Advancing Pharmaceutical Processes and Tools for Improved Health Outcomes

Tagelsir Mohamed Gasmelseid
International University of Africa, Sudan

A volume in the Advances in Medical Technologies and Clinical Practice (AMTCP) Book Series

Medical Information Science
REFERENCE
An Imprint of IGI Global

Published in the United States of America by
Medical Information Science Reference (an imprint of IGI Global)
701 E. Chocolate Avenue
Hershey PA, USA 17033
Tel: 717-533-8845
Fax: 717-533-8661
E-mail: cust@igi-global.com
Web site: http://www.igi-global.com

Library of Congress Cataloging-in-Publication Data

Library of Congress Cataloging-in-Publication Data

Names: Gasmelseid, Tagelsir Mohamed, 1967- editor.
Title: Advancing pharmaceutical processes and tools for improved health outcomes / Tagelsir Gasmelseid, editor.
Description: Hershey, PA : Medical Information Science Reference, [2016] | Includes bibliographical references and index.
Identifiers: LCCN 2016000675| ISBN 9781522502487 (hardcover) | ISBN 9781522502494 (ebook)
Subjects: LCSH: Pharmacy--Research. | Pharmaceutical industry--Technological innovations. | Pharmaceutical services. | Pharmacy informatics.
Classification: LCC RS122 .A29 2016 | DDC 615.10285--dc23 LC record available at http://lccn.loc.gov/2016000675

This book is published in the IGI Global book series Advances in Medical Technologies and Clinical Practice (AMTCP) (ISSN: 2327-9354; eISSN: 2327-9370)

British Cataloguing in Publication Data
A Cataloguing in Publication record for this book is available from the British Library.

Advances in Medical Technologies and Clinical Practice (AMTCP) Book Series

Srikanta Patnaik
SOA University, India
Priti Das
S.C.B. Medical College, India

ISSN: 2327-9354
EISSN: 2327-9370

Mission

Medical technological innovation continues to provide avenues of research for faster and safer diagnosis and treatments for patients. Practitioners must stay up to date with these latest advancements to provide the best care for nursing and clinical practices.

The **Advances in Medical Technologies and Clinical Practice (AMTCP) Book Series** brings together the most recent research on the latest technology used in areas of nursing informatics, clinical technology, biomedicine, diagnostic technologies, and more. Researchers, students, and practitioners in this field will benefit from this fundamental coverage on the use of technology in clinical practices.

Coverage

- Patient-Centered Care
- E-Health
- Medical Informatics
- Biometrics
- Telemedicine
- Clinical Nutrition
- Biomechanics
- Neural Engineering
- Diagnostic Technologies
- Nursing Informatics

Titles in this Series

For a list of additional titles in this series, please visit: www.igi-global.com

Virtual Reality Enhanced Robotic Systems for Disability Rehabilitation
Fei Hu (University of Alabama, USA) Jiang Lu (University of Houston -Clear Lake, USA) and Ting Zhang (University of Alabama, USA)
Medical Information Science Reference • copyright 2016 • 383pp • H/C (ISBN: 9781466697409) • US $210.00 (our price)

Software Innovations in Clinical Drug Development and Safety
Partha Chakraborty (Cognizant Technology Solutions, India) and Amit Nagal (GVK Bioscience, India)
Medical Information Science Reference • copyright 2016 • 305pp • H/C (ISBN: 9781466687264) • US $225.00 (our price)

Handbook of Research on Holistic Perspectives in Gamification for Clinical Practice
Daniel Novák (Czech Technical University in Prague, Czech Republic) Bengisu Tulu (Worcester Polytechnic Institute, USA) and Håvar Brendryen (Norwegian Center for Addiction Research, University of Oslo, Norway)
Medical Information Science Reference • copyright 2016 • 636pp • H/C (ISBN: 9781466695221) • US $340.00 (our price)

Optimizing Assistive Technologies for Aging Populations
Yosry S. Morsi (Swinburne University of Technology, Australia) Anupam Shukla (ABV - Indian Institute of Information Technology and Management Gwalior, India) and Chandra Prakash Rathore (Oracle India Private Limited, India)
Medical Information Science Reference • copyright 2016 • 494pp • H/C (ISBN: 9781466695306) • US $220.00 (our price)

Modern Concepts and Practices in Cardiothoracic Critical Care
Adam S. Evans (Icahn School of Medicine at Mount Sinai, USA) Gregory E. Kerr (Weill Cornell Medical College, USA) Insung Chung (Icahn School of Medicine at Mount Sinai, USA) and Robin Varghese (Icahn School of Medicine at Mount Sinai, USA)
Medical Information Science Reference • copyright 2015 • 1110pp • H/C (ISBN: 9781466686038) • US $415.00 (our price)

Recent Advances in Assistive Technologies to Support Children with Developmental Disorders
Nava R. Silton (Marymount Manhattan College, USA)
Medical Information Science Reference • copyright 2015 • 424pp • H/C (ISBN: 9781466683952) • US $210.00 (our price)

Table of Contents

Detailed Table of Contents

The pharmaceutical industry is often characterized as a research-driven sector because of its exceptionally high ratio of R&D inputs to sales. Development of novel drugs is very difficult because of several issues including heavy investment, high risks, and long development cycle. Government plays an important role in regulating the development of the pharmaceutical industry. This is true for all phases in pharmaceutical development: from R&D to market. This chapter will focus on the discussion of prospective high-tech areas, development strategies, and innovation resources in the pharmaceutical industry. Expert opinions were analyzed based on the conditions in China's biopharmaceutical sector. Policy recommendations are provided to support technological innovation.

Open Access (OA) is reshaping the world by redefining the scholarly communication methods with focus on building a knowledge society. It has the power to democratize the knowledge by removing hurdles from free access to scholarly works while encouraging knowledge sharing. Institutional Repositories (IRs) play a vital role in the OA movement by facilitating the Green Route to Open Access. In India, some elite educational and research institutes such as the Indian Statistical Institute, some CSIR Laboratories, IITs, and IIMs have taken significant initiatives in building IRs. In addition to theses, a few Universities have also taken the initiative and now have their own IRs as their proudest possession. However, many Indian Irs, once functional are in bad shape and a few have been closed. This paper deals with IR initiatives in the discipline of Medicine and Biology including. The paper discusses their scope, collection strength, growth rate and their current status.

Web 2.0 technologies are being rapidly integrated in higher education, which dramatically influences the ways learners approach and use information. Knowledge transfer has evolved into a two-way process. Users no longer simply consume and download information from the web; they create and interact with it. Several theoretical works were developed in order to discuss the possibilities of integration of Web 2.0 tools in Pharmacy, Medicine, Allied Health, Nursing and many other Biomedical Areas. Other works have started gathering qualitative and quantitative evidence of the importance of Web 2.0 tools in the learning process. By performing this integrative review, this paper will provide an overview of what is being done in biomedical and pharmaceutical education, and elaborate some of the potential opportunities and challenges that these applications present. With this updated review we hope to give our contribution to consolidate research in this promising area.

This contribution examines solutions how Information Quality (IQ) dimensions as a framework along with Radio Frequency Identification (RFID) and the Electronic Product Code Information System (EPCIS) as tools may improve needed drug trackability and traceability capabilities in the pharmaceutical industry (PI). For years counterfeit drugs have been impacting the industry and putting patients' health in danger. We analyze applications, methods and practices in the improvement of the quality of drug tracking and tracing. The potential of IQ, RFID, EPCIS and related applications and technologies suggest and design corresponding information and materials flows. This research presents examinations, reviews and recommendations and utilizes two theoretical frameworks: Transaction Cost Theory and Collective Action Theory. This setting may be viewed as a large, complex and international web of corporations, legislation, regulatory efforts, compliance regimes, manufacturers, wholesalers, pharmacies, importers as well as rapidly advancing technologies and applications.

Polypharmic patients are patients who chronically use five or more medicines. The number of polypharmacy patients continues to increase even though it is a risk factor for morbidity and mortality. A medication review is an important measure to mitigate medication risks. It is known to effectively reduce the number of drug related problems per (polypharmic) patient. STRIP is a Dutch method to perform a structured medication review. Based on this method, the STRIPA(ssistent) tool is developed. However, whether or not this app is considered useful by the healthcare professional is not known yet. In order to assess

bundlet (SWNT/NT-BUD) and droplet (fullerene) models predict dissimilar behaviours. Interaction-energy parameters are derived from C60. An NT-BUD behaviour or further is expected. Solubility decays with temperature result smaller for GR/GR-BUD than SWNT/NT-BUD than C60 in agreement with lesser numbers of units in clusters. Discrepancy between experimental data of the heat of solution of fullerenes, CNT/NT-BUDs and GR/GR-BUDs is ascribed to the sharp concentration dependence of the heat of solution. Diffusion coefficient drops with temperature result greater for GR/GR-BUD than SWNT/NT-BUD than C60 corresponding to lesser number of units in clusters. Aggregates (C60)13, SWNT/NT-BUD7 and GR/GR-BUD3 are representative of droplet, bundlet and columnlet models.

This chapter provides discussion of the visual ways of learning basic physical and chemical concepts related to symmetry and the crystalline structures. All kinds of symmetrical structures are present in substances and their various molecular compositions that are researched in fields related to pharmacology. Great part of technologies, methodologies, tools, and applications require knowledge visualization skills to understand and present concepts and processes. Exploration of science-based concepts and nature-related processes supports attaining visualization literacy, which is needed for explaining physical and chemical notions, clinical procedures, and publicizing clinical and mobile medical informatics. This chapter discusses the ways of preparing to this task artists, graphic designers, computer graphics students, as well as people in charge of hospitals, medical centers, and pharmaceutical industries who hire designers. The chapter offers exercises in visualization of scientific concepts by providing two projects about basic science-related themes: (1) Symmetry and pattern in animal world: geometry and art, and (2) Crystals and crystal caves. Each project invites the reader to create visual presentation of the theme.

Preface

There has been a growing concern for the improvement of pharmaceutical services provided by healthcare institutions. This concern is also shared by other stakeholders including patients, regulatory organizations (such as Food and Drug Authorities and their pharmacovigilance centers), pharmaceutical companies, insurance companies and research institutions. While the concerns held by companies tend to be driven by financial drivers, the attention of healthcare institutions stems from their social and institutional obligations towards the provision of patient-oriented pharmaceutical services. While such drivers tend to significantly shape the processes adopted by each stakeholder, their collective effort is the main critical success factor for effective planning and provision of pharmaceutical care services at local and national settings. Within this context, interventions tend to be oriented toward the following areas:

1. Medication-specific Pharmacoinformatics tools that aim at enhancing interactions as well as co-operation between physicians and pharmacists. While such tools are basically oriented towards improving information accessibility and sharing among healthcare professionals, they also play significant roles in improving patient outcomes and patient-physician interactions. The functionality and usefulness of the entire tools are contingent upon medication processes as well the technological platforms adopted.
2. Institutional and organizational dimensions directed towards examining policy and decision making and supporting strategic interventions.

This contributions included in this handbook reflects these two areas. Leong Chan and Dan Liu focused on the roles to be played by policy planning in supporting technological innovations in the pharmaceutical industry. The growing attention being made for planning originates from the intensive competitiveness experienced in the Chinese pharmaceutical sector, the high R&D investments and the imperativeness of using advanced biotechnologies. By focusing on the prospective "prioritized" technology areas and innovation resources, the authors discussed potential technology development strategies, innovation pathways and biosimilars.

Chakravarty analyzed the main open Access Initiatives in Medical Biology by studying Institutional Repositories in India. By focusing on the roles to be played by new information technologies and related access frameworks, the author discussed the importance of developing innovative models that enhance haring of scholarly information. The shared literature may include research papers, primary data and other evidence, creative activity and other products of research and scholarship including across institutions and audiences. The author's recommendation to use open access information in medical biology domains has been driven by the high cost of subscription to international databases, research-specific complications and access-related difficulties. Angelo Jesus & Maria João Gomes examined and reviewed the use of web 2.0 in pharmaceutical education.

Dinah M. Mande and Rolf T. Wigand examined the context of major information quality issues and dimensions in the identification and tracking of drugs within the pharmaceutical industry. They used two technologies (mainly Radio Frequency Identification [RFID] and the Electronic Product Code Information System [EPCIS]) as tools to develop a framework that aims ate the improvement of needed drug track-ability and traceability capabilities in the Pharmaceutical Industry (PI).

The work of Marco Spruit et al examined the complications faced by polypharmacy patients who use multiple drugs and proposed the use of STRIPA as a tool to perform a structured medication review in The Netherlands. The authors focused on examining the drivers and functionalities of the tool in relation with other tools and collected some data to prove its usefulness and acceptability by healthcare professionals.

Eduardo Contreras and Gustavo Puente examined the context of identifying Rheumatic diseases by general physicians (Rheumatology specialty) and pointed out the main challenges being faced. The authors conceptualized the context of disease identification by coupling "Clinical Semiology" and "software agents' technologies". The overall functionality and design pathways of the diagnostic tool proposed by the authors is based on the heuristic compilation of the tacit knowledge of rheumatologists. Such knowledge is taken as a key element in identifying and designing necessary rheumatic inflammatory and non-inflammatory conditions.

The work of Gasmelseid focused on the use of intelligent agent-based Pharmacoinformatics frameworks for the improvement of pharmaceutical care in response to the expansion of drug-related problems especially adverse drug reactions (ADRs). Such frameworks have the potential in providing necessary decision support for the identification, analysis and reporting of ADRs. The author proposed an intelligent multi-agent decision support system that includes a process model, a multi-agent architecture and an integrated data processing model with clear description of agent functionalities. The model reflects three main modules: a data capture and update module, diagnosis module and a pharmaceutical care and drug monitoring module. The study also reflected on the practical and managerial environment of the model and the basic considerations to be taken into account.

Francisco Torrens and Gloria Castellano addressed the clusters origin of solvation features of C-nanostructures in organic solvents. Their use of different mathematical computations shed light on the approaches that can be used in handling pharmaceutical integration.

Anna Ursyn discussed the possibilities of making visualizations about connection between science, computing, and art. The author proposed two projects about science-related themes:

1. Symmetry and pattern in animal world – geometry and art, and
2. Crystals and crystal caves.

Projects are aimed at supporting the readers' skills in visualization and scientific illustration for the purpose of medicine and pharmacology applications.

Chapter 1
Policy Planning to Support Technological Innovation in the Pharmaceutical Industry

Leong Chan
Portland State University, USA

Dan Liu
Portland State University, USA

ABSTRACT

The pharmaceutical industry is often characterized as a research-driven sector because of its exceptionally high ratio of R&D inputs to sales. Development of novel drugs is very difficult because of several issues including heavy investment, high risks, and long development cycle. Government plays an important role in regulating the development of the pharmaceutical industry. This is true for all phases in pharmaceutical development: from R&D to market. This chapter will focus on the discussion of prospective high-tech areas, development strategies, and innovation resources in the pharmaceutical industry. Expert opinions were analyzed based on the conditions in China's biopharmaceutical sector. Policy recommendations are provided to support technological innovation.

1. THE PHARMACEUTICAL MARKET IN CHINA

China's pharmaceutical market is one of the most dynamic in the world. It grew 22% in 2010 to US$116 billion and ranked the fifth largest in the world (Bieri, 2012). With an average annual growth rate above 20% from 2005 to 2010, it is set to overtake Japan as the world's second largest market by 2015 (Giniat, Fung, Weir, & Meyring, 2011). Due to the economic recession in the Western countries, the Chinese pharmaceutical market is steadily moving up toward the leading position globally. However, the Chinese pharmaceutical industry faces huge challenges in the area of technological innovation. Breakthrough technological innovation from the domestic Chinese pharmaceutical sector is rarely seen for decades. Although China is a major exporter of pharmaceuticals, it is specialized in the production of crude drug substances and low-tech generics, rather than novel drugs.

DOI: 10.4018/978-1-5225-0248-7.ch001

Figure 1. Innovative and generic drug market share in China 2010
Adapted from IMAP (2012).

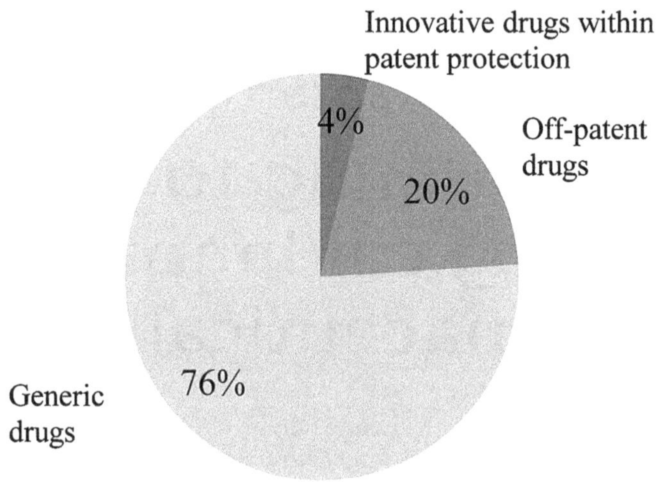

The Chinese pharmaceutical market is highly fragmented and very different from the market in developed countries. In 2010, generic drugs had about 76% of the entire pharmaceutical market in China, while only 4% of the market was comprised of innovative drugs still under patent protection. The remaining 20% of the market consisted of off-patent drugs (Figure 1) (Bieri, 2012). The generic drugs market has the largest segment and has mostly been controlled by domestic products. However, the profit margin is low due to intense competition. The innovative drug market has the smallest segment and is dominated by imported products, particularly those produced by MNCs. For the off-patent drug segment, both imported and domestically-produced branded drugs compete to survive.

1.1 The Biopharmaceutical Sector

China's pharmaceutical industry consists of three major sectors:

1. Chemical pharmaceuticals;
2. Biopharmaceuticals; and
3. Traditional herbal medicines.

Although the chemical pharmaceutical technologies have been regarded as the industrial foundations in the last century, biopharmaceutical technologies have been emerging as a prospective area with huge growth potential. Many leading chemical pharmaceutical companies have already tapped into the biotechnology area. There has been a paradigm shift in industrial R&D from high-risk synthetic pharmaceuticals towards R&D in biopharmaceuticals. The top chemical pharmaceutical companies spent tens of billions of dollars to acquire biotechnology companies and in-licensing deals. Pfizer, Roche, Lilly, Astra Zeneca, Glaxo Smith Kline (GSK) and Bristol Myers Squibb have all underlined their strong commitment and highlighted their biological projects in R&D pipeline (Maggon, 2007).

The Chinese biopharmaceutical sector has been developing rapidly in recent years. Since this is still a new area with good prospects, both established pharmaceutical companies and startup firms are trying to profit from the expanding market. Although the overall innovation capability of domestic players is not very strong, research in some specialty areas has already caught up with the level of leading countries. However, it is generally accepted that the Chinese biopharmaceutical industry still needs to increase its competitiveness globally, especially in high technology areas.

1.2 Policy Factors

The pharmaceutical sector is among the mostly regulated areas in healthcare. The innovation capability of the industry is largely influenced by macro factors including regulations, economy, demographics, and technology level. For example, the reimbursement issue needs to be considered in the Chinese market. The government introduced the National Essential Drug List (NEDL) in 2009 to set the range of reimbursement and lower drug prices for the general public. NEDL sets upper limits of the retail prices for the drugs on the list. Innovative drugs are usually not included on the reimbursement list. These drugs are graded at the highest price in the market. Since no imbursement are available, the innovative drugs have a very limited market size. As illustrated in Figure 1, the market share was only 4% in 2010 (Bieri, 2012). Therefore policy factors can have significant impact on the direction of the pharmaceutical industry and its development. Successful technology policy relies on a better understanding of both global and domestic market environments. Decision makers need to evaluate related issues and adjust their strategy accordingly. In order to build a competitive and innovative pharmaceutical sector, both foreign factors and domestic settings call for effective strategic orientation to adapt to global market competition.

2. TECHNOLOGIES, STRATEGIES, AND RESOURCES

Innovation is the driving force for pharmaceutical development. Technological competitiveness and innovation are vital for the fast-developing biopharmaceutical industry in China (Han, 2009; Zhouying, 2005). Due to historical reasons, the technology level of the Chinese biopharmaceutical sector remains less competitive globally, and it still faces challenges including weak innovative capacity and lack of R&D investment. Due to the high investment risk and long development cycle, the biopharmaceutical sector relies heavily on regulations and support from governments. Strengthening technological competitiveness and building up innovative capabilities are primary concerns of industry as well as policy makers.

2.1 Prospective Technology Areas

It is important to identify global technology trends and adapt to local capabilities and needs. Choosing the right technology areas and guiding investment are major topics in technology policy. While it is unrealistic for the Chinese biopharmaceutical industry to excel in all high technology areas, it is more realistic to focus on key areas where the country has potential capabilities to achieve competitive advantages. From the perspective of industrializing countries, the appropriate technology can offer windows of opportunity to catch-up with leading countries. In other words, China should look into the global technology frontiers and seize the opportunities for catching up.

To represent the global technology trends and emerging areas, this research starts from the findings from technology forecasting reports published by international organizations including the United Nations (UN) and Organization for Economic Co-operation and Development (OECD). The UN has published the

Table 1. Top ten biotechnologies for improving health in developing countries

Rank	Biotechnology
1	Modified molecular technologies for affordable, simple diagnosis of infectious diseases
2	Recombinant technologies to develop vaccines against infectious diseases
3	Technologies for more efficient drug and vaccine delivery systems
4	Technologies for environmental improvement (sanitation, clean water, bioremediation)
5	Sequencing pathogen genomes to understand their biology and to identify new antimicrobials
6	Female-controlled protection against sexually transmitted diseases, both with and without contraceptive effect
7	Bioinformatics to identify drug targets and to examine pathogen-host interactions
8	Genetically modified crops with increased nutrients to counter specific deficiencies
9	Recombinant technology to make therapeutic products more affordable
10	Combinatorial chemistry for drug discovery

Source: UNESCO (2006).

research results in a report titled Top Ten Biotechnologies for Improving Health in Developing Countries (Table 1) (Daar, et al., 2006). More recently in 2009, OECD published the forecasting report Human Health Biotechnologies to 2015, which is based on the conditions of its member countries (Arundel, Sawaya, & Valeanu, 2009) (OECD, 2009).

The available research indicates that different countries have different needs for technologies due to various developmental conditions (Chin, 2008). As an emerging nation, which walks in between the developed and developing cohort, China needs to identify prospective technology areas based on its needs and capabilities. The following technology areas and definitions were extracted from the forecasting reports from the OECD and UN (Daar, et al., 2006; Arundel, et al., 2009; OECD, 2009).

- **Recombinant Therapeutic Proteins:** Therapeutic proteins are used to treat many non-communicable diseases. These technologies provide affordable and sustainable sources for treatment of chronic disease.
- **Recombinant Vaccines against Infectious Diseases:** Vaccines produced using recombinant DNA technology. The products can be used to effectively treat infectious diseases.
- **Monoclonal Antibody Technology:** Monoclonal antibodies (mAb) can be used for therapeutic treatment and diagnostic tests. Many therapies are undergoing clinical trials. Most are concerned with immunological and oncology targets.
- **Stem Cell and Tissue Engineering:** These technologies involve techniques that replace or act directly on cells and tissues in the body. The treatment repairs tissues damaged from injuries and diseases.
- **Gene Therapy:** This technology involves the treatment of a disease by introducing a new gene into a cell. It either uses or acts directly on nucleic acids, which are the molecules that serve as the building blocks for DNA and RNA.
- **Antisense Therapy:** Antisense drugs are being researched to treat a wide range of diseases such as cardiovascular diseases, asthma, and arthritis. There are currently more than 30 anti-sense therapies in clinical trials.

- **RNAi (Ribonucleic Acid Interference):** This includes all entries for products which act therapeutically via an RNA interference mechanism. There have been a great number of research activities in this new area. Most proposed clinical uses are aimed at treating infections.
- **Nanobiotechnology for Efficient Drug and Vaccine Delivery:** This type of technology aims for improved drug delivery systems from the convergence between biotechnology and nanotechnology.
- **Synthetic Biology:** The design and construction of new biological parts, devices and systems that do not exist naturally; The redesign of existing biological systems to perform specific tasks.
- **Bioinformatics to Identify Drug Targets and Examine Pathogen-Host Interactions:** These technologies cover the manipulation and analysis of large datasets of genetic and health information.
- **Pharmacogenetics:** This technology identifies inherited differences (variation) between individuals in drug metabolism and response. It can be applied in clinical trials and prescribing practices.
- **Gene Sequencing:** Sequencing of pathogen genomes provides ways to identify new antimicrobials. These technologies can accelerate the process of drug discovery and fight against infectious diseases.
- **Biotechnology Diagnostics:** This technology includes both in vitro diagnostics and in vivo diagnostics. Modified molecular technologies provide affordable and simple diagnosis of infectious diseases.

The prospective technology areas in the biopharmaceutical sector are extracted from available foresight reports. There are two major reasons for these technology areas to be validated. Firstly, these reports were published a few years ago, and they need to be updated and validated according to recent development. The OECD forecasting report was published in 2009. If we consider the publication lag, the research should be done between 2008 and 2009. Secondly, the OECD report represented the findings from a club of developed countries, while this research focuses on emerging economies. As a result, the identified technology areas were sent to expert panel for validation. The finalized list is illustrated in Table 2.

Table 2. Prospective technology areas (T_k)

$\mathbf{T_1}$: Recombinant Therapeutic Proteins
$\mathbf{T_2}$: Recombinant Vaccines
$\mathbf{T_3}$: Monoclonal Antibody Technology
$\mathbf{T_4}$: Cell and Tissue Engineering
$\mathbf{T_5}$: Gene Therapy
$\mathbf{T_6}$: Antisense Therapy
$\mathbf{T_7}$: RNAi
$\mathbf{T_8}$: Nanobiotechnology
$\mathbf{T_9}$: Synthetic Biology
$\mathbf{T_{10}}$: Bioinformatics
$\mathbf{T_{11}}$: Pharmacogenetics
$\mathbf{T_{12}}$: Gene Sequencing
$\mathbf{T_{13}}$: Biotechnology Diagnostics

2.2 Development Strategies

The strategies define how technologies should be developed and implemented. As an industrializing country, China faces the decisions of "Make" or "Buy", or somewhere in between (White, 2000). According to the findings from the literature review section, the following strategies are defined to describe the situation:

- **Indigenous Innovation:** This strategy relies on the host country's local technology base and available innovation resources to build up indigenous competence (Jin, 2005; Lazonick, 2004; Yang & Shu, 2005).
- **Imitative Innovation:** Also known as re-innovation in literature, it is based on imitation, adaptation, and improvement of the original innovators' technology (Cheng & Shiu, 2008; Mukoyama, 2003).
- **Collaborative Innovation:** This strategy means the participants cooperate and develop new ideas altogether. Competitors may share resources and work together toward innovation (Yang & Shu, 2005; X. Wang & Li, 2007).
- **International Technology Transfer:** This includes technology import and acquisitions. This is a fast track to save valuable time and resources during the catching-up process (Meyer, 2001; Salicrup & Fedorkova, 2006).

2.4 Innovation Resources

Under the condition of a transitional economy, China's National Innovation system carries some characteristics from both a market economy and centrally-planned system. Here we need to identify the key contributors toward technology development and innovation in the Chinese biopharmaceutical sector. Subsidies and favorable policy measures should be designed and prioritized to strengthen the performance of effective innovators. The following innovation resources have been identified by the literature review.

- **State-Owned Enterprises (SOEs):** SOEs are medium- to large-sized companies left by the centrally planned system. These companies constitute the main production capacity of the Chinese pharmaceutical industry, but most of them specialize in low-tech generics drugs. Compared with foreign counterparts, domestic pharmaceutical companies are weaker in terms of technology level and research capabilities (White, 2000; K. Wang, Hong, Marinova, & Zhu, 2009; Nolan & Yeung, 2001).
- **High-Tech Small-to-Medium Enterprises (SMEs):** These smaller companies have emerged since the 1980s, when the government started to allow private ownership of companies. Many small dedicated biotechnology firms (DBFs) belong to this category. They probe into potential technology areas with the purpose of obtaining leadership status in some niche sub-sectors (Jiang, Wang, & Yan, 2001; Boutellier & Ullman, 2007; K. Wang, et al., 2009).
- **Multinational Company and Subsidiaries (MNCs):** Currently, many top MNCs have established subsidiaries in China. These large American and European pharmaceutical companies have dominant innovative capability in most technological areas. They act as technology leaders in both production and R&D activities in the Chinese pharmaceutical sector (Han, 2009; Ghauri & Rao, 2009).

- **Contract Research Organizations (CROs) and Contract Manufacture Organizations (CMOs):** These organizations provide services for both foreign and domestic companies. Through learning-by-doing from leading innovators, CROs and CMOs have shown increasing capabilities in developing advanced technologies and manufacturing practice aligning to international standards (Lawrence, 2005; Singh, 2006).
- **University Research Programs (URPs):** Some top research universities are emerging forces in pharmaceutical innovation, and they have been producing more publications and patents in recent years. Not only do these research universities innovate through laboratories, but they also cultivate talented young students for the domestic pharmaceutical industry (K. Wang, et al., 2009; Boutellier & Ullman, 2007).
- **Equity Joint Ventures (EJVs):** This is a common way for foreign companies to enter the Chinese biopharmaceutical sector, especially during the 1990s. Two or more investors share the ownership and control over the equity, property (including IP), and operation (Boutellier & Ullman, 2007; Lee, 2008).
- **Public Research Institutes (PRIs):** PRIs and national R&D laboratories are owned and managed by government departments. These organizations carry out research projects according to government instructions (K. Wang, et al., 2009; Boutellier & Ullman, 2007).
- **Foreign R&D Centers (FR&D):** In recent years, some foreign invested R&D centers have been established in China. The biopharmaceutical sector is one of the target areas. This has also happened in India in recent years. Foreign R&D Centers are capable of carrying out comprehensive research to develop new medicines at the innovation frontiers (Lawrence, 2005; Gassmann, Reepmeyer, & Zedtwitz, 2008).

2.5 Expert Panel and Data Collection

The expert panel included policy-makers from government agencies and technology management experts from the health industries. Some of the experts were stakeholders from the domestic and foreign organizations. These included domestic enterprises, top multinational pharmaceutical companies and foreign research institutions. Foreign stakeholders were included because of their technology strength and their long-term investment stakes in the Chinese pharmaceutical market. Domestic biopharmaceutical companies were selected to represent local perspective in research and development. The idea here is to reach a "Win-Win" situation in technology development.

A panel of 20 experts participated in the research process. The expert panel was divided into three subgroups to match the purpose of this research. Subgroup-G experts have backgrounds from various government agencies. All of them are senior officials or researchers from the National Medical Policy Research Center, MOH, State Food & Drug Administration, and Center of Drug Evaluation. Subgroup-F experts have foreign backgrounds, and represent the interests of various foreign organizations, which include two foreign research institutions and three of the top-ten multinational enterprises (Roche, Sanofi-Aventis, and GSK). Subgroup-L consists of local experts from the domestic industry and research organizations (non-government & without foreign backgrounds), which include bio-tech SMEs, industrial associations, and public research institutions. Since the experts come from diverse sources, they can provide valuable judgment from different perspectives.

Three sets of research instruments were developed to quantify the relative weights of criteria. The confidence level measurement used Likert scaling. Confidence scores with medium or above scores from the experts are acceptable. The pair-wise comparison method was applied for quantifying experts' judgments. Different sets of instruments were sent to and filled out by different experts according to their expertise and areas of specialty. During this process, necessary information was conducted through emails and phone calls. Upon receiving all of the completed research instruments from experts, the results were calculated either in matrix or in vector formats. Some important indicators and values such as the inconsistencies and disagreements were evaluated. In this research, such issues were resolved after several iterations of communication with the experts.

3. ANALYSIS AND DISCUSSION

3.1 Prospective Technology Areas

With achieving technological competitiveness and sustained innovation as the mission, this research examines a number of prospective technology areas in the biopharmaceutical industry. Although the experts come from different backgrounds, they have reached a high level of agreement in their judgments. The results can be classified under three categories: high >10%, medium 6%-10%, and low 1%-5% (Figure 2).

3.1.1 High Priority Technology Areas

The "high" category is defined where the contribution is larger than 10%. These are the technology areas that should be China's highest priorities for R&D. The recommended areas include T1 Recombinant Therapeutic Proteins, T2 Recombinant Vaccines, and T3 Monoclonal Antibody Technology.

For T1 Recombinant Therapeutic Proteins, the preferred strategy is imitative innovation. The ideal innovation resources include: high-tech SMEs, university research programs, and public research institutes. The development of recombinant proteins is well known to have relied on blockbuster revenue, primarily derived from 16 brands. For example, Amgen's blockbusters Enbrel, Neulasta and Aranesp

Figure 2. Prioritized technology areas

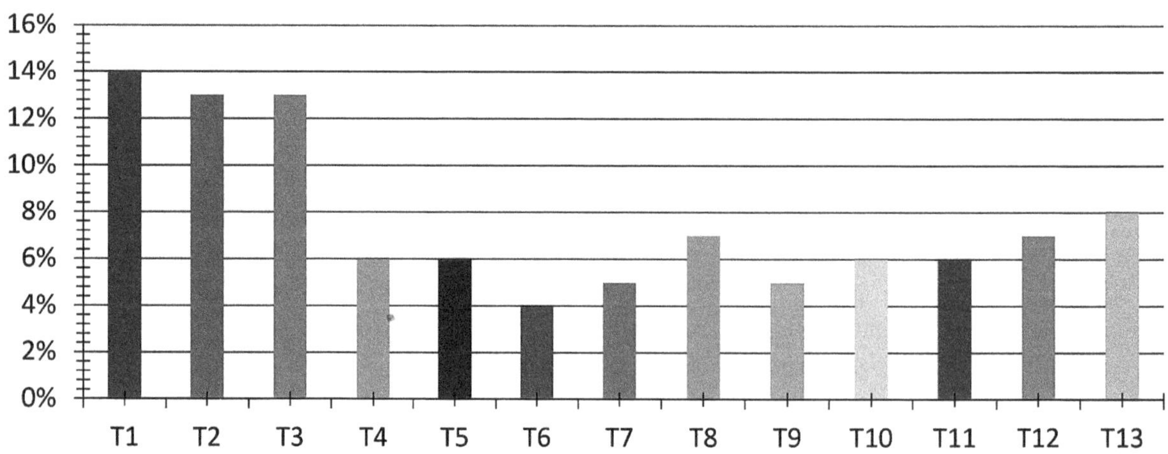

are the sector's key leading blockbusters, with combined sales of more than US$10.8billion in 2010. The expected growth during 2012-2014 will be around 13% (Maheshwari, October 2011). Hematology, diabetes, endocrinology and oncology are the most valuable therapy areas for recombinant proteins. Taking the imitative innovation strategy means that China should focus on developing biosimilars in this area; and biotech SMEs, Universities, and Research Institutes should play more important roles.

For T2 Recombinant Vaccines, the preferred strategy is indigenous innovation. The recommended innovation leaders include: MNCs and subsidiaries, high-tech SMEs, and foreign R&D centers. Vaccines are among the most lucrative segments in the global pharmaceutical market. With an average growth of over 13% during 2009-2012, the global market for human vaccines is forecasted to reach US$32 billion by the year 2017 (GIA, 2012a). The US and EU are the two largest vaccine markets in the world. The vaccine market in China has the potential to record phenomenal growth in the coming years. The growth rate will be around 20% and its size will reach CNY 12 billion by 2013 (WiCON, January 2012). The quantity and the variety of vaccines produced in China are similar to those of developed countries, but China needs to improve the production capability and critical technology in order to produce higher quality vaccines (Han, 2009). Taking the indigenous innovation strategy indicates that China should develop more novel products, and focus on the improvement of local competence. For such a purpose, innovation resources like multinationals and subsidiaries, biotech SMEs, and Foreign-invested R&D centers can take the lead.

For T3 Monoclonal antibody technology, the preferred strategy is imitative innovation. The recommended innovation leaders include: high-tech SMEs, university research programs, and public research institutes. In the global market, the antibody bandwagon has been joined by 200 companies with hundreds of new projects and targets that have attracted billions of dollars in R&D investment, acquisitions and licensing deals leading to monoclonal antibody (Maggon, 2007). The total global monoclonal antibody (mAb) sales are forecasted to reach US$49 billion by 2013. The "big five" mAbs – Avastin, Herceptin, Rituxan, Humira, and Remicade – have dominated the market, cornering almost 80 percent of sales (Bird, 2008). There has been a great gap between China and Western countries in the research, development and manufacture of monoclonal antibody drugs (Hu, Ma, & Zhang, 2006). Since the imitative innovation strategy has a higher priority, China should focus on the development of biosimilars, especially for the blockbuster drugs. Ideally, biotech SMEs, Universities, and Research Institutes should play more important roles in this process.

3.1.2 Medium Priority Technology Areas

The "medium" category is defined as being where the contribution ranges from 6% to 10%. These technology areas are recommended as medium priorities for China to carry out R&D. This list consists of seven technology areas including T4 Cell and tissue engineering, T5 Gene therapy, T8 Nanobiotechnology, T10 Bioinformatics, T11 Pharmacogenetics, T12 Gene sequencing, and T13 Biotechnology Diagnostics.

For T4 Cell and tissue engineering, the preferred strategy is imitative innovation. The recommended innovation leaders include: high-tech SMEs, university research programs, and public research institutes. According to market reports, there are more than 60 tissue engineering products in the global market and about 30 in clinical trials. China's biomedical materials industry is largely driven by foreign technology, and domestic companies accounted for a mere 3% of global market share in 2011 (Hvistendahl, 2012). By following the imitative innovation strategy, China should focus on catching up with the advanced countries. The results suggest that biotech SMEs, Universities, and Research Institutes should play more important roles.

For T5 Gene therapy, the preferred strategy is imitative innovation. The recommended innovation leaders include: high-tech SMEs, university research programs, and public research institutes. Gene therapy is a high-tech area with very few available products. However, there about 80 gene therapies are in clinical trials (Arundel, et al., 2009). Many experts believe that gene therapy will play a significant role in future medical treatment. The research recommends imitative innovation strategy, indicating that China should focus on learning from advanced countries. The results also suggest that biotech SMEs, Universities, and Research Institutes play more important roles.

For T8 Nanobiotechnology, the preferred strategy is imitative innovation. The recommended innovation leaders include: high-tech SMEs, university research programs, and public research institutes. The applications of nanobiotechnology in the biomedical field are principally directed towards development of novel drug delivery systems. According to a market report in 2012, the global nanobiotechnology market will reach $6.0 billion by 2017 (GIA, 2012b). As a catching up country in this area, it is recommended that China follow the imitative innovation strategy. The results also suggest that biotech SMEs, Universities, and Research Institutes should play more important roles in the process.

For T10 Bioinformatics, the preferred strategy is indigenous innovation. The recommended innovation leaders include: MNCs and subsidiaries, high-tech SMEs, and foreign R&D centers. The applications in bioinformatics are increasingly powerful, allowing researchers to garner more knowledge about more complex organisms and systems. The worldwide bioinformatics market was estimated at US$3.0 billion in 2010, and the applications will continue to have very rapid growth to 2015 (Arundel, et al., 2009). Today, bioinformatics research in China still lags behind the best in the world. There are relatively few applications for drug discovery in the domestic market (Wei & Yu, 2008). Since the indigenous innovation strategy is recommended, China should focus on developing new applications in bioinformatics. To support this strategy, innovators such as MNCs, high-tech SMEs, and Foreign R&D centers can take the leading roles.

For T11 Pharmacogenetics, the preferred strategy is imitative innovation. The recommended innovation leaders include: high-tech SMEs, university research programs, and public research institutes. Pharmacogenetics is a prospective field that could lead to personalized medicines. According to market reports, the worldwide pharmacogenetics market was estimated at US$3.7 billion in 2009, but there will be a limited number of new pharmacogenetic products arriving on the market by 2015 (Arundel, et al., 2009). The experts recommend the imitative innovation strategy, indicating that China should focus on catching up with advanced countries. The results also suggest that biotech SMEs, universities, and research institutes should play more important roles in this process.

For T12 Gene sequencing, the preferred strategy is imitative innovation. The recommended innovation leaders include: high-tech SMEs, university research programs, and public research institutes. Applications of gene sequencing technology will help researchers to find genes associated with human disease. China is catching up rapidly in the field of gene sequencing. The acquisition in 2012 of a California-based DNA sequencing company by a Chinese firm (BGI) led to wide concerns. Some American scientists, politicians and industry executives said the takeover represented a threat to American competitiveness in DNA sequencing (Pollack, December 30, 2012). So far, the experts in this research still recommend an imitative innovation strategy, where biotech SMEs, Universities, and Research Institutes should take the lead in China's catching-up process.

For T13 Biotechnology Diagnostics, the preferred strategy is imitative innovation. The recommended innovation leaders include: high-tech SMEs, university research programs, and public research institutes. The diagnostics market was estimated at $ 52.4 billion in 2012, and it is expected to grow at a rate of 7% during 2012-2017 (RNCOS, 2013). Roche Diagnostics is the dominant leader with 20% market share. Nine of the world's top 15 firms are based in the United States. Other firms are based in either Europe or Japan. In this research, the experts recommend an imitative innovation strategy for China. The country should focus on catching up with advanced countries. The results suggest that biotech SMEs, universities, and research institutes play more important roles in this process.

3.1.3 Low Priority Technology Areas

The "low" category is where the contribution is equal to or less than 5%. These technology areas are regarded as having lower priorities for China. This list consists of three technology areas including T6 Antisense therapy, T7 RNAi, and T9 Synthetic biology. These technology areas have one thing in common: there are still very few or even no approved applications in the market today. However, these drugs have good prospects in that some candidate drugs are already in clinical trials (Arundel, et al., 2009). The related studies in China are still in initial stages. The experts recommend an imitative innovation strategy for all three technology areas. China should focus on catching up with advanced countries. The results also suggest that biotech SMEs, universities, and research institutes should play more important roles in this process.

3.1.4 Summary on Prospective Technology Areas

The research results of technology areas highlights the directions for investment and improvement. For most of the above discussed high-tech areas, the Unites States is the dominant leader worldwide. Research has shown that American academic publications and patents comprise more than 40% of the world. China and some major European countries belong to the second tier in these areas. As a latecomer country in the biopharmaceutical industry, China's innovation capabilities have been steadily growing since the mid 1990s. However, China's technology level is still lagging behind the world's leading standard, and the country needs to take a learning position as discussed in the above analyses. In order to accelerate the catching-up process, the government's role of long-term investment in these identified areas cannot be overemphasized.

3.2 Technology Development Strategies

The relative contribution of the strategies to the innovation mission is presented in Figure 3.

The results in Figure 3 show that technology development strategy should focus more on imitative innovation (33%). Indigenous innovation (26%) is regarded as the second best option to improve national competitive capability and to obtain high industry value for China. Collaborative innovation ranked third at 23%. Lastly, international technology transfer contributed 17% to the overall mission.

Any static strategy is often less effective for latecomer countries since competitive advantage could be accumulated through multiple approaches or sources. From an industrial perspective, a country's technology strategies need to be dynamic for its unique but changing developmental contexts. Therefore, policy makers should adopt a comprehensive approach to technology strategy using all possible

Figure 3. Contribution of strategies to overall innovation

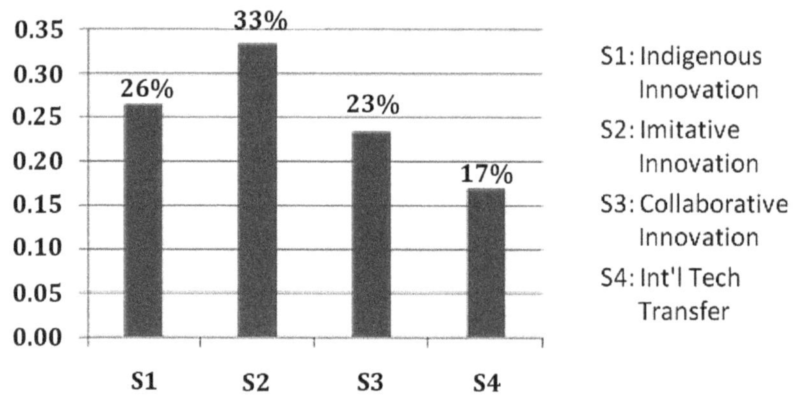

resources, and engage various stakeholders in the process of technology development. Such an approach entails decision makers becoming more involved in governance initiatives to improve the innovation environment, or scaling their influence over the relevant high-tech areas for the long-term innovation goals. This involves balancing various technology development strategies to build up industrial innovation capacity for competitiveness and future success.

3.2.1 Imitative Innovation and Biosimilars

Accumulation of technological capacities to compete in the global market has become a major concern for China. The research brings to light that imitative innovation is still the best option to achieve such a purpose under the current conditions. The experts' judgments give high priority to imitative innovation (33%) in the development of biopharmaceutical technologies. This conforms to the fact that technology leaders in high-tech areas are mostly foreign enterprises, which mainly belong to the United States and Western Europe. If the latecomers want to catch up with the technological frontiers, their strategies are likely to start from imitation. This has been the case for many of the East Asian economies – first for Japan, then for Taiwan, Korea and Singapore – and now for China (Kim, 1997). The results of this research indicate that China's biopharmaceutical industry is at the stage of learning from advanced countries.

When discussing imitative innovation in the biopharmaceutical industry, biosimilars are topics that cannot be circumvented. Novel biologics are noted for their high production cost and expensive purchase prices. Biosimilars bring clear potential for payers in the emerging pharmaceutical or "pharmerging" markets, such as Brazil, India and China (IMS, 2011). Developing biosimilar products is also a relatively low-risk strategy for newcomers entering the health biotech space and generating short-term revenues (Frew, et al., 2008). Of the approximate 150 approved originator biologic drugs on the market today, almost half of them have lost or are close to losing their patent protection. This provides an external condition for cheaper biosimilar products to enter the market and be available for consumers. However, under the current registration regime, biosimilar drugs and new biologic drugs are not treated with any differences in China. Both applications require the same process for clinical trials. Although the United States does not currently have related regulations, India and the European Union have developed abbreviated approval processes for biosimilar products (IMS, 2011). China should consider adopting similar approaches to remove or lower the legislative hurdles for the development of biosimilars.

3.2.2 India's Experience in Pharmaceutical Development

The rise of India's pharmaceutical industry in the last few decades may provide some insight for other emerging economies. The Indian government adopted an imitative strategy in the pharmaceutical industry during the 1970s (Janodia, Sreedhar, Ligade, Pise, & Udupa, 2008). The weak IPRs regime fostered the development of domestic technology capabilities in that period. If an Indian domestic company could merely modify the manufacturing process of a foreign medicine, the company was allowed to produce the same product without patent infringement (N. Kumar, 2002). This strategy established low cost leadership advantage among local companies and increased domestic social welfare due to the lowered drug prices. It was remarkable achievement that new drugs can be introduced to India only within four to five years after their introduction in foreign countries. However, the negative effect was that most MNCs chose to leave the Indian market for afraid of patent infringement.

As a large number of medicines went off patent protection during the late 1980s, Indian medicines further experienced a rapid growth of exports to the world market. Moreover, an interesting effect was that the increased technological capabilities of Indian pharmaceutical companies have brought back the FDI from the Western developed countries. New joint ventures or R&D centers were setup mainly draw upon trained manpower and research infrastructure available in the country, despite the fact that the Indian patent regime did not provide strong patent protection (Nagesh Kumar & Aggarwal, 2005). By introduction and assimilation of advanced technologies from abroad, the Indian pharmaceutical companies have emerged as competitive suppliers in the world.

The Indian experience highlights the fact that government strategy may lead to industrial success. However, India's imitative strategy can no longer be duplicated by other emerging economies because the global innovation environment has changed significantly in recent years. Many countries have joined WTO and endorsed the TRIPS Agreement, which requires strengthened IPR protection. Therefore, direct replication of foreign products will not be applicable, and even India has to make a change in recent years. Nowadays, emerging economies should focus on re-innovation or imitative innovation, which is beyond pure replication. In summary, the Indian experience in pharmaceutical development strengthened the findings in this research.

3.2.3 Indigenous Innovation, Collaborative Innovation, and Novel Drugs

Indigenous innovation strategy ranks second at 26%, followed by collaborative innovation strategy at 23%. These strategies cannot be overlooked as optional choices for China's current technology capabilities. China should try to develop its indigenous strengths and also collaborate with leading countries. In most technology areas, China belongs to the second cohort among the worldwide biopharmaceutical communities. Therefore, the country cannot afford to totally rely on indigenous innovation. The open-door policy in the last 30 years has proven that foreign elements are extremely important resources for the local industries. China would not have achieved the current technology level if the doors were closed for FDI and MNCs. As a long term goal, indigenous innovation strategy should be encouraged in China. Past experiences from other countries have repeatedly demonstrated that an emerging economy will ultimately move from the imitative stage to the innovative stage (Kim, 1997).

Although the collaborative innovation strategy did not rank the highest for any specific technology areas, it has been regarded as an increasingly important strategy in recent years. In reality, some industrial players have come up with new channels to develop novel drugs through bridging collaborative and indigenous innovation. For example, Chinese-based SinoPharm struck a cooperation deal with American-based Harbor Biosciences to develop novel drugs in the area of therapeutic protein drugs. The two sides share resources in terms of financial investment, technology know-how, and research facilities. Regarding the research outcomes, SinoPharm has exclusive rights in China, while Harbor Biosciences maintains rights in other countries. This type of collaboration is very attractive to both parties: the foreign player benefits from lowered investment and more research resources, while the Chinese player is able to gain access to advanced R&D techniques and essentially obtain their very own novel drugs (Bioassociate, December 2012; Nasto, December 2011).

3.2.4 International Technology Transfer

Compared with other strategies, international technology transfer ranks relatively low in the overall contribution toward mission. MNCs and Foreign R&D Centers are the major contributors for this strategy. The Chinese industries have benefited substantially from international technology transfer deals during the 1980s and 1990s. The MNCs transferred many technologies which helped China to upgrade its industries. However, as local technology capability matured, the reliance on foreign technologies decreased. In recent years, there have been many cases where foreign companies began acquiring domestic firms or technologies (Xia, 2008). International technology transfer deals no longer travel in one direction from abroad to local, but also from home to abroad. Nevertheless, in high-tech areas such as the biopharmaceutical industry, foreign players still play more important roles in international technology transfer in China. From a government perspective, China needs to establish effective macro-level technology policies to guide and promote technology transfer activities in high-tech areas.

3.2.5 Disagreement on Technology Development Strategies

The experts' judgments recorded a relatively high disagreement regarding the strategies. Experts with different backgrounds suggested different strategies for China to follow. Government experts have higher expectations for imitative innovation or indigenous innovation. Indeed, due to noticeable gaps between China and the global innovation frontiers, it is more realistic to expect some learning and imitative activities. This process happened to other countries during their catching up stages, such as Japan in the 1960's and South Korea in the 1970s. India's pharmaceutical industry grew very fast during the 1970's when the government adopted similar strategies to imitate foreign drugs (Janodia, et al., 2008). However, in today's more globalized environment, China's strategy should focus on re-innovation or imitative innovation, which goes beyond simple imitation. The domestic industrial experts have suggested more on imitative innovation strategy, indicating the catching up trends in the industry. They also have higher expectations for collaborative innovation, which confirms that collaborations are necessary for domestic companies. Although the foreign experts suggested relying more on collaborative innovation, the reality is that some MNCs have limited collaborations with domestic players. Their major concerns are loss of technology edge or IPR. However, the results have demonstrated that foreign players have the motive or interests to collaborate with domestic players.

The global biotech communities are well aware of the advantages and attractions China can offer. As China's economy grows and incomes rise, it can be anticipated that people will increase their spending on healthcare and medicines. This brings great opportunities not only for domestic companies, but also for multinationals. The government should welcome global partners to jointly share the development and prosperity of the nation's emerging industry. To facilitate internationalization of China's biopharmaceutical industry, policy makers should plan to establish ideal conditions for attracting foreign innovators. Promoting research collaboration between domestic companies and their overseas counterparts will benefit the technology learning process, as well as the mission of sustained industrial growth. For its innovation goals to be met, China needs to have more integrated strategies for technology development. Past experiences revealed that the country needs to be more integrated into the global innovation networks. This means keeping an open-door policy and encouraging foreign investment in high-tech areas.

3.3 Supportive Innovation Resources

The relative contribution of resources toward innovation is presented in Figure 4.

From the results shown in Figure 4, we can observe that A4 High-tech SMEs contributed about 18% and ranked first toward the overall mission, indicating its important position in technological innovation. A8 MNCs ranked second in overall contribution at 15%. Lastly, A3 SOEs are ranked surprisingly low with 9% in overall contribution, indicating an area for improvement in terms of innovative performance.

This research provides the Chinese biopharmaceutical industry a performance report of various innovators with regard to their contribution toward global strategies and technology objectives. This will assist policy makers in determining which infrastructure items require improvement or investment. Based on the feedback from result validation, the research suggests improving the conditions and environment

Figure 4. Contribution of resources to mission

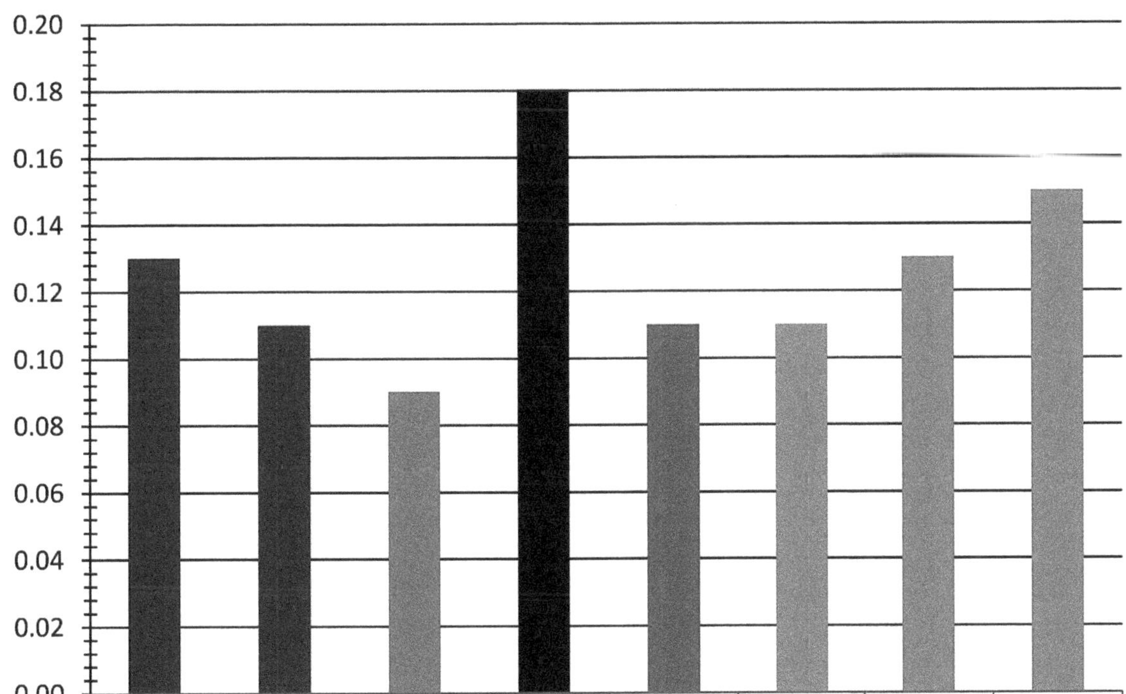

	A1: URPs	A2: PRIs	A3: SOEs	A4: SMEs	A5: EJVs	A6: CROs	A7: FR&D	A8: MNCs
■ Series1	0.13	0.11	0.09	0.18	0.11	0.11	0.13	0.15

for innovation. The result analyses indicate that High-tech SMEs are the most important contributors for China's biopharmaceutical industry in the current development stage. The second group of important contributors is considered to be the MNCs and subsidiaries. The Foreign R&D Centers and University Research Programs tie for the third place toward mission. These important innovation resources will be discussed in this section.

3.3.1 High-Tech SMEs

Owing to the narrowed gaps of competitive advantages in recent years, many emerging biotech SMEs have entered the race for technology development. These companies have certain advantages over large established enterprises, including greater flexibility, better efficiency, less bureaucracy, and profit-seeking behaviors which allow them to succeed in the fast-changing markets. Many biotech SMEs in the Chinese biopharmaceutical sector share similar advantages and traits. For example, they are more successful in some specialized high-tech areas, and most of them are very eager or active in collaborative innovations with other players. This is mainly due to the reality that SMEs are usually not strong as standalone innovators. They need to search for complementary resources to cover their deficiencies in certain aspects.

Despite a clear evidence of progress in recent years, biotech SMEs still have some key issues to be addressed. One of the main challenges faced by biotech SMEs is to obtain funding, not only for their business purposes but also for their R&D activities. With lower research inputs, most biotech SMEs are not well prepared to compete globally in many high-tech fields. The government should provide services to expand biotech SMEs' networking with other players so that interactions can create a synergy where knowledge, expertise, and experiences are shared. This is also a measure to avail biotech SMEs with complementary resources and related activities that they lacked. Moreover, the government has the ability to alleviate biotech SMEs' tax burdens, and induce them to invest more in R&D activities with incentives.

3.3.2 Multinational Companies and Subsidiaries

MNCs' technological strength, institutional heritage, and their global coverage generate specific advantages for their operations in the Chinese biopharmaceutical sector. MNCs are in a better competing position because they are better endowed with both R&D capacities and funding capital. Researchers found that the presence of foreign pharmaceutical firms can make a number of important contributions to the success of the industrial networks of the host country. For example, these firms have better expertise in developing and protecting intellectual property with high commercial potential, they have well-established marketing and distribution channels, and they are experienced in both shaping and working within strict regulatory guidelines (Chiesa & Chiaroni, 2005). Domestic players in the host country may benefit from technology spillover through MNCs' demonstration effects, labor turnover, and overall industrial structure upgrading (both upstream and downstream) (Fan, 2003; Liu & Buck, 2007). These are essential factors to build up a better innovation ecosystem for the biopharmaceutical industry in China.

The research results have revealed that there are certain disagreements on whether MNCs collaborate with other players. According to the feedback from result validation, the contributions of MNCs toward collaborative innovation are uneven. One expert claimed that her company has initiated many partnership programs in recent years, which indicates higher contribution. On the other hand, several other experts argued that their collaborative programs are limited in clinical trials which do not contribute much to innovation. Most of the MNCs in China focus on the collaborations with domestic players in the late-stage clinical trials, which are required by domestic regulations. The Chinese government does not allow First-in-Human clinical research for foreign drugs in China. Even if the drug is approved for marketing in a foreign country, the drug company still needs to restart all three phases of clinical trials locally. Although some MNCs also collaborate with Chinese universities and research institutes in early-stage discoveries, the cases are relatively few, and the MNCs would eventually acquire the research outcomes to enrich their own product pipelines. Therefore, the government should provide more favorable policies to facilitate and support collaboration between MNCs and domestic players, especially in early-stage drug discovery. This presupposes that both sides will benefit from cooperation or even competition with their counterparts. The trend of globalization and industrial liberalization needs such a change in attitude or business orientation where even competitors can cooperate with each other to achieve mutual success (Awuah & Amal, 2011). From a foreign perspective, MNCs also need to achieve a better balance between their demands in exploiting the potential Chinese market and utilizing the plentiful local resources.

3.3.3 University Research Programs

According to the results, the contribution of University Research Programs ties with that of Foreign R&D Centers for third place. This is mainly due to a university's higher contribution toward collaborative innovation and imitative innovation. University programs are more oriented toward basic research. Many biotech startups and spinoffs were created to take advantages of discoveries in academic research. MNCs collaborate with top Chinese universities in early-stage drug discoveries. Many universities also have academic connections with foreign research institutions. Moreover, university partnerships are not limited to R&D. They are important for training future talents with advanced research techniques, as well as providing companies the opportunity to recruit a highly qualified workforce (Frew, et al., 2008). In the past two decades, Chinese universities have trained about 100,000 biotech researchers. Nearly 1,000 universities and colleges in China offer biology-related courses, and more than 500 universities and colleges offer biology-related programs. More than 20,000 university students graduated each year in biology-related fields before 2006 (Zhou, January 2007). Universities provide the foundations for China to catch up with developed countries.

3.3.4 Foreign R&D Centers

Foreign R&D Centers also contribute the third highest toward the overall mission. This is largely due to their contributions toward international technology transfer and indigenous innovation. In recent years, more and more foreign-invested R&D centers have been established in China. This is a new approach to R&D in that it builds Chinese portals to the global biotech communities. On the one hand, these innovation centers have stringent ties with research resources in their own countries. On the other hand, they hire and train many domestic scientists and researchers to carry out local R&D projects. The vast population and different disease patterns in China provide a convenient condition for the application of new technologies. There are plenty of opportunities for both early-stage drug discovery and late-stage clinical trials. These organizations bring some of the latest research techniques, routines, and practices to China. Many of these research centers focus on finding specific biomarkers and genes related to diseases that are more prevalent in China and Asian regions. This is an important aspect because they contribute to indigenous innovation as well as technology transfer.

3.3.5 Other Innovation Resources

Although the research can be utilized to identify a single best contributor toward the mission, the research goal is far beyond such a finding. The technology development process in reality involves complementary technologies which can be developed by different innovators in the innovation systems. In other words, the current research is not merely for choosing the only innovator for each candidate technology. For example, when the results suggested high-tech SMEs as the best contributing resource toward the overall mission of competitiveness and innovation, it should not be interpreted that other innovators are excluded from the development strategy; rather, they just contribute less than the ideal option but still remain as contributors. Because rankings provide guidelines and direction for the design of technology policies to leverage investment input, such as time, effort, human capital, and related monetary support.

4. POLICY RECOMMENDATIONS

Based on the above discussions and experts' feedback, major areas of policy recommendations include:

1. Establishing a clear vision for technology objectives and implementation strategies in the biopharmaceutical sector;
2. Creating actors or organizations with responsibility for promoting or executing such strategies;
3. Nurturing a collaborative environment and innovation ecosystem;
4. Providing service platforms, education, and training; and
5. Legislating support and financial incentives for innovators at a micro or organizational level.

4.1 Establishing a Clear Vision for Technology Objectives and Strategies

In the development of high-tech industries in emerging economies, the synergies of two major aspects need to be considered, which will reap some of the benefits of globalization while remaining responsive to local market needs. Technology objectives and strategies serve to provide direction and schemes for the making of innovation policy within the sectoral regime. This research suggests that emerging economies must set clear technology goals before formulating their global strategies. The decision makers have the roles of clarifying the vision for those undertaking risky R&D activities in the high-tech areas, determining proper regulations, and industrial standards to maintain the desired boundaries of research activities, and making sure that an effective innovation environment is provided with appropriate market mechanisms. The concern is to enhance competitiveness for the industry's present needs in globalization and local applications.

A major achievement of the findings is that the experts with diverse backgrounds have reached unanimous agreement for prospective technology areas in China. This means that the identified high-tech areas are in need of attention not only for the local biotech community to upgrade but also for the overseas stakeholders to invest. These high priority areas include Recombinant Therapeutic Proteins, Recombinant Vaccines, and Monoclonal Antibody Technology. As a large country, China should not give up developing other technologies, but the focus here is to make priorities based on both global trends and local needs. The long-term objective is to develop a full-fledged biopharmaceutical sector that ranks among the higher end of the global industrial value chain.

4.2 Creating Actors and Organizations to Improve Innovation Capacity

Global strategies often require substantial supporting resources during the implementation processes. This research suggests that policy makers actively consider the strategic need and ensure adequate resources are provided. The results bring to light that China should strive to build an enterprise-centered innovation system for the biopharmaceutical sector. The change from a structural and institutional context of scientific research to an entrepreneurial mode may be beneficial for the biopharmaceutical industry in the long run. This concern needs to be taken for policy making.

To build a more competitive biopharmaceutical Industry, the government can strengthen the development of enterprises by enhancing the factors that lead them to business success. Research shows that institutional support, especially support from the government, can play a major role in SMEs' competitiveness development within the domestic market (Awuah & Amal, 2011). The actions can include executing an elite promotion program to support a number of flagship enterprises, and motivate them toward innovation. In order to stimulate innovation in the identified priority high-tech areas, policies can be designed to encourage Chinese scientists overseas to return to China and contribute their experience and expertise to the local industry. Domestic scientists should be encouraged to participate in the commercialization process or even to become technological entrepreneurs themselves. Policy responses should be developed with regard to supporting innovation and investment in high-tech areas.

4.3 Nurturing a Collaborative Environment and Innovation Ecosystem

A major concern of the government is to improve the innovation ecosystem in the biopharmaceutical sector. The research results showed that both indigenous innovation and collaborative innovation are contributing strategies for the biopharmaceutical sector in China. Under the present conditions and status of China's biopharmaceutical sector, imitative innovation is the preferred strategy. However, increasing international collaboration, as well as appropriate competition, is crucial for technological innovations and scientific breakthroughs. With increased partnerships and collaborations, both domestic and foreign actors cooperate to carry out scientific research that leads to innovations.

The situation of limited collaborations between foreign and domestic biotech companies has been discussed in the literature. There are suggestions that foreign firms should adjust their IP strategies and change from a defensive position of filing and enforcing patents to a more active exploitation of the commercial value of their technologies in China (Li, Meijer, Duysters, & Rochemont, 2011). First, large MNCs usually have a broad range of non-commercialized patents where they are willing to open their innovation processes through patent transactions. Chinese firms may have opportunities either to buy or in-license some of the foreign state-of-the-art technologies. Second, smaller or medium-sized foreign firms usually have more specific and state-of-the-art technologies. They may consider out-licensing such technologies before their IPs are infringed by Chinese firms (Li, et al., 2011).

The foreign firms should be encouraged to employ an open innovation strategy. In China's context, the government can play a decisive role in promoting innovation activities in the high-tech areas. Measures can include setting up appropriate institutional frameworks, opening up more business opportunities, and providing generous incentives for various innovators. Policy makers must have a strong commitment to establish a more innovative environment for various players. Promoting the development of new ventures and encouraging foreign investment from MNCs are not only very necessary, but also represent the current needs and future trends for high-tech development.

4.4 Providing Service Platforms, Education, and Training

In order to support innovation, the government can provide various services and facilities. These include clinical trial facilities and R&D service platforms that align with international best practices. The government should promote the development of biotech contract research services for clinical trial phase I & II, and establish biotech contract manufacturing services for clinical trial phase III. The government can also provide support through education, information propaganda, and procurement programs. The need for high quality scientists and engineers arises from both industry and research institutes. The investment in the higher education system should keep up with rapidly growing demand. The availability of human resources will ultimately contribute to the increase of industrial competitiveness.

4.5 Legislating Support and Financial Incentives for Innovators

The nature of innovation networks in China is different from those in developed countries such as the United States and EU. The infrastructure and supporting systems are mainly policy driven. Favorable policies and economic support from the government are key factors to drive continuous growth in the long run. Substantial investments may be required in reforming established SOEs, cultivating startups, and supporting various academic/research institutes. Another area that needs improvement is the efficiency of the drug evaluation process and reviewing mechanism. The bureaucratic nature of decision making due to the involvement of government agencies has deterred the progress of technological development and cooperation. Such barriers should be simplified or removed completely. For example, the registration procedures for biosimilars could be simplified according to related practices in advanced foreign countries.

Policies can serve to arouse the market participants' interests in high-tech areas through fiscal incentives. Such measures may include offering support to SMEs in preferred research areas through investment incentives or low-interest loans. The government can provide startups with special funding to lower their burden of initial sunk cost. Lastly, the competitiveness of companies also depends on how well the companies handle their networks of complementary relationships with other players in the biotech sector. Institutional setup, active collaboration, and financial support from the government will enhance the companies' capabilities with regard to innovation and competitiveness.

REFERENCES

Arundel, A., Sawaya, D., & Valeanu, I. (2009). *Human health biotechnologies to 2015.* OECD.

Awuah, G. B., & Amal, M. (2011). Impact of globalization: The ability of less developed countries' (LDCs') firms to cope with opportunities and challenges. *European Business Review*, *23*(1), 120–132. doi:10.1108/09555341111098026

Bieri, C. (2012). *Global Pharma & Biotech M&A Report.* IMAP.

Bioassociate. (2012). *Pharmerging markets: China - the next major innovative pharma market.* Bioassociate Consulting & Management Ltd.

Bird, J. (2008). *Monoclonal Antibodies Report: 2008 Update.* Academic Press.

Boutellier, R., & Ullman, F. (2007). China's unique position in discovery and preclinical research. *Drug Discovery Today*, *12*(1-2), 4–7. doi:10.1016/j.drudis.2006.11.009 PMID:17198968

Cheng, C. J., & Shiu, E. C. C. (2008). Re-innovation: The construct, measurement, and validation. *Technovation*, *28*(10), 658–666. doi:10.1016/j.technovation.2007.08.002

Chiesa, V., & Chiaroni, D. (2005). *Industrial Clusters in Biotechnology: Driving Forces, Development Processes and Management Practices*. London: Imperial College Press.

Chin, C. D. (2008). Biotechnology for global health: Solutions for the developing world. *Consilience*, *1*, 1–12.

Daar, A. S., Thorsteinsdóttir, H., Martin, D. K., Smith, A. C., Nast, S., & Singer, P. A. (2006). *Top ten biotechnologies for improving health in developing countries*. UNESCO.

Fan, E. X. (2003). Technological spillovers from foreign direct investment. *Asian Development Review*, *20*(1), 34–56.

Frew, S. E., Sammut, S. M., Shore, A. F., Ramjist, J. K., Al-Bader, S., Rezaie, R., & Singer, P. A. et al. (2008). Chinese health biotech and the billion-patient market. *Nature Biotechnology*, *26*(1), 37–53. doi:10.1038/nbt0108-37 PMID:18183014

Gassmann, O., Reepmeyer, G., & Zedtwitz, M. v. (2008). *Leading Pharmaceutical Innovation: Trends and Drivers for Growth in the Pharmaceutical Industry*. Berlin: Springer. doi:10.1007/978-3-540-77636-9

Ghauri, P. N., & Rao, P. M. (2009). Intellectual property, pharmaceutical MNEs and the developing world. *Journal of World Business*, *44*(2), 206–215. doi:10.1016/j.jwb.2008.05.008

GIA. (2012a). *Human Vaccines: A Global Strategic Business Report*. Global Industry Analysts, Inc.

GIA. (2012b). *Nanobiotechnology: A Global Strategic Business Report*. Global Industry Analysts, Inc.

Giniat, E., Fung, P., Weir, A., & Meyring, N. (2011). *China's Pharmaceutical Industry - Poised for the Giant Leap*. KPMG Advisory (China) Limited.

Han, P. (2009). China's growing biomedical industry. *Biologicals*, *37*(3), 169–172. doi:10.1016/j.biologicals.2009.02.010 PMID:19427231

Hu, X., Ma, Q., & Zhang, S. (2006). Biopharmaceuticals in China. *Biotechnology Journal*, *1*(11), 1215–1224. doi:10.1002/biot.200600083 PMID:17089435

Hvistendahl, M. (2012). China's Push in Tissue Engineering. *Science*, *338*(6109), 900–902. doi:10.1126/science.338.6109.900 PMID:23161989

IMS. (2011). *Shaping the biosimilars opportunity: A global perspective on the evolving biosimilars landscape*. IMS Health.

Janodia, M. D., Sreedhar, D., Ligade, V. S., Pise, A., & Udupa, N. (2008). Facets of Technology Transfer: A Perspective of Pharmaceutical Industry. *Journal of Intellectual Property Rights*, *13*, 28–34.

Jiang, Y., Wang, Y., & Yan, X. (2001). Chinese pharmaceutical companies: An emerging industry. *Drug Discovery Today*, *6*(12), 610–612. doi:10.1016/S1359-6446(01)01842-6 PMID:11408192

Jin, C. (2005). *Towards Indigenous Innovation: Pathways for Chinese Firms. Workshop of Technology Innovation and Economic Development.*

Kim, L. (1997). *Imitation to Innovation: The Dynamics of Korea's Technological Learning*. Cambridge, MA: Harvard Business School Press.

Kumar, N. (2002). *Intellectual property rights, technology and economic development: Experiences of Asian countries.* RIS Discussion Paper, 25.

Kumar, N., & Aggarwal, A. (2005). Liberalization, outward orientation and in-house R&D activity of multinational and local firms: A quantitative exploration for Indian manufacturing. *Research Policy*, *34*(4), 441–460. doi:10.1016/j.respol.2005.01.010

Lawrence, R. (2005). New opportunities in Asia: A focus on India and China. *Drug Discovery Today*, *10*(2), 89–91. doi:10.1016/S1359-6446(04)03315-X PMID:15718156

Lazonick, W. (2004). Indigenous innovation and economic development: Lessons from China's leap into the information age. *Industry and Innovation*, *11*(4), 273–297. doi:10.1080/1366271042000289360

Lee, P. (2008). Opening the door to the Chinese pharmaceutical market. *Chemistry Today*, *26*, 76–81.

Li, Y., Meijer, E., Duysters, G., & Rochemont, M. (2011). Patent transactions with China in a new era: A European perspective. *Journal of Knowledge-based Innovation in China*, *3*(2), 136–156. doi:10.1108/17561411111138973

Liu, X., & Buck, T. (2007). Innovation performance and channels for international technology spillovers: Evidence from Chinese high-tech industries. *Research Policy*, *36*(3), 355–366. doi:10.1016/j.respol.2006.12.003

Maggon, K. (2007). Monoclonal antibody gold rush. *Current Medicinal Chemistry*, *14*(18), 1978–1987. doi:10.2174/092986707781368504 PMID:17691940

Maheshwari, S. (October 2011). *Global protein therapeutics market: Beefing up towards futuristic growth.* Pharmaphorum.

Meyer, A. D. (2001). Technology transfer into China: Preparing for a new era. *European Management Journal*, *19*(2), 140–144. doi:10.1016/S0263-2373(00)00088-8

Mukoyama, T. (2003). Innovation, imitation, and growth with cumulative technology. *Journal of Monetary Economics*, *50*(2), 361–380. doi:10.1016/S0304-3932(03)00005-9

Nasto, B. (2011, December). Biotech in China: Special feature on China's emerging biotech industry. *Nature Biotechnology*, 1–16.

Nolan, P., & Yeung, G. (2001). Big business with Chinese characteristics: Two paths to growth of the firm in China under reform. *Cambridge Journal of Economics*, *25*(4), 443–465. doi:10.1093/cje/25.4.443

OECD. (2009). *The Bioeconomy to 2030: Designing a Policy Agenda*. Organization for Economic Co-Operation and Development.

Pollack, A. (2012, December 30). U.S. clears DNA firm's acquisition by Chinese. *The New York Times*.

RNCOS. (2013). *In-Vitro Diagnostics Market Analysis to 2017*. RNCOS Industry Research Solutions.

Salicrup, L. A., & Fedorkova, L. (2006). Challenges and opportunities for enhancing biotechnology and technology transfer in developing countries. *Biotechnology Advances*, *24*(1), 69–79. doi:10.1016/j.biotechadv.2005.06.004 PMID:16098701

Singh, R. (2006). Clinical research in China and India: A paradigm shift in drug development. *Drug Discovery Today*, *11*(15/16), 675–676. doi:10.1016/j.drudis.2006.06.009 PMID:16846793

Wang, K., Hong, J., Marinova, D., & Zhu, L. (2009). Evolution and governance of the biotechnology and pharmaceutical industry of China. *Mathematics and Computers in Simulation*, *79*(9), 2947–2956. doi:10.1016/j.matcom.2008.09.001

Wang, X., & Li, J. (2007). Innovation network in harvest. Paper presented at the International Conference on Technology Innovation, Risk Management, and Supply Chain Management.

Wei, L., & Yu, J. (2008). Bioinformatics in China: A personal perspective. *PLoS Computational Biology*, *4*, 1–11.

White, S. (2000). Competition, capabilities, and the make, buy, or ally decisions of Chinese State-Owned firms. *Academy of Management Journal*, *43*(3), 324–341. doi:10.2307/1556398

WiCON (2012). *Pharma China 2012*. WiCON International Group LLC.

Xia, Q. V. (2008). Analysis of future directions and growth of biotechnology in China. *Asia Biotech*, *12*, 1–6.

Yang, J., & Shu, W. (2005). On the choice of technological innovation strategy of Chinese enterprises at present. *China-USA Business Review*, *4*(10), 63–66.

Zhou, Y. (2007, January). Opportunities in Biopharmaceutical Outsourcing to China. *BioProcess International*, 16–23.

Zhouying, J. (2005). Globalization, technological competitiveness and the 'catch-up' challenge for developing countries: Some lessons of experience. *International Journal of Technology Management and Sustainable Development*, *4*(1), 35–46. doi:10.1386/ijtm.4.1.35/1

Chapter 2
Open Access Initiatives in Medical Biology: A Study of Institutional Repositories in India

Rupak Chakravarty
Panjab University, India

ABSTRACT

Open Access (OA) is reshaping the world by redefining the scholarly communication methods with focus on building a knowledge society. It has the power to democratize the knowledge by removing hurdles from free access to scholarly works while encouraging knowledge sharing. Institutional Repositories (IRs) play a vital role in the OA movement by facilitating the Green Route to Open Access. In India, some elite educational and research institutes such as the Indian Statistical Institute, some CSIR Laboratories, IITs, and IIMs have taken significant initiatives in building IRs. In addition to theses, a few Universities have also taken the initiative and now have their own IRs as their proudest possession. However, many Indian Irs, once functional are in bad shape and a few have been closed. This paper deals with IR initiatives in the discipline of Medicine and Biology including. The paper discusses their scope, collection strength, growth rate and their current status.

1. INTRODUCTION

The journey of man from a nomadic society to knowledge society is very interesting and perhaps never ending. Free flow of information has played a pivotal role in this journey. But the journey is still incomplete as there exits a thirst – thirst for knowledge and is still unquenched. To facilitate a free flow of information, especially scholarly information is very crucial. At the same time it is a challenge as financial aspects are also involved in this. Scholarly communication has the potential to make this world a better place to live in as it is based upon the philosophy of equal access to knowledge. Open Access is augmenting and enriching the scholarly communication cycle and making the world a global village.

DOI: 10.4018/978-1-5225-0248-7.ch002

Scholarly communication can be understood as the system through which research and other scholarly works are created, evaluated, disseminated to the scholarly community, and preserved for future use. It is the cyclical process It is a cyclical process in which content is generated, reviewed, disseminated, acquired, preserved, discovered, accessed, and assimilated for the advancement of scholarship. The assimilation can potentially lead to generation of new content and thus start a new iteration of the process (or life cycle).

2. OPEN ACCESS AND THE TWINS

Open access (OA) means unrestricted online access to peer-reviewed scholarly research. Open access is primarily intended for scholarly journal articles, but is also provided for a growing number of theses, book chapters, and scholarly monographs. Open access comes in two degrees: gratis OA, which is free online access, and libre OA, which is free online access plus some additional usage rights.

The Budapest statement defined open access as - "There are many degrees and kinds of wider and easier access to this literature. By 'open access' to this literature, we mean its free availability on the public internet, permitting any users to read, download, copy, distribute, print, search, or link to the full texts of these articles, crawl them for indexing, pass them as data to software, or use them for any other lawful purpose, without financial, legal, or technical barriers other than those inseparable from gaining access to the internet itself. The only constraint on reproduction and distribution, and the only role for copyright in this domain, should be to give authors control over the integrity of their work and the right to be properly acknowledged and cited."

The Bethesda and Berlin statements add that for a work to be open access, users must be able to "copy, use, distribute, transmit and display the work publicly and to make and distribute derivative works, in any digital medium for any responsible purpose, subject to proper attribution of authorship."

The two ways authors can provide OA are Green and Golden Routes/Roads. In green OA, authors self-archive their journal articles in an OA repository (IR). In the Golden OA approach, the author decides to submit the research papers in a selected repository, making it freely available over the world wide web. In Gold Open Access, the end users i.e. The readers are not supposed to pay any subscription cost, but the author has to pay the Article Processing Charges (APCs). The third approach, which is also becoming an accepted model is Hybrid OA which is most commonly associated with Gold Open Access. In this OA model is a mix of subscription charges and publication fees. In this model, some articles are available only to subscribers, while others are made available at no charge to anyone searching the web. Authors pay an additional fee for the open access option. They may do this because open access is a requirement of their research funding agency. Or they may do it so that non-subscribers can access their article for free. In this model, only the articles for which the authors have covered APCs are available for free.

According to Bo-Christer "in just over two years [2009-2012] the number of journals from major publishers offering hybrid Open Access has more than doubled, from approximately 2,000 to over 4,400. Since the overall numbers of journals from these publishers has remained on the same level, the hybrid share has risen from 25% to around 50% of all eligible journals."

3. PHARMACOLOGY

Pharmacology is the branch of medicine and biology concerned with the study of drug action, (Vallance P & Smart TG, January 2006) where a drug can be broadly defined as any man-made, natural, or endogenous (within the body) molecule which exerts a biochemical and/or physiological effect on the cell, tissue, organ, or organism. More specifically, it is the study of the interactions that occur between a living organism and chemicals that affect normal or abnormal biochemical function. If substances have medicinal properties, they are considered pharmaceuticals. IUPAC defines pharmacology as Science of drugs, including their origin, composition, pharmacokinetics, pharmacodynamics, therapeutic use, and toxicology.

4. RESEARCH PROBLEM

New technologies, specifically ICT, and innovative models are enhancing the shared model of scholarly information. The shared literature may include research papers, primary data and other evidence, creative activity and other products of research and scholarship including across institutions and audiences. However, on account of the exponential price rise in journal subscription, libraries, especially in underdeveloped and developing countries were finding it difficult to serve their readers adequately and qualitatively. They are still finding their budget either shrinking or stagnant. The problem is further aggravated as the funders and taxpayers do not have access to the research they pay for. Producing a research paper may require time and money. Scholars are devoid of scholarly content and open access to scholarly literature is the only solution to this problem. Also, there is need of conducting research in the discipline of Medicine and Biology repositories in India.

Figure 1. France circa 1913: R. Dubois anesthetizing machine
(Photo by Boyer/Roger Viollet/Getty Images).

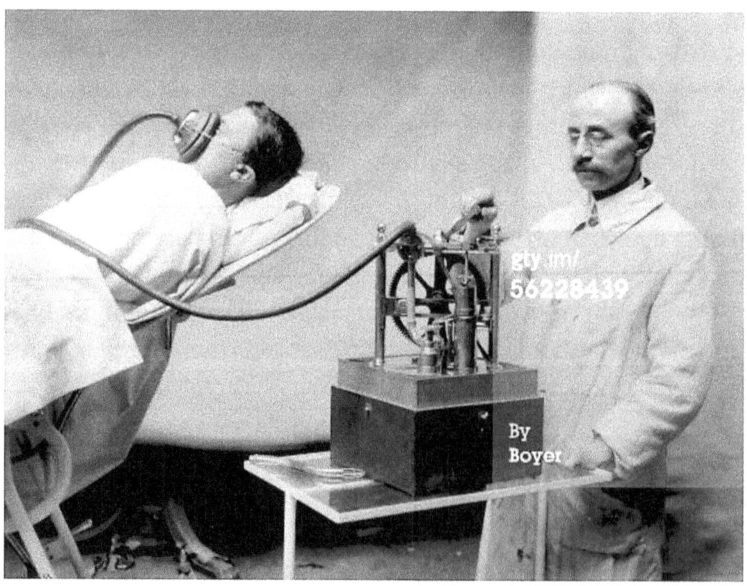

5. LITERATURE REVIEW

Halliday (2001) in his paper titled "Scholarly Communication, Scholarly Publication and the Status of Emerging Formats," explored the role of scholarly communication and provided a working definition of scholarly publication consisting of a list of criteria, which may be used to analyse the degree to which emerging formats can be categorized as scholarly publications. The paper also helped to identify the means by which they may be supplemented so that their status may be promoted to that of the 'scholarly publication.'

Ghosh (2006) in his paper "Open Access and Institutional Repositories: A Developing Country Perspective – A Case Study of India," discussed the concept of scholarly communication as a method wherein the author facilitates the availability and distribution of scholarly communication freely, as a means and effort to solve the problem of inaccessibility, primarily due to financial constraints, particularly in the developing countries. He mentioned that in India there has been a gradual realization of the usefulness of open access among various institutions. Various Open Access Initiatives have been undertaken and are operational. Many are in the developmental stage. Some Initiatives have also been taken in the area of metadata harvesting services, particularly publicly funded ones. He claimed that the future of open access in India is dependent upon a proper policy and developing a proper framework: in the implementation of open access, LIS Professionals should play a proactive role in the growth of collections in Institutional Repositories. The paper provides an overview about the present state of Open Access Initiatives by various institutions of the country.

Koganuramath (2006) discussed the future plans towards implementation of Institutional Repositories in JNU Central Library and also described some of the available Institutional Repository (IR) software. He stressed that IR is the one and only way to make the whole world know the esteem of a university. Being a premier university in teaching and research programmes, IR aims to manage networked information services, research work for benefits of faculty, students, and research scholars.

Chakravarty (2006) in the paper titled "Institutional Repositories: A Perspective for the Universities," discussed the current situation of scholarly communication. According to him, the world is witnessing a sea change in the area of scholarly communication. Perhaps the control over scholarly communication has started a gradual shift from commercial publishers to academic organizations and many authors initiatives in the area of Open Access. OA is perhaps opening up the major barriers that higher education institutes and libraries face, especially when it comes to escalating journal prices and shrinking budgets. Institutional Repositories are one of the two most powerful tools to empower and strengthen the open access movement. Universities and other academic institutions of the developed countries are already reaping the rich benefits of IRs. He emphasizes that the technology is free, the software is available free of cost, and the universities are also having the necessary infrastructure for implementing institutional repositories at their premises. The only required link that is missing is the awareness and willingness. He argues that it is high time that Indian universities make a decision and a strong commitment to develop institutional repositories and convince the faculty members and research scholars to deposit papers in digital archives. He also considered that IR may also contain learning objects in digital formats, thus facilitating IT-enabled pedagogy in Indian Universities.

Rao (2007) in the paper "Institutional Repository: A Key Role for Libraries," mentioned that institutional repository is a digital archive, owned and maintained at institutional level. It is a tool for collecting, storing, and disseminating information to advance scholarly communication. This paper also mentioned essential elements of institutional repositories. He also discussed the challenges for creating institutional repositories at a national level. The key role of the libraries in successfully implementing the institutional repositories is discussed.

Vishala (2007) in the article "Building Institutional Repository: Role of the Library" emphasized in an intellectual life and output of an institution. Institutional repositories are being recognized as an essential infrastructure for scholarship in the digital world. The paper seeks to provide an overview of institutional repositories and their benefits to the institutions and also describes the role of library in building institutional repositories.

Singh (2007) in this article titled "Open Source Software: A Comparative Study of Greenstone and DSpace," provided the information that software is available free of cost with source code for anyone to use. Open source software reduces the cost of creating digitization. Among open source software, Greenstone and DSpace are becoming more popular in India. This paper deals with comparisons of these two popular open source softwares. This paper helps the professionals who are planning to create and help the professionals who are planning to create an institutional repository.

Potel (2007) in his paper titled "Institutional Digital Repositories/e-Archives: INFLIBNET Initiatives in India," the author describes the technological advances, storing knowledge in less-expensive ways to create institutional preservations. INFLIBNET (Information and Library Network centre) decided to opt for DSpace for institutional repository and archive its publications, conference proceedings, and lecture notes. Tghe paer narrated the practical experiences and provides an overview of INFLIBNET's Institutional Repository and Digital Archive-India developed for the Indian academic and research community to archive their intellectual work.

Gyatri (2007) in her paper titled "Sharing Digital Resources with an Institutional Repository," emphasised on the need for storing, reusing, and sharing these digital information resources for knowledge sharing and conservation of time and effort. She discussed the concept of IR as "collection of digital objects" is where digital content assets are stored, searched, and retrieved for later use. Institutional repositories are being developed by academic institutions for the purpose of knowledge sharing, reuse, and preservation. This paper discusses the DSpace open source digital repository software, implemented at ICFAI Business School, Ahmadabad, and highlighted the need for information sharing.

Khan (2008) in his paper titled "An Assessment on Present Situation of Institutional Digital Repositories in India: A Study," highlighted the present status of Institutional Repository (IR) in India by its collection type, subject coverage, and total number of digital repository collection available to the academic community as open sources. For this purpose, the directory of Open Access Repositories (Open DOAR) and the Registry of Open Access Repositories (ROAR) have been consulted. This paper focuses to develop humanities and social sciences, institutional repositories in India. Results have been shown by collection type and subject coverage.

Nazim (2008) in his paper titled "Open Access Journals and Institutional Repositories: Practical Need and Present Trends in India," described trends in open access publishing in India. Data was collected from directors of Open Access Journals and Institutional Repositories. The URL of each Institutional Repository and Open Access Journal was visited to collect relevant data and included information from earlier studies. Case study method was used to know the trends of open access publishing in India. The data is analysed based on certain parameters, such as number of Institutional Repositories and Open Access Journals, number of documents, software used, types of documents, etc. Among the top 25 Open Access publishing countries, India ranks 12th for the overall number of journals, but drops to 18th for journals with online content. However, its positioning in the list of open access journals is fifth. At present, India ranks 12th in the list of countries with registered interoperable archives in the Registry of Open Access Repositories (ROAR).

Kamila (2009) in her paper titled "Institutional Repository Project in India," discussed the concept of the Institutional Repository, its relevance, merits, software requirements, and current trends in India, with special reference to the initiatives at Burdwan University, Burdwan.

Antelman (2010) in his article titled "Do Open-Access Article Have a Greater Research Impact?," Focussed at four disciplines at varying stages of adoption of open access—philosophy, political science, electrical and electronic engineering, and mathematics—to see whether they have a greater impact as measured by citations in the ISI web of science database when their authors make them freely available on the Internet. The finding revealed that, across all four disciplines, freely available articles do have a greater research impact. Shedding light on this category of open-access reveals that scholars in diverse disciplines are adopting open-access practices and being rewarded for it.

Dev (2010) in his article titled "Institutional Repository," urged to create a way for organizations to provide users with a method for self-service preservation and dissemination of digital users with a method for self-service preservation and dissemination of digital information. He found that, many organizations have reported difficulty convincing users to voluntarily deposit material into their repositories. The paper discussed the experience of The University of Rochester and carried out studies of its faculty and students to determine the reasons for their lack of interest in using the repository. Driven by these user needs, the University of Rochester created institutional repository which offers tools for authoring, collaboration, portfolios, and analytics—providing value throughout the authoring and academic lifestyles.

Bhat (2010) in his paper titled "Open Access Repositories in Computer Science and Information Technology: An Evaluation," the author explained evaluation of nine Open Access Repositories in the field of computer science IT. The repositories were identified from the open DOAR, Directory of Open Access Repositories (http://www.opendoar.org/), the findings revealed that most of the repositions are maintained by 1-2 faculty members on a part-time basis. Most of the repositories follow policies for selection of content and submission of documents have provision for withdrawal of content by the authors and voluntary faculty deposit policies. Backups are taken for short-term preservation of the content, but still no strategies for long-term preservation of the content. He observed that authors are responsible for ensuring copyright compliance of their articles. Most repositories have feedback from users. A few repositories provide access statistics. The paper provided recommendations for the establishment and management of open access repositories.

Nazim, M., and Mukherjee, B. (2011). The article discusses the status of institutional repositories (IRs) in Asia, with information on the effect of information communication technologies (ICT) on academic publishing. Topics include the efforts to launch an Open Access (OA) model that utilizes the Internet to broaden access to information resources and publications, the definition of IRs as a platform through which the research output of a specific university or research organization is available, and the growth and distribution of IRs throughout Asian countries based upon the survey results.

Jain, P. (2011). The paper reviewed the recent literature about institutional repositories (IRs) including the benefits and possible obstacles of setting up an IR. It discussed librarians' and authors' participative roles and open access. The findings reveal that in spite of all the obstacles to successful implementation, including associated negative perceptions, IRs have been increasingly recognised as a vital tool for scholarly communication and an important source of institutional visibility and a viable source of institutional knowledge management. The paper aimed at institutions with low-use repositories and at institutions considering initial development of an IR. The paper outlines the implications for IR practice for different groups, namely authors, librarians and academic administrative staff. It could, therefore, be used to persuade and influence different sets of stakeholders at institutions with under-populated or embryonic IRs, about the value of open access, the importance of depositing material and the potential functionality afforded by IR packages.

Krishnamurthy, M. M., and Kemparaju, T. D. (2011). The purpose aimed to study the IRs in use in Indian universities and research institutes. Repositories in various institutions in India were accessed and described in a standardised way. The 20 repositories studied covered collections of diverse types. Most of these collections have unique content. Originality/value - The goal of this study is to study the IR software and data based on the content type, metadata and characteristics. The paper also describes the collections and some important observations from this study.

Sawant, S. (2012). The study aimed to investigate various issues concerning the management of institutional repositories (IRs) developed in India. The survey method was used with the help of a web questionnaire. The questionnaire was e-mailed to the entire population i.e., all IRs identified in India. It was observed that 79 per cent of the institutions had used the DSpace IR software package. The respondents considered end-user interface to be the top-ranking IR-system feature. It was found that all IRs supported text (HTML, Postscript, PDF, Spreadsheet etc.) file formats. Half of the respondents marked bitstream copying as a long-term preservation strategy. Almost all institutional repositories were OAI-PMH compliant.. The study identified the existence of 16 functional IRs, some of which were not registered in any of the directories such as ROAR or Open DOAR.

Roy, B., Mukhopadhyay, P., and Biswas, S. (2012). The article presents the results of an analytical study investigating institutional digital repositories (IDRs) in India. An overview of the goals of IDRs and their association with the open access movement has been discussed. It was observed that the IRs are developing at a reasonably rapid pace in India and that they have the potential to be a very useful resource is commented on. Detailed information on the structure, policies, and capabilities of the 60 Indian IDRs studies is also offered and commented on.

Nagra, K. A. (2012). The author observed that IRs are increasingly becoming an essential component of academic institutions. An institutional repository showcase scholarly and research output to the wider community, and significantly helps in institutional advancement and outreach. It has the benefit of sharing and marketing research with others. It also provides an opportunity to raise the profile and brand awareness of an institution, faculty and students to the global community. Building an effective and successful institutional repository for an academic institution requires careful planning and enthusiasm from the institution's community. This article describes the detailed steps based on the review of the literature and the author's experience in creating institutional repositories. The libraries need to make critical decisions and choices before designing or redesigning institutional repositories. Among them are the funding, staffing, technology, metadata, content and operations to build, manage and run the institutional repositories successfully.

(Johnson, 2002; Nagahban, 2010). A study of nine libraries in the National Capital Region of India was conducted to: identify the benefits of IRs, learn the satisfaction level of users with respect to IR facilities, identify the incentives for publication in IRs, and identify the appropriate policies to be adopted by institutions for implementing IRs. The results of the responses from the 496 respondents indicate that: 1) Most Faculty members and Research scholars in this study indicated that they do not publish their research in IRs. 2) Opinions about the benefits of IRs are user-specific. For instance, a substantially larger percentage of those Faculty members and Research scholars in the study who do use IRs are Most satisfied with potential benefits of IR while a relatively higher percentage of Students fall in the Satisfied category. 3) The benefits of IRs ranked higher, although this varied by the different user groups was wider readership. Students ranked this and "quality aspect" – the opportunity to improve the quality of one's work through the provision of feedback from other researchers. 4) A serious concern about publishing in IRs is the potential for plagiarism and overall lower control over one's work. A second concern is the potential loss of content in IRs, which are often not archived as well as scholarly journals are. The findings suggest that institutions need to give due attention to policies related to two aspects of publication, especially those related to quality and copyright issues and to the academic value of research output. Other policy-related topics include a citation in another publication with due acknowledgement, inclusion in indexing systems for retrieval, interoperability with other IRs, and Permanent storage.

Haridasan, S., and Rani, K. (2012). The article offers information on the research conducted to evaluate the Institutional Repositories (IRs) initiatives in Indian universities. It mentions that the Electronic Thesis or Dissertation (ETDs) is the most commonly deposited material in IRs. It revealed that Dspace is the commonly used software and English is the most prominent language used by all IRs and the popularity about IRs can be increased through postings in list serves, web search engines and metadata harvesting services.

Fralinger, L., and Bull, J. (2013). The authors attempted to identify factors that might affect the international usage of US IRs as part of assessment efforts to determine an IR's return-on-investment. A survey was disseminated to IR administrators asking for demographic information, international usage counts for website hits and downloads, and any internationalization efforts connected to the IR in order to determine any influencing factors on an IR's international usage. While many IRs reported various rates of international usage, the largest group of respondents did not report an international usage rate for both page hits and downloads, despite overwhelmingly expressing an importance of international traffic to their IR and parent institution. Research limitations/implications – It is not clear if this non-reporting of international usage could be due to ignorance, apathy, or lack of technological support on the part of the IR administrators. Practical implications – Determining international usage as a part of an IR assessment might be problematic or even impossible for many US IRs. Originality/value – This study suggests that many IR administrators either do not know, do not care, and/or cannot record international usage data for their respective IRs, which could hinder determining an international return-on-investment for the IR.

Inefuku, H. W. (2013). In the interest of providing access to and preserving scholarship produced by faculty, students and staff, many universities have developed and implemented institutional repositories. Repositories are often organized into communities that correspond to campus units, including departments, research centres and institutes, and administrative offices. Universities, however, frequently undergo academic restructuring, which can make repository organization out of sync with university organization. This article addresses the impact of academic restructuring on repositories and looks to practices from archival arrangement and description to create repository organizational structures and community descriptions flexible enough to reflect the university organization, despite changes arising from academic restructuring.

Ruiz-Conde, E., and Calderón-Martínez, A. (2014). Are institutional repositories mere warehouses for digital documents or are they in fact establishing themselves as a rigorous option for the spread of scientific knowledge? This study analyses the competitive environment of the Top100 university repositories, defined as leaders in terms of market participation and penetration. The study also analyses the basic functionalities of preservation and diffusion of academic production through factors related to the prestige of the repositories and of the institutions that operate them. The results show that repositories with a larger digital academic supply are associated with the production of demonstrated scientific rigour.

Nwadiuto Igwe, K. (2014). The paper is a blueprint designed for the development of the country, with emphasis on making it one of the top developed economies by the year 2020. The study showed that knowledge/information (its generation, acquisition, utilization and possible application) was not accorded the due attention as a tool for the vision's success. Added to the above statement is the fact that the research findings and intellectual output of higher education institutions (HEIs) and research-based institutes in the country mostly in the form of grey literature (i.e. Unpublished technical/research reports, manuals, theses and dissertations, monographs, etc.) gather dust in their institutions of origin, with little or no visibility, thereby affecting accessibility, utilization and application. This paper advocated for a national policy for the development and management of open access repositories in institutions, where the research findings and intellectual outputs of these institutions will be domiciled, which will be easily accessible online for utilization and possible application in various sectors of the economy for the realization of vision 20: 2020.

6. INSTITUTIONAL REPOSITORY (IR): AN INTRODUCTION

Clifford Lynch defines IR as " a university-based Institutional Repository is a set of services that a university offers to its members of its community for the management and dissemination of digital materials created by the institution and its community members. It is most essentially an organizational commitment to the stewardship of these digital materials, including long-term preservation where appropriate, as well as organization and access or distribution." Foster and Gibbons describe IR as "an electronic system that captures, preserves, and provides access to the digital work products of a community." Wikipedia defines IR as "an online locus for collecting and preserving in digital form the intellectual output of an institution, particularly a research institution. For a university, this would include materials such as research journals, articles before (pre-prints) and after (post-prints) undergoing peer review and digital versions of theses and dissertations, but it might also include other course notes, or learning objects."

6.1 Types of Material

Types of resource material include blue print, pre-print, post-print, working paper, thesis/dissertation, technical report, conference papers, progress reports, research articles, A-V chips, teaching material, diversity of contents and language used in full texts undergoing peer review and digital versions of course notes. The contents can be classified, restricted, or unrestricted/open.

6.2 Characteristics of Institutional Repositories

- Promote self-archiving.
- Are institutionally defined rather than subject based.
- Have wider distribution.
- Increases dissemination, collect content in a single location.
- Are supported by Scholarly Communication.

The following characteristics of Institutional Repository (IRs) are outlined by Crow (2002) Shearer (2003) and Ware (2004):

- **Digital:** IR collect mainly digital information. This is different to the mandate of the university archives which is to collect all types of content related to the university
- **Institutionally Defined:** Unlike the traditional discipline-specific and digital libraries, institutional repositories capture the research of the entire population of the institution
- Institutional repositories collect scholarly material exclusively. In this context, scholarship includes (preprints, post prints, technical reports, etc.)
- **Cumulative and Perpetual:** This refers to the long-term nature of data preservation and accessibility through IR
- **Open Access:** A key defining feature of IR is the free and open access to their content
- **Interoperability:** This refers to the fact that IRs can be used by both data providers (who expose data in various forms) and service providers (who process and add value to the data)
- **Self-Archiving:** Most IRs requires the author or someone associated with the author to deposit the content directly.

6.3 Need and Advantage

Institutional Repositories facilitate innovation, wider access, visibility to research output, preserve institutional heritage, reduce publication delay, faster communication, increase citation, strength research, easy access to grey literature, persistent URL, research work archive. While the most popular open source and hosted applications share the advantages that institutional repositories bring to institutions, such as the increased visibility and impact of research output, interoperability and availability of technical support, institutional repository advocates tend to favour open source solutions for the reason that they are by their nature more compatible with the ideology of the freedom and independence of the internet from commercial interests. On the other hand, some institutions opt for outsourced commercial solutions.

In her briefing paper on open access repositories, advocate Alma Swan lists (Swan, Alma. "Open Access institutional repositories: A Briefing Paper". *Open Scholarship*. Retrieved 24 September 2013.) The following as the benefits that repositories bring to institutions:

- Opening up outputs of the institution to a worldwide audience;
- Maximizing the visibility and impact of these outputs as a result;
- Showcasing the institution to interested constituencies – prospective staff, prospective students and other stakeholders;
- Collecting and curating digital output;
- Managing and measuring research and teaching activities;
- Providing a workspace for work-in-progress, and for collaborative or large-scale projects;
- Enabling and encouraging interdisciplinary approaches to research;
- Facilitating the development and sharing of digital teaching materials and aids, and
- Supporting student endeavours, providing access to theses and dissertations and a location for the development of e-portfolios.

The Institutional Repository will:

- Help to develop National Research Repository infrastructure by setting up, populating, and linking individual repositories.

- Stimulate development of services that draw on research information made available through the repository infrastructure.
- Provide a window that gives open access to improve the sponsoring institution visibility and status.
- Support the Open Access model of publication.

Institutional Repositories are an online locus for collecting, preserving, disseminating in digital form. Institutional Repositories are intellectual research output of an institution. It manages and disseminates its research material created by community members. Librarian plays a role of a leader in planning and building these repositories. They act as experts in the collection, preserve for document and providing digital information.

7. RESEARCH METHODOLOGY

The present study uses *Open*DOAR for generating a list of IRs in the field of Medicine and Biology. *Open*DOAR is a project to provide a comprehensive and authoritative list of repositories categorised by several criteria like country, software, subject, etc. It has two subject categories related to the scope of the present study. One is Biology and Biochemistry and the other is Health and Medicine. Indian IRs under these two subject categories were explored for further study. Eight IRs was listed under each subject category respectively, with minor overlapping. The total number of relevant IRs after combining the results of the two lists and removing the overlapping comes out to be 12. For the purpose of data analysis last five year data were collected from individual repositories.

Table 1. List of Indian IRs in medical biology

SN	Repository Name	No. of Records	Software	Status
1	CMFRI Digital Repository	9661	EPrints	Operational
2	Digital Knowledge Repository of Central Drug Research Institute (CDRI)	796	DSpace	Operational
3	ePrints@NII	10	EPrints	Broken
4	EPrints@NIRT	830	EPrints	Operational
5	Eprints@SBT MKU	89	EPrints	Broken (data till October 2011)
6	Indian Institute of Petroleum (IIP) Institutional Repository	454	DSpace	Broken
7	Institutional Repository@CSIO	311	EPrints	Operational
8	IR@NPL	996	EPrints	Operational
9	Eprints @MDRF	676	Eprints	Operational
10	Open Access Repository: Indian Academy of Sciences	91745	Eprints	Operational
11	KNOOR: Knowledge Repository Open Network	911	DSpace	Operational
12	OpenMED@NIC	2904	Eprints	Operational
Total		**108502**		

8. DATA ANALYSIS

Out of these twelve repositories, 3 were found to be broken, i.e. non-functional. The Open Access Repository of Indian Academy of Sciences (publications of Fellows) has maximum number of records – 91745. Table 2 also reveals that the total research output under OA by these IRs in Medicine and Biology including Pharmacology is 108502.

Indian IRs has a share of 5.84% of the total IRs in Biology and Biochemistry whereas its contribution is 3.2% in the discipline of Health and Medicine. Table 2 also reveals that Indian IRs in Medicine and Biology at global level is 4.12%.

Figure 2 reveals that 15% Indian IRs are dedicated to Medicine and Biology and 85% are dedicated to other disciplines.

Figure 3 depicts that The Open Access Repository of Indian Academy of Sciences has the highest share of scholarly outputs in Medicine and Biology. While the CMFRI Digital Repository has 9661 records followed by OpenMED@NIC which has only 2904 records, although it specifically deals in Medical Sciences.

DSpace is an open-source digital asset management system originally created by developers from MIT and HP Labs in 2002. EPrints is a free and open-source software package originally developed by researchers at the University of Southampton School of Electronics and Computer Science in 2000 (making it the oldest of the platforms in this report). It was designed specifically for archiving research papers, theses and teaching materials, though it can accept any content. The other popular IR software includes Fedora, Figure 4 reveals that 80% of the Indian IRs on selected discipline are running on Eprints software whereas only 20% of the IRs are based on Dspace.

Table 2. Share of Indian IRs in medicine and biology at global level

SN	Discipline	World	India
1	Biology and Biochemistry	137	8 (5.84%)
2	Health and Medicine	251	8 (3.2%)
	Total	388	16 (4.12%)

Figure 2. Medicine and biology IRs of India

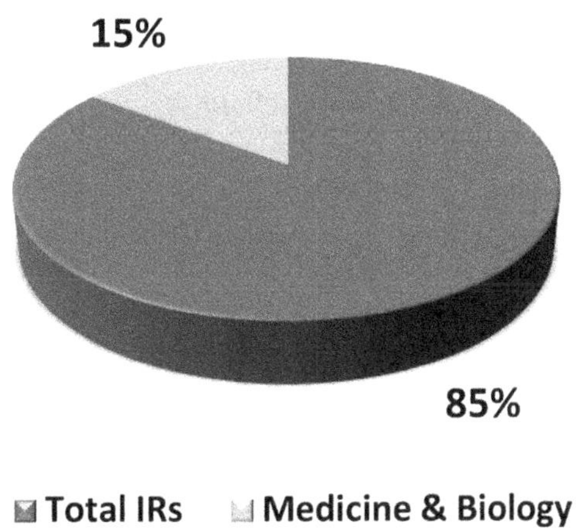

Figure 3. Share of scholarly literature by Indian IRs

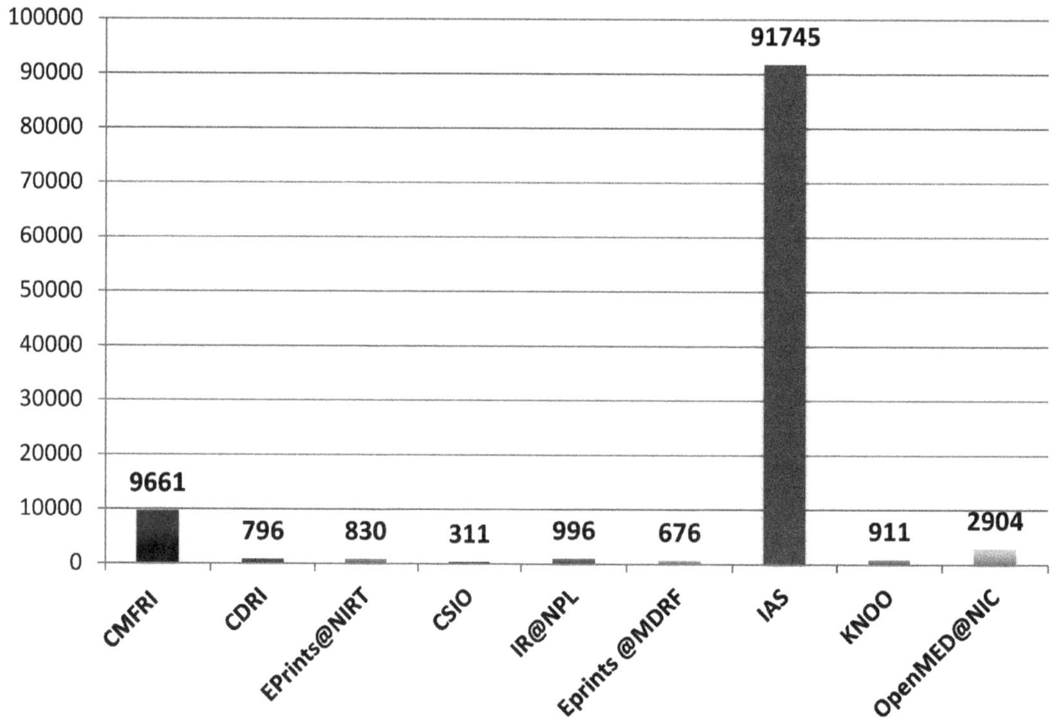

Figure 4. IR software preference

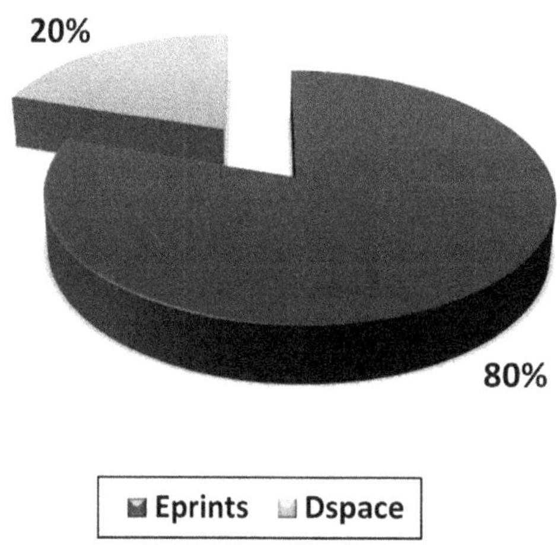

9. INDIAN IR IN MEDICINE AND BIOLOGY

9.1 Eprints@CMFRI

Eprints@CMFRI is the Open Access Institutional Repository of Central Marine Fisheries Research Institute. The research outputs of CMFRI include: journal papers, conference papers, reports, theses, patents etc. - are uploaded/self-archived by CMFRI scientists who do research on fisheries and related areas. Interested users can freely download and use documents as most of them are directly accessible and full-text downloadable. 'Request Copy' forms can be used for documents to which direct full-text download is restricted due to publisher embargo.

Figures 5 and 6 reveal that there is a significant rise in the number of scholarly content from 2011 which continued up to 2012. After July 2012 it has maintained a constant growth rate with the minor upward trend. The figure also reveals that the overall growth rate is over 1737 items per year.

9.2 Digital Knowledge Repository of Central Drug Research Institute (DKR@CDRI)

CDRI is considered to be a pioneer multidisciplinary research organization in the field of biomedical research where all the infrastructure and expertise are available to develop a drug right from its concept to market. The very latest techniques and methodologies are employed for developing drugs, diagnostics and vaccines to combat diseases prevalent among mankind in general and Indian population in particular. For administrative and scientific purposes the Institute's manpower has been grouped into 17 R&D divisions and few divisions providing technical and scientific support. In order to carry out work in cohesion and in a focused environment, the research works have been grouped into following six major research areas. DKR@CDRI strives to collect, preserve and disseminate different institutional publications (journal articles, conference proceeding articles, technical reports, thesis, dissertations, etc.). Users can search, browse and access publications of CDRI from this collection.

Figure 5. Growth rate of CMFRI IR

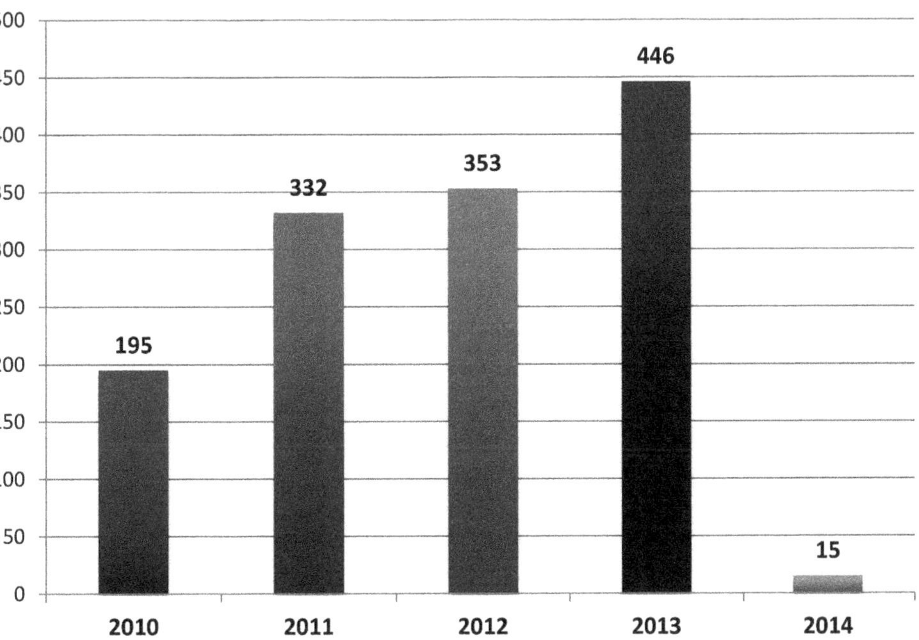

Figure 6. Growth statistics of CMFRI IR 2011 onwards

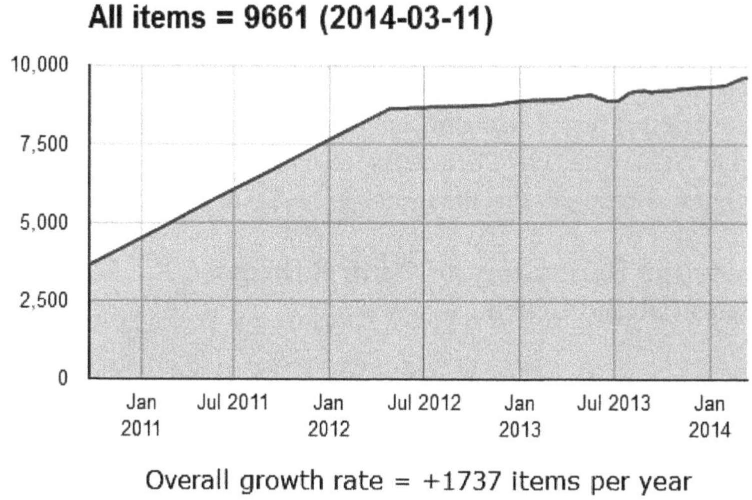

Figure 7. Growth rate of DKR@CDRI

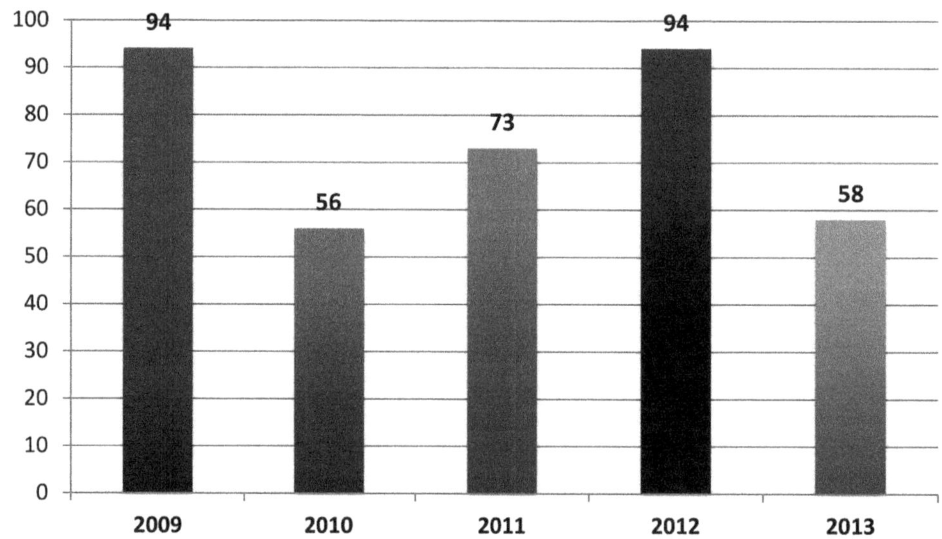

Figure 7 indicates that the highest number of e-prints were archived in the year 2009 and 2012 (94 each). However, there is a downfall in the number of records in the year 2013 as only 58 records were added during this year.

Figure 8 reveals that there is a rise in self-archiving from 2012 and is mentioned upto January 2014. This is a positive sign as more scholarly work is now available under open access. The figure also reveals that the overall growth rate is over 275 items per year.

Figure 8. Growth statistics of DKR@CDRI 2012 onwards

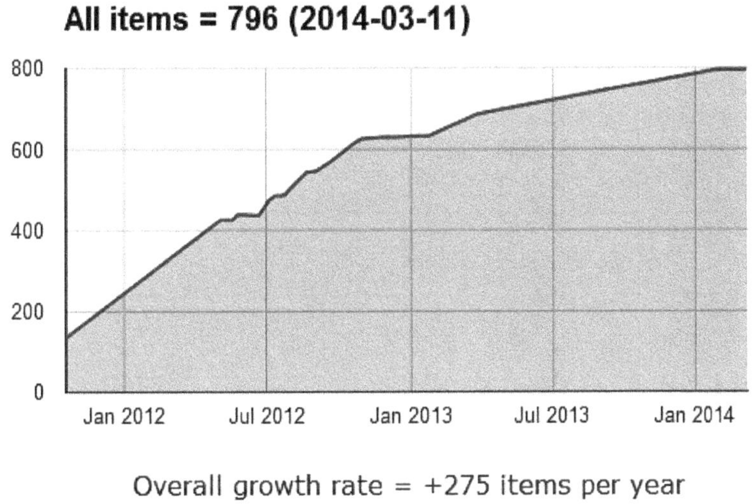

9.3 EPrints@NIRT

The National INstitute for Research in Tuberculosis (Formerly Tuberculosis Research Centre (TRC)), a permanent institute, under the Indian Council of Medical Research (ICMR), is an internationally recognized institution for Tuberculosis (TB) research. It is a Supranational Reference Laboratory and a WHO Collaborating Centre for TB Research and Training. Recently, an International Centre for Excellence in Research (ICER) in collaboration with NIH was established at the Centre. The centre evolved supervised intermittent chemotherapy, especially in urban areas and large cities subsequently. Later, treatment for approximately 6 months with powerful drugs has been shown to be equally effective and these regimens were evaluated under field conditions.

This is the Institutional Repository of the National Institute for Research in Tuberculosis (NIRTIR) Chennai, India. A scholarly archiving facility for NIRT scientific community, where the institute's intellectual output is preserved, searched and shared. The archive runs on EPrints, an open source software package. We expect the service to facilitate long-term preservation of our research output and provide easy access to these publications, as well as improve their impact. Further, by making our research findings freely available, we hope to contribute to the expanding knowledge base in the global fight against tuberculosis.

The data collected from EPrints@NIRT presents a very sad picture as we can see that after the year 2007 there is a steep fall in the deposit rate (Figure 9). To get the idea of post 2012, the IR growth chart from OpenDOAR was consulted which is reproduced in Figure 10.

Figure 10 reveals that the self-archiving momentum picked up from August 2013 and continued aupto December 2013 after which it become steady.

9.4 EPrints@CSIO

Central Scientific Instruments Organisation (CSIO), a constituent unit of Council of Scientific & Industrial Research (CSIR), is a premier national laboratory dedicated to research, design and development

Figure 9. Growth rate of EPrints@NIRT

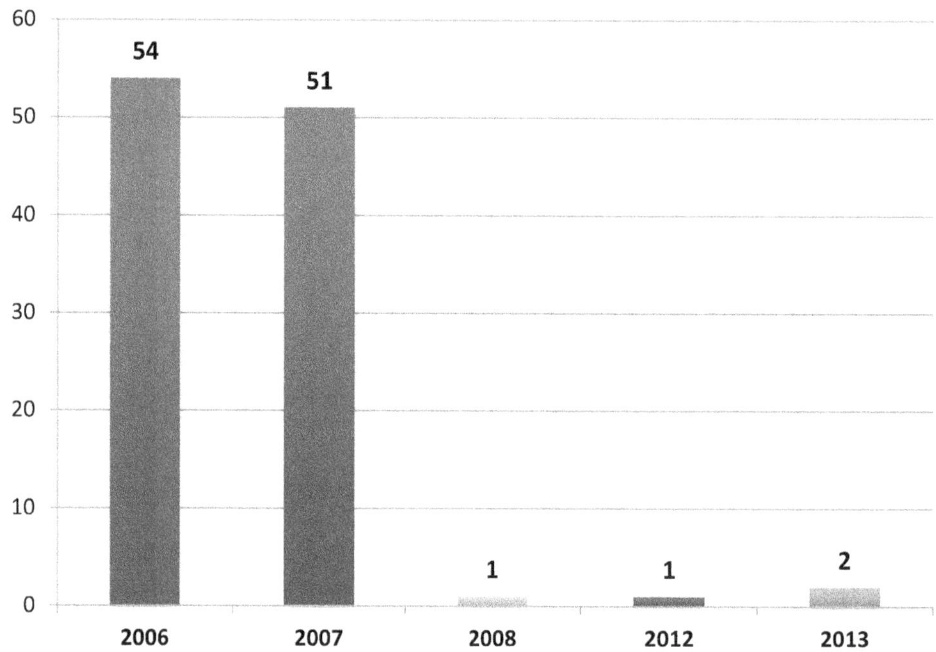

Figure 10. Growth statistics of EPrints@NIRT 2013 onwards

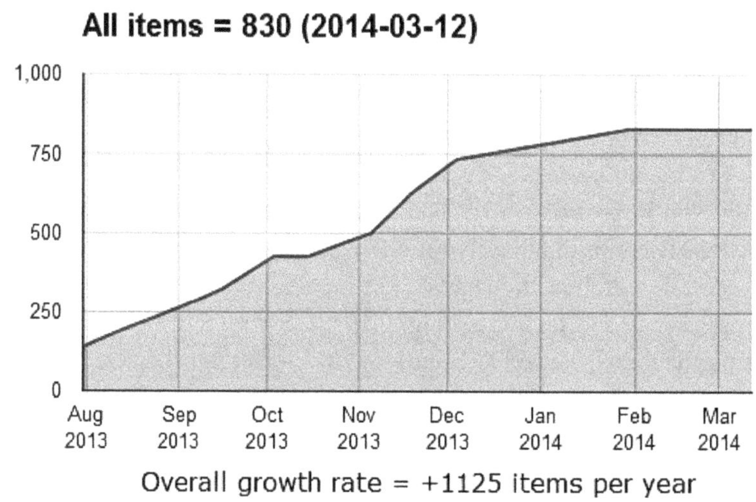

of scientific and industrial instruments. It is a multi-disciplinary and multi-dimensional apex industrial research & development organisation in the country to stimulate growth of Instrument Industry in India covers wide range and applications.

CSIO's Institutional Repository is the digital archive of the research output of our scientists. This knowledge base covers journal articles, conference papers, technical reports, presentation/lectures, pre-prints, Thesis, images etc. One can browse the documents by author, division, subject, date and document

Figure 11. Growth rate of EPrints@CSIO

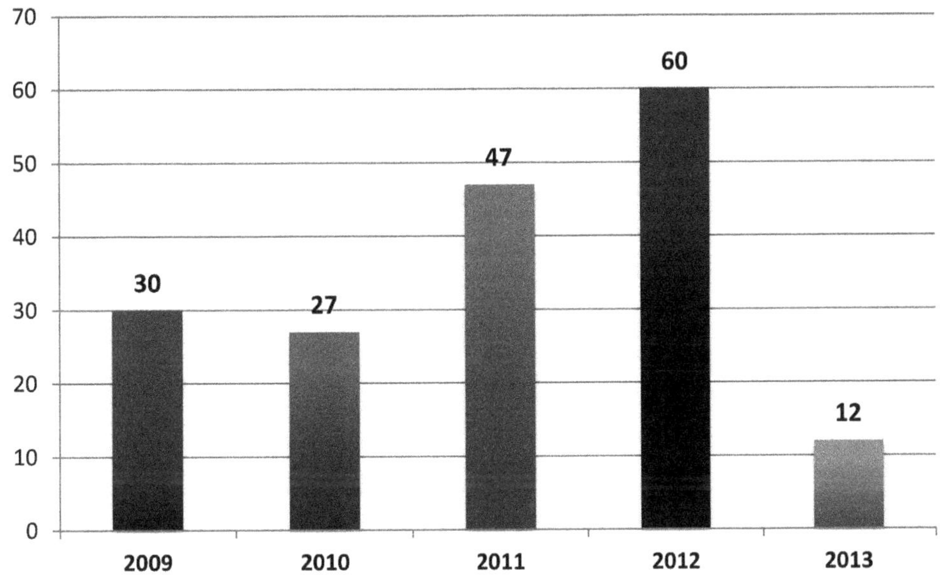

Figure 12. Growth statistics of EPrints@CSIO 2013 onwards

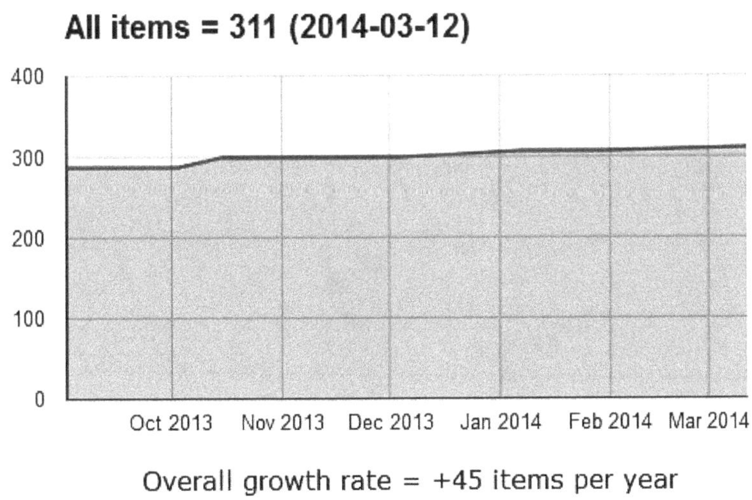

type. Both simple and advanced search facilities have been provided. CSIO scientists are welcome to submit their publications on their own by registering.

Figure 11 reveals that the highest deposit ware made in the year 2012 whereas only 27 submissions were made in 2010. In the year 2013, the IR has observed lower self-archiving rate. For post 2012 self-archiving rate the OpenDOAR date is presented below.

Figure 12 reveals that the archiving rate is almost constant from September 2013 to March 2014 and the overall growth rate is above 45 per year.

9.5 Eprints@SBT MKU

This is official website of School of Biotechnology (SBT), Madurai Kamraj University (MKU). This repository provides access to the research output of the institute. Users may set up Atom or RSS feeds to be alerted to new content. The software used is EPrints, the subjects covered are biology and biochemistry, content covered are articles and language used is English. As the repository is now in broken stage data is available till 2011 only. Figure 13 shows the growth rate of collection in SBT MKU EPrint@ NML from 2006 to 2011.

Figure 13. Growth rate of EPrints@SBT MKU

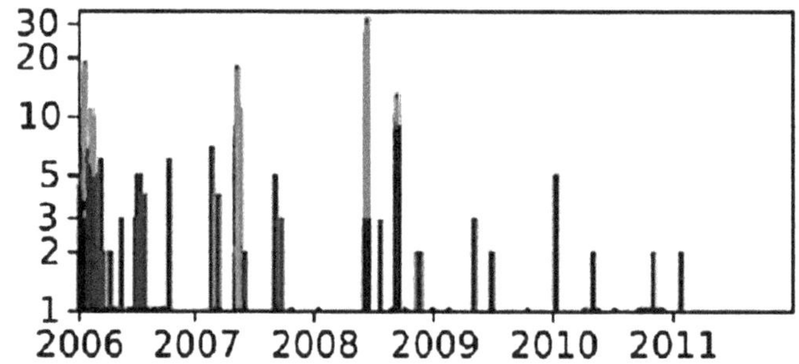

Figure 14. Growth rate of IR@NPL

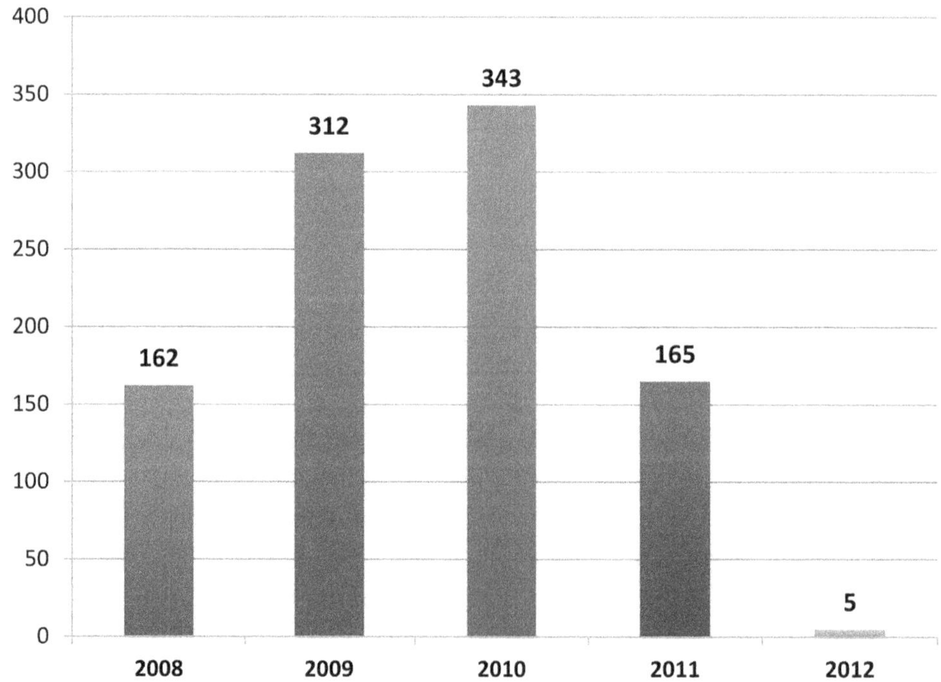

9.6 IR@NPL (National Physical Laboratory)

IR@NPL has been established/maintained by the National Physical Laboratory (NPL), New Delhi, India. It is an online digital archive of research outputs of Scientists/ Researchers of the NPL. The purpose of IR@NPL is to capture, preserve and provide free worldwide access to the intellectual outputs of the NPL in the form of journal articles, research papers, conference papers, technical reports, preprints, technical bulletin, annual reports, golden jubilee document, sameeksha etc. IR@NPL can be accessed globally, but submission of documents to this repository is reserved only for the NPL community. Deposits in this repository can be browsed by author, subject, year and document type, also have simple and advance search facility.

Figure 14 reveals that the highest archiving was done in the year 2010 (343) followed by the year 2009 (312), 2011 (165) and 162 in the year 2008. Lowest submission was observed in the year 2012.

Figure 15 reveals that there has been a constant submission rate after August 2013 and is maintained till March 2014. According to the above graph, the overall growth rate is more than 74 items per year.

9.7 Eprints @MDRF

The prevalence of Diabetes Mellitus is growing in epidemic proportions all over the world, particularly in India. It is now known that India has the highest number of diabetic subjects, in the world even higher than China and USA. India has the highest prevalence rates of diabetes i.e. about 20% of the total diabetic population in the world. In 1996, Dr. V. Mohan and Dr. M. Rema felt the need for an exclusive Diabetes Research Foundation to do original research work with the excellent infrastructural facilities. This resulted in the setting up of the Madras Diabetes Research Foundation (MDRF)

MDRF scientists do research on Diabetes and related fields and publish the outputs in professional journals and conferences. With a view to eliminating barriers to access our local research outputs, this open access repository has been set up. Here MDRF scientists archive post-prints of the research papers soon after it is accepted by journal publishers. In some cases published version is archived with "Request

Figure 15. Growth statistics of IR@NPL 2013 onwards

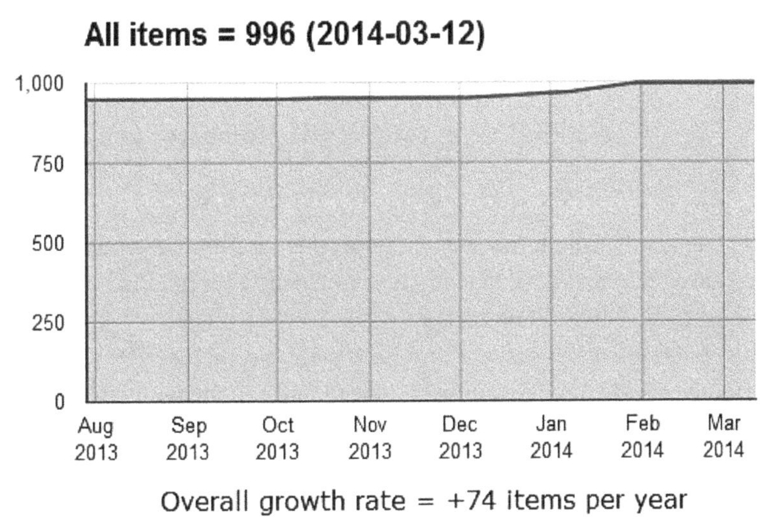

Figure 16. Growth rate of Eprints @MDRF

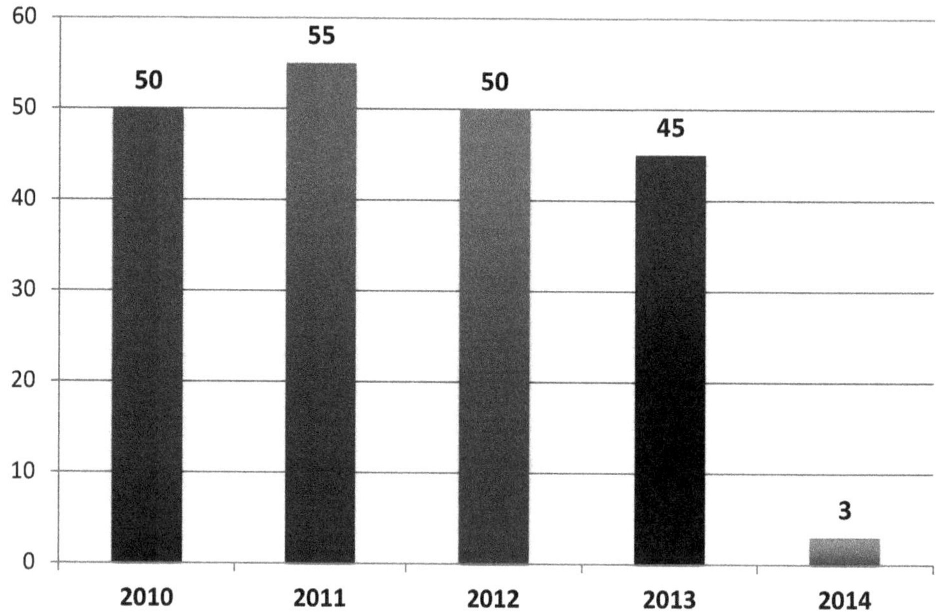

Figure 17. Growth statistics of Eprints @MDRF 2013 onwards

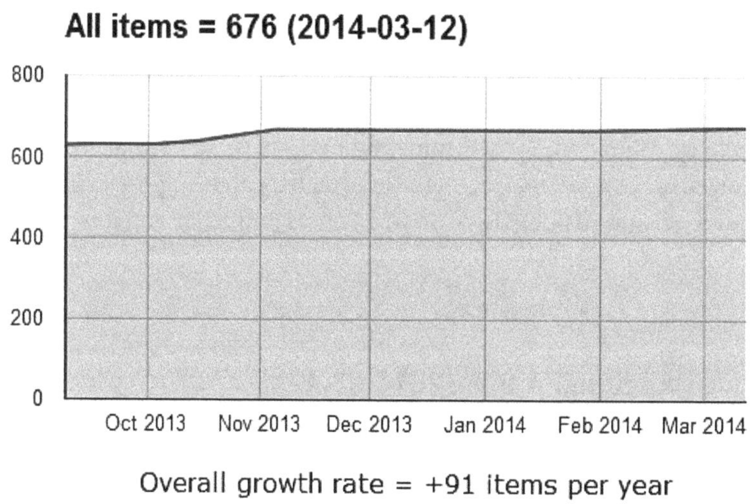

Copy" facility. Users can send "article request" through this facility. Dr. MGR Medical University and University of Madras have recognized MDRF as a research centre for the PhD program. Eprints@MDRF holds all the theses guided by MDRF scientists

Figure 16 displays the deposit rate of pre and post prints in the MDRF repository during the year 2010 to 2014. The figure reveals that the submission rate has neither witnessed a steep rise or downfall in the submission rate. However, only three submissions were being made at the time of writing the paper (March 2014).

Figure 17 depicts the growth rate of research publications after October 2013. The overall growth rate is above 90 items per year.

9.8 Open Access Repository: Indian Academy of Sciences (IAS)

The Indian Academy of Sciences was founded and registered as a society in 1934 with the aim to promote the progress and uphold the cause of science, both in pure and applied branches. We strive to meet our objectives through original research and dissemination of scientific knowledge to the community via our meetings, discussions, seminars, symposia and publications. The academy as on date has around 1,500+ Fellows, out of which, past members account for 600-700. It is estimated that the total number of articles published so far by all fellows in various national and international publications could be around 100,000.

Publications of the IAS fellows repository collects, preserves and disseminates in digital format the research output created by the fellows of the Indian Academy of Sciences. It enables the Academy community to deposit their preprints, postprints and other scholarly publications, and organizes these publications for easy retrieval. While Publications of the IAS fellows can be accessed by anybody, submission of documents to this repository is limited to the fellows of the Academy only. Publications of the IAS fellows repository are running on EPrints open archive software, a freely distributable archive system available from eprints.org. Publications of the IAS fellows complies with the Open Archives Initiative (OAI) framework allowing publications to be easily indexed by web search engines and other indexing services.

Figure 18 depicts that there is a sharp decline in the growth of IR in terms of archived papers. Number of records submitted in the year 2010 were 3001 which decreased to 2351 in the year 2011 and this trend continued in 2012 (753 records) and 2013 (99 records).

Figure 19 reveals that the overall growth rate of the IR is about 19500 items per year.

Figure 18. Growth rate of IAS repository

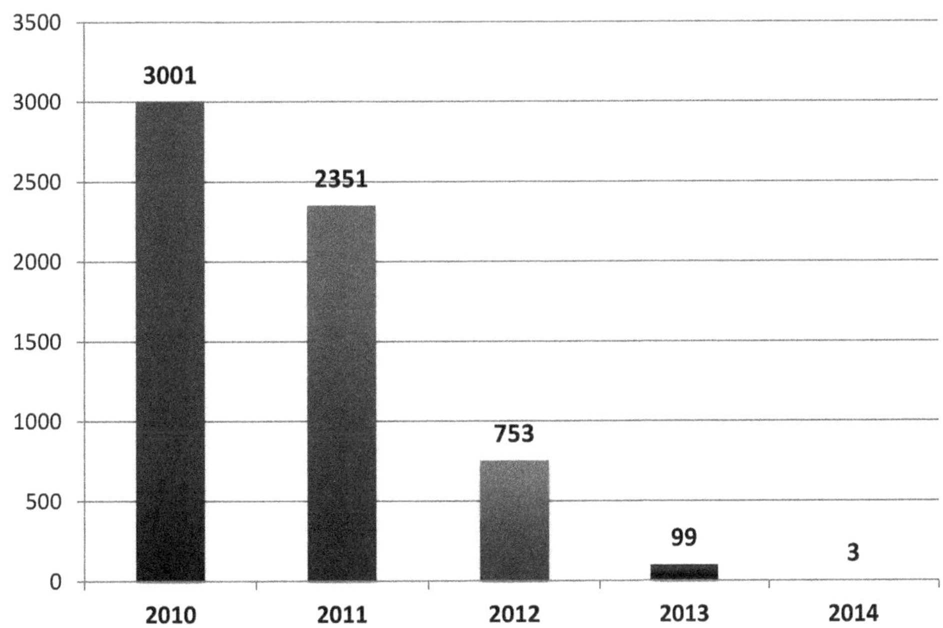

Figure 19. Growth statistics of Indian Academy of Sciences repository

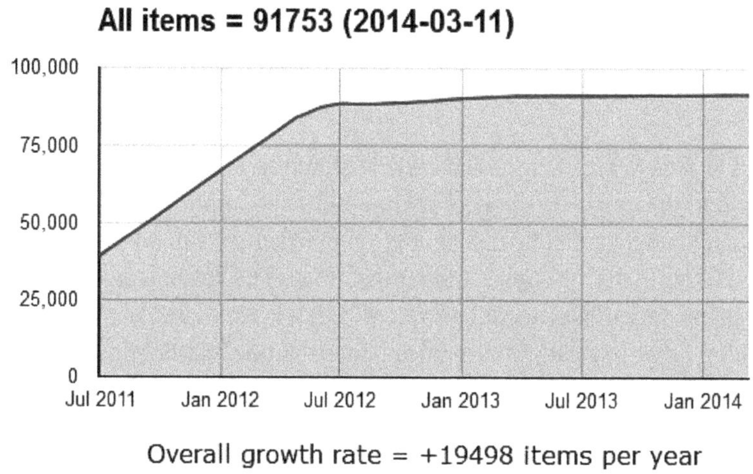

Figure 20. Growth rate of KNoor

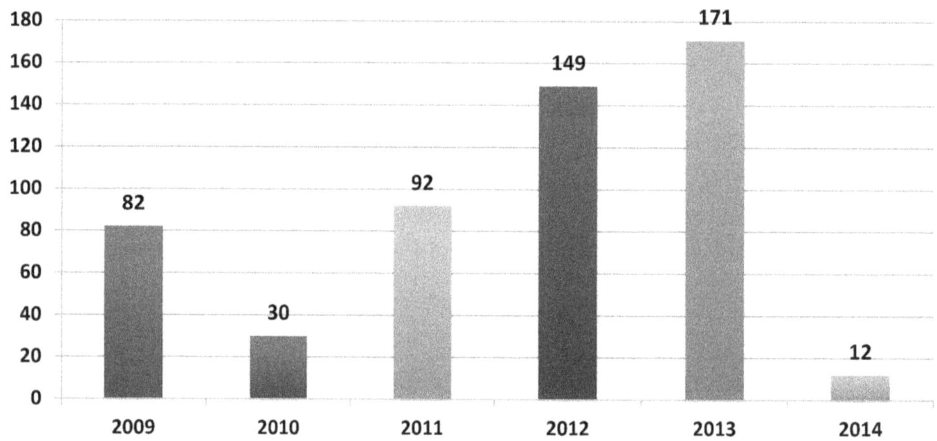

9.9 KNoor: Knowledge Repository Open Network

Knoor - Knowledge Repository Open Network is a joint Initiative of Premier Institutes of Jammu and Kashmir (J&K)". This repository is funded by University Grants Commission (UGC), the apex body of the Indian Higher Education system under the UGC-Major Project "E-repository for J&K". It provides access to the research output from research institutes of the J&K (viz. S.K Institute of Medical Sciences, S.K University of Agricultural Science & Technology and University of Kashmir). The interface is available in English. Many items are available as abstracts only. Users may set up RSS feeds to be alerted to new content. It provides an interface for the Faculty members and researchers, for creating their own collections in the IR. It also maintains multilingual Phd Theses. Authors are supposed to mail their contributions for uploading in the Repository (peer reviewed papers only OR pre-prints) for their free, open access and high visibility. Authors are also required to check whether concerned publisher permits self-archiving at SHERPA/RoMEO's publisher copyright policies. The service is designed, maintained

Figure 21. Growth statistics of KNoor 2012 onwards

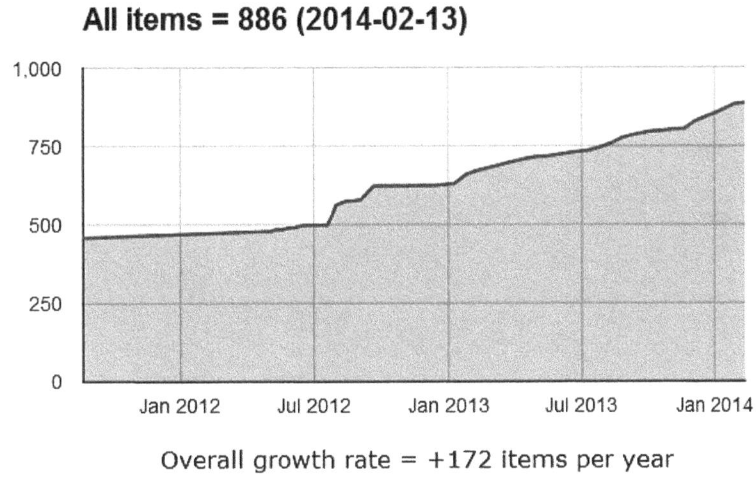

Figure 22. Growth rate of OpenMED@NIC IR

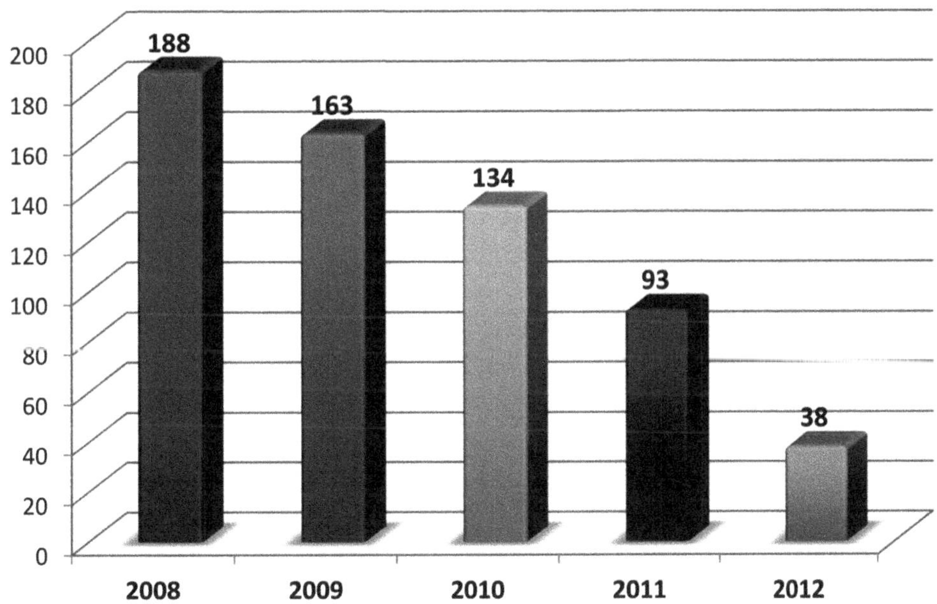

and regulated by the Department of Library and Information Science (University of Kashmir) with infrastructure support from Directorate of IT&SS, University of Kashmir.

Figure 20 reveals that the growth in the number of IT records has been consistent from 2011 onwards. It decreased from 82 (2009) to 30 (2010) but after that it gained momentum and there has been a constant growth in archiving by the researchers and scientists.

Figure 21 reveals that the self-archiving rate has increased from 2012 to 2014 and the overall growth rate is over 170 items per year.

Figure 23. Growth statistics of OpenMED@NIC 2009 onwards

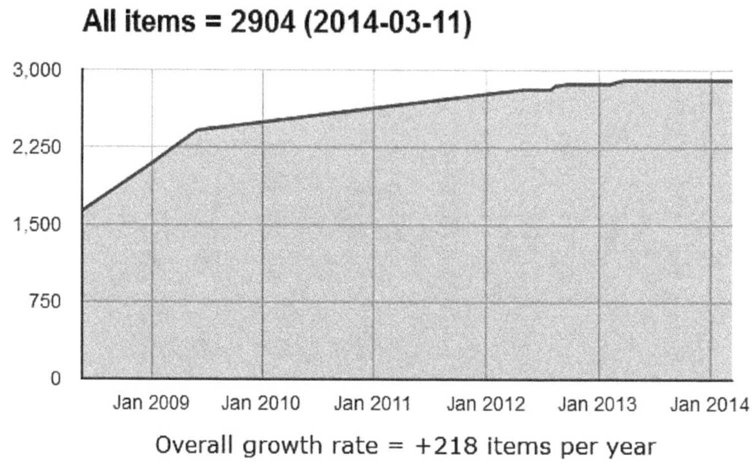

9.10 OpenMED@NIC

OpenMED is an Open Access Archive in the area of Medical and Allied Sciences, including biomedical, Medical Informatics, Dental, Nursing and Pharmaceutical Sciences. It is international in scope and includes both published (post-prints) and unpublished (pre-prints) peer-reviewed documents having relevance to research in these disciplines. It is hosted by Bibliographic Informatics Division, National Informatics Centre (India) using the GNU EPrints Archive Software. The major objective of OpenMED is to provide a free facility for authors to self-archive their publications for making them available under open access and improve their impact. It encourages authors to make their work visible, searchable and usable by potential users. The metadata (core descriptive information about the documents) is expected to be aggregated by various service providers and search engines compliant with OAI-PMH (Open Archives Initiative Protocol for Metadata Harvesting). It would help providing the authors the maximum exposure as well as easy and free access to the users. Submitted documents are placed into the submission buffer and becomes part of the OpenMED@NIC archive on their acceptance.

Figure 22 reveals that there is continuing decline in paper archiving as it was 188 in the year 2008 which came down to 38 in the year 2012.

Figure 23 reveals that the overall growth rate for OpenMED repository is 218 items per year.

10. CONCLUSION

An Institutional repository is gradually becoming the life and blood of a research organization as it provides the appropriate platform to showcase the research productivity. At the same time it also serves the humanity by facilitating open access to quality literature. It protects and preserves the intellectual capital of its research scientists.

Number of IRs in the field of medicine and biology specially covering Pharmacology as a subject is very low. Although there are many premier institutions undertaking R&D work in this domain but not all of them are having dedicated Irs. Moreover, the existing IRs lack in collection, excluding a few of them.

In such a scenario, the authors/scientists should be motivated for motivated for self- archiving. If self-archiving does not attract their attention, the approach of mediated-archiving should be facilitated. Other factors which need to be taken care of include quality, adequate budget for establishing new IRs as well as maintaining existing IRs, proper marketing and promotion of IRs, HR training for metadata creation and metadata harvesting, etc. More IRs in universities should be created, as it will enhance research and growth of the Nation's Research output. IRs should be designed to support and achieve the needs and goals of the institution, as it is the tangible way for the institution to communicate with the rest of the world.

Open Access is very important in the discipline of Pharmacology as humans, howsoever developed and advanced can't win immortality and will continue to suffer from diseases. There is an urgent requirement of free scholarly communication in this field so as to provide state-of-the-art solutions for well being of the society. It should be free from all barriers and should explore optimum benefits of Open Access either by following the golden route (OA Journals) or the green route (self-archiving in IRs)

REFERENCES

Bethesda Statement on Open Access Publishing. (2003, June 20). Retrieved March 11, 2014, from http://legacy.earlham.edu/~peters/fos/bethesda.htm

Björk, B.-C. (2012). The hybrid model for open access publication of scholarly articles: A failed experiment? *Journal of the American Society for Information Science*, *63*, 1496–1504. doi:10.1002/asi.22709

BOAI10. (2002, February 14). *Budapest Open Access Initiative | Read the Budapest Open Access Initiative*. Retrieved March 11, 2014, from http://www.budapestopenaccessinitiative.org/read

Carver, B. (2003). *Creating an institutional repository: A role for libraries*. Retrieved from http://marylaine.com/exlibris/xlib 181.html

Chakravarty, R. (2011). Self-archiving in open access institutional repositories: Whose court is the ball in? *International Journal of Information Research*, *1*(1).

Chakravarty, R., & Mahajan, P. (2011). Open access journal initiatives in India. *International Journal of Information Dissemination and Technology*, *1*(1).

Fralinger, L., & Bull, J. (2013). Measuring the international usage of US institutional repositories. *OCLC Systems & Services*, *29*(3), 134–150. doi:10.1108/OCLC-10-2012-0039

Haridasan, S., & Rani, K. (2012). Institutional Repository Initiatives in Indian Universities: An Evaluative Study. *Journal of Library and Information Science*, *37*(2), 71–87.

Inefuku, H. W. (2013). Whatever Happened to Art and Design?: Using Archival Practice to Manage the Impact of Academic Restructuring on Institutional Repositories. *Journal of Library Administration*, *53*(4), 209–222. doi:10.1080/01930826.2013.865383

Internet Repository. (2010). In *Wikipedia*. Retrieved December 17, 2010, from http://en.wikipedia.org/w/index.php?title=Institutional_repository&oldid=402861088

Jain, P. (2011). New trends and future applications/directions of institutional repositories in academic institutions. *Library Review*, *60*(2), 125–141. doi:10.1108/00242531111113078

Jeffery, K. G. (2006). Open Access: An Introduction. *Ercim News, 64*.

Joint, N. (2006). Institutional repositories, self archiving and the role of the library. *Library Review*, *55*(2), 81–84. doi:10.1108/00242530610649576

Krishnamurthy, M. M., & Kemparaju, T. D. (2011). Institutional repositories in Indian universities and research institutes: A study. *Program: Electronic Library & Information Systems*, *45*(2), 185–198. doi:10.1108/00330331111129723

Lynch, C. A. (2003). *Institutional repositones: Essential infrastructure for scholarship in the digital age*. Retrieved from http:// DSpace.uniroma2.it/DSpace/bitstream/2108/261/1/ir.html

Nagra, K. A. (2012). Building Institutional Repositories in the Academic Libraries. *Community & Junior College Libraries*, *18*(3/4), 137–150. doi:10.1080/02763915.2012.799028

Nazim, M., & Mukherjee, B. (2011). Status of Institutional Repositories in Asian Countries: A Quantitative Study. *Library Philosophy & Practice*, 60-75.

Nwadiuto Igwe, K. (2014). Open Access Repositories in Academic and Research Institutions for the Realization of Nigeria's Vision 20: 2020. *International Journal Of Information Science & Management*, *12*(1), 33–46.

Open Access. (2006). In *Wikipedia*. Retrieved 26 December 2006 from en.wikipedia.org/wiki/Open_access

Open Doar. (2008). *Introduction to the project*. Retrieved from http://www.opendoar.org/about.html

Roy, B., Mukhopadhyay, P., & Biswas, S. (2012). An Analytical Study of Institutional Digital Repositories in India. *Library Philosophy & Practice*, 1-13.

Ruiz-Conde, E., & Calderón-Martínez, A. (2014). University institutional repositories: Competitive environment and their role as communication media of scientific knowledge. *Scientometrics*, *98*(2), 1283–1299. doi:10.1007/s11192-013-1159-5

Sawant, S. (2012). Management of Indian institutional repositories. *OCLC Systems & Services*, *28*(3), 130–143. doi:10.1108/10650751211262128

Scholarly Communication. (2012). In *Wikipedia*. Retrieved March 9, 2012, from http://en.wikipedia.org/w/index.php?title=Scholarly_communication&oldid=486006103

Schöpfel, J., & Prost, H. (2013). Degrees of secrecy in an open environment. The case of electronic theses and dissertations. *ESSACHESS - Journal for Communication Studies, 6*(2).

Suber, P. (2003). *Removing the barriers to research: An introduction to open access for librarians*. Retrieved 26 December 2006 from http://www.earlham.edu/~peters/writing/acrl.htm

Vallance, P., & Smart, T.G. (2006). The future of pharmacology. *British Journal of Pharmacology*, *147*(S1), S304–7. doi:10.1038/sj.bjp.0706454

Vert, M., Doi, Y., Hellwich, K., Hess, M., Hodge, P., Kubisa, P., & Schué, F. et al. (2012). Terminology for biorelated polymers and applications (IUPAC Recommendations 2012). *Pure and Applied Chemistry*, *84*(2), 377–410. doi:10.1351/PAC-REC-10-12-04

Wasan, S., & Chakravarty, R. (2013). Digital Repositories as Harbingers of Open Access in India: A Study. In T. Ashraf & P. Gulati (Eds.), *Design* (pp. 191–218). Hershey, PA: Development, and Management of Resources for Digital Library Services; doi:10.4018/978-1-4666-2500-6.ch017

Chapter 3
Web 2.0 Tools in Biomedical and Pharmaceutical Education:
Updated Review and Commentary

Ângelo Jesus
Instituto Politécnico do Porto, Portugal

Maria João Gomes
Universidade do Minho, Portugal

ABSTRACT

Web 2.0 technologies are being rapidly integrated in higher education, which dramatically influences the ways learners approach and use information. Knowledge transfer has evolved into a two-way process. Users no longer simply consume and download information from the web; they create and interact with it. Several theoretical works were developed in order to discuss the possibilities of integration of Web 2.0 tools in Pharmacy, Medicine, Allied Health, Nursing and many other Biomedical Areas. Other works have started gathering qualitative and quantitative evidence of the importance of Web 2.0 tools in the learning process. By performing this integrative review, this paper will provide an overview of what is being done in biomedical and pharmaceutical education, and elaborate some of the potential opportunities and challenges that these applications present. With this updated review we hope to give our contribution to consolidate research in this promising area.

INTRODUCTION

Internet social applications, normally referred as Web 2.0 tools, are making their way in the new teaching paradigms of higher education. Since their early development, primarily for entertainment and social communication within the general population, applications such as blogs, social video sites, and virtual worlds (Barsky, 2006) are being adopted by higher education institutions in a vast range of scientific areas (Boulos, Maramba, & Wheeler, 2006). It has been argued that Web 2.0 technologies have the potential to change the education of healthcare professionals, from a didactic one way process to a collaborative and participative process, empowering the student to be an equal participant in the learning

DOI: 10.4018/978-1-5225-0248-7.ch003

process(Ward, Moule, & Lockyer, 2009). Web 2.0 applications appear to offer exciting new ways to teach, however, research into the use and evaluation of Web 2.0 tools in Biomedical and Pharmaceutical Education is still in its infancy, and the current pedagogic evidence about these tools is still lacking (Boulos et al., 2006)(Cain & Fox, 2009). In specific areas such has Pharmacy and Allied Health education, the examples are not wide, but the increasing use of teaching methodologies such as case based learning (Jesus, Cruz, & Gomes, 2011), might contribute to the implementation and dissemination of these tools. Wikis, blogs, podcasts and other tools have already been mentioned in some papers regarding higher education (Poonawalla & Wagner, 2006) (Kamel Boulos & Wheeler, 2007). These Web 2.0 technologies are in fact emerging as platforms to enable or encourage students to collaboratively create and share their own insights into current and emerging themes within their education. The number of tools and users are increasing and finding a place in healthcare management, education and practice (Ward et al., 2009). While it seems Web 2.0 might offer the potential for online learning to support pedagogy, in higher education there is little understanding of how and where it is being used to support biomedical education (Ward et al, 2009). This study is aimed to explore what specific Web 2.0 tools are being used, with what purposes and in what contexts.

In this updated review we will also focus on the most recent evidence about the use o Web 2.0 tools in pharmaceutical education, and share our own experiences on this topic.

METHODOLOGY

The study design is descriptive (MacMillan & Shumaker, 1997) and adopted the format of an integrative review (Cooper, 1984) since the objective was to make a synthesis of results (secondary analysis) from previous studies (primary analysis), in order to respond to new questions, new hypotheses and to verify or establish new relationships (Fortin, 2009). It is well documented that research reviews are considered research of research, and therefore should meet the same standards as primary research in methodological accuracy (Cooper, 1998). Cooper (1998) has delineated the process of conducting a research review as encompassing a problem formulation stage, a literature search stage, a data evaluation stage, a data analysis stage, and a presentation stage.

Problem Formulation Stage

This study primarily aims to characterize the use of Web 2.0 tools in biomedical and pharmaceutical education between the years of 2004 to 2014, and gather data to assess the potential educational values and hurdles to overcome. To constitute the categories of analysis we kept in mind the specificity of the object. Accordingly, the following variables were considered:

1. Year of publication,
2. Type of web 2.0 tool,
3. Biomedical area,
4. Type of publication,
5. Type of article and, finally,
6. Type of empirical study.

Table 1. Brief research design descriptions of retrieved studies

Type of Article		Brief Description
Empirical Study	Quasi-experimental	It is similar to an experimental design, where strict control and randomization of treatment conditions are not possible.Usually the groups of subjects are already created by some previous criteria whic are independent from the aims of the research study.This is typical in educational settings
	Survey	Structured questions given to participants, by means of interviews or questionnaires.
	Qualitative Study	Family of methods involving an interpretive, naturalistic approach of the subject.
	Mixed Study	Applies principles of Qualitative and Quantitative research
	Case Study	A form of qualitative research, the case study looks intensely at an individual or small group, drawing conclusions only about that participant or group and only in that specific context.
	Action Research	Action research involves utilizing a systematic cyclical method of planning, taking action, observing, evaluating (including self-evaluation) and critical reflection prior to planning the next cycle.
Theoretical/Reflection		Include expert opinions, reflections and theoretical aspects on the topic.
Instructional Design and Assessment Studies (IDEAS)		Includes articles that describe new courses, parts of courses, integration of selected competencies across the curriculum, assessment of instructional outcomes,the use of technologies and new delivery methods
Study Protocol		Described, proposed or ongoing research, providing a detailed account of the hypothesis, rationale, and methodology of the study

For the variable "biomedical area" we started with 5 main categories – Medicine and Dental Medicine; Nursing; Pharmacy and Pharmacology, Allied Health and Other Topics. For the variable "type of publication" we considered three categories – journal article, proceedings and other. For the variable "type of article," we adapted the proposals of Coutinho (2008) and Poirier et al (2009), and considered four main categories – theoretical/ reflection, empirical, instructional design and assessment studies (IDEAS)[1], and study protocol descriptions. Finally, in the variable "type of empirical study" we started with an initial range of six categories adapted from the proposal of Gomes & Coutinho (2008): quasi - experimental, survey, case study, mixed study, action research and qualitative study. During the data analysis we didn't found any paper that falls in the categories "case studies" and "action research". Brief research design descriptions of retrieved studies are summarized in Table 1. Classification and definition are based in the proposals of O'Brien (2001), Poirier et al., (2009) and Coutinho (2011).

Literature Search Stage

Well-defined literature search strategies are critical for enhancing the accuracy of any type of review because incomplete and biased searches result in an inadequate database and the potential for inaccurate results (Cooper, 1998). Ideally, all of the relevant literature on the problem or topic of interest is included in the review; yet obtaining this literature can be challenging and costly (Jadad, Moher, & Klassen, 1998). Our literature search consisted mainly on electronic search of published articles and congress proceedings, in international educational databases and university repositories. Published journal articles and conference proceedings, between 2004 and 2014, relating to the use of Web 2.0 tool in Biomedical and Pharmaceutical Higher Education were searched in the databases ERIC; EBSCO Academic Search Complete, MEDLINE, PubMed, JSTOR, CINAHL, Science Citation Index Expanded, Social Sciences Citation Index and Conference Proceedings Citation Index. Keywords, utilized in the search, consisted in: "blog", podcast", "virtual worlds", "Wiki", "Docs"," Web 2.0", " Health Education"," Biomedical

Education", "Medicine Education", "Nurse Education" and "Pharmacy Education" in several combinations. The titles, abstracts, and keywords of each of these articles were reviewed by two members of the research team, to assess their relevance to the research at hand. The references of the selected articles were also surveyed for appropriate articles. Computerized databases are efficient and effective; however, limitations associated with inconsistent search terminology and indexing problems may yield only about 50% of eligible studies (Whittemore & Knafl, 2005). Thus, and following the recommendations of Conn, Valentine, Cooper, & Rantz (2003), other approaches to searching the literature were included, such as, journal hand searching and searching research registries. A total of 55 articles were selected for further analysis.

Data Evaluation

The final sample for this integrative review included empirical studies, theoretical reports instructional design and assessment studies, and study protocols. Empirical reports included a wide variety of methods. Due to the diverse representation of primary sources, reports were coded according to two criteria relevant to this review: methodological or theoretical accuracy and data relevance on a 2-point scale (high or low). Since there are limited studies on this topic, an in order not to compromise the review, no report was excluded, based on this data evaluation rating system; however, the score was included as a variable in the data analysis stage. High relevance was only attributed to empirical studies.

Data Analysis

Data analysis required that the data from primary sources was ordered, coded, categorized, and summarized into a unified and integrated conclusion about the research problem (Cooper, 1998). Data analysis comprised of data coding from primary sources to simplify, abstract, focus, and organize data into a manageable framework (Whittemore & Knafl, 2005). Relevant data of each subgroup classification was extracted from all primary data sources and compiled into a matrix. The matrix was created according to the assumptions and categories mentioned earlier. This approach provides succinct organization of the literature, which facilitates the ability to systematically compare primary sources on specific issues, variables, or sample characteristics.

Presentation

Conclusions are reported in table form. Explicit details from primary sources and evidence to support conclusions will be provided to demonstrate a logical chain of evidence, allowing the reader of the review to ascertain that the conclusions do not exceed the evidence (Oxman, 1994).

PROSPECTS

By performing this integrative review, this chapter will provide an overview of what is being done in biomedical and pharmaceutical education, and elaborate on some of the potential opportunities and challenges that these applications present. We hope to give our contribution to consolidate research in this promising area.

Table 2. Array of literature

Author	Year of Publication	Web 2.0 Tool	Biomedical Area	Type of Publication	Type of Article	Type of Empirical Study
(Bouldin, Holmes, & Fortenberry, 2006)	2006	Blog	Pharmacy	Article	Instructional Design and Assessement	
(Boulos, Maramba, & Wheeler, 2006.)	2006	Wiki + Blog + Podcast	Other Topics	Article	Theoretical/ Reflection	
(Sandars, 2006)	2006	Wikis + Blogs	Medicine	Article	Theoretical/ Reflection	
(Boulos, Hetherington, & Wheeler, 2007)	2007	Virtual Worlds	Medicine	Article	Theoretical/ Reflection	
(Mathieu, 2007)	2007	Wiki + Blog + Podcast	Nursing	Article	Theoretical/ Reflection	
(Oomen-Early & Burke, 2007)	2007	Blog	Other Topics	Article	Empirical	Survey
(Chretien, Goldman, & Faselis, 2008)	2008	Blog	Medicine	Article	Instructional Design and Assessement	
(Forbes & Hickey, 2008)	2008	Podcast	Nursing	Article	Empirical	Survey
(Geyer, Beylefeld, & Alwyn, 2008)	2008	Podcast	Medicine	Proceedings	Empirical	Survey
(Goldman, Cohen, & Sheahan, 2008)	2008	Blog	Other Topics	Article	Empirical	Survey
(Kaldoudi, Bamidis, Papaioakeim, & Vargemezis, 2008)	2008	Web 2.0	Medicine	Proceedings	Theoretical/ Reflection	
(Pilarski, Alan Johnstone, Pettepher, & Osheroff, 2008)	2008	Podcast	Medicine	Article	Empirical	Survey
(Sandars, Homer, Pell, & Croker, 2008)	2008	Web 2.0	Medicine	Article	Empirical	Survey
(Skiba, 2008)	2008	Microblogging	Nursing	Article	Theoretical/ Reflection	
(Sumer, 2008)	2008	Web 2.0	Medicine	Article	Theoretical/ Reflection	
(Billings, 2009)	2009	Wikis + Blogs	Nursing	Article	Theoretical/ Reflection	
(Bratsas, Kapsas, Konstantinidis, Koutsouridis, & Bamidis, 2009)	2009	Web 2.0	Medicine	Proceedings	Theoretical/ Reflection	
(Cain & Fox, 2009)	2009	Web 2.0	Pharmacy	Article	Theoretical/ Reflection	
(Grassley & Bartoletti, 2009)	2009	Wikis + Blogs	Nursing	Article	Theoretical/ Reflection	

continued on following page

Table 2. Continued

Author	Year of Publication	Web 2.0 Tool	Biomedical Area	Type of Publication	Type of Article	Type of Empirical Study
(Hansen, Murray, & Erdley, 2009)	2009	Virtual Worlds	Nursing	Article	Theoretical/ Reflection	
(Honey, Diener, Connor, Veltman, & Bodily, 2009)	2009	Virtual Worlds	Nursing	Proceedings	Instructional Design and Assessement	
(Meade, Bowskill, & Lymn, 2009)	2009	Podcast	Pharmacology	Article	Empirical	Survey
(Miller, Bookstaver, & Norris, 2009)	2009	Wikis	Pharmacy	Article	Instructional Design and Assessement	
(Phadtare, Bahmani, Shah, & Pietrobon, 2009)	2009	Docs	Allied Health and Nursing	Article	Empirical	Quasi-Experimental
(Rogers, 2009)	2009	Virtual Worlds	Nursing	Proceedings	Instructional design and assessement	
(Rosh, Jones, & Wahl, 2009)	2009	Web 2.0	Medicine	Article	Experiment Report	
(Shantikumar, 2009)	2009	Podcast	Medicine	Article	Empirical	Survey
(Woulfe, Williams, & Ryan, 2009)	2009	Wikis	Pharmacy	Proceedings	Empirical	
(Archambault et al., 2010)	2010	Wikis	Allied Health and Medicine	Article	Study Protocol	
(Bristol, 2010)	2010	Microblogging	Nursing	Article	Theoretical/ Reflection	
(Chu, Young, Zamora, Kurup, & Macario, 2010)	2010	Web 2.0	Medicine	Article	Theoretical/ Reflection	
(Melissa Tan, K. Ladyshewsky, & Gardner, 2010)	2010	Blog	Allied Health	Article	Empirical	Qualitative Study
(O'Sullivan & McGlynn, 2010)	2010	Web 2.0	Allied Health	Proceedings	Instructional design and assessement	
(Mattheos, Schoonheim, Klein, Walmsley, & Chapple, 2010)	2010	Web 2.0	Dental Medicine	Article	Theoretical/ Reflection	
(Varga-Atkins, Dangerfield, & Brigden, 2010)	2010	Wikis	Medicine	Article	Empirical	Mixed Study
(Younger, 2010)	2010	Wikis	Nursing	Article	Theoretical/ Reflection	
(Schreiber, Fukuta, & Gordon, 2010)	2010	Vodcast	Medicine	Article	Empirical	Experimental
(Wiecha, Heyden, Sternthal, & Merialdi, 2010)	2010	Virtual Worlds	Medicine	Article	Empirical	Survey

continued on following page

Table 2. Continued

Author	Year of Publication	Web 2.0 Tool	Biomedical Area	Type of Publication	Type of Article	Type of Empirical Study
(Mirk, Burkiewicz, & Komperda, 2010)	2010	Wikis	Pharmacy	Article	Empirical	Survey
(Boutin & Therriault, 2011)	2011	Wiki	Allied Health	Proceedings	Instructional Design and Assessement	
(Fischer, Haley, Saarinen, & Chretien, 2011)	2011	Blog	Medicine	Article	Empirical	Quasi-Experimental
(Fox & Varadarajan, 2011)	2011	Microblogging	Pharmacy	Article	Empirical	Survey
(George & Dellasega, 2011)	2011	Web 2.0	Medicine	Article	Empirical	Mixed Study
(Llambí et al., 2011)	2011	Wikis	Medicine	Article	Empirical	Quasi-Experimental
(Meade, Bowskill, & Lymn, 2011)	2011	Podcast	Pharmacology	Article	Empirical	Qualitative
(Ros-Rodriguez et al., 2011)	2011	Wiki	Pharmacy	Proceedings	Instructional Design and Assessement	
(Sparks, O'Seaghdha, Sethi, & Jhaveri, 2011)	2011	Web 2.0	Medicine	Article	Theoretical/ Reflection	
(Stiffler, Stoten, & Cullen, 2011)	2011	Podcast	Nursing	Article	Empirical	Survey
(Zanussi, Paget, Tworek, & McLaughlin, 2011)	2011	Podcast	Medicine	Article	Theoretical/ Reflection	
(Pierce & Fox, 2012)	2012	Podcast	Pharmacy	Article	Instructional Design and Assessment	
(Bussières, Métras, & Leclerc, 2012)	2012	Microblogging	Pharmacy	Article	Instructional Design and Assessment	
(Miller, Norris, & Bookstaver, 2012)	2012	Wikis	Pharmacy	Article	Empirical	Survey
(Stewart, Panus, & Hagemeier, 2013)	2013	Podcast	Pharmacy	Article	Instructional Design and Assessement	
(Camiel, Goldman-Levine, Kostka-Rokosz, & McCloskey, 2014a)	2014	Microblogging	Pharmacy	Article	Instructional Design and Assessement	
(Camiel, Goldman-Levine, Kostka-Rokosz, & McCloskey, 2014b)	2014	Microblogging	Pharmacy	Article	Empirical	Survey

RESULTS PRESENTATION AND DISCUSSION

Year of Publication

The documents retrieved were published over a time period between the end of 2004 and the third quarter of 2014 (see Table 2). The year which shows the highest number of publications, was 2009 with 13 documents, followed by the years 2010 and 2011 with ten to eleven records.

Web 2.0 Tools

One of the main aspects to consider in the literature search, and data evaluation, was the Web 2.0 tool utilized in the studies (Table 3). Taking into account the diversity of tools available, and the possible combinations, our option was not to previously set categories, but to categorize, according to the data evaluation. During the data analysis process we accounted for 9 categories: "Web 2.0"; "Podcast/Vodcast"; "Wikis"; "Blogs"; "Second Life"; "Twitter®"; "Wikis + Blog", "Wikis + Blog + Podcast" and finally "Google Apps® and Similar". The "Web 2.0" category was considered due to the number of theoretical and empirical works being published, regarding the general application of Web 2.0 tools in biomedical sciences, or the application of more than three specific tools in the same context.

Biomedical Area

Medicine is the largest biomedical area represented in this study. This result may be explained by the large literature being published by medical professionals, the large number of medical faculties, and also because other biomedical areas, like Allied Health and Pharmacy, represent a younger existence when compared with traditional Medical Education. During the development of the study we coded two new variables – "Allied Health and Nursing"; "Allied Health and Medicine"- which are related to experiments conducted with both kind of professionals (Table 4). In this updated review we focused in newer studies regarding pharmaceutical education.

Table 3. Number of articles according with each Web 2.0 tool

Web Tool	Total Number of Articles	Type of Articles			
		Empirical	Theoretical/ Reflection	Instructional Design / Assessment Studies	Study Protocol
Web 2.0	11	2	7	2	0
Podcast /Vodcast	11	8	1	2	0
Wikis	10	5	1	3	1
Blogs	6	4	0	2	0
Virtual Worlds	5	1	2	2	0
Twitter/Microblogging	6	2	2	2	0
Wikis + Blog	3	0	3	0	0
Wikis + Blog + Podcast	2	0	2	0	0
Google Apps/Similars	1	1	0	0	0

Type of Publication

Regarding the variable type of publication, journal articles are the most common source and medium of dissemination of work being done in Biomedical Education with Web 2.0 tools, consisting on a total of 40. This may be due to two major factors. Firstly, conference proceedings, by themselves, do not constitute a characteristic and homogeneous group of scientific work that can be easily accessed through the web, mainly because only a few organizations with long-standing organization history of scientific conferences have indexing. Additionally, situations where proceedings are published (full text) in supplements of related Journals, we choose to categorize them as Journal Articles.

Type of Article and Type of Empirical Studies

As previously stated, four main categories were considered, for the variable type of article: theoretical/ reflection; empirical study; instructional design and assessment study, and study protocol (Table 5). The empirical studies relating to the topic in question, were categorized accordingly with the followed methodology. The following table summarizes the results. Surveys are the most common type of empirical study.

Table 4. Number of articles according to biomedical area

Biomedical Area	Total Studies Retrivied
Medicine and Dental Medicine	21
Pharmacy and Pharmacology	15
Nursing	11
Allied Health	3
Allied Health and Medicine	1
Allied Health and Nursing	1
Other Topics	3
Total number of Articles	55

Table 5. Number of articles according with type of article and type of empirical study

Type of Article	Number of Articles	Type of Empirical Study			
		Quasi-Experimental	Survey	Qualitative	Mixed Study
Empirical	24	4	15	2	3
Theoretical/ Reflection	18				
Instructional Design and Assessment Studies	12				
Study protocol	1				

RESULTS BY WEB 2.0 TOOL USED

Web 2.0

Web 2.0 technologies are being rapidly integrated in higher education, which dramatically influences the ways learners approach and use information. Knowledge transfer has become a two-way process. Users no longer simply consume and download information from the web; they create and interact with it (Rosh et al., 2009). Several theoretical works were developed in order to discuss the possibilities of integration of Web 2.0 tools in Pharmacy (Cain & Fox, 2009), Medicine (Chu et al., 2010), Allied Health (O'Sullivan & McGlynn, 2010) and many other Biomedical Areas. Other works have started gathering evidence of the importance of Web 2.0 tools for students and even show some quantitative and qualitative results. Initial works like the one from Sandars et al., (2008) propose a very straightforward approach. With a survey type study, they aimed to determine the nature and extent of the use of social software by first year medical students. From a sample of 212 students, they discovered that over 90% used instant messaging and 70% social networking sites. There was no significant difference between males and females. Blogs were read by about a fifth of the students and 8% wrote their own blogs. A fifth of males stated that they were users of media sharing and contributed to wikis. Social bookmarking was rarely used by either sex. On a more complex approach George & Dellasega (2011) used and evaluated the integration of new social media tools into the curricula of two graduate-level medical humanities electives offered to 4th year students at Penn State College of Medicine. Five social media tools were selected - Twitter®, YouTube®, Flickr®, Blogger® and Skype® - to promote student learning. At the conclusion of each course, students provided quantitative and qualitative course evaluation. Results demonstrate that students gave high favorability ratings to both courses, and expressed that the integration of social media into coursework augmented learning and collaboration. Others identified challenges such has demands on time, concerns about privacy and lack of familiarization with technology. These reports represent an integrated approach to the use of Web 2.0 technologies in Biomedical Higher Education. Reports relating to the use of specific tools are presented below.

Podcast and Vodcast

Much that is written about podcasting refers to its ability to enhance convenience, flexibility and accessibility to learning (Nathan & Chan, 2007). It appears that when one examines the purpose behind the use of podcasting, it falls into three broad categories: enhancing the flexibility of learning, increasing accessibility to learning (particularly in relation to enabling mobile access) and enhancing the student's learning experience (particularly in on campus courses through the use of more blended learning experiences) (McGarr, 2009). We found 10 articles relating to the use of podcast/vodcast in biomedical education. Six of them were surveys. Meade et al. (2009) tried to provide an evaluation of the usefulness of pharmacology podcasts as a supplementary learning tool for Non Medical Prescribing Students. The focus of this evaluation was on student use of the podcasts, perceptions of the usefulness and any potential impact on students' pharmacological knowledge. The podcasts were made available to students through WebCT® and the authors tracked two successive cohorts of nurse prescribing students (n = 69). Survey data, investigated reasons for podcast use and perceived usefulness of podcasts as a learning tool. Their results showed that not only 91% of students accessed at least one podcast but also that 93% of students used the podcasts to revisit a lecture. Only 22%

used the podcasts because they had missed a pharmacology session. Most students (81%) generally listened to the entire podcast rather than specific sections and most (73%) used them while referring to their lecture handouts. More than 90% of students found the podcasts helpful as a learning tool, as a revision aid and in promoting their understanding of the subject. Of the 69 students, 64 completed the pharmacology exam. In order to examine any impact of podcasts on student knowledge, their exam results were compared with those of two historical cohorts who did not have access to podcasts (n = 70). Evaluation of range of marks obtained, mode mark and mean mark suggested improved knowledge in students with access to podcasts compared to historical cohorts of students who did not have access to pharmacology podcasts(Meade et al., 2009). Also in the UK, but with medical students, Shantikumar (2009) aimed to investigate medical students' perceptions of a series of enhanced podcasts for revision of contents. Thirteen audiovisual podcasts covering general surgery were developed, consisting of a PowerPoint slideshow with a voiceover narrative. A questionnaire was distributed to 211 final year medical students, two months after the podcast became available. Results showed that students who used the resource felt that the enhanced podcasts were straightforward to access, were a useful learning supplement, and felt that similar resources for the remainder of the undergraduate medical syllabus would be useful for revision purposes (Shantikumar, 2009). Similar results were described by Pilarski et al (2008) in first year medical students. According to their results, the availability of lecture recordings aided medical students in their studies and reduced stress and anxiety. Student response to the recordings was positive, and no negative outcomes were noted by students or faculty (Pilarski et al., 2008).

Analogous methods were used in Nursing Education (Forbes & Hickey, 2008), (Stiffler et al., 2011). In both cases, the authors provided an overview of the potential uses and techniques for implementing podcasting in nursing education, and described students perceptions and attitudes. The results are in consonance with the above-mentioned studies. Although these are promising results we must not be drawn to generalize their conclusions. First of all, the survey type study does not provide all relevant information about context on the application of podcasts. Secondly, student performance in these cases, except Meade et al (2009), is not categorized or even measured in a consistent way. However, student's perceptions and responses to these implementations are crucial for their success, but more detailed, in-depth studies are required in order to truly account for an impact in biomedical and pharmaceutical education. This will imply bigger samples and longer studies.

We only came across with one qualitative study exploring the use of podcast in biomedical education. Meade et al. (2011) conducted semi-structured interviews with seven non-medical prescribing students. Their results determined that students used podcasts for revisiting lectures, preparing for exams, to clarify or revise specific topics and, to a lesser extent, to catch up on a missed lecture. Barriers to podcast use were identified, mainly around technological issues. Lack of experience and lack of access to suitable technology to access podcasts proved a barrier for some students. Students found that using podcasts allowed them to have greater control over their learning and to gauge their learning needs, as well as helping them build their understanding of a complex topic (Meade et al., 2011). The qualitative methods provide a depth of understanding of issues that is not possible through the use of statistically based, quantitative investigations. The qualitative approach can be a useful tool since it centralizes and places primary value on how students understand experience and operate with web 2.0 tools. The results of qualitative studies can be of interest to educators and policy makers when design a strategy for the implementation of such tools in their specific academic context.

Initially we did not consider a category for vodcasts, however, in the array of literature, one study referring to vodcast in medical education was found (Schreiber et al., 2010) and one in Pharmacy (Pierce & Fox, 2012). In the first study the authors compared information recall and student experience and preference after live lectures and video podcasts in undergraduate medical education. A crossover trial was performed with a total of 100 students. The first group attended a live lecture on arthritis and then a video podcast on vasculitis, while the second group attended a live lecture on vasculitis and then a video podcast on arthritis. The study hypothesis was that knowledge after vodcast would not be significantly different from knowledge after live lectures. The primary outcome was the score on the multiple-choice questionnaire after each intervention. This study found that students showed similar information recall after video podcasts and live lectures, but preferred live lectures. Students appreciated the convenience and control over vodcasts, but found them, generally, less engaging. Design strengths in this study included testing the students only after the interventions, to avoid influencing the educational intervention with the test itself and keeping the two interventions as similar as possible using the same speaker, the same PowerPoint® slides, and the same verbal information. Authors, point out however, that selection bias might have been present, since participation was out of hours and voluntary and also a significant dropout rate occurred.

Pierce and Fox (2012), promoted vodcast of lectures prior to the scheduled class and then discussed interactive cases of patients with end-stage renal disease in class. Students' performance on the final examination significantly improved compared to performance of students the previous year who completed the same module in a traditional classroom setting. Althought these results are very promissing, we must consider taht this experiment was based in a "Flipped Classroom" model, which may also contribute to the performance improvement.

These studies open possibilities for future interventions, and although conclusions were not always favorable to the Web 2.0 tool (Stewart et al., 2013), several factors were identified as confounding factors. We suggest that any pursue in this type of designs take in consideration the duration of the intervention, the possible selection bias and of course a better isolation of threats to the internal validity. The incorporation of podcasts in biomedical and pharmaceutical education, namely in support of a mobile form of learning, may have potential, but the reason given to justify its use should be student led, and not technology led. Educators need to be cautious of the claims made in relation to new and emerging technologies, particularly in the assumptions made by proponents of the technologies (McGarr, 2009). The development of future policies in institutions should be cognisant of the limitations of the technology in supporting a more mobile form of learning. If used to provide access to records of previous lectures, or to provide summaries and syntheses of course material, careful consideration should be given to how the students should use the material. If seen as the primary source of material and learning by the students, this type of technology may not contribute to the students' learning experience. So it is important to ensure that students possess the study skills needed to use the support material effectively.

Blogs

Blogs may serve as a potential teaching tool, as well as an important channel for Biomedical Education. Previous research has shown that blogging, integrated into curriculum, may increase student's motivation, critical thinking, class interaction, and student's course satisfaction (Beldarrain, 2006; Olofsson, Lindberg, & Hauge, 2011). Blogging may even propel students to take social and political actions outside of the virtual environment (Oomen-Early & Burke, 2007). Concerning the use of blogs alone, in biomedical and pharmaceutical education, a total of 6 articles were retrieved. Regarding instructional design and assessment, two examples were found, in Pharmacy (Bouldin et al., 2006) and Medicine (Chretien et al., 2008), respectively. Bouldin et al., (2006) applied blogs to a reflective journaling exercise in a communication course during the second-professional year at the University of Mississippi School of Pharmacy, in order to encourage students to reflect on course concepts and apply them to environments outside the classroom, and to assess their communication performance. Seventy-eight students enrolled in the study. Two blog entries per week were required for full credit. Blogs were evaluated at three points during the entire term, and at the end of the course, students evaluated the assignment using a survey. Results showed that the assignment contributed to student learning and increased awareness level for approximately 40% of the class, however, there were some limitations. Because of the proximity to the end of the semester, student responses may have been influenced by the total curricular workload; moreover this evaluation was only taken at one point in time.

Chretien et al.(2008) performed a similar study with 91 medical students during a basic medicine clerkship rotation at an academic institution over a one-year period. Students were asked to contribute with reflective postings to the class blog during their rotation (each rotation was four weeks long). They were able to read each other's postings and leave feedback in a comment section. An instructor provided feedback on entries, aimed to stimulate further reflection. Students could choose anonymous names if desired. One hundred and seventy seven posts were written, and only a minority was not reflective. The instructor's feedback stimulated additional reflection and one-third of students left feedback comments. Both studies show that it is possible to foster reflective thinking by the use of blogs in such different biomedical areas as Pharmacy and Medicine and even provide opportunities to support student professional development. It is notable that the number of students in each study is quite considerable, a quality not often seen in these types of studies. However, the number of tutors /instructor wasn't proportional, and this caused an incomplete and not as frequent feedback. Additionally, both studies were conducted in only one institution, which may limit generalization. Nevertheless, as instructional design and assessment studies, these studies show and describe thoroughly the process by which any educator can implement and assess the opportunities promoted by blogs in their classrooms, or even during clinical practice.

Regarding the survey type study, Goldman et al (2008) evaluated whether "seminar blogs" enhanced learning in a graduate-level introductory public health school class. An online survey was conducted and the main questions were about blog use, perceived impact on learning, and interactions with classmates. The survey also asked about the "ease of blogging" compared with speaking in class, the development of group dynamics, potential usefulness of blogging in student's work, and advantages and disadvantages of seminar blogs. Results generally favor the study objective, however several limitations are addressed. First of all, the study was conducted in a post-graduate setting, where subject's age and maturity may (or may not) contribute for the study objectives. As referred by the authors, the short time frame (8 weeks) may have been too brief to fully determine the developing of group dynamics or social presence. And finally we should keep in mind that this a descriptive study based on student self-report of effects on learning and participation and only with a comparison group would we be able to assess the impact of the blogs on specific measurable educational outcomes.

Oomen-Early and Burke (2007) also conducted a survey to explore faculty and students' perceptions of blogging as an effective teaching and learning tool within the online health education classroom. The researchers measured if blogging increased students' interaction, engagement, and feelings of connectedness with peers. Results are favorable to the study objective, but again we are dealing with a small sample of students. This study, shows however a great deal of care in developing a survey (validated by expert and tested in a group of students), and points out several limitations and challenges regarding the use of online free blog hosting services. Using a much more robust study design Fischer et al. (2011), conducted a comparative study of student reflective writings, produced using either a blog format or a traditional essay to assess differences in content, depth of reflection and student preference. Students were quasi-experimentally assigned to one of the two groups, and although no significant difference was determined in themes addressed, or levels of reflection, the authors showed that it's possible, and quite achievable to promote the use of blogs in biomedical teaching. The study suggests that the use of blogs, in no way diminishes the ability for reflective writing. Moreover the fact that blogs support features such as, image, sound and video, can in fact enhance the participation and may even encourage those who feel greater reluctance in participate in these reflective activities. Nevertheless one should not draw hasty conclusions, since more studies in different knowledge areas, with a greater number of students and over a longer period than are necessary.

We also came across with a qualitative study, conducted by Melissa Tan et al. (2010) who aimed to determine the impact of blogging on the clinical reasoning of physiotherapy students. With a sample of 83 students, the authors divided them randomly in 8 groups. Each group had a tutor/moderator (faculty member), and all students were required to contribute to and participate in reflective practice discussions within their blog group for the duration of their clinical program with at least one post and comments on two of their group members' original posts per week. Data was retrieved and analyzed using NVivo®. About 475 blog posts were analyzed. Results demonstrated that the students applied a range of clinical reasoning skills categorized as collaborative; diagnostic; ethical; interactive; and procedural reasoning. Of these, 90% were procedural, ethical and interactive reasoning. The authors concluded, even though the students were dispersed across facilities and changed locations every four-five weeks, that blogging provided these physiotherapy students with opportunities to explore clinical reasoning dilemmas, in a safe space. The blog enabled the students to maintain connections with their peers and supported them in their fieldwork practice. This apparently promoted the development of metacognition at both an individual and collective level. This research indicates a promising result, that blogging can be used to promote learner self-discovery and to assist self regulated learning in biomedical science. Evidence suggests that blogs help to create connections between students with diverse opinions and interests. This encourages critical thinking and teaches the value of respect towards other students' points of view. Discussions in blogs may also promote higher levels of thinking, because people can think before answering back. They can be used for individual students, or for groups, fostering collaborative learning. Some limitations were however addressed by the authors, of the studies reviewed, but they were attained as ways to improve the experience, and none of them could theoretically diminish the main conclusions.

Wikis

Wikis are increasingly being used as means of sharing and communicating medical knowledge. The Medpedia Project, is a good example of an up-to-date, unbiased medical information repository, maintained by health experts from all around the world, and of free access (Park, Crocker, Nussey, Springate, & Hutchings, 2010). The use of wikis is quite common in other educational levels (K-12). But their potential is already being discovered by faculty members. Wikis (as tools for group authoring and collaboration) have substantial applications. Often group members collaborate on a document by emailing to each member of the group a file that each person edits on their computer, and some attempt is then made to coordinate the edits so that everyone's work is equally represented; using a wiki pulls the group members together and enables them to build and edit the document on a single, wiki page (Parker & Chao, 2007). Students can even use a wiki to develop research projects, with the wiki serving as ongoing documentation of their work, build a collaborative annotated bibliography or develop map concepts. Wikis can also be useful for brainstorming, and editing a given topic, to produce a linked network of resources. Teachers can also use wikis as a knowledge base, enabling them to share reflections and thoughts regarding teaching practices, documentation or presentation tool in place of conventional software.

The use of wikis alone, is one of the most documented, with a total of 10 articles retrieved, ranging from theoretical works (Younger, 2010), to instructional design and assessments' (Miller et al., 2009; Ros-Rodriguez et al., 2011; Boutin & Therriault, 2011), surveys (Mirk et al., 2010; Miller et al., 2012) study protocols (Archambault et al., 2010) and mixed studies (Varga-Atkins et al., 2010) . There are even studies associating Wikis to Blogs (Billings, 2009; Sandars et al., 2008) and Podcast (Mathieu, 2007; Boulos et al., 2006) however, all of them are theoretical works, in several biomedical areas. Even teachers are discussing the use of wikis to promote and update study materials for their students (Rodríguez, Carrillo, & Alvarez, 2010). Focusing on instructional design and assessments', Miller et al. (2009) described their use of wikis in Advanced Pharmacy Practice, with an overall student answers with more 90% of respondents either strongly agreeing or agreeing that wikis helped reinforce concepts and ideas and that they enjoyed working with wikis. Within the same study type, Ros-Rodriguez et al. (2011) recently published their experience in the development of a Virtual Classroom of Pharmacology. Using wiki utilities the students had to transform an initial pdf outline to a wiki format. The students had to create or edit one wiki for each lesson outline and to include hyperlinks for the main concepts to additional material located in the "Virtual Classroom of Pharmacology" and had also to create a glossary with those terms. The student involvement in the task led them to relate, integrate and assimilate the knowledge about those particular aspects of pharmacology. Besides, the authors conclude that the generation of these materials reinforced the learning and promoted work as a team. Also in instructional design and assessments, Boutin & Therriault (2011) presented a pedagogical experience in an Occupational Therapy program that used a wiki to enable the learner to become a" knowledge producer and actor of hi/her learning"(Boutin & Therriault, 2011). The authors targeted three courses, each with different objectives intended for the wiki. Following application of the wiki, the degree of satisfaction of students and professors and the impact thereof on the acquisition of skills was evaluated through an online survey. The published results, from both student and faculty, seem to indicate that using a wiki as a learning tool contributes to the acquisition of knowledge and the development of skills in this new competency-based program of occupational therapy.

On a different approach, Llambí et al. (2011) developed a quasi-experimental model in which they designed and implemented a blended-learning course on tobacco cessation. Three hundred thirty-five health professionals participated in the course. Of these, 145 attended the on-site workshop, 216 participated in the online activities, and 109 completed both phases. When comparing pre and post test scores, the authors found a statistically significant difference among those who completed both phases compared with those who only did the online phase ($p = .003$ and $p = .009$, respectively), demonstrating that wikis, could be a useful tool in promoting continuing education activities for health professionals. We must however mention that tests were completed by a minority of the participants, since fewer than half performed the pretest and fewer than a third completed both and allowed paired analysis. The fact that tests were not compulsory became a weakness of this study. Using a mixed approach, Varga-Atkins et al. (2010) tried to discover whether the use of wikis could enhance medical students' development of professionalism. Using both survey and interviews, 32 students were questioned. The authors described that the wikis helped students finding, compiling, evaluating and sharing resources related to their learning objectives on professionalism. Secondly, it was demonstrated by the study that the online wiki spaces could have a role in enhancing students' development of "professionalism in a wider sense". Although, using quite a small sample and limited time span, this is a pioneer study, which can be a launch base for more broad research in several subjects of biomedical education. More outcomes are expected as Archambault et al. (2010) develops also a mixed study on the use of wikis. The authors published the study protocol for the mixed study, that aims to develop and test the metrological qualities of a questionnaire for the assessment of healthcare professionals' intentions and the determinants of those intentions to use wiki-based reminders that promote best practices in trauma care. This promising work may contribute even further for the use of wikis in biomedical and health education.

The only survey type study regarding wikis in biomedical education at this point is the research conducted by Woulfe et al. (2009). The authors applied a "wiki approach" to Master of Pharmacy students at The University of Sydney. Normally these students were required to submit a series of group-generated PBL cases. However faculty members often reported a fragmented approach and a lacking in "patient-focus". After a successful pilot study in 2007, the authors introduced a wiki format, with a view to encourage a whole patient approach to the task. A total of 38 students were divided into 6 tutorial groups in each semester, producing one PBL case summary per group per week for a total of 10 weeks. In the first semester, students were required to submit their case summaries on paper. In the second Semester, they were required to write their case summaries in a wiki. The results are not favorable to the study objective. The technical difficulties reported by students appear to have greatly affected the quality of their case summaries, as well as influenced their perceptions of the wiki. This increases the awareness of educators for the necessity for training in the use of the wikis, before instructional changes.

We also introduced a "wiki iniciative" in our own pharmacotherapy course (Jesus, Gomes, & Cruz, 2012, 2013), mainly as an e-portfolio for the students, in Wikispaces. The choice of platform was based on the capabilities provided by the same level of editing (integrating text, images, links, videos, audio, documents, presentations and other features embed), the issue of privacy and access and log edits, and you can follow in detail the contributions of each group member. Students should produce a new page for each case would be presented with only one case per week. In total 20 were developed e-portfolios with different characteristics. A systematic analysis of the contributions made by the students allowed the observation of commitment and motivation from them and formed the basis of a quantitative evaluation.

Microblogging

Microblogging is the practice of posting small pieces of digital content on the Internet (EDUCAUSE, 2009). Microblogging has become popular among groups of friends and even professional colleagues who frequently update content and follow each other's posts, creating a sense of online community (EDUCAUSE, 2009). Although literature about this Web 2.0 tool is still in its infancy, we uncovered two theoretical studies in nursing education(Skiba, 2008; Bristol, 2010), and four studies relating the use of Twitter® in Pharmacy Education (Fox & Varadarajan, 2011; Bussières et al., 2012; Camiel et al., 2014b, 2014a). The first study describes implementation and effectiveness assessment of Twitter® use, in encouraging interaction between faculty members, guests, and students in a pharmacy course taught simultaneously on 2 campuses. Students were required to tweet a minimum of 10 times over several class sessions. The course instructor and guest professionals also participated but no limit of tweets was established for them. One hundred forty-three students were enrolled in the course: 119 in one campus and 24 at the satellite campus. Course requirements were the same across campuses. The Twitter® activity, was worth 2% of the grade. Students were also given an opportunity to earn 5 bonus points by completing an online, anonymous evaluation of the Twitter® activity. The results presented by Fox and Varadarajan (2011), indicate that more than eighteen hundred tweets were made by students, guests and the instructor. Students tweeted most frequently with each other and found value in reading each other's tweets. From the 131 students that completed the optional evaluation survey 71% percent indicated that Twitter® was distracting, 69% believed it prevented note taking, and more than 80% indicated that it facilitated class participation and allowed an opportunity to voice opinions. These results led the researchers to conclude that the use of Twitter® in pharmacy courses (or any other biomedical course) balanced the potentially positive aspects of the technology, such as increased interaction among students, with potentially negative aspects, such as the interruptive nature of Twitter® use and the large volume of tweets generated by assignment. Far from extolling the benefits of Twitter®, this study shows clearly and objectively the results obtained, even if they do not translate into a steadfast discovery the potential of microblogging. The study design, the number of professionals involved and the methodological description shows a high robustness, which can definitely be a starting point for other researches or experiments in the biomedical classroom.

In 2012, Bussières and colleagues (2012), examined the use of Twitter® in a first-year pharmacy law course, as a backchannel (an electronic discussion that occurs simultaneously in real time during a lecture or conference where students may post questions, comments, or respond to other posts). The authors compared the frequency of questions either through Twitter® or hand-raising. Of 200 students, only 8 asked questions using Twitter® vs 30 questions with hand-raising. Despite the authors' feelings that the use of Twitter® was a failure, students reported that backchannel allowed those students to express themselves who otherwise might not have. The idea of using Twitter® as a backchannel was also pursued by Camiel and colleagues (2014a). In their paper, the authors describe the use of Twitter® in a "large required, lecture-based, team-taught nonprescription drugs/self-care products course". The students were remineded students at the beginning of each lecture that backchannel was being used. The feed was continuasly monitored, and after class, responses to unanswered questions were posted on the course learning-management system discussion board. The author promoted a survey at the end of the year, where 266 students completed an anonymous voluntary survey, where thirty-nine percent (and 40% neutral) felt that being anonymous to their classmates was important to them. In their written comments, a number of students commented positively about not needing to raise their hand to ask a question in front of the entire class.

In more recent article, the authors (Camiel et al., 2014b) explore the use of Twitter® for the building of a personal learning network. Although the initial results showed a high percentage of students (36%) that were not confortable using Twitter, and even considering that the use of a personal learning network was a new concept, the final results were very promising, with more than 50% students planning to continue using Twitter® as a social network.

One must consider that as diverse as microblogging may be, it has some shortcomings. Although microblogging promotes writing and editing as fun activities, and therefore may be used in informal learning (Ebner, Lienhardt, Rohs, & Meyer, 2010) there is the possibility that students get distracted and carried away during ongoing lessons due to microblogging with their mobile phones. Not only is this addiction harmful to their education but it is also disrespectful to the teachers or lectures. Further, educators insist that due to microblogs being limited to adopting 140 characters for message updates, it could potentially lead to bad grammar usage by students (Grosseck & Holotesku, 2008).

Google Apps and Similars

The use of Google Apps or other similar applications isn't particularly widespread. Only one study (Phadtare et al., 2009), using GoogleDocs® was retrieved. With a heterogeneous sample of 48 participants (from a medical, nursing and physiotherapy background from US and Brazil) randomly assigned to two groups, the authors proposed to compare on-line vs. traditional classroom-based methods for teaching scientific writing. In the on-line group, participants used virtual communication, GoogleDocs® and standard writing templates, and in the standard group participants received standard instruction without the aid of virtual communication and writing templates. Outcome variables consisted of manuscript quality and self-reported participant satisfaction. The quality of each manuscript was evaluated according to well-defined parameters using a scale. The obtained results suggest that the on-line scientific writing group performed significantly better than the standard writing guidance group in terms of writing quality. They also reported greater overall satisfaction and a greater number of participant-mentor communication events. The authors agree that the use of GoogleDocs® enhanced participants' familiarity with this method of collaboration, as well as improved the mentors' efficiency. Although being a study with several statistical procedures and high strength study design, Phadtare et al. (2009) also address the problems associated with using imperfect measurement scales, which can be prone to subjective bias. The authors also reflected on the lack of tools to objectively evaluate manuscript quality and participants' self-assessment, which makes it difficult to interpret the results of this type of study. Several other factors are also pointed out, as possible influences, including participants' previous experience with on-line courses, their grades, computer competency, and interaction with the system in question, as well as the instructor skills and the presence or absence of supervision.

Virtual Worlds

Virtual worlds, such as Second Life®, are described as online computer-based simulations where the user is given the impression of being in another place/location through replications of real life objects(Honey et al., 2009). Second Life® enables learners to manipulate information and synchronously interact with other people, via a digital representation, also known as an avatar (Rogers, 2009). Virtual worlds, like Second Life®, have the ability to create an artificial social structure where problem-based scenarios can be created and developed, allowing students to actively (co) construct mental models of technical and interpersonal skills (Dickey, 2005). While some literature discusses the uses of Second Life® in positive terms (Boulos et al., 2007; Kirriemuir, 2008, Hansen et al., 2009) citing its ability to support interaction and collaborative learning there is little research about actual projects. Our research retrieved one instructional design and assessment study, one qualitative study and one survey type study. Rogers, (2009) proposed to discover how virtual world simulations can assist nursing students developing characteristics and skills essential to future roles as healthcare professionals. In order to accomplish his goal, the author compiled a series of investigative interviews that researched the attitudes and experiences of a sample of sixteen nursing students enrolled in a Bachelor of Nursing Program, who were previously exposed to six simulated clinical scenarios created in Second Life®. Students were placed into groups based only on their year level and exposed to the simulation in separate locations to replicate the intended purpose of the simulation (Rogers, 2009). The students agreed that the simulation in Second Life® assisted them in the development of technical and non-technical skills and in most cases the students found the technology quite easy to use. Also, the online simulation enabled the students to collaborate and solve the scenarios as a team (even though they were isolated in the real world), allowing students to actively co-construct technical and interpersonal skills (Rogers, 2009).

Also in nursing education Honey et al. (2009) conducted a pilot study based around a simulation of a woman having a postpartum hemorrhage. The focus of the scenario was on recognizing abnormal presentation in a postpartum assessment, nursing interventions for hemorrhage, and communication among health care professionals. Although the project led to significant publication on technical issues and simulation development in Second Life, results about student's participation and assessment are scarce and insufficient for a detailed analysis. With a more structured approach, but with a different objective Wiecha et al (2010), proposed, designed and delivered a pilot postgraduate medical education program in Second Life. They enrolled and trained 14 primary care physicians in an hour-long, highly interactive event in the virtual world on the topic of type 2 diabetes. Participants completed surveys to measure change in confidence and performance on test cases to assess learning. The post survey also assessed participants' attitudes toward the virtual learning environment. On a seven-point Likert scale, participants' mean reported confidence increased from pre to post online event with respect to insulin selection, initiation and dose adjustment on test cases, the percent of participants providing a correct insulin initiation plan increased from 60% pre to 90% post. All participants agreed that this experience in Second Life® was an effective method of medical education, that the virtual world approach was superior to other methods of online education. Although these last conclusions are in fact favorable, most of the results retrieved show that insufficient evidence is still available about the potential of Virtual Worlds in Biomedical education.

There are also potential problems with using a virtual world as an educational tool. It's not intuitive, it contains many types of environments, some of which are inappropriate for educational and professional uses. Non-student avatars may appear in student locations, and some of those avatars can manipulate and even act aggressively toward others. Nevertheless, where problems arise, the teacher must see ways to overcome them. The careful planning of activities, and an on-demand technical support, are undoubtedly two features to keep in mind while developing an online educational experience in virtual worlds. Further research is also necessary to investigate the attitudes of students towards technology-supported learning and to identify and further explain links between the students' personal online practices and their experiences in Virtual Worlds. Moreover it is advisable to demonstrate real, quantifiable outcomes (with large number of students) of these types of interventions in order to objectively assess their impact on student learning.

FUTURE RESEARCH DIRECTIONS

Despite the great deal of opportunities for the implementation of Web 2.0 applications within higher education, there are several challenges that are not permitting a widespread use. These include finding faculty members who are willing to accept change and adopt new pedagogical approaches. Increasing faculty members' acceptance of these new tools may require efforts at improving the perceived usefulness and benefits of the applications (Cain & Fox, 2009). Institutions have great deal of power on these challenges, and the role of leadership and management is crucial in promoting short-term workshops/seminars to increase informatics literacy and promote pedagogical innovations/adaptations. Another aspect to consider, especially in survey studies is standardization. Standardization provides survey research with some of its great strengths, because standardization makes it possible to seriously consider surveys as measurement tools (Beatty, 2001). Studying students is always difficult and always involves subjectivity, but the potential for measurement gives survey research great scientific credibility. It also allows for efficiency in handling data, provides for clear quality control criteria, and minimizes random error introduced by the interviewer (Beatty, 2001). Validating these instruments can take long periods of time but it will benefit outcomes and the validity of the conclusions.

All the conclusions and results presented in this chapter are about professionals who are trained in several types of methodologies and study designs. They must know how to perform and assess studies like clinical trials and other experimental studies that have strong design strength. Similarly, studies performed in order to assess the impact of Web 2.0 tools in biomedical and pharmaceutical education need to follow the same principle. We are aware that we cannot be so straightforward in Educational Sciences, but there are options and ways to better control and predict threats to the study conclusions, giving more strength to our results. IDEAS studies and surveys have their role in creating evidence, but quasi-experimental studies and qualitative studies can provide a stronger, wider and more reliable research. The privacy issue has also arisen during the use of Web 2.0 tools, with focus in wikis and blogs. The issue can be easily controlled, if blogs are password protected, or accessed by invitation only, and if wikis are developed inside LMS Systems, like Moodle, and accessible only to members. This will also avoid vandalism and promote student participation (Boulos et al., 2006). Nevertheless an active tutor role is required in order to readily access, edit or comment on new materials posted by students. Last but not least, the technical difficulties felt by students in some of the studies referred are an alert that, although we live in a society of information, maybe we need a little more care in educating our students about these technologies in order to truly benefit from them.

Indeed, Web 2.0 tools and applications have a powerful place in instruction and in the classroom, but caution is warranted regarding the strategies used. Instead of beginning instructional planning with these tools in mind, the learning objectives and instructional strategies should guide the adoption process. An awareness of the different tools, gives faculty members additional mechanisms from which to choose, but selecting an appropriate instructional strategy is even more important.

REFERENCES

Archambault, P. M., Légaré, F., Lavoie, A., Gagnon, M.-P., Lapointe, J., St-Jacques, S., & Pham-Dinh, M. et al. (2010). Healthcare professionals' intentions to use wiki-based reminders to promote best practices in trauma care: A survey protocol. *Implementation Science; IS*, *5*(1), 45. doi:10.1186/1748-5908-5-45 PMID:20540775

Barsky, E. (2006). Introducing Web 2.0: RSS trends for health librarians. *Journal of Canadian Health Library Association*, *27*(1), 7–8. doi:10.5596/c06-001

Beatty, P. (2001). Standardization in Survey Research: How Did it Get Here and How Far Should it Go? In *Bulletin of the International Statistical Institute. 53rd Session Proceedings*. Seoul, South Korea: International Statistics Institute.

Beldarrain, Y. (2006). Distance Education Trends: Integrating New Technologies to Foster Student Interaction and Collaboration. *Distance Education*, *27*(2), 139–153. doi:10.1080/01587910600789498

Billings, D. M. (2009). Wikis and blogs: Consider the possibilities for continuing nursing education. *Journal of Continuing Education in Nursing*, *40*(12), 534–535. doi:10.3928/00220124-20091119-10 PMID:20000260

Bouldin, A. S., Holmes, E. R., & Fortenberry, M. L. (2006). "Blogging" About Course Concepts: Using Technology for Reflective Journaling in a Communications Class. *American Journal of Pharmaceutical Education*, *70*(4), 84. doi:10.5688/aj700484 PMID:17136203

Boulos, M. N. K., Hetherington, L., & Wheeler, S. (2007). Second Life: An overview of the potential of 3-D virtual worlds in medical and health education. *Health Information and Libraries Journal*, *24*(4), 233–245. doi:10.1111/j.1471-1842.2007.00733.x PMID:18005298

Boulos, M. N. K., Maramba, I., & Wheeler, S. (2006). Wikis, blogs and podcasts: A new generation of Web-based tools for virtual collaborative clinical practice and education. *BMC Medical Education*, *6*(1), 41–41. doi:10.1186/1472-6920-6-41 PMID:16911779

Boutin, S., & Therriault, P. (2011). The learner as a participant in the construction of content through the use of a wiki in a competency-based program of occupational therapy. In *INTED 2011 Proceedings*. Valencia: IATED. Retrieved from http://library.iated.org/view/BOUTIN2011THE

Bratsas, C., Kapsas, G., Konstantinidis, S., Koutsouridis, G., & Bamidis, P. D. (2009). A semantic wiki within moodle for Greek medical education. In *22nd IEEE International Symposium on Computer-Based Medical Systems, 2009. CBMS 2009* (pp. 1–6). IEEE. doi:10.1109/CBMS.2009.5255417

Bristol, T. J. (2010). Twitter: Consider the possibilities for continuing nursing education. *Journal of Continuing Education in Nursing*, *41*(5), 199–200. doi:10.3928/00220124-20100423-09 PMID:20481418

Bussières, J.-F., Métras, M.-É., & Leclerc, G. (2012). Use of Moodle, ExamSoft, and Twitter® in a First-Year Pharmacy Course. *American Journal of Pharmaceutical Education*, *76*(5), 94. doi:10.5688/ajpe76594 PMID:22761535

Cain, J., & Fox, B. I. (2009). *Web 2.0 and Pharmacy Education*. Academic Press.

Camiel, L. D., Goldman-Levine, J. D., Kostka-Rokosz, M. D., & McCloskey, W. W. (2014a). *Twitter® as a Medium for Pharmacy Students' Personal Learning Network Development*. Currents in Pharmacy Teaching and Learning. doi:10.1016/j.cptl.2014.04.008

Camiel, L. D., Goldman-Levine, J. D., Kostka-Rokosz, M. D., & McCloskey, W. W. (2014b). Twitter® as an In-Class Backchannel Tool in a Large Required Pharmacy Course. *American Journal of Pharmaceutical Education*, *78*(3), 67. doi:10.5688/ajpe78367 PMID:24761028

Chretien, K., Goldman, E., & Faselis, C. (2008). The reflective writing class blog: Using technology to promote reflection and professional development. *Journal of General Internal Medicine*, *23*(12), 2066–2070. doi:10.1007/s11606-008-0796-5 PMID:18830767

Chu, L. F., Young, C., Zamora, A., Kurup, V., & Macario, A. (2010). Anesthesia 2.0: Internet-based information resources and Web 2.0 applications in anesthesia education. *Current Opinion in Anaesthesiology*, *23*(2), 218–227. doi:10.1097/ACO.0b013e328337339c PMID:20090518

Conn, V. S., Valentine, J. C., Cooper, H. M., & Rantz, M. J. (2003). Grey literature in meta-analyses. *Nursing Research*, *52*(4), 256–261. doi:10.1097/00006199-200307000-00008 PMID:12867783

Cooper, H. M. (1998). Synthesizing research: A guide for literature reviews. *Sage (Atlanta, Ga.).*

Coutinho, C. (2008). Web 2.0: uma revisão integrativa de estudos e investigações. In Actas do Encontro sobre Web 2.0. Braga: Universidade do Minho, CIEd.

Coutinho, C. (2011). *Metodologia de Investigação em Ciências Sociais e Humanas: Teoria e Prática.* Almedina.

Dickey, M. D. (2005). Three-dimensional virtual worlds and distance learning: Two case studies of Active Worlds as a medium for distance education. *British Journal of Educational Technology*, *36*(3), 439–451. doi:10.1111/j.1467-8535.2005.00477.x

Ebner, M., Lienhardt, C., Rohs, M., & Meyer, I. (2010). Microblogs in Higher Education – A chance to facilitate informal and process-oriented learning? *Computers & Education*, *55*(1), 92–100. doi:10.1016/j.compedu.2009.12.006

EDUCAUSE. (Ed.). (2009). *7 Things You Should Know About Microblogging.* Retrieved from http://net.educause.edu/ir/library/pdf/ELI7051.pdf

Fischer, M. A., Haley, H., Saarinen, C. L., & Chretien, K. C. (2011). Comparison of blogged and written reflections in two medicine clerkships. *Medical Education*, *45*(2), 166–175. doi:10.1111/j.1365-2923.2010.03814.x PMID:21208262

Forbes, M. O., & Hickey, M. T. (2008). Podcasting: Implementation and evaluation in an undergraduate nursing program. *Nurse Educator*, *33*(5), 224–227. doi:10.1097/01.NNE.0000334775.98018.e8 PMID:18769328

Fox, B. I., & Varadarajan, R. (2011). Use of Twitter® to Encourage Interaction in a Multi-campus Pharmacy Management Course. *American Journal of Pharmaceutical Education*, *75*(5), 88. doi:10.5688/ajpe75588 PMID:21829262

George, D. R., & Dellasega, C. (2011). Use of social media in graduate-level medical humanities education: Two pilot studies from Penn State College of Medicine. *Medical Teacher*, *33*(8), e429–e434. doi:10.3109/0142159X.2011.586749 PMID:21774639

Geyer, H., Beylefeld, A., & Alwyn, H. (2008). To Podcast or not to Podcast? Students' Feedback on a Different Learning Experience in Histology. In R. Williams & D. Remenyi (Eds.), *The Proceedings of the 7th European Conference on e-Learning.* Agia Napa, Cyprus: Academic Conferences Limited.

Goldman, R. H., Cohen, A. P., & Sheahan, F. (2008). Using Seminar Blogs to Enhance Student Participation and Learning in Public Health School Classes. *American Journal of Public Health*, *98*(9), 1658–1663. doi:10.2105/AJPH.2008.133694 PMID:18633075

Gomes, M. J., & Coutinho, C. (2008). Meta-análise da investigação realizada no âmbito do mestrado em Tecnologia Educativa da UM. In *As TIC na Educação em Portugal. Concepções e Práticas* (pp. 60–70). Porto: Porto Editora.

Grassley, J. S., & Bartoletti, R. (2009). Wikis and blogs: Tools for online interaction. *Nurse Educator*, *34*(5), 209–213. doi:10.1097/NNE.0b013e3181b2b59b PMID:19726963

Grosseck, G., & Holotesku, C. (2008). Can we use Twitter® for educational activities? In *4th Scientific Conference eLSE "elearning and Software for Education"*. Retrieved from http://www.scribd.com/doc/2286799/Can-we-use-Twitter-for-educational-activities

Hansen, M. M., Murray, P. J., & Erdley, W. S. (2009). The potential of 3-D virtual worlds in professional nursing education. *Studies in Health Technology and Informatics*, *146*, 582–586. PMID:19592909

Honey, M. L., Diener, S., Connor, K., Veltman, M., & Bodily, D. (2009). Teaching in Virtual Space: An interactive session demonstrating Second Life® simulation for haemorrhage management. In *Same places, different spaces*. Retrieved from. http://www.ascilite.org.au/conferences/auckland09/procs/honey-interactivesession.pdf

Jadad, A. R., Moher, D., & Klassen, T. P. (1998). Guides for reading and interpreting systematic reviews: II. How did the authors find the studies and assess their quality? *Archives of Pediatrics & Adolescent Medicine*, *152*(8), 812–817. doi:10.1001/archpedi.152.8.812 PMID:9701144

Jesus, Â., Cruz, A., & Gomes, M. J. (2011). Case Based, Learner Centered Approach to Pharmacotherapy. In *Proceedings from EDULEARN 2011* (pp. 6074–6080). Barcelona: IATED.

Jesus, Â., Gomes, M. J., & Cruz, A. (2012). *A B-learning strategy for Therapeutics at the Bachelor Level*. Presented at the FIP World Centennial Congress of Pharmacy and Pharmaceutical Sciences, Amsterdam: International Pharmaceutical Federation.

Jesus, Â., Gomes, M. J., & Cruz, A. (2013). *Case Based Learning Digital - Proposta para Estruturação da Formação*. Presented at the XII Congresso Internacional Galego-Português de Psicopedagogia, Braga.

Kaldoudi, E., Bamidis, P., Papaioakeim, M., & Vargemezis, V. (2008). *Problem-Based Learning via Web 2.0 Technologies*. IEEE; doi:10.1109/CBMS.2008.136

Kamel Boulos, M. N., & Wheeler, S. (2007). The emerging Web 2.0 social software: An enabling suite of sociable technologies in health and health care education. *Health Information and Libraries Journal*, *24*(1), 2–23. doi:10.1111/j.1471-1842.2007.00701.x PMID:17331140

Kirriemuir, J. (2008). Second Life® in higher education, medicine and health. *Health Information on the Internet*, (64), 6–8.

Llambí, L., Esteves, E., Martinez, E., Forster, T., García, S., Miranda, N., & Margolis, A. et al. (2011). Teaching tobacco cessation skills to Uruguayan physicians using information and communication technologies. *The Journal of Continuing Education in the Health Professions*, *31*(1), 43–48. doi:10.1002/chp.20100 PMID:21425359

Mathieu, J. (2007). Blogs, podcasts, and wikis: The new names in information dissemination. *Journal of the American Dietetic Association*, *107*(4), 553–555. doi:10.1016/j.jada.2007.02.027 PMID:17383254

Mattheos, N., Schoonheim-Klein, M., Walmsley, A. D., & Chapple, I. L. C. (2010). Innovative educational methods and technologies applicable to continuing professional development in periodontology. *European Journal of Dental Education*, *14*, 43–52. doi:10.1111/j.1600-0579.2010.00624.x PMID:20415976

McGarr, O. (2009). A Review of Podcasting in Higher Education: Its Influence on the Traditional Lecture. *Australasian Journal of Educational Technology*, *25*(3), 309–321.

Meade, O., Bowskill, D., & Lymn, J. S. (2009). Pharmacology as a foreign language: A preliminary evaluation of podcasting as a supplementary learning tool for non-medical prescribing students. *BMC Medical Education*, *9*(1), 74. doi:10.1186/1472-6920-9-74 PMID:20021673

Meade, O., Bowskill, D., & Lymn, J. S. (2011). Pharmacology podcasts: A qualitative study of non-medical prescribing students' use, perceptions and impact on learning. *BMC Medical Education*, *11*(2). Retrieved from http://www.biomedcentral.com/1472-6920/11/2 PMID:21223547

Melissa Tan, S., & Ladyshewsky, K.,, R., & Gardner, P. (2010). Using blogging to promote clinical reasoning and metacognition in undergraduate physiotherapy fieldwork programs. *Australasian Journal of Educational Technology*, *26*(3), 355–368.

Miller, A. D., Bookstaver, P. B., & Norris, L. B. (2009). *Use of Wikis in Advanced Pharmacy Practice Experiences*. Academic Press.

Miller, A. D., Norris, L. B., & Bookstaver, P. B. (2012). Use of wikis in pharmacy hybrid elective courses. *Currents in Pharmacy Teaching and Learning*, *4*(4), 256–261. doi:10.1016/j.cptl.2012.05.004

Mirk, S. M., Burkiewicz, J. S., & Komperda, K. E. (2010). Student perception of a wiki in a pharmacy elective course. *Currents in Pharmacy Teaching and Learning*, *2*(2), 72–78. doi:10.1016/j.cptl.2010.01.002

Nathan, P., & Chan, A. (2007). Engaging Undergraduates with Podcasting in a Business Subject. In ICT: Providing choices for learners and learning. Singapore.

O'Brien, R. (2001). An Overview of the Methodological Approach of Action Research. In R. Richardson (Ed.), Theory and Practice of Action Research. Brazil: Universidade Federal da Paraíba. Retrieved from http://www.web.ca/~robrien/papers/arfinal.html

O'Sullivan, S., & McGlynn, H. (2010). Use of e-portfolios and web 2.0 tools in the assessment of group work activities in the undergraduate biomedical science classroom. In *EDULEARN10 Proceedings*. Barcelona: IATED. Retrieved from http://library.iated.org/view/OSULLIVAN2010USE

Olofsson, A. D., Lindberg, J. O., & Hauge, T. E. (2011). Blogs and the Design of Reflective Peer-to-Peer Technology-Enhanced Learning and Formative Assessment. *Campus-Wide Information Systems*, *28*(3), 183–194. doi:10.1108/10650741111145715

Oomen-Early, J., & Burke, S. (2007). Entering the Blogosphere: Blogs as Teaching and Learning Tools in Health Education. *International Electronic Journal of Health Education*, *10*, 186–196.

Oxman, A. D. (1994). Checklists for review articles. *BMJ : British Medical Journal*, *309*(6955), 648–651. doi:10.1136/bmj.309.6955.648 PMID:8086990

Park, C. L., Crocker, C., Nussey, J., Springate, J., & Hutchings, D. (2010). Evaluation of a Teaching Tool--Wiki--in Online Graduate Education. *Journal of Information Systems Education*, *21*(3), 313–321.

Parker, K. R., & Chao, J. T. (2007). Wiki as a teaching tool. *Interdisciplinary Journal of E-Learning and Learning Objects*, *3*, 57–72.

Phadtare, A., Bahmani, A., Shah, A., & Pietrobon, R. (2009). Scientific writing: A randomized controlled trial comparing standard and on-line instruction. *BMC Medical Education*, *9*(1), 27. doi:10.1186/1472-6920-9-27 PMID:19473511

Pierce, R., & Fox, J. (2012). Vodcasts and Active-Learning Exercises in a "Flipped Classroom" Model of a Renal Pharmacotherapy Module. *American Journal of Pharmaceutical Education*, *76*(10), 196. doi:10.5688/ajpe7610196 PMID:23275661

Pilarski, P. P., Alan Johnstone, D., Pettepher, C. C., & Osheroff, N. (2008). From music to macromolecules: Using rich media/podcast lecture recordings to enhance the preclinical educational experience. *Medical Teacher*, *30*(6), 630–632. doi:10.1080/01421590802144302 PMID:18677662

Poirier, T., Crouch, M., MacKinnon, G., Mehvar, R., & Monk-Tutor, M. (2009). Updated Guidelines for Manuscripts Describing Instructional Design and Assessment: The IDEAS Format. *American Journal of Pharmaceutical Education*, *73*(3), 55. doi:10.5688/aj730355 PMID:19564998

Poonawalla, T., & Wagner, R. F. (2006). Assessment of a blog as a medium for dermatology education. *Dermatology Online Journal*, *12*(1), 5. PMID:16638373

Rodríguez, M. L. G., Carrillo, A. R., & Alvarez, A. M. R. (2010). Aplicación de la filosofía ¿wiki¿ a la actualización del material docente en el área de farmacia y tecnología farmacéutica. *Ars Pharmaceutica*, *51*(4), 187–194.

Rogers, L. (2009). Simulating clinical experience: Exploring Second Life® as a learning tool for nurse education. In *Same places, different spaces. Proceedings ASCILITE Auckland 2009*. Retrieved from http://www.ascilite.org.au/conferences/auckland09/procs/rogers.pdf

Ros-Rodriguez, J., Encinas, T., Picazo, R.A., Labadia, A., Artalejo, A.R., Gutiérrez-Martin, Y., ... Gilabert, J.A. (2011). Use of Wikis as collaborative tools in a B-learning course of Pharmacology. In *INTED 2011 Proceedings*. Valencia: IATED. Retrieved from http://library.iated.org/view/ROSRODRIGUEZ2011USE

Rosh, A., Jones, K., & Wahl, R. (2009). Receiving: The Use of Web 2.0 to Create a Dynamic Learning Forum to Enrich Resident Education. *Academic Emergency Medicine*, *16*, S274–S275. doi:10.1111/j.1553-2712.2009.00392_2.x

Sandars, J. (2006). Twelve tips for using blogs and wikis in medical education. *Medical Teacher*, *28*(8), 680–682. doi:10.1080/01421590601106353 PMID:17594577

Sandars, J., Homer, M., Pell, G., & Croker, T. (2008). Web 2.0 and social software: The medical student way of e-learning. *Medical Teacher*, *30*(3), 308–312. doi:10.1080/01421590701798729 PMID:18608950

Schreiber, B. E., Fukuta, J., & Gordon, F. (2010). Live lecture versus video podcast in undergraduate medical education: A randomised controlled trial. *BMC Medical Education*, *10*(1), 68. doi:10.1186/1472-6920-10-68 PMID:20932302

Shantikumar, S. (2009). From lecture theatre to portable media: Students' perceptions of an enhanced podcast for revision. *Medical Teacher*, *31*(6), 535–538. doi:10.1080/01421590802365584 PMID:18937140

Skiba, D. J. (2008). Nursing education 2.0: Twitter® & tweets. Can you post a nugget of knowledge in 140 characters or less? *Nursing Education Perspectives*, *29*(2), 110–112. PMID:18459627

Sparks, M. A., O'Seaghdha, C., Sethi, S. K., & Jhaveri, K. D. (2011). Embracing the Internet as a Means of Enhancing Medical Education in Nephrology. *American Journal of Kidney Diseases: The Official Journal of the National Kidney Foundation*. doi:10.1053/j.ajkd.2011.06.009

Stewart, D. W., Panus, P. C., & Hagemeier, N. E. (2013). An analysis of student performance with podcasting and active learning in a pharmacotherapy module. *Currents in Pharmacy Teaching and Learning*, *5*(6), 574–579. doi:10.1016/j.cptl.2013.07.004

Stiffler, D., Stoten, S., & Cullen, D. (2011). Podcasting as an Instructional Supplement to Online Learning. *CIN: Computers, Informatics. Nursing*, *29*, TC84–TC88. doi:10.1097/NCN.0b013e3181fc3fdf

Sumer, S. (2008). Web 2.0 and Radiology. *Internet Journal of Radiology, 8*(2). Retrieved from http://www.ispub.com/journal/the_internet_journal_of_radiology/volume_8_number_2_11/article/web_2_0_and_radiology.html

Varga-Atkins, T., Dangerfield, P., & Brigden, D. (2010). Developing professionalism through the use of wikis: A study with first-year undergraduate medical students. *Medical Teacher*, *32*(10), 824–829. doi:10.3109/01421591003686245 PMID:20854158

Ward, R., Moule, P., & Lockyer, L. (2009). Adoption of Web 2.0 Technologies in Education for Health Professionals in the UK: Where are we and why. *Electronic Journal of E-Learning*, *7*(2), 165–172.

Whittemore, R., & Knafl, K. (2005). The integrative review: Updated methodology. *Journal of Advanced Nursing*, *52*(5), 546–553. doi:10.1111/j.1365-2648.2005.03621.x PMID:16268861

Wiecha, J., Heyden, R., Sternthal, E., & Merialdi, M. (2010). Learning in a Virtual World: Experience With Using Second Life® for Medical Education. *Journal of Medical Internet Research*, *12*(1), e1. doi:10.2196/jmir.1337 PMID:20097652

Woulfe, J., Williams, K., & Ryan, G. (2009). Evaluating pharmacy students ' wiki-based collaboration. In R. Atkinson & C. McBeath (Eds.), *Same places, different spaces. Proceedings ascilite Auckland 2009* (pp. 1197–1199). Auckland: The University of Auckland, Auckland University of Technology, and Australasian Society for Computers in Learning in Tertiary Education (ASCILITE). Retrieved from http://www.ascilite.org.au/conferences/auckland09/procs/

Younger, P. (2010). Using wikis as an online health information resource. Nursing Standard, 24(36), 49–56.

Zanussi, L., Paget, M., Tworek, J., & McLaughlin, K. (2011). Podcasting in medical education: Can we turn this toy into an effective learning tool? *Advances in Health Sciences Education: Theory and Practice*. doi:10.1007/s10459-011-9300-9 PMID:21544550

KEY TERMS AND DEFINITIONS

Blog: A kind of journal published on the World Wide Web consisting of posts typically displayed in reverse chronological order so the most recent post appears first. Blogs are usually the work of a single individual, occasionally of a small group, and often are themed on a single subject.

Microblogging: Microblogging is the practice of posting small pieces of digital content on the Internet.

Podcast: A type of digital media consisting of an episodic series of audio files subscribed to and downloaded through web syndication or streamed online to a computer or mobile device.

Virtual Worlds: Described as online computer-based simulations where the user is given the impression of being in another place/location through replications of real life objects.

Vodcast: A type of digital media consisting of an episodic series of video files subscribed to and downloaded through web syndication or streamed online to a computer or mobile device.

Web 2.0: Web 2.0 is a loosely defined intersection of web application features that facilitate participatory information sharing, interoperability, and user-centered design.

Wiki: A website whose users can add, modify, or delete its content via a web browser using a simplified markup language or a rich-text editor. Most are created collaboratively.

ENDNOTE

[1] Coutinho (2008) proposes four categories: theoretical/ reflection, empirical, reports of educational experiences and study protocols. Poirier et al. (2009) presents a more structured approach to reports of educational experiences – called Instructional Design and Assessment Studies (IDEAS).

Chapter 4
Information Quality Issues in the Identification and Tracking of Drugs within the Pharmaceutical Industry

Dinah M. Mande
University of Arkansas – Little Rock, USA

Rolf T. Wigand
University of Arkansas – Little Rock, USA

ABSTRACT

This contribution examines solutions how Information Quality (IQ) dimensions as a framework along with Radio Frequency Identification (RFID) and the Electronic Product Code Information System (EPCIS) as tools may improve needed drug trackability and traceability capabilities in the pharmaceutical industry (PI). For years counterfeit drugs have been impacting the industry and putting patients' health in danger. We analyze applications, methods and practices in the improvement of the quality of drug tracking and tracing. The potential of IQ, RFID, EPCIS and related applications and technologies suggest and design corresponding information and materials flows. This research presents examinations, reviews and recommendations and utilizes two theoretical frameworks: Transaction Cost Theory and Collective Action Theory. This setting may be viewed as a large, complex and international web of corporations, legislation, regulatory efforts, compliance regimes, manufacturers, wholesalers, pharmacies, importers as well as rapidly advancing technologies and applications.

INTRODUCTION

Information Quality (IQ) as a discipline is applicable to and encompasses any form of organization. The pharmaceutical industry, the focal setting of this research, is no exception. Since the pharmaceutical industry is an information-intensive industry utilizing extensively both structured and unstructured data to create, sell and distribute drugs, IQ is of paramount importance in determining the identity and

DOI: 10.4018/978-1-5225-0248-7.ch004

authenticity of drugs and their ingredients. For all parties involved, the safety of drugs is often taken for granted when visiting the pharmacy to purchase medicine that cures our various ills. However, the problem of counterfeit drugs and drug ingredients has made and is continuing to make its way into the U.S. drug supply chain at an alarming and ever increasing rate. We explore drug counterfeit Information Quality-related issues in the pharmaceutical industry and how they can be, at least partially, addressed by embracing Radio Frequency Identification (RFID) and related technologies and applications together with various regulatory schemes addressing this dilemma.

Information Quality as a field offers important dimensions that make information resources useful for planning and decision-making. For this chapter, the focus of our research setting is the pharmaceutical industry. Many IQ dimensions are crucial to numerous industries, but counterfeiting in the pharmaceutical industry is cheating and defrauding this industry by engaging in unauthorized manufacturing, packaging or the illicit commercialization of drugs being protected by trademarks, patents or copyrights.

The pharmaceutical industry has been continually fighting battles on many fronts including the adoption of pricy new technology in order to reduce drug counterfeiting or to improve supply chains. Counterfeit products, expired products that are re-entering the distribution chain as well as costs to guard for these illicit practices have been among the battles the pharmaceutical industry has been fighting for many years. Moreover, the industry is faced with patient safety issues as well as packaging practices being designed for the market rather than patients' use.

LITERATURE REVIEW

This literature review addresses several bodies of literature. They include identity management and authentication, dimensions of Information Quality, various technologies and applications such as RFID, counterfeiting in general, as well as the practice of counterfeiting in the pharmaceutical industry.

In the following section we provide an overview of identity management.

Identity Management

Identity management is simply a tool to help identify a person, place, or thing (Millet & Holden, 2003). It can be used within multiple areas, including technology and security. Confirming the identity of a person is a common event, such as showing an ID when attempting to cash a check at the bank. For many products, identities are not as essential. For example, the manufacturer of nails does not necessarily care where the metal comes from. But, the beef industry needs to know the origin of a cow. These examples show how identity management can come in multiple forms, but its importance depends on the field in which it is utilized.

Along with identity management, authentication is a means of confirming something or someone to be authentic (Millet & Holden, 2003). It can also confirm the identity of a person or tracing the path of an item back to its origin. To authenticate an item, its attributes are compared with known characteristics of its creator. For example, a painting from Picasso can be authenticated by art experts by examining the attributes of the work. If they match known patterns, such as style, time of creation, etc., then the painting can be called authentic.

Identity management is a key concern when dealing with authenticity of a drug or drug component. The dilemma with counterfeit drugs and drug components is that they look authentic, i.e. their identity seems to be the same as they look real, i.e. just like the real thing, including the packaging, the design, logos as well as printing on the package. Another focal concern with counterfeit drugs is the information, the lack of information associated or provided with the drug as well as counterfeit drug.

The next section addresses dimensions of Information Quality.

Dimensions of Information Quality

Identity management and authentication have been fundamental components of Information Quality. Over the years, IQ researchers have used various methods to manage and authenticate information in an attempt to improve quality. In the well known Wang and Strong (1996) framework a set of IQ dimensions allows researchers to quantify or ascertain the notion of quality of information. Treating information in the pharmaceutical industry (PI) as a product itself, we apply IQ dimensions to the worldwide pharmaceutical industry counterfeit drug problem. We are guided by the Wang and Strong framework to examine the current state of anti-counterfeiting efforts by both the pharmaceutical industry and governmental regulatory agencies. And, in turn, they provide an approach to a potential solution in addressing this worldwide problem using RFID and related technologies and applications.

Although the creation and selling of drugs have been around for a very long time, Information Quality is a relatively new aspect in the pharmaceutical industry. In this research, the authors utilize the existing Wang/Strong IQ dimensions. Those 16 dimensions provide a valuable framework in which researchers can classify various Information Quality problems (Wang & Strong, 1996). Although most, if not all, of the IQ dimensions could theoretically be applied to the pharmaceutical industry for counterfeiting issues, the authors chose only eight dimensions as the most relevant and applicable of the issues studied in this research. Those dimensions are: accuracy, completeness, consistency, amount of data, accessibility, timeliness, believability, and security. A definition of each dimension and its applicability to this research is provided below.

In the following, we provide IQ dimensions, their definitions and applicability to this research in the pharmaceutical industry (Wang & Strong, 1996). The eight IQ dimensions utilized are:

- **Accuracy:** Defined as the extent to which data is correct and reliable. Accuracy of components, ingredients, and the amounts of drugs is of utmost importance for all parties involved.
- **Completeness:** Defined as the extent to which data is not missing and is of sufficient breadth and depth for the task at hand. Data and information about drugs and their ingredients are assumed to be complete.
- **Consistency:** Defined as the extent to which data is presented in the same format. Consistency in the manufacture of drugs is required and assumed.
- **Amount of Data:** Refers to the quantity of data. The amount of data has to be adequate.
- **Accessibility:** The extent to which data are available or easily and quickly retrievable. Accessibility of drug information is essential for tracking and tracing efforts.
- **Timeliness:** Defined as the extent to which the data is sufficiently up-to-date for the task at hand. Information about the drug has to be available in a timely fashion.
- **Believability:** Extent to which information is regarded as true and credible. Believability is a fundamental aspect of trust in drugs expected by patients and doctors alike.
- **Security:** Extent to which access to information is restricted appropriately to maintain its security. Security and safety of all drugs are paramount.

The applicability of these dimensions to this research is discussed throughout the chapter. Next we present a discussion of quality attributes of the actual physical product, i.e. the drug or medicine itself.

Due to its impact on people's health, the pharmaceutical industry (PI) has characteristics of regulated industries and follows the requirements common to R&D firms or chemical laboratories. Although the PI experienced a number of quality problems such as the product should be put in a package that contains accurate or precise information about the medication and at the same time be appealing to the client. Measurement methods require control; packaging must promote inalterability of The source of quality in the PI can be found in both policy and attitude which can influence sources of quality exhibited by management and personnel as in the characteristics of the supply chain, the product design, and the human resources management. Incorporating technology into the process allows greater consistency and less variability in critical attributes of the pharmaceutical product.

Next, we are providing a brief overview of radio frequency identification as well as the Electronic Product Code Information Services and how these technologies and applications relate to our research. Both of these technologies are viewed as essential tools in the effort of tracking and tracing of drugs and as helpful and promising tools to identify counterfeit drugs.

RFID, EPCIS, and a RFID-Based Infrastructure

Before addressing counterfeiting in the PI and related Information Quality issues, a brief introduction to radio frequency identification (RFID) is in order. First, RFID is not a new technology. It was developed in World War II as a means of identifying friendly airplanes from enemy airplanes using radar over a battlefield (History of RFID, 2010). More recently, RFID has been adopted for use by supply chain management and retail giants such as Walmart and Target. Products can be tagged with a RFID chip inserted or built into the products at the manufacturer. Products are typically tagged with a RFID chip at the palette, case and individual product item levels as needed. This tag may allow the purchaser of the product to ensure the identity and authenticity of the product during the transportation process or once it arrives at its destination.

RFID tags work by one of two methods – active or passive. An active tag has a battery, while a passive tag does not. Passive tags get their power from an interrogator (e. g., a reading device). It generates an electro-magnetic field providing sufficient power to transmit and read the recorded information on the tag, i.e. the RFID chip. Due to the higher cost of an active tag, passive tags currently dominate the market. Almost any information may be recorded on these tags and many companies include, e.g., a unique serial number on each tag. These serial numbers make it possible to verify a product's identity and authenticity, date of manufacture, name of manufacturer, etc. Such information could potentially save lives (such as with certain pharmaceutical products) as products move through the supply chain and are distributed. The creation, adoption, and standardization of RFID products by companies have come about relatively slowly, but increasingly yet deliberately companies are using and adopting the technology. RFID technology is predicted to grow into a $2.1 billion industry by 2016 (History of RFID, 2010). The cost of individual RFID chips when mass produced in very large quantities are today below US$ 0.07.

In the following section, the leading organization developing RFID standards, i.e. EPCglobal, is discussed and provides additional details on RFID standards, their usage and adoption. Most importantly, we are attempting to paint a picture how EPCglobal's role-out of Electronic Product Code Information Services (EPCIS) provides a suitable underlying infrastructure for the tracking and tracing of products, here pharmaceutical products or drugs in particular.

EPCglobal had been working for many years on the bar code (EPCglobal, 2010), as well as its successor, Radio Frequency Identification (RFID). In these efforts Electronic Product Code Information Services (EPCIS) was developed and has since become a widely used standard. EPCglobal is spearheading the development of industry-driven standards for the Electronic Product Code™ (EPC) in support of RFID in today's fast-moving and information-rich trading networks. EPCglobal's goal is increased visibility and efficiency throughout the supply chain and higher quality information flow between companies and their respective trading partners (EPCglobal, 2010).

EPCglobal describes this effort as follows:

The goal of EPCIS is to enable disparate applications to leverage Electronic Product Code (EPC) data via EPC-related data sharing, both within and across enterprises. Ultimately, this sharing is aimed at enabling participants in the EPCglobal Network to gain a shared view of the disposition of EPC-bearing objects within a relevant business context. ... The EPC Information Service approach will define a standard interface to enable EPC-related data to be captured and queried using a defined set of service operations and associated EPC-related data standards, all combined with appropriate security mechanisms that satisfy the needs of user companies. In many or most cases, this will involve the use of one or more persistent databases of EPC-related data though elements of the Services approach could be used for direct application-to-application sharing without persistent databases. With or without persistent databases, the EPCIS specification specifies only a standard data sharing interface between applications that capture EPC-related data and those that need access to it. It does not specify how the service operations or databases themselves should be implemented. This includes not defining how the EPCISs should acquire and/or compute the data they need, except to the extent the data is captured using the standard EPCIS capture operations. The interfaces are needed for interoperability, while the implementations allow for competition among those providing the technology and EPC Information Service. – (EPC Information Services, 2010).

We can see how EPCglobal's Electronic Product Code Information Services (EPCIS) provides a suitable infrastructure potentially enabling the tracking and tracing of products, here pharmaceutical products or drugs in particular.

One vendor, IBM, has developed an elaborate framework within which RFID-based tracking and tracing, but also identity management and authentication is envisioned. IBM's RFID Information Center (RFIDIC) (IBM RFID Information Center, 2010) is based on EPCglobal's Electronic Product Code Information Services (EPCIS) standard specification. Accordingly, this RFID Information Center makes possible the tracking of uniquely identifiable, i.e. serialized, product throughout the supply chain. Despite the name, the RFID Information Center is said to be sensor-agnostic, implying it recognizes product serialized with RFID, barcode and/or 2D barcode.

IBM's RFIDIC Shipment Verification feature (IBM RFID Information Center, 2010) offers an automated means to track shipments and confirm receipts. Automating these processes helps reduce product losses and the labor needed to resolve discrepancies and errors in shipment or receipt processes.

The following steps (see Figure. 2) demonstrate the Shipment Verification model (Coase, 1937):

1. The retail distribution center ships totes to the retail pharmacy. Object events with a business step of *shipping* are sent to the RFID Information Center for the retail distribution center.

Figure 1. RFID information center system

Source: IBM RFID Information Center. Retrieved on February 16, 2016, from http://en.wikipedia.org/wiki/IBM_RFID_Information_Center.

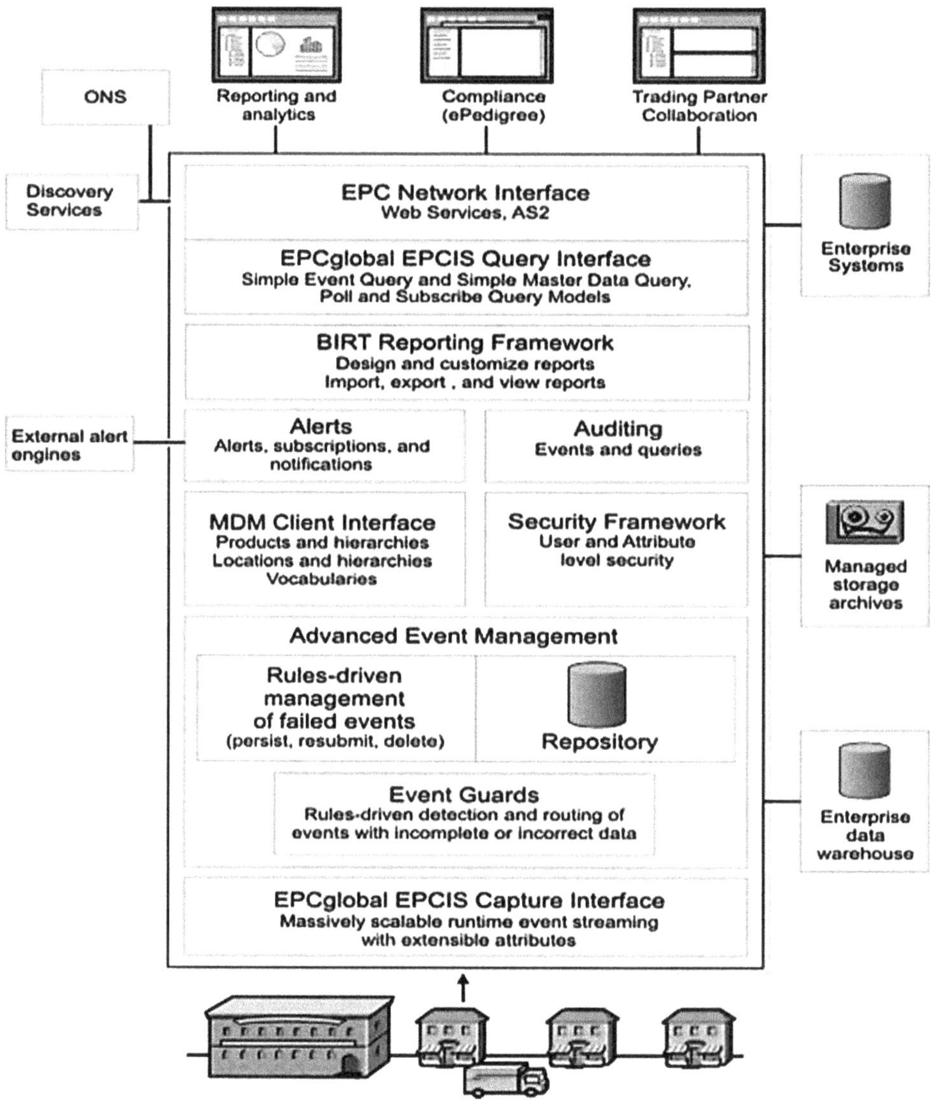

2. The retail distribution center responds to the subscription query from the retail pharmacy by sending a list of the totes that were shipped. These Advanced Shipping Notices (ASNs) provide advanced visibility of shipments in the form of Dashboards. The receiver's *Expected Receipts* dashboard highlights all in-transit totes so that labor can be scheduled for expected receipts. The shipper's *Shipment Verification* dashboard highlights all shipped totes that are eventually received to provide visibility of the downstream inventory. Electronic Product Codes (EPCs) in the delayed status (those not received in the expected time frame) can be investigated and the problems resolved.
3. The retail pharmacy receives the totes. Object events with a business step of *receiving* are sent to the RFIDIC for the retail pharmacy.

Figure 2. RFIDIC shipment verification feature
Source: IBM RFID Information Center. Retrieved on February 16, 2016, from http://en.wikipedia.org/wiki/IBM_RFID_Information_Center.

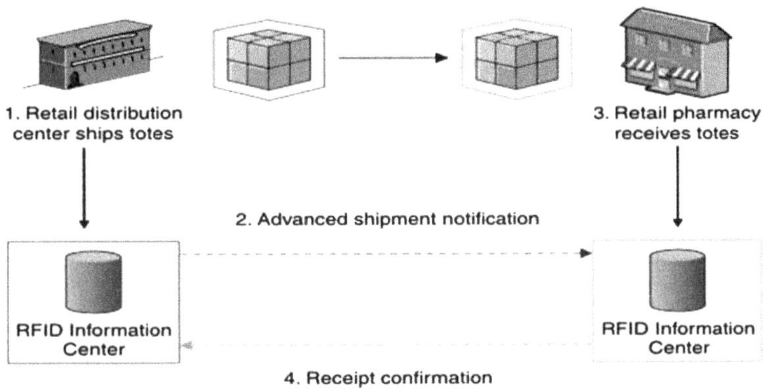

4. The retail pharmacy responds to the subscription query from the retail distribution center by sending a list of the totes received.

This EPCIC infrastructure must be viewed here in light of the motivation, the underlying interests and incentives of counterfeiters. We demonstrated earlier that counterfeiting is indeed financially rather enticing and attractive, especially vis-à-vis the penalties which are on balance relatively low. Organizationally and also physically we claim that there is relative easy access and entry to legitimate supply chains (from manufacturers through pharmacies and drug stores) within the PI, especially when considering the low level of supply chain controls on both the commercial and the customs sides. This occurs in spite of considerable uptake in control efforts by participants in the supply chain including such organizations as the Food and Drug Administration (FDA). This latter point will be addressed and demonstrated separately in a subsequent section.

Considerable gains have been made in supply chains in regards to visibility, transparency and control. These desirable features are now built into supply chain management software and have become standard features welcomed by numerous players within supply chains. Often such features make such manufacturing processes as *just-in-time delivery* possible in the first place. In many cases the practice of warehousing has become a concept from the past and in a way the trucks on the road and the containers on trains in transit to the manufacturer have become the substitutes of warehouses. Catching counterfeiters in such supply chain settings, however, is a difficult task. It requires deliberate cooperation and collaboration among almost all stakeholders in the supply chain, including such organizations as Customs and the FDA, in the design of architectures and processes as well as standard daily operations, the processing of transactions (physical (the handling of products) and digital (the processing of documents, payments, etc.).

Thus we realize that all of the above steps and measures are needed and require intensive cooperation and collaboration to guard supply chains. In turn, such raised complexity and risks would raise the cost of counterfeiting considerably. Further below we will take a look at process changes that would increase the risk for counterfeiters. The next section presents two theoretical lenses via which this research is understood and explained conceptually.

Finally, in this literature review, it is important to have an understanding of counterfeiting problems and practices in the pharmaceutical industry. Such an overview and review are presented in the next section.

Counterfeiting and the Pharmaceutical Industry

The pharmaceutical industry is of course not the only business with counterfeiting problems. In 2009 U.S. officials captured a total of about $260 million in fake products (Johnson, 2010). This is not just a problem in the United States but this phenomenon is observable worldwide.

Chinese criminal gangs are the biggest purveyors of fake products in the United States, accounting for about 80% by value of the counterfeit goods seized last year, according to U.S. government data. Footwear tops the list of fake goods, followed by consumer electronics, luxury goods, and pharmaceutical products (Johnson, 2010).

The single, biggest sweep against counterfeit goods had taken place in December 2009, when federal officials confiscated about $26 million worth of fake toys, Christmas ornaments, perfume, and electronics (Johnson, 2010). Since then, U.S. officials said they made their biggest-ever seizures of counterfeit goods in April 2010 in two operations yielding more than $240 million in total as part of a broader federal offensive against the buying, selling and transacting of pirated products (Johnson, 2010). In this large-scale sting operation, federal, state and local law enforcement officials (including U.S. Immigration and Customs Enforcement, or ICE, at the Department of Homeland Security), part of the National Intellectual Property Rights Coordination Center, confiscated about $40 million worth of counterfeit goods, including fake Rolex watches, Coach handbags, and Nike shoes, as well as pirated DVDs and fake pharmaceutical products, in a sweep of more than 30 U.S. cities. In addition, as part of a separate, long-running investigation, federal officials confiscated $200 million in fake goods made in Asia and smuggled through the port of Baltimore. Immigration and Customs Enforcement announced the double-barreled operation on April 26, 2010 to coincide with World Intellectual Property Day (Johnson, 2010).

This massive effort to reduce and yet better to eliminate the trafficking of fake goods, reminds the world business community that fake products, in turn, steal jobs, creativity, benefits derived from entrepreneurship and innovation, it funds organized crime and there is a serious risk to public safety. The latter point hits home quickly when looking at the statistics pertaining to fake automotive replacement parts such as fake breaks or even fake aircraft replacement parts. Starting April 26, 2010, the Naval Criminal Investigative Service and the Defense Criminal Investigative Service started targeting counterfeit goods that could get into the military supply chain. The U.S. General Services Administration targets fake goods in the federal civilian supply chain. Additionally, federal, state, and local officials have created more than 20 "IP theft enforcement teams" to target the traffic of fake goods nationwide, ICE officials have announced (Johnson, 2010).

In the past, some terrorist groups such as Hezbollah have tried to use the sale of counterfeit goods as a way to finance terrorism, according to the Federal Bureau of Investigation. More recently though counterfeit goods seized in the latest sweeps appeared to be linked not to terrorism, but to organized criminal gangs seeking to make easy money. So far, the trade in fake goods has been largely physical—shipping containers packed with knock-off sneakers and handbags, later smuggled to bazaars and neighborhood stores. According to ICE, the "next big frontier" is Web sites offering pirated movies, music, and pharmaceutical products (Johnson, 2010). A major effort examining web sites offering pharmaceutical products has been carried out by the National Association Boards of Pharmacy (NABP) (2011). These impressive efforts are specifically highlighted later in this chapter.

It is readily evident that counterfeiting of goods is a major worldwide problem and it must be faced by the U.S. just the same. Counterfeiting is especially problematic when dealing with pharmaceutical products such as drugs and when patients' safety, well-being and even their very lives may be at stake.

Various stakeholders within the pharmaceutical industries and their supply chains have become very concerned about such practices and have joined efforts to curtail the production and distribution of counterfeit drugs. Such efforts require cooperation and collaboration, and they suggest deliberate collective action among the relevant players in this setting, including governments and regulators, pharmaceutical manufacturers, wholesalers, importers, pharmacies as well as patients.

The following example illustrates the above described issues and dilemmas, especially when the lives of people are at risk. This case highlights the international nature of supply chains, concerns with identification and authentication as well as intentional fraud.

A Poison's Path

Here we present one case of trafficking in fake goods, specifically a pharmaceutical intermediate product (glycerin) produced to be used eventually in the manufacture of various pharmaceutical products, here specifically, heparin. This brief case study illustrates well how this setting of distributing fake products very quickly is international in nature, how at the core of the issue is identity management and authentication as well as how consequences of such fraud may result in the disfigurement and even over 80 deaths in patients worldwide.

In 2007, a popular blood thinning drug, heparin, was found to contain a mysterious ingredient. This ingredient was only discovered after multiple cases of severe allergic reactions, and even deaths, occurred upon taking the drug. Forensic tests and examinations were conducted to determine the cause. The FDA determined that some heparin contained a substance mimicking the real drug, but was a counterfeit instead (Bogdanich & Hooker, 2007). The FDA linked 19 deaths in the U.S., and hundreds of severe allergic reactions directly to this counterfeit drug (Information on Heparin, 2010). Through intense investigation the journey of the contaminated heparin was traced back to its origin - to a small factory in Hengxiang, China (Bogdanich & Hooker, 2007; see Figure. 4). The chemicals used were prime ingredients for anti-freeze. They were sold, instead, as a type of glycerin used in the production of heparin. For each step in the transactions as the fake product was sold, separate and unique certificates assuring the product's identity and authenticity were issued (see Figure 5). Chinese authorities arrested the person believed to be responsible, Wang Guiping, and sentenced him to life in prison in September of 2009 (Bogdanich & Hooker, 2007). This is, however, just one of many examples of how a counterfeit drug was discovered in the international as well as U.S. pharmaceutical supply chain.

This literature review has addressed several major areas this research is guided by and will deal with. They are identity management, the dimensions of Information Quality, RFID as well as associated technologies and applications and, lastly, we presented an overview of counterfeiting issues and problems in the pharmaceutical industry,

The next section presents a theoretical perspective through which to view the proposed research.

THEORETICAL PERSPECTIVE OF THE RESEARCH

The present research efforts we view through one major theoretical lens: Transaction Cost Theory. This theory lends itself nicely to understand and explain the present, already described setting. The vast number of transactions carried out in the pharmaceutical industry, especially those pertaining to sales and distribution, clearly may be viewed as applicable transactions within a Transaction Cost Theory

Figure 5. Findings from the NABP review of 7,430 Internet drug outlets
Source: National Association of Boards of Pharmacy (NABP). Internet Drug Outlet Identification Program. Mount Prospect, IL: NABP, January 2011.

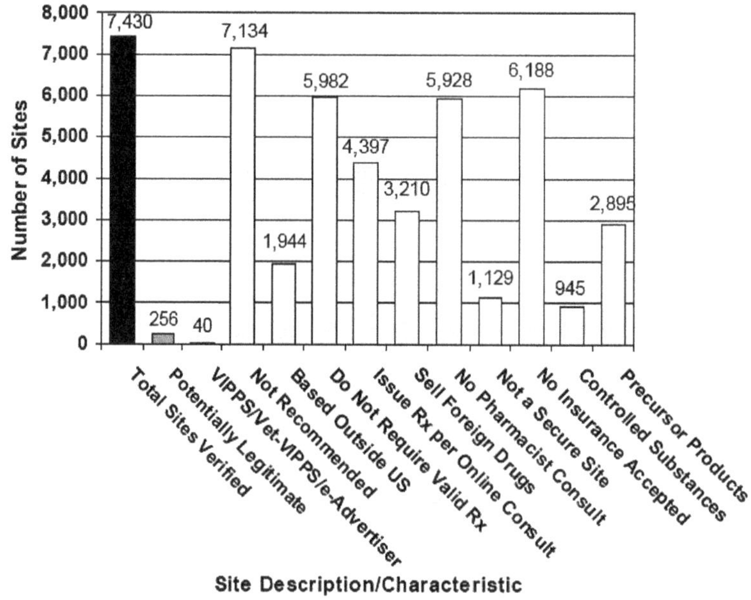

Table 1. Events, trail of transactions, and consequences in the Heparin case

Ref.	Action	Result
1	Glycerin created in Hengxiang, China	Sold a syrup containing diethyfene glycol, a toxic industrial solvent and prime ingredient in some antifreeze, to Beijing broker using forged documents showing syrup was 99.5% pure glycerin.
2	Shipped from port in Shanghai to Barcelona, Spain	Purchased by a new broker, assuming original documentation was authentic, gave additional stamp of authenticity. Resold to broker in Panama.
3	Shipped 46 barrels to Colon, Panama	Panamanian government assumed barrels certificates of authenticity were real and purchased for medicinal use
4	Shipped by truck to Panama City, Panama	Government officials used syrup in 260,000 bottles of medicine
5	Medicine dispersed throughout Panama's medical community and pharmacies	At least 100 deaths in Panama are blamed on this counterfeit glycerin, causing numerous severe allergic reactions, with some patients suffering permanent damage, i.e. facial paralysis. In China, at least 81 deaths were blamed on the drug
6	After thorough worldwide Investigation, contaminated drugs linked to Chinese citizens; Wang Guiping	Receives life sentence in Chinese prison for his crimes

framework. In order to introduce and apply measures of oversight and control, various regulatory efforts, the interaction and cooperation among a vast number of stakeholders (drug manufacturers, distribution companies, shipping companies, logistics firms, conventional drug stores, but also entities such as the FDA and Customs) is needed to make such integrated efforts and tracking possible. This theoretical approach is briefly presented below.

Figure 3. A poison's path
Source: A Poison's Path from China to Medication in Panama. New York Times, May 6, 2007, p.1. Retrieved on February 16, 2016, from http://www.nytimes.com/ref/world/20070506_POISON_GRAPHIC.html.

Figure 4. False certificates of authentication were issued for each step the fake product was sold
Source: Bogdanich, W. and Hooker, J. A Poison's Path. The New York Times, May 6, 2007, p. 1.

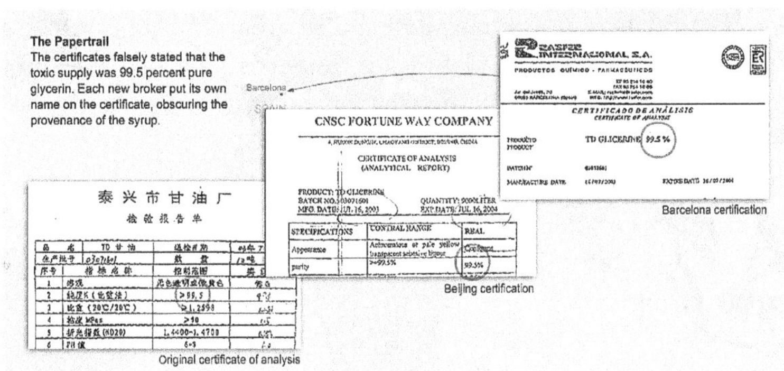

Transaction Cost Theory

Transaction costs are costs incurred in making an economic exchange (Coase, 1937; Wigand et al., 1997; Williamson, 1979). A number of different kinds of transactions can be identified:

Search and information costs are costs such as those arising in determining that the required good is available on the market, which firm has the lowest price, etc. Bargaining costs are the costs required to come to an acceptable and negotiated agreement with the other party in the transaction, drawing up an appropriate contract, etc. Policing and enforcement costs are the costs of making sure the other party in the transaction sticks to the terms of the contract, and taking appropriate actions (often through the legal system).

Transaction costs consist of costs incurred in searching for the best supplier/partner/customer, the cost of establishing a maximally developed contract, and the costs of monitoring and enforcing the implementation of the contract, i.e. transaction.

Transaction cost theorists assert that the total cost incurred by a firm can be grouped largely into two types—transaction costs and production costs. Transaction costs, often known as coordination costs (Williamson, 1979), are well defined as the costs of "all the information processing necessary to coordinate the work of people and machines that perform the primary processes, "whereas production costs include the costs incurred from "the physical or other primary processes necessary to create and distribute the goods or services being produced" (Williamson (1997) cited in Wigand et al., 1999, p.37).

As described above, the Transaction Cost Theory perspective lends itself well to view the activities, i.e. transactions, as a focal theoretical and organizational point when exploring and examining this research within the pharmaceutical industry.

RESEARCH METHODS

This section describes the background of the research conducted for the study, followed by an account of the methods utilized and how they and the underlying procedures were followed.

This research attempted to acquire or generate supporting data retrieved from multiple sources (see below), including the Food and Drug Administration (FDA), the Federal Bureau of Investigation (FBI), the World Health Organization (WHO) and other organizations, as well as several news sources such as the *Wall Street Journal* and *New York Times*. In addition, three in depth interviews were conducted with three industry experts. Their names and organizational affiliations are not revealed here to protect their respective identity. These sources provide a plethora of supporting data and background information. In numerous ways, they should illustrate and illuminate the underlying issues and problems of counterfeiting of drugs that were explored.

The collection and retrieval of data for research purposes for this chapter can be seen from different perspectives and foci as well as levels. These perspectives are from (1) an industry-level, (2) a regulatory and enforcement, (3) an independent, non-partisan and (4) an international perspective. Briefly, this researcher will describe each:

1. **The View of Counterfeit Activity in the Pharmaceutical Industry from an Industry-Level Perspective:** The pharmaceutical manufacturing industry develops and produces many thousands of medicinal products for diagnostic, preventive, and therapeutic uses, saving the lives of millions of people and reduces suffering from illness and help to recover patients' productive lives. The U.S. pharmaceutical industry enjoys worldwide prominence through research and development (R&D) of new drugs, yet in spite of the expenditure and investments of billions of dollars, these may eventually yield fewer than 100 new prescription medicines annually (Pharmaceutical and Medicine Manufacturing, 2011). R&D laboratories perform the work of drug discovery and development. They tend to be located separately from manufacturing plants, although some are integrated with production plants. Once a drug passes animal and clinical tests, the Food and Drug Administration (FDA)'s Center for Drug Evaluation and Research (CDER) is required to review the drug's performance on human patients before approval for commercial use. The entire laborious process, ranging from the first discovery of a promising compound to FDA approval, may well take over a decade and may cost often many hundreds of millions of dollars. From an industry-level research perspective it seems to be possible to secure Information Quality-related data pertaining to the tracking and tracing of counterfeit pharmaceutical products. It seems unlikely to receive such data voluntarily from individual firms, but industry groups and associations such as the Pharmaceutical Research and Manufacturers of America (PhRMA) (950 F St. NW., Washington, DC 20004, Internet: http://

www.phrma.org) may collect and maintain such data. We explored this industry-level resource together with others.

2. **The View of Counterfeit Activity in the Pharmaceutical Industry from a Regulatory and Enforcement Perspective:** Much of counterfeit activity is being monitored and captured, at least as much as possible, by such agencies as the Food and Drug Administration (FDA), ICE as well as the Federal Bureau of Investigation (FBI). Both organizations capture data and issue reports on drug counterfeiting and distribution activities. Each one does so from a somewhat different perspective (e.g., the FDA focuses on health and assurance of drug quality perspectives whereas the FBI focuses more on capturing individuals and identifying organizations engaged in illegal drug counterfeiting activities, just one of several counterfeiting activities), nevertheless all are interested in diminishing and, ideally, eradicating all counterfeiting activities. The aim of the research for this chapter under this perspective is in identifying and locating available secondary data and reports addressing counterfeiting activities over time and to examine if Information Quality issues and attributes are being addressed and are captured.
3. **The View of Counterfeit Activity in the Pharmaceutical Industry from Independent, Non-Partisan Perspectives:** We contacted two independent, non-profit organizations active within the pharmaceutical industry. They are the Drug Information Association and the Center for Medicine in the Public Interest. The Drug Information Association (DIA) (800 Enterprise Road, Suite 200, Horsham, PA 19044, Internet: http://www.diahome.org), a neutral, non-profit organization. The DIA is a professional association with nearly 18,000 members working in every facet of the discovery, development, and life cycle management of pharmaceuticals, medical devices, and related products. Similarly, the Center for Medicine in the Public Interest (CMPI) (757 Third Avenue, 20th Floor, New York, NY 10017, Internet: http://cpmi.org) is a nonprofit, non-partisan organization promoting innovative solutions that advance medical progress, reduce health disparities, extend life and make health care more affordable, preventive and patient-centered. The CMPI also provides the public, policymakers and the media a reliable source of independent scientific analysis on issues ranging from personalized medicine, food and drug safety, health care reform and comparative effectiveness. The CPMI has specifically addressed the problem of counterfeit drugs in several reports and efforts. This researcher anticipated that both, the DIA and the CMPI, may have looked at or encountered issues surrounding Information Quality dimensions and attributes in their pursuit to stem the flow of counterfeit drugs. Moreover, we were hopeful that these organizations would be willing to share such data.
4. **The View of Counterfeit Activity in the Pharmaceutical Industry from an International Perspective:** Supply chains within the pharmaceutical industry easily cross international boundaries and thus the underlying problem is no longer just one nation's problem. It has indeed become an international problem. The World Health Organization (WHO) (Avenue Appia 20, CH-1211 Geneva 27, Switzerland, Internet: http://www.who.int), a United Nations organization agency, is responsible for providing leadership on global health matters, shaping the health research agenda, setting norms and standards, articulating evidence-based policy options, providing technical support to countries and monitoring and assessing health trends. The WHO has actively addressed the problem of counterfeit drugs for many years and has asked for large-scale international actions to stop or at least reduce this problem. Every country is actually involved in counterfeit medicines, estimated to make up ten per cent of the global pharmaceutical trade. According to a report by the US-based Center for Medicines in the Public Interest (2011), counterfeit drug sales leaped

> to $75 billion in 2010, a 92 per cent jump from 2005. WHO officials created a global task force (International Medical Products Anti-Counterfeiting Taskforce (IMPACT)) focusing on legislation, law enforcement activities, trade, risk communications and innovative technology solutions that should curb the flow and sale of counterfeit drugs. Solutions would include public-private initiatives utilizing new and innovative new technologies that can detect counterfeits. WHO set up the world's first web-based system for tracking the activities of illicit drug traders in the Western Pacific Region. The Rapid Alert System (RAS) communications network relays reports on the distribution of counterfeit medicines to appropriate authorities so they can respond quickly. Information on fake drug identity and distribution needs to be shared at national and international levels among drug regulatory authorities, customs and police organizations, pharmaceutical companies, non-governmental organizations and consumer groups.

These researchers contacted the WHO to see if counterfeit drug data may be available and accessible. Moreover, such data could be examined and analyzed to address several of the above mentioned Information Quality dimensions, attributes, issues and features.

RESULTS

Here we are reporting our results and findings. Right away it should be stated that the effort to locate actual data on the problem of counterfeit medicine was not successful in that the respective agencies were unwilling to share such data. It is thus almost superfluous to state that no data addressing specific IQ dimension were found either. There are, unquestionably, efforts underway by government agencies to collect such data. This seems to be done through sampling efforts as the task to examine every good, shipment, parcel, pallet and container is too vast of an undertaking. Unfortunately, at least for this researcher, no such data are being reported or made available publicly. Often, it seems, formal investigative and/or legal efforts are undertaken and, understandably, during this process no such data can be shared, as the data themselves may have become legal evidence. On the other hand, the FDA, e.g., has issued various annual reports on counterfeit drugs, yet the last one issued dates back to 2006. This researcher was unable to receive an answer why such annual reports are no longer being produced. One may only surmise various underlying political and other undertones that may be potential explanations. It is also understandable that there must be concerns about ongoing legal pursuits and other forms of litigation. In our request for such information, it was clearly stated that we are seeking only publicly available data and that we have no interest in any sort of interference in existing legal proceedings.

Numerous efforts were made via formal letter, by phone as well as e-mail in contacting representatives from the FBI, FDA, ICE and other agencies. Some such agencies responded, some never did. Those who responded informed these researchers that such requested information is available on the agency's respective website. The websites, however, had been examined previously in depth and no such data were available which was the very reason to write the agency heads in the first place. We were also informed of our rights to request information and data utilizing the Freedom of Information Act. In conclusion, in the effort to seek suitable Information Quality data in the context of counterfeit drugs, it must be stated that this was a most frustrating and unsatisfactory experience. Consequently, in this section we present some informative secondary source data and efforts surrounding Information Quality dimensions and issues.

In the following, results of two studies are reported that address in part Information Quality issues and problems (the NABP study of 7,430 Internet drug outlets and the international efforts of the WHO IMPACT program). Both studies and efforts demonstrate numerous Information Quality issues and concerns of the counterfeit drug problem.

Results of a Study of 7,430 Internet Drug Outlets by the NABP: An Effort in Information Quality

The National Association of Boards of Pharmacy (NABP) released in 2011 a study examining 7,430 Internet drug outlets that was initiated in 2009. This massive effort has become the NABP's Internet Drug Outlet Identification Program and, for all practical purposes, the effort is a very large scale attempt to examine the Information Quality aspects presented to the consumer or public by these Internet drug websites. As will be shown below, this research examined many of the key components of Information Quality such as accuracy, completeness, timeliness, believability, security and many others as identified in the well known Wang and Strong (1996) scheme or by McGilvray (2008).

The NABP, founded in 1904, is an impartial professional organization supporting the state boards of pharmacy in creating uniform regulations to protect public health (see: http://www.nabp.net). The organization aims to ensure public health and safety through its pharmacist license transfer and pharmacist competence assessment programs as well as its various accreditation programs. The member boards of pharmacy are grouped into eight districts including all 50 U. S. states, the District of Columbia, Guam, Puerto Rico, the Virgin Islands, eight Canadian provinces and New Zealand. NABP's mission statement asserts that "NABP is the independent, international, and impartial association that assists its member boards and jurisdictions in developing, implementing and enforcing uniform standards for the purpose of protecting the public health" (http://www.nabp.net, last accessed on September 5, 2014).

As of December 17, 2010 NABP (NABP, 2011) had conducted initial reviews and verified via a subsequent review, 7,430 Internet drug outlets selling prescription medications. This number was up from 5,449 sites by the end of 2009, a 36% increase. 7,134 (96.02%) of these 7,240 sites were found to be operating out of compliance with state and federal laws and/or NABP patient safety and pharmacy practice standards. These 7,134 sites are listed as 'Not Recommended' in the "Buying Medicine Online" section, under 'Consumers' on the NABP website. This 'Not Recommended' list has grown 37% from 5,226 sites posted at the end of 2009. It should be noted that sites listed as 'Not Recommended' include those found not to be in compliance at the time of review, but they may since have been deactivated. In all, this massive effort evaluating 7,430 Internet drug outlets was indeed an impressive institutional undertaking by the 'NABP applying numerous Information Quality criteria to classify and rate each outlet.

Overall, the results reported above do not paint a pretty picture. The underlying problems with Internet drug outlets create a most frustrating environment within which to regulate, monitor and police such outlets by the NABP but also the FDA, ICE, FBI, American Medical Association (AMA) and other regulatory and enforcement agencies. There is, nevertheless, also a bright sight: it can be recognized that stakeholders committed to patient safety, cleaning up the Internet and identifying and shutting down rogue drug sellers has made significant progress in 2010 (NABP, 2011, p. 4). Such progress reflected increasing momentum and included actions ranging from voluntary actions by Internet electronic commerce providers, to legislative efforts, various enforcement actions as well as public awareness campaigns. Once again we observe that the benefits of concerted collective action of stakeholders and industry players can indeed make a major difference in tackling this problem. – This topic of collective action is also the focus of a later section in this chapter.

Although the overall total number of Internet drug outlets have increased, the number of sites reviewed by the NABP and verified as 'Not Recommended' during 2010 declined when compared to 2009. The NABP (2011, p. 5) reports this change as a shift in focus, recognizing that reputable search engines, domain name registrars, and other Internet commerce and patient advocacy stakeholders have increased their efforts to stop rogue drug sellers. As a partial reaction, the NABP has reallocated resources to identifying legitimate Internet pharmacies and other prescription drug-related entities, in support of its Verified Internet Pharmacy Practice Sites^{CM} (VIPPS®) and Veterinary-Verified Internet Pharmacy Practice Sites^{CM} (Vet-VIPPSCM) accreditation programs and e-Advertiser Approval^{CM} Program Web sites under review following "the same basic standards for legitimate online practice whether operating suspiciously or being considered for accreditation or approval (NABP, 2011, p. 5)."

Accordingly, the growth of the 'Not Recommended' list has slowed over 2010. NABP reported 1,926 additional sites to its 'Not Recommended' list in 2010 – a 51.57% decrease from the 3,977 sites added to the list in 2009 – but still clearly reflecting the ongoing practice and severe problem.

At the time of this research the 7,134 Internet drug outlets currently listed as 'Not Recommended' on the NABP Web site are characterized as follows:

- 1,944 have a physical address located outside of the U.S.
- 4,005 sites do not provide any physical address.
- 5,982 do not require a valid prescription.
- 4,397 issue prescriptions per online consultation or questionnaire only.
- 3,210 offer foreign or non-FDA-approved drugs.
- 5,928 do not offer medical consult with a pharmacist.
- 1,129 do not have secure sites.
- 6,188 do not accept insurance.
- 2,429 have server locations in foreign countries.
- 6,722 appear to be affiliated with a network.
- 945 dispense controlled substances.
- 2,895 sell precursor products.

It should be noted that the above characteristics, descriptions and analysis address many dimensions and features of Information Quality such as accuracy, completeness, timeliness believability, security and many others as identified by Wang and Strong (1996) or by McGilvray (2008). Of the total 7,430 sites examined, 256 (3.45%) merely seem to be potentially legitimate, meaning they meet NABP program criteria that could be verified solely by examining the sites visually, up 15% from the 223 potentially legitimate sites discovered in 2009.

Merely forty (0.54%) of the 7,430 reviewed Internet drug outlets have been accredited through NABP's VIPPS or Vet-VIPPS programs, or approved through the NABP e-Advertiser Approval Program, up 111% from the then 19 entities that were approved in 2009 by these programs. As asserted by the NABP, these increases reflect the staff's redoubled focus on Internet pharmacies and related sites seeking accreditation or approval through NABP (2011). Additional information is provided in Figure 5.

The next section looks at the counterfeit drug problem and some associated Information Quality issues at the international level.

Results from the World Health Organization (WHO) and Its Project IMPACT Task Force: Drug Counterfeiting at the International Level

The World Health Organization, a United Nations agency based in Geneva, addresses the drug counterfeit problem at the international level. Counterfeit medicines are a worldwide problem and they are found everywhere. Such *medicines* may range from random mixtures of harmful toxic substances to inactive, ineffective preparations. Almost always, the source of a counterfeit drug is unknown and its content unreliable at best. In all countries counterfeit drugs are always illegal. As anywhere else, they can result in treatment failure or even death. The WHO made it as one of its goals, yet eliminating counterfeit drugs is a considerable public health challenge.

The WHO provides some key facts characterizing the international drug counterfeit problem (World Health Organization, 2011):

- Counterfeit medicines are medicines that are deliberately and fraudulently mislabeled with respect to identity and/or source.
- Use of counterfeit medicines can result in treatment failure or even death.
- Public confidence in health-delivery systems may be eroded following use and/or detection of counterfeit medicines.
- Both branded and generic products are subject to counterfeiting.
- All kinds of medicines have been counterfeited, from medicines for the treatment of life-threatening conditions to inexpensive generic versions of painkillers and antihistamines.
- Counterfeit medicines may include products with the correct ingredients or with the wrong ingredients, without active ingredients, with insufficient or too much active ingredient, or with fake packaging.

Responding to the growing public health crisis of counterfeit drugs, in February 2006, the World Health Organization launched the International Medical Products Anti-Counterfeiting Taskforce (IMPACT). At its core, IMPACT aims to build coordinated networks across and between countries in 0rder to halt the production, trading and selling of fake medicines around the globe. IMPACT is a partnership comprised of all the major anti-counterfeiting players, including: international organizations, non-governmental organizations, enforcement agencies, pharmaceutical manufacturers associations and drug and regulatory authorities. WHO provides direct country and regional support for strengthening medicines' regulation.

To fight counterfeit medicines effectively, a range of stakeholders – not just health professionals – is needed. As already mentioned, in 2006 WHO helped to create the International Medical Products Anti-Counterfeiting Taskforce, or IMPACT. The aim is to involve a range of stakeholders in collaborative efforts to protect people from buying and taking counterfeit medicines. To prevent the manufacture and distribution of counterfeit medicines, IMPACT's areas of focus are:

- Legislative and regulatory infrastructure,
- Regulatory implementation,
- Enforcement,
- Technology,
- Communication.

Table 2. Examples of counterfeit medicine around the globe

Counterfeit Medicine	Country/Year	Report
Anti-diabetic traditional medicine (used to lower blood sugar)	China, 2009	Contained six times the normal dose of glibenclamide (two people died, nine people hospitalized)[1]
Metakelfin (antimalarial)	United Republic of Tanzania, 2009	Discovered in 40 pharmacies: lacked sufficient active ingredient[2]
Viagra & Cialis (for erectile dysfunction)	Thailand, 2008	Smuggled into Thailand from an unknown source in an unknown country[3]
Xenical (for fighting obesity)	United States of America, 2007	Contained no active ingredient and sold via Internet sites outside the USA[4]
Zyprexa (for treating bipolar disorder and schizophrenia)	United Kingdom, 2007	Detected in the legal supply chain: lacked sufficient active ingredient[5]
Lipitor (for lowering cholesterol)	United Kingdom, 2006	Detected in the legal supply chain: lacked sufficient active ingredient[6]

IMPACT Planning Group members have taken note of the recent discussions of the World Health Organization's (WHO) Working Group on Substandard/Spurious/Falsely-Labelled/Falsified/Counterfeit (SSFFC) Medical Products, which recognized that the work undertaken thus far by the International Medical Product Anti-Counterfeiting Task Force (IMPACT) has delivered valuable results to countries around the world.

The Report of the Working Group on SSFFC Medical Products recognizes that WHO's key role in protecting public health has to include the fight against SSFFC medical products, notably in the areas of information exchange and awareness raising, in developing/updating/promoting norms and standards, and providing technical support to countries to build and further strengthen national regulatory infrastructures and capacity.

Counterfeiting is greatest in regions where regulatory and enforcement systems for medicines are weakest. In African countries, and in parts of Asia, Latin America, and countries in transition, a much higher percentage of the medicines on sale may be counterfeit.

Not only is there a huge variation between geographic regions in terms of incidence of counterfeit medicines, variation can also be significant within countries: for example, between urban and rural areas, and between cities. Table 2 presents examples of this depicting the mentioned considerable variety.

PROJECT IMPACT FINDINGS

The WHO's Project IMPACT has been active since 2006 and reports impressive results in its worldwide efforts (IMPACT, 2011, pp. 102-103). The range of counterfeit and illicit medicines belongs to wide therapeutic categories. 19 of the 25 countries reported the name of the medicines seized. Analysis of these data leads to the findings below.

The number in brackets indicates the number of times the product has been cited by the participating countries (IMPACT, 2011, pp. 102-103):

- Life-style products as erectile dysfunction medicines and female libido pills [80], have been mentioned 104 times by 17 countries. For weight loss medicines [19], authorities found Reductil and Sibutramine, its active pharmaceutical ingredient. Lida Dai Dai Hua, which claim to be a slimming formula made only from Chinese herbs, was also seized in Switzerland, Belgium and Australia. The last type of life-style products seized is hair loss medicines [5].

- Hormones and steroids have been mentioned 56 times by 10 countries. Human Growth Hormones, Testosterone and anabolic steroids are very popular drugs in the world of bodybuilding. Some diuretics which are included on the World Anti-Doping Agency's banned drug list due to their alleged use as a masking agent for other drugs were also seized during the operation. This may suggest the fact that some competitive sportsmen could also use the Internet in order to buy these illegal medicines.
- Narcotics and psychotropic drugs have been mentioned 32 times by 11 countries. They consist mainly of benzodiazepines [12], antidepressant [5] and narcotic analgesic [3] as Vicodin or Codeine.
- Amphetamines and precursors were also seized. In three cases, the precursors have been identified as ephedrine. The similarity in chemical structure to the amphetamines has made ephedrine a sought-after chemical precursor in the illicit manufacture of methamphetamine and methcathinone. Consequently, it would be interesting to analyze further the potential links between narco-trafficants and counterfeiters.
- Pain killers [12], antibiotics [11], contraceptives, blood pressure medication, anti-fungal were also amongst the medicines seized during Operation Pangea II. When the origin of the products was reported, it appears that most of the medicines seized at postal hubs were .com. During web monitoring and investigation on websites, some participating countries analyzed and reported in which country the illicit websites were hosted. At least 41 different countries were identified.

This allows criminals to cover their tracks and make investigations more complex and difficult. After identifying the host country of the website, international collaboration will be essential for law enforcement agencies to manage to shut down the site, notably due to the absence of harmonization between the laws of different countries. Criminals will always try to establish their websites where Internet legislation is weak or permissive.

In several cases, it appears that multiple websites, located in different places and offering medicines to customers from different countries, are associated to the same criminal organization.

- In Austria, law enforcement authorities identified seven illegal websites with the same IP address and hosted by the same provider. Investigation seems to confirm links with Germany and United Kingdom, where a suspect managed a company who owns around 85 domains.
- In Germany, customs officials are investigating a complex of illicit websites selling pharmaceuticals: at least three different websites could be linked and connections are suspected to link with Italy, United Kingdom and France.

In several countries, authorities seized food supplements or herbal products which were found, after analysis, to contain undeclared active pharmaceutical ingredients (API).

- In Australia, authorities seized some capsules which claim to be natural herbal formulas for weight loss but which contain in fact an important dose of Sibutramine, the active pharmaceutical ingredient (API) of weight loss medicines which can only be supplied on a doctor's prescription.

- In Israel, some food supplements were found to contain also Sibutramine or Sildenafil (API for erectile dysfunction medicines).
- In New Zealand, seven consignments of products purporting to be purely herbal were found to contain undeclared western medicines. Finally, in Singapore, most of the packages seized contained health products suspected of being adulterated with western medicines.

Patients with allergies, high blood pressure, heart problems, kidney diseases or who are pregnant could suffer serious side effects. This can be aggravated by the fact that the API added discretely in so-called "natural" products can be over-dosed. In Australia, some capsules were found to contain 24.1mg of Sibutramine while the genuine product contain only between 10 and 15mg of the same active pharmaceutical ingredient.

This problem should be taken very seriously as it has increased significantly over the last two years. Undeclared API mixed with herbal products has already engendered several deaths, notably in Singapore, Hong Kong and France.

Another impressive result that can be reported by the IMPACT effort (2011) is the joint development of an extensive questionnaire entitled, "Data Collection Tool for Review of National Situations Concerning Counterfeit Medicines" (IMPACT, 2011, pp. 129-138). Via this questionnaire it may be possible to collect and tabulate national data across the globe along many important aspects and issues of counterfeit medicines. Establishing such a questionnaire make it possible to compare nationally collected data with other nations' data. This, in turn, establishes baselines, benchmarks as well as standards making comparison possible and meaningful. This questionnaire was developed relatively recently. Thus no data collected can be reported yet.

In the following section a Discussion of these findings and results is presented.

DISCUSSION

Counterfeiting and piracy cost the U.S. economy billions of dollars annually, have led to the loss of hundreds of thousands of jobs, and needlessly expose consumers to dangerous and defective products. Counterfeit drugs are just a subset of these costs and problems. It was reported earlier that in February 2008, investigations by the Panamanian government revealed that at least 174 people were poisoned and of those 115 people died as a result of counterfeit glycerin that was used to make cough syrup. The government of Panama inadvertently purchased diethylene glycol, a substance commonly used in antifreeze, which was falsely certified as medical-grade glycerin (Bogdanich, 2008). This example as well as many others demonstrate that the consequences of counterfeit drug manufacturing and distribution can be quite deadly. Thc World Health Organization asserts that up to ten percent of medicines worldwide are counterfeit—a deadly hazard that could be costing the pharmaceutical industry $46 billion a year (Balfour et al., 2005).The U.S.-based Center for Medicine in the Public Interest predicted that counterfeit drug sales will reach $75 billion globally in 2010, an increase of more than 90% from 2005 (Counterfeiting Facts and Stats, 2009). These astonishing statistics speak on their own and one does not need to be too persuasive to convince interested readers and researchers that counterfeit drugs constitute an enormous problem worldwide.

The research for this chapter made an attempt to examine the underlying issues of competitive changes in the industry at many levels including the medical, legal, policy, legislative as well as investment areas, the rising costs of prescription medication has been at the forefront of business, politics and the economy. Moreover, the purpose of this research is to examine specific counterfeiting issues in the pharmaceutical industry and to answer the question if Information Quality dimensions and RFID technology serve as a framework and technology, respectively, to reduce drug counterfeiting practices in the pharmaceutical industry. More specifically, our focus is on the tracking and tracing of drug practices in the pharmaceutical supply chain.

We reported in some detail on a study of 7,430 Internet-based drug outlets conducted by the National Association of Boards of Pharmacy (NABP) (2011). Several dimensions of Information Quality were utilized to analyze this specific research setting. This study shows that the joint community of concerned regulators and stakeholders still has quite a long way to go. Unquestionably, considerable progress was made in 2010 to make the Internet a safer place for prescription medications as well as the patients consuming those. The NABP reports increasing support from Internet commerce providers, legislators, policymakers as well as advocacy groups to reduce and eliminate rogue drug sellers. Increased attention was also reported from Internet pharmacies seeking to distinguish themselves as trustworthy businesses and websites. NABP continues to educate the public through its $AWAR_XE$ program informing consumers of dangerous and illicit practices of rogue Internet drug outlets. In turn, patients are provided with information to make informed decisions about their online drug purchases.

U. S. federal law prohibits the importation of prescription drugs from foreign sources, yet some services openly approve such importation, irregardless that many of these sites masquerade as legitimate pharmacies. Many act as if they are Canadian pharmacies, sell medications not approved under Canadian regulations and, moreover, many have no ties to Canada whatsoever.

More specifically, the NABP (2009, p. 4) reports:

According to an FDA statement before the Nevada State Board of Pharmacy in 2006, "evidence shows there are weaknesses in the oversight of the drug distribution system by foreign governments for drugs that are imported into the U.S. We have found that although 'Canadian pharmacies' purport to dispense drugs that are FDA approved, generally the drugs, in fact, are not. Rather, the dispensed drugs are of unknown quality and country of origin.

Cyveillance, an online risk monitoring and management firm in Arlington, VA, in2005 discovered approximately 11,000 Web sites that were designed to appear as Canadian pharmacy sites. Of these sites, most redirected users to 1,009 sites that actually sell prescription drugs, and of those, only 214 were actually based in Canada. Cyveillance also found many of the online pharmacy sites claiming to be based in Canada were actually registered to individuals or companies in other countries, such as Australia, the Czech Republic, El Salvador, Germany, Mexico, and Vietnam, the *Washington Post* reported on June 14, 2005.

Following the death of the Canadian woman who ingested counterfeit drugs she bought online, Canadian Pharmacists Association Executive Director Jeff Poston was quoted in a news release as saying, "[m]any Internet pharmacies claim to be Canadian but in fact can be based anywhere in the world. A Canadian flag is no guarantee – nor can the origin and safety of drugs bought online be guaranteed."

Counterfeit and substandard drugs are becoming more common in the global marketplace. The World Health Organization has estimated that more than 50% of drugs purchased over the Internet from sites that conceal their physical address are counterfeit.

Problems of authenticity also occur in association with domain name practices. Here, the NAPB (2009, pp. 5-6) reports:

The U.S. Government Accountability Office (GAO) undertook a study to determine just how accurate domain name records are. Citing the results of this study in its 2005 report to the U.S. House of Representatives' Subcommittee on Courts, the Internet, and Intellectual Property, GAO estimates that 2.31 million domain names (5.14%) were registered with data that appeared obviously and intentionally false in one or more of the required contact information fields. Examples of such entries included "(999) 999-9999" for a telephone number, "asdasdasd" for a street address, and "XXXXX" for a postal code. The GAO report, entitled Internet Management: Prevalence of False Contact Information for Registered Domain Names, also states that 1.64 million domain names (3.65%) were registered with incomplete data in one or more of the required fields. In total, GAO estimates that 3.89 million domain names (8.65%) had at least one instance of patently false or incomplete data in the required WhoIs contact information fields.

The report further notes that,

… although registrants are required to provide accurate contact information during the domain name registration process, they may supply false or incomplete information in order to hide their identities or to shield themselves from being contacted by members of the public.

More recently, in a June 24, 2008 statement before the US House Committee on the Judiciary Subcommittee on Crime, Terrorism, and Homeland Security, Christine N. Jones, general counsel and corporate secretary of the Go Daddy Group, Inc, corroborated this finding. The Go Daddy State of the Internet: NABP Position Paper Group consists of eight Internet Corporation for Assigned Names and Numbers (ICANN)-accredited registrars, including GoDaddy.com, and manages some 30 million domain names. Jones says, "bad actors typically do not want to pay extra to hide their WHOIS data when they are probably going to provide false WHOIS data, anyway. Most online pharmacies do not have privacy protection on them. More often than not, the registrant simply provides false, but typically valid looking, WHOIS data, upon registration."

Jones also acknowledges that the registrar does not verify registration information for domain names; the process is conducted electronically, and the information is never viewed by a person. "The domain name registration system is entirely automated. There is no human intervention into the process," she says.

Bob Parsons, CEO and founder of GoDaddy.com writes in a March 23, 2005 blog post, "often times the information within the [WhoIs] database is inaccurate. Inaccurate information happens mostly because some registrants who want to achieve anonymity – for a myriad of reasons, some of which are despicable – provide false information to begin with. . . . There's often no way to track down a registrant who provided false information when registering their domain name."

In September 2003, Benjamin Edelman, then a research fellow at the Berkman Center for Internet & Society at Harvard Law School, testified to similar findings before the U.S. House of Representatives' Subcommittee on Courts, the Internet, and Intellectual Property. "As the DNS [domain name system] is currently structured, registrants are under only an honor system to provide accurate WhoIs data. Meanwhile, it makes no economic sense for registrars to enforce WhoIs accuracy," says Edelman, now an assistant professor at the Harvard Business School. "The result is that in terms of accuracy, when compared with other compilations of public data (such as driver's licenses and trademark registrations), the WhoIs database is substantially fiction. Despite years of inquiry by this subcommittee, in addition to numerous ICANN working groups and other discussions, intentionally invalid WhoIs data remains widespread."

As already noted above, this researcher's effort to locate actual and reliable data on the problem of counterfeit medicine and drugs was quite impossible. Through interviews with industry experts, newspaper reports and reading between the lines we are aware that, unquestionably, efforts are underway by government agencies to collect such data. This is accomplished through sampling efforts as the task to examine every good, shipment, parcel, pallet and container is far too vast as well as costly of an undertaking.

Such typical investigative efforts through sampling methods are well described by Giuliani Partners (2004, pp. 5-7), provide representative snapshots, and, according to several interviews with industry experts, do not seem to differ too much today. Although some of this information dates back to 2003 and 2004, it was impossible to provide more current similar such detailed information. According to Giuliani Partners (2004, p. 5) weaknesses in the existing system threaten the quality and integrity of the nation's drug supply: "Despite best efforts, the evidence we have seen thus far supports the notion that the drug supply is indeed vulnerable."

Some examples are as follows (Giuliani Partners, 2004, pp. 5-7).

Random Examinations Conducted by the FDA and U.S. Customs and Border

Protection

The FDA and U.S. Customs and Border Protection conducted a number of random inspections or "blitzes" at several mail ports in the fall and early winter of 2003:

- In the first inspection, 1,153 drug products were examined and 1,019 or 88% were not approved by the FDA; the drugs came from countries such as India, Thailand, and the Philippines.
- In the second exam, 1,982 parcels were examined and 1,728 or 87% were not approved; 16% of those shipments were from Mexico.
- Many of the drugs examined during these visits were non-FDA approved for many reasons, including:
    - Improper labeling, e.g., there were no instructions for proper use;
    - The presence of controlled substances;
    - Potentially recalled drugs, e.g., drugs that had been withdrawn from the market for safety reasons;
    - Animal drugs not approved for human use;
    - Drugs requiring risk management and/or restricted distribution (e.g., initial screening or periodic monitoring); drugs with clinically
    - Significant drug interactions; or drugs requiring careful dosing; and
    - Required special storage conditions for certain drugs were violated.

Portal Visits

In order to gain an appreciation for the scope of the problem, United States mail facilities were visited to observe the volume and nature of the packages allegedly containing prescription drugs entering the United States. A number of the observations follow.

John F. Kennedy Airport Mail Facility

At the invitation of United States Senator Norm Coleman, former New York City Mayor Rudolph W. Giuliani and former New York City Police Commissioner, Bernard B. Kerik, accompanied the Senator on a visit in March, 2004 to the US Mail facility located at JFK Airport. Customs officials advised that approximately 40,000 packages of suspected drug shipments are received each day from the postal service for review and inspection. Based upon information, the FDA focuses on "countries of interest" and visually inspects 500 to 700 parcels per day. Thus, the majority of packages are sent on to the addressee uninspected. The following was learned:

- Drugs purported to be Xanax, Valium (Diazepam), Lorazapam, Vicodin (all controlled substances) and Lupron were observed; there were numerous packages from the Netherlands, Brazil, Pakistan, as well as other countries.
- Many of the drugs contained in the parcels were non-FDA approved because they were inappropriately packaged, expired, mislabeled or otherwise noncompliant.
- The sheer volume of shipments overwhelms Customs and FDA; FDA has only 6 staff members assigned to JFK.
- Although much of what is inspected is non-FDA approved, few parcels are actually detained. The processing requirements to detain a shipment are cumbersome and time consuming. The rules require the FDA to send a notice to the addressee of the package. If the person does not respond or the response is insufficient, the package must then be returned to the sender (manufacturer). This process varies significantly from the way controlled substances or narcotics are handled. Such drugs can be destroyed without further processing.

Miami International Mail Branch Facility Visit in March 2003

Giuliani Partners was provided with a Congressional staff report regarding a similar review of the Miami facility in March 2003. The findings of the bipartisan Congressional report were consistent with the findings of this review:

- Congressional staff witnessed "thousands of shipments of foreign drugs" being processed; the packages were from countries such as Honduras, Costa Rica as well as Great Britain; and the packages purportedly contained "valium" (diazepam), Reteina (Ritalin), Zolipedem, and Ciprofloxacin.
- The volume of drugs coming through the mail facilities is too great to allow for any meaningful inspection.
- Parcels are only visually inspected; there is no testing as to the quality or integrity of the product.
- FDA and Customs detain very limited numbers of questionable drugs coming into the facility because of the cumbersome nature of the detention process.

In addition to weaknesses at the entry points of supply chains as reported above, there are also weaknesses observable within the supply and distribution chains themselves. At first glance it appears that U. S. distribution chains are straightforward for prescription medicines, i.e. manufacturers sell their products to stores and retail pharmacies. Those, in turn, dispense medications to patients submitting prescriptions. Giuliani Partners (2004, p. 4) report that it is not until the system is studied in greater detail that the observer begins to appreciate both the complexities and vulnerabilities of the distribution chain and the potential for exploitation or abuse.

According to Giuliani Partners (2004, p. 4) factors characterizing the supply chain and contributing to the counterfeit problem are as follows:

- Wholesalers or distributors are primarily regulated by the states with no uniform standards across state borders. States have a comparatively small number of investigators to monitor the licensed wholesalers; thus, given the sheer number of wholesalers, oversight is minimal.
- There are thousands of "secondary" pharmaceutical wholesalers in addition to McKesson, AmerisourceBergen and Cardinal Health (the "big three") involved in the distribution of prescription medicines. As reported in the industry (e.g., Giuliani Partners, 2004), there are more than 6,500 small wholesalers nationwide.
- There is no uniform mechanism, i.e., a chain of custody or "pedigree," to track the medicine from point of manufacture to point of sale. It should be noted that the FDA has not implemented the pedigree requirement that was mandated by law in 1988.
- Repackaging is a vulnerable point in the process and can provide an opportunity for counterfeit or non-FDA approved products to compromise the system.

Challenges are also faced in terms of oversight and enforcement. Giuliani Partners (2004, pp. 4-5) report that there are challenges associated with the oversight and enforcement of our current laws as well with ensuring that medicines being purchased or sold in the U. S. are FDA-approved, safe and effective.

- The current volume of parcels of drugs coming into this country through the mail (it is estimated to be more than 10 million packages annually) and the increasing volume of internet purchases make meaningful inspection by the FDA almost impossible.
- The FDA has less than 100 investigators to deal with drug importation issues nationwide, and its investigative authority is limited relative to its ever-increasing law enforcement responsibilities. For example, the FDA has no administrative subpoena authority in order to facilitate the conduct of its investigations; thus it must either partner with another investigative agency or request subpoenas from the local United States Attorney's office.
- Investigating and prosecuting counterfeit drug cases or illegal internet sales cases are not, with few exceptions, a priority for the federal or state law enforcement agencies.
- The penalties are comparatively low for engaging in this kind of activity – the current penalties for FDA violations are approximately three years.
- The technologies being advanced as mechanisms to ensure an imported drug shipment is safe and effective are not foolproof, and, in some instances, not yet available.
- Electronic Track and Trace – most agree that these technologies, e.g., using bar coding or radio frequency identification (RFID) chips that could track drug products in real time throughout the system and then provide an electronic pedigree, are still very costly when available.
- Counterfeit resistant technologies that include covert and overt packaging and labeling techniques, such as holograms, watermarks, color shifting inks or fluorescent inks, as well as chemical agents, are widely used by the industry already. However, they can be easily duplicated and, therefore, must be changed on a periodic basis.

- "Unit of Use" packaging, which is a container closure system designed to hold a specific quantity of drug product for a specific use and dispensed to a patient without any modification except for appropriate labeling, does eliminate the need for some repackaging; however, there are packaging and cost issues for the manufacturers, and some drugs do not lend themselves to such packaging.
- Authentication testing, while not a technology *per se*, is also an option when determining the integrity of a pharmaceutical product. It is a complicated, time consuming and costly process, however, and can be performed only by the original manufacturer. There are no available tests that can be conducted "in the field" to ascertain whether a product is real or fake.

The above factors, among others, make the counterfeit drug business a high profit, low risk business, including those involved in circumventing the laws in supplying medicines outside the traditional distribution chain. Consequently, such high profit and low risk business may be appealing to organized crime and terrorist organizations.

Unfortunately, in general no data on counterfeit drugs are being reported or made available publicly. Thus the above description and data provided by Giuliani Partners provide important and unique insights and specifics. As already mentioned, often, formal investigative legal efforts are undertaken simultaneously and during this process no such data can be shared, as the data themselves may take on the nature of legal evidence. Puzzling, however, is that the FDA, e.g., has issued various annual reports on counterfeit drugs, yet the last one issued dates back to 2006. It is difficult to comprehend why such reports stopped, even though the counterfeit drug problem has been increasing since. It was impossible to receive an explanation why such annual reports are no longer being produced. One may only surmise various underlying political and other undertones and currents that may be potential explanations. It is also understandable that there must be concerns about ongoing legal pursuits and other forms of litigation. In our request for such information, it was clearly stated that we are seeking only publicly available data and that we have no interest in any sort of interference in existing legal proceedings.

As stated in the result section, numerous efforts were made in contacting representatives from the FBI, FDA, ICE and other agencies. Some agencies responded, some never did. Those who responded suggested that requested information is available on the agency's respective website, but such information is not available which was the reason we contacted the agencies in the first place.

In conclusion, in the effort to seek suitable Information Quality data in the context of counterfeit drugs, it must be stated that this was a most frustrating and unsatisfactory experience. As a result we presented some secondary source data and efforts surrounding Information Quality-near dimensions and issues.

CONCLUSION

The authors examined Information Quality issues in the identification and tracking of drugs within the pharmaceutical industry. Identification, authentication and tracking are viewed as fundamental components of Information Quality. The specific field of study addressed in this setting is counterfeit drugs, a dilemma of paramount importance for public health and the well-being and safety of patients. The authors advocate RFID and related technologies, including EPCglobal's Electronic Product Code Information Services (EPCIS) and IBM's RFID Information Center system that, in turn, provide a suitable and fitting infrastructure for the tracking and tracing of uniquely identifiable, i.e. mass-serialized, products throughout the supply chain.

As a reflective and concluding effort, the authors present here an additional theoretical perspective, i.e. in addition to the already earlier described Transaction Cost Theory perspective, utilizing Collective Action Theory to view the present research setting conceptually. Several regulatory efforts and compliance regimes are recognized and a call for collective action for all stakeholders within the supply chain in the pharmaceutical industry is advanced.

The authors recognize that the pharmaceutical industry is a large complex web of corporations, legislation, regulatory efforts, compliance regimes, manufacturers, wholesalers, pharmacies, importers as well as rapidly advancing technologies and applications. Counterfeit drugs are a major problem for public health, the concern of physicians trying to help as well as the well-being and safety of patients. Drug counterfeiting no longer is a concern in a given country or region; it is truly an international problem as drug suppliers around the globe feed their raw and intermediate products into the global drug supply chain. Considerable gains have been made in supply chains in regards to visibility, transparency and control, thus contributing to the potential exposure of counterfeiting practices.

This situation, however, is yet more difficult when considering the comments of one reader (Dr. Marv Shepherd) to a *Wall Street Journal* article entitled, "China never investigated tainted Heparin, says Probe" (Mundy, 2010):

I'm not surprised. When I was in China in 2008 and again in 2009 talking to Chinese about the counterfeit drug problem, one Chinese FDA official told me and I quote: "... our drug exports are not our problem they are your (U.S.) problem." This summed it up very well for me.

Catching counterfeiters in such supply chain settings, however, is a difficult task. It requires cooperation and collaboration among almost all stakeholders together with such organizations as Customs, ICE and the FDA in the design of architectures and processes as well as standard daily operations, including the processing of transactions (physical (the handling of products) and digital (the processing of documents, payments, etc.). Although the authors presented a number of very promising technical solutions such as RFID, Electronic Product Code Information Services (EPCIS) together with the RFID Information Center system as well as the various ePedigree efforts that collectively indeed can function as a suitable infrastructure providing tracking and tracing capabilities, the solution, however, is not solely a technical one. The counterfeit drug problem is also an organizational and a people problem. This, in turn, requires cooperation and collaboration, i.e. indeed collective action, among all the many stakeholders within the supply chain. It is clear that this is an industry-wide as well as an interorganizational concern requiring coordinated efforts, i.e. collective action, including industry, government, third-party representatives as well as supranational organizations.

For this we need to embrace Collective Action Theory described below.

The Theory of Collective Action

Collective Action refers to the pursuit of a common goal by more than one person. Presumably the achievement of the goal will then benefit all of society (e.g., Sandler, 1992). The term dates back to some of the work by Vilfredo Pareto in the 1930ies and Mancur Olson (1965, 1971) as Olson applied this concept to economics subsequently in his *The Logic of Collective Action: Public Goods and the Theory of Groups*. Collective action problems arise when each individual in a group pursues a rational strategy, yet the collective outcome is bad for all of those same individuals, thus, in effect, creating "collective irrationality"

(Wheelan, 2011).To an extent Ronald Coase (Coase, 1937) should be mentioned in this context as well in that he provided in his classic *The Nature of the Firm* the concept of transaction costs making possible the measurement of the size of firms as well as the problem of social cost (Coase, 1960). Accordingly, transaction costs, especially those pertaining to the cost of organizing of such collective action, for a majority attempting to achieve the utility of the goal (typically a public good) are disproportionately higher than the transaction costs for a small minority. An additional problem of collective action is the benefit gained by those who do not participate in its achievement. This is generally referred to as the *free rider problem*, elegantly explained by Vilfredo Pareto (1935). The concept of collective action has been used extensively also by several scholars in the standards evolution, standards diffusion as well as standards adoption literature (e.g., Markus, Steinfield, Wigand & Minton; Wigand, Steinfield & Markus).

When applying Collective Action Theory to the counterfeit drug problem within the PI, one can readily recognize numerous weaknesses, problems and dilemmas. When collective action is maximized in an economic sense, we quickly realize that this is indeed an extreme ideal case, rather unlikely to occur with such perfection in the real world. Group settings—and the players within the counterfeit drug setting attempting to reduce counterfeit drugs can be seen as such a group, actually once aggregated a rather large group—tend to create free-rider problems, especially as the group gets larger. Realizing though that any society has no choice but to engage in collective action in certain settings and certain levels (in every village, county, state, nation as well as the entire planet) how can we then overcome collective action problems and shortcomings? Some of these have been tried in the PI. Often these solutions are informal and voluntary, yet others necessitate the action of an authority (such as the ICE, FBI and FDA) with coercive powers (typically governments, i.e. their agencies).

Wheelan (2011, pp. 130-131) but also others have identified the following operative scheme to reduce a collective action problem. Each may vary of course depending on the nature of the situation:

- **Regulate:** Individuals can be enticed to act in a way leading to best collective outcome such as the setting and enforcing of quotas (e.g., fishing quotas or requiring mandatory union dues). There are situations in which self-regulation (e.g., within an industry) does indeed work, but often such schemes are wishful thinking as this is may be too reminiscent of the fox guarding the hen house, but here appears to be agreement industry-wide that these "successes" are by far insufficient. The problem here is also, once again, that no data are publicly available making these "successes" assessable and analyzable as well as fully comparable. One also recognizes the underlying political difficulties and consequences if such data would be publicly available. A solid case in point is the earlier reported NABP effort of examining Internet drug outlets: As soon one such outlet is shut down, it is indeed possible for the same organization to open another website the very next day, although under a different name while using the very same web pages, software, databases, etc. but the NABP has been successful to an extent in encouraging, e.g., online service providers to engage in collective action practices by not providing services to obvious entities clearly engaged in selling counterfeit drugs.
- **Use Government to Provide the Good:** Some level of government can provide a good or service, at times this may actually create a monopoly. This may include such activities as maintaining a park, fire-fighting, snow removal. In return, governments collect then taxes from those benefitting. This then eliminates the problem of free-riding. An example of this in the PI is the government's current proposal to engage in high-risk research and development (R&D) of much needed drugs in areas in which the private sector is unwilling to pursue such R&D efforts while there is general

agreement that such efforts may result in likely success. Accordingly, the Federal Government is ready to spend billions of dollars on a drug development center to help create medicine (Harris, 2011) through which this center is to take on promising drug exploration that the private sector is unwilling to undertake.

- **Privatize the Resource:** When resources are available, they may be public goods or government may privatize these goods for use or distribution. Governments may stipulate, however, how the resource is to be used. Differences in utilization could be observed if there is just one vs. several/many owners of that resource: If overfishing is a concern, a single fisherman in such a market would be very concerned not to overfish, as in the long-run the fisherman would work himself out of business. If many fishermen fish in this market, it is highly likely that individually they will care much less about overfishing, i.e. individually they will tend to catch as much fish as possible on each fishing attempt. Similar behavior is observable by ranchers in overgrazing situations who lease land from the government for grazing their herds. In the PI some such behavior is observable, e.g., when a high demand for a vaccine may exist yet the supply is regionally, nationally or internationally limited and production is delayed such as during national immunizations efforts administered and recommended by the Centers for Disease Control.
- **Rely on a 'Privileged Group':** A privileged group is a subset of a larger group whose private incentives are such that they will cover the costs required to support a group activity, even if all other potential beneficiaries choose to free-ride. Several examples come to mind: A dedicated group of neighbors may create a 'neighborhood watch' organization. This will benefit the entire neighborhood including those neighbors who do not contribute time or other resources to this effort. One very grade-conscious student may deliver the vast amount of work needed on a team project, benefitting all other team members who choose to watch and enjoy MTV instead. Similar behavior is observable in standards development efforts at the industry level. In the PI it is conceivable that such a privileged group could be created by industry players where each player contributes considerable resources (time, money, etc.) to stem the flow of counterfeit drugs, yet there are likely to be also players who choose not to participate, benefit from the overall effort by free-riding. An example of such a "privileged group" is the Global Security Team of Pfizer, Inc. This Team is assembled by Pfizer and includes former employees of the FBI, Homeland Security, the Drug Enforcement Administration and various former narcotics agents. The group works with local enforcement agencies to identify and prosecute international drug counterfeiters. Mr. John Clark of Pfizer, Inc. heads the Global Security Team.
- **Provide 'Selective Incentives':** One may wonder why individuals as well as firms contribute to NPR or PBS when they easily cloud free-ride. Surely there are some who contribute to get that special coffee mug or tote bag coming with a contribution to show other that they contributed. Other organizations may offer magazines or discounts (e.g., AARP with a magazine and insurance discounts) as incentives. Free riders still can enjoy the benefits of listening free to NPR, but there are also these added benefits for those who pay their share. This may include also the psychological gratification of having contributed, one benefit of altruism. In addition, such contributions are usually tax-deductible for both individuals and firms. In the PI setting it may be conceivable that the Federal Government encourages strongly efforts and practices that will reduce the distribution of counterfeit drugs by encouraging certain procedures and practices. These, however, may be costly or require costly initial investments. As such a selective incentive the Federal Government

might offer a certain amount of these investments or all to become recognized tax write-offs for firms. Such a tax write-off is likely to stimulate firms to participate in these encouraged efforts.

- **Develop Informal Solutions:** Groups throughout history have always engaged in some form of collective action to solve problems. Often there is no other choice, it seems. In doing so, they often find informal solutions how to solve such problems and to find ways of cooperation and collaboration. Examples are solving crime problems in neighborhoods or creating a communal garden. In the PI several examples can be reported in which selected industry players get together informally to make a contribution stemming the flow of counterfeit drugs. One such example is the aforementioned Global Security Team by Pfizer, Inc.

The discussion in this Conclusion section and insights derived from it demonstrates that groups such as those involved in the reduction of counterfeit drugs do not always perform and behave such that welfare of the group or society is maximized. At times puzzling results are possible in that rational individuals as well as firms seeking to advance their own interests may end up making themselves worse off. This in and by itself is a powerful, yet puzzling finding. The problem of collective action has been tackled by virtually every society throughout the history of human civilization as collective action is a central ingredient in communal decision making. Public policy and its decision makers pursue goals to create policies and institutions that minimally at least bring about better collective outcomes for society. Accordingly and as demonstrated, collective action is a fitting framework and conceptual lens through which to view, examine and explain the setting of counterfeit drug practices within the pharmaceutical industry.

ACKNOWLEDGMENT

This research was funded in part by the National Science Foundation's Virtual Organizations as Sociotechnical Systems (VOSS) program (Award Numbers: IIS-0838402 and IIS-0838231), as well as the Jerry L. Maulden/Entergy Fund at the University of Arkansas at Little Rock. Any opinions, findings, and conclusions or recommendations expressed in this material are those of the authors and do not necessarily reflect the views of the funding agency. The researchers gratefully acknowledge their support.

REFERENCES

Akerlof, G. (1970). The market for lemons: Quality uncertainty and the market mechanism. *The Quarterly Journal of Economics*, *84*(3), 488–500. doi:10.2307/1879431

Balfour, F., Barrett, A., Brady, D., Capell, K., Magnusson, P., Matlack, C., . . . Wheatley, J. (2005). "Fakes!". *Business Week*. Retrieved September 5, 2014, from http://www.businessweek.com/magazine/content/05_06/b3919001_mz001.htm

Barnett, J. M. (2005). Shopping for Gucci on Canal Street: Reflections on Status Consumption, Intellectual Property, and the Incentive Thesis. *Virginia Law Review, 91*(6), 1381-1428.

Bickman, L., & Rog, D. J. (Eds.). (2009). Applied Social Research Methods. Thousand Oaks, CA: SAGE.

Bogdanich, W. (2008). Panama Releases Report on '06 Poisoning. *The New York Times*. Retrieved on September 6, 2014, from http://www.nytimes.com/2008/02/14/world/americas/14panama.html?ref=health

Bogdanich, W., & Hooker, J. (2007, May 6). A Poison's Path. *The New York Times*.

Brown, S. A. (1997). *Revolution at the Checkout Counter: The Explosion of the Bar Code*. Cambridge, MA: Harvard University Press.

Buller, D. B., & Burgoon, J. K. (1996). Interpersonal Deception Theory. *Communication Theory*, *6*(3), 203–242. doi:10.1111/j.1468-2885.1996.tb00127.x

Center for Medicines in the Public Interest. (2011). *Intellectual Property Rights Issues and Imported Counterfeit Good*. Center for Medicine in the Public Interest. Retrieved on September 5, 2014, from http://cmpi.org/in-the-news/testimony/

Chakraborty, G., Allred, A., Sukhdial, A. S., & Bristol, T. (1997). Use of negative cues to reduce demand for counterfeit products. In M. Brucks & D. J. MacInnis (Eds.), *Advances in Consumer Research* (Vol. 24). Provo, UT: Association for Consumer Research.

Chaudhry, P. E., & Walsh, M. G. (1996). An assessment of the impact of counterfeiting in international markets: The piracy paradox persists. *The Columbia Journal of World Business*, *3*(1), 34–49. doi:10.1016/S0022-5428(96)90039-3

Coase, R. (1937). The Nature of the Firm. *Economica, 4*, 386–405.

Coase, R. (1960). The Problem of Social Cost. *The Journal of Law & Economics*, *3*(1), 1–44. doi:10.1086/466560

Colvin, R. (1999). Innovative technologies help thwart counterfeiting. *Modern Plastics*, *76*(7), 57–58.

Cordell, V. V., Wongtada, N., & Kieschnick, R. L. Jr. (1996). Counterfeit purchase intentions: Role of lawfulness attitudes and product traits as determinants. *Journal of Business Research*, *35*(1), 41–53. doi:10.1016/0148-2963(95)00009-7

Counterfeiting Facts and Stats. (2009). *Protection from Brand Infection*. CMO Council. Retrieved on July 21, 2011, from http://www.cmocouncil.org/programs/current/protection/protection_couterfeit_stats.asp

DeMatos, C. A., Ituassu, C. T., & Rossi, C. A. V. (2007). Consumer attitudes toward counterfeits: A review and extension. *Journal of Consumer Marketing*, *24*(1), 36–47. doi:10.1108/07363760710720975

Doney, P., & Cannon, J. P. (1997). An examination of the nature of trust in buyer-seller relationships. *Journal of Marketing*, *61*(2), 35–51. doi:10.2307/1251829

Ekman, P. (1992). *Telling Lies: Clues to Deceit in the Marketplace, Politics, and Marriage*. New York: Norton.

EPC Information Services (EPCIS) Version 1.0.1 Specification. (2010). Retrieved on September 5, 2014, from http://www.epcglobalinc.org/standards/epcis

EPCglobal. (2010). Retrieved on September 7, 2014, from http://www.epcglobalinc.org/about/

Giuliani Partners LLC. (2004). *Examination and Assessment of Prescription Drug Importation from Foreign Sources to the United States. Interim Findings*. Author.

Grazioli, S., & Jarvenpaa, S. (2003). Consumer and business deception on the Internet: Content analysis of documentary evidence. *International Journal of Electronic Commerce*, *7*(4), 93–118.

Grossman, G. M., & Shapiro, C. (1988a). Counterfeit product trade. *The American Economic Review*, *78*(1), 59–75.

Grossman, G. M., & Shapiro, C. (1988b). Foreign counterfeiting of status goods. *The Quarterly Journal of Economics*, *103*(1), 79–100. doi:10.2307/1882643

Ha, S., & Lennon, S. J. (2006). Purchase intent for fashion counterfeit products: Ethical ideologies, ethical judgments, and perceived risks. *Clothing & Textiles Research Journal*, *24*, 297–315.

Harris, G. (2011, January 23). Federal Research Center Will Help Develop Medicines. *New York Times,* p. 1.

Hilton, B., Choi, C. J., & Chen, S. (2004). The ethics of counterfeiting in the fashion industry: Quality, credence and profit issues. *Journal of Business Ethics*, *55*(4), 343–352. doi:10.1007/s10551-004-0989-8

History of RFID. (2010). *RFID Journal*. Retrieved on September 6, 2014, from http://www.rfidjournal.com/article/articleview/1338/1/129/

Hyman, R. (1989). The psychology of deception. *Annual Review of Psychology*, *40*(1), 133–154. doi:10.1146/annurev.ps.40.020189.001025

IBM RFID Information Center. (2010). Retrieved on September 8, 2014, from http://en.wikipedia.org/wiki/IBM_RFID_Information_Center

IMPACT. (2011). *The Handbook: Facts-Activities-Documents 2006-2010*. Rome: Agenzia Italiana del Farmaco.

Information on Heparin. (2010). US Food and Drug Administration. Retrieved on September 6, 2014, from http://www.fda.gov/Drugs/DrugSafety/PostmarketDrugSafetyInformationforPatientsandProviders/UCM112597

International Anti-Counterfeiting Coalition. (2008). *Get real: The truth about counterfeiting*. Retrieved on September 5, 2014 from www.iacc.org/aboutcounterfeiting/the-truth-about-counterfeiting.php)

Johnson, K. U. S. (2010, April 30). Seizes Big Batches of Fake Goods. *Wall Street Journal*.

Johnson, P. E., Grazioli, S., Jamal, K., & Berryman, R. G. (2001). Jamal, K. & Berryman, R.G. Detecting deception: Adversarial problem solving in a low base-rate world. *Cognitive Science*, *25*(3), 355–392. doi:10.1207/s15516709cog2503_2

King, B., & Zhang, X. Securing the pharmaceutical supply chain using RFID. In *International Conference on Multimedia and ubiquitous Engineering*. doi:10.1109/MUE.2007.186

Kirmani, A., & Rao, A. R. (2000). No pain, no gain: A critical review of the literature on signaling unobservable product quality. *Journal of Marketing*, *64*(2), 66–79. doi:10.1509/jmkg.64.2.66.18000

Klein, S., Higgins, A., Kipp, A., & Mangan, A. (2010). Drug Living Lab – Information Infrastructures to Increase the Security of Pharmaceutical Supply Chains. Unpublished.

Kumar, R. (2005). *Research Methodology*. Thousand Oaks, CA: SAGE.

Lazear, E. (1995). Bait and switch. *Journal of Political Economy*, *103*(4), 813–830. doi:10.1086/262004

Lightner, N. J., & Eastman, C. (2002). User preference for product information in remote purchase environments. *Journal of Electronic Commerce Research*, *3*(3), 174–186.

Markus, M. L., Steinfield, C. W., Wigand, R. T., & Minton, G. (2006). Standards, Collective Action and IS Development-Vertical Information Systems Standards in the Home Mortgage Industry. *MIS Quarterly. Special Issue on Standards and Standardization*, *30*, 439–465.

McGilvray, D. (2008). *Executing Data Quality Projects: Ten Steps to Quality Data and Trusted Information*. Burlington, MA: Morgan Kaufman.

Millett, L. I., & Holden, S. H. (2003, November-December). Authentication and its privacy effects. *IEEE Internet Computing*, *7*(6), 54–58. doi:10.1109/MIC.2003.1250584

Mundy, A. (2010, July 22). China never investigated tainted Heparin, says probe. *The Wall Street Journal*, p. A7.

National Association of Board of Pharmacies (NABP). (2009). *State of the Internet: NABP Position Paper on the Continued Proliferation of Rogue Internet Drug Outlets*. Mount Prospect, IL: NABP.

National Association of Boards of Pharmacy (NABP). (n.d.). Retrieved from http://www.nabp.net, last accessed on May 28, 2011.

National Association of Boards of Pharmacy (NABP). (2011). *Internet Drug Outlet Identification Program: Progress Report for State and Federal Regulators*. Mount Prospect, IL: NABP.

Nia, A., & Zaichkowsky, J. L. (2000). Do counterfeits devalue the ownership of luxury brands? *Journal of Product and Brand Management*, *9*(7), 485–497. doi:10.1108/10610420010351402

Olsen, J. E., & Granzin, K. L. (1992). Gaining retailers' assistance in fighting counterfeiting: Conceptualization and empirical test of a helping model. *Journal of Retailing*, *68*, 90–109.

Olson, M. (1971). The Logic of Collective Action: Public Goods and the Theory of Groups (rev. ed.). Cambridge, MA: Harvard University Press.

Pareto, V. (1935). *The Mind and Society*. New York: Harcourt, Brace.

Penz, E., & Stottinger, B. (2005). Forget the "real" thing—take the copy! An explanatory model for the volitional purchase of counterfeit products. *Advances in Consumer Research. Association for Consumer Research (U. S.)*, *32*(1), 568–575.

Pharmaceutical and Medicine Manufacturing. (n.d.). Bureau of Labor Statistics, U. S. Department of Labor. Retrieved on September 5, 2014, from http://www.bls.gov/oco/cg/cgs009.htm#addinfo

Prescription Drug Marketing Act (PDMA) of 1987 (P.L. 100-293, 102 Stat. 95).

Sandler, T. (1992). *Collective Action: Theory and Applications*. Ann Arbor, MI: University of Michigan Press.

Timeline of RFID. (2010). Retrieved on September 6, 2014, from http://www.google.com/#q=Timeline+for+RFID&hl=en&sa=X&rlz=1R2SKPB_enUS374&tbs=tl:1,tl_num:100&ei=R9VETKb4CML38AbCyoGTBw&ved=0CIgBEMsBKAQ&fp=1&bav=on.2,or.r_gc.r_pw.&cad=b

Tom, G., Garibaldi, B., Zeng, Y., & Pilcher, J. (1998). Consumer demand for counterfeit goods. *Psychology and Marketing*, *15*(5), 405–421. doi:10.1002/(SICI)1520-6793(199808)15:5<405::AID-MAR1>3.0.CO;2-B

U.S. Customs and Border Protection. (2009). *Intellectual property rights—Seizure statistics: Fiscal year 2009*. Retrieved September 12, 2014, from www.cbp.gov/linkhandler/cgov/trade/priority_trade/ipr/pubs/seizure/fy09_stats.ctt/fy09_stats.pdf)

Wang, R. Y., & Strong, D. M. (1996). Beyond Accuracy: What Data Quality Means to Data Consumers. *Journal of Management Information Systems*, *12*(4), 5–34. doi:10.1080/07421222.1996.11518099

Wee, C. H., Tan, S. J., & Cheok, K. H. (1995). Non-price determinants of intention to purchase counterfeit goods. *International Marketing Review*, *12*(6), 19–46. doi:10.1108/02651339510102949

Wheelan, C. (2011). *Introduction to Public Policy*. New York: W. W. Norton.

Wigand, R. T., & Benjamin, R. I. (1996) Electronic Commerce: Effects on Electronic Markets. *Journal of Computer-Mediated Communication*, *1*(3). Retrieved on September 5, 2014, from http://shum.huji.ac.il/jcmc/jcmc.html

Wigand, R. T., Picot, A., & Reichwald, R. (1997). *Information, Organization and Management: Expanding Markets and Corporate Boundaries*. Chichester, UK: Wiley.

Wigand, R. T., Steinfield, C. W., & Markus, M. L. (2005, Fall). IT Standards Choices and Industry Structure Outcomes: The Case of the United States Home Mortgage Industry. *Journal of Management Information Systems*, *22*(2), 165–191.

Williamson, O. (1979). Transaction-Cost Economics: The Governance of Contractual Relations. *The Journal of Law & Economics*, *22*(2), 233–261. doi:10.1086/466942

World Health Organization (WHO). (2010). *Medicines: Counterfeit Medicines*. Fact Sheet No. 175. Retrieved on September 10, 2014, from http://www.who.int/mediacentre/factsheets/fs275/en/index.html

Zou, J., Zhong, W., & Ward, R. K. (2006) A novel digital watermarking method for commercial bills based on a class of orthogonal functions. In *IEEE International Symposium on Signal Processing and Information Technology*. Vancouver: IEEE Computer Society. doi:10.1109/ISSPIT.2006.270766

ENDNOTES

1. Deadly counterfeit diabetes drug found outside China's Xinjiang, China View, 5 February 2009.
2. Tanzania Food and Drugs Authority
3. Center for Combating Counterfeit Drug, Thailand
4. US Food and Drug Administration
5. The Medicines and Healthcare Products Regulatory Agency, United Kingdom
6. The Medicines and Healthcare Products Regulatory Agency, United Kingdom

Chapter 5
STRIPA:
The Potential Usefulness of a Medical App

Floor Aarnoutse
Utrecht University, The Netherlands

Ronald Batenburg
Utrecht University, The Netherlands

Cassandra Renes
Utrecht University, The Netherlands

Marco Spruit
Utrecht University, The Netherlands

ABSTRACT

Polypharmic patients are patients who chronically use five or more medicines. The number of polypharmacy patients continues to increase even though it is a risk factor for morbidity and mortality. A medication review is an important measure to mitigate medication risks. It is known to effectively reduce the number of drug related problems per (polypharmic) patient. STRIP is a Dutch method to perform a structured medication review. Based on this method, the STRIPA(ssistent) tool is developed. However, whether or not this app is considered useful by the healthcare professional is not known yet. In order to assess this, a systematic literature study is conducted. In addition, an effectiveness study design is described. The results show that there is indeed a need for medication reviews and Dutch healthcare professionals are likely to adopt new technologies, an effectiveness study based on a randomized controlled trial is necessary to assess the effectiveness of STRIPA.

INTRODUCTION

In the Netherlands around 10% of the pharmacy visitors are polypharmic patients, which mean they chronically use five or more medicines (KNMP, 2013). Research showed that the number of polypharmacy patients continues to increase and that it is a known risk factor for morbidity and mortality (Hajjar, Cafiero & Hanlon, 2007). In the Netherlands alone, polypharmacy costs society between 103 and 229 million euros (Zorginstituut Nederland, 2013). Polypharmacy can possibly lead to dangerous combinations of drugs, which can be harmful for the patients. Not only can certain drug-drug interactions be harmful, they can also neutralize the active substances in one another. The chronic use of multiple drugs increases patients' risks to experience adverse effects, under-prescription, overtreatment, and decreased drug adherence (Meulendijk et al., 2013). Besides that, using multiple drugs also leads to an increased

DOI: 10.4018/978-1-5225-0248-7.ch005

chance of hospitalization (Meulendijk, 2012). Therefore it is of importance that general practitioners (GPs) scan for drug-drug interactions. This can be formally done via a periodic medication review, by which GPs together with pharmacists have to review their polypharmic patients' medicine use.

In the Netherlands there were two methods for medication reviews: the Prescribing Optimization Method (POM) and 'Gebruik-Indicatie-Veiligheid-Effectiviteit' (GIVE). There was a need for a unified method and that is why based on the POM method, STRIP (Systematic Tool to Reduce Inappropriate Prescribing) was developed (TPO, 2012). STRIP is a step-by-step method that aims at assisting GPs and pharmacists with determining the optimal medication for polypharmic patients (Meulendijks, 2013) and can be used in software for healthcare professionals. The rise of mobile technology brought exponential growth of software use by healthcare professionals (O'Hagan, 2012).

STRIPA, short for STRIP-Assistant, is a STRIP-based app for GPs and pharmacists to use when making a medication review for polypharmic patients. For these medication reviews STRIPA offers several functionalities:

- An overview of the suffered conditions, diseases and medication for each patient.
- Linking the prescribed medication to the associated diseases.
- Advice when to start new medication.
- Advice when to stop current medication.
- Linking side effects to the associated medication.
- Showing drug-drug interactions.
- Advice about medication dosage.

These functionalities will be further elaborated on in the STRIPA section.

In 2009 almost a third of all pharmacists did not do a single medication review (Inspectie voor de Gezondheidszorg, 2009). It is hypothesized that this is the case because it takes a lot of time for the pharmacists as well as for the GPs. STRIPA can help these healthcare professionals with making faster and more structured medication reviews. However, it is unknown if GPs and pharmacists see the potential benefits and are willing to adopt STRIPA as a support tool. Therefore the main research question for this paper is:

RQ 1: What is the potential usefulness of STRIPA for conducting medication reviews?

This research question will be answered based on several subquestions. First of all (i) 'why is medication reviewing important?'. Answering this question will lead to a better understanding of the field. The second sub question (ii) is 'what can STRIPA offer?'. Answering this questions will lead to a better understanding of the app. Sub question three (iii) is 'are there any apps similar to STRIPA?'. This will deepen the understanding about STRIPA and help positioning it within the market. The fourth sub question (iv) is 'how can an effectiveness study for STRIPA be designed?. By designing an effectiveness study, a foundation for future research is built.

The remainder of this paper is structured as follows; in the next section a systematic literature review will be conducted in order to give a state-of-the-art overview of the current situation regarding medication reviews, drug-drug interactions, polypharmic patients and the usage of e-health by healthcare professionals. This will give an answer to the first sub question. In the third section STRIPA will be introduced, in order to give an in-depth understanding of the app and therefore answer sub question number two. In the subsequent section STRIPA will be compared to existing apps, which answers sub question three. Thereafter an effectiveness study design will be presented. This will answer sub question five. The article concludes with a discussion, conclusion and future research.

1. SYSTEMATIC LITERATURE REVIEW

A systematic literature study (SLR) is carried out in order to identify and select research relevant for STRIPA. This section will start with the explanation of the process of the SLR. After that the result of the SLR in the form of related literature is presented.

1.1 SLR Procedure

The SLR used in this study is based on the study of Liberati et al. (2009) and is therefore divided in four successive phases; identification, screening, eligibility, included. Figure 1 in Appendix 1 gives an overview of the phases along with the yielded results per phase.

- **Identification:** Two databases, Google Scholar and PubMed, are searched for relevant articles. PubMed comprises a large number of citations for biomedical literature from MEDLINE, life science journals and online books and thus is purely focussed on medical literature. Google scholar indexes articles across an array of publishing formats and disciplines and includes most peer-reviewed online journals of Europe and America. In addition to that Scholar also includes many scholarly publishers, books and other non-peer reviewed journals. By not limiting the search to only the more on-subject database, PubMed, a far larger number of articles can be searched through for the SLR.

Two reviewers independently scanned the electronic records to identify potential articles. In Appendix 2, Table 1, the used search keys and their underlying relationship can be found. These search keys are selected based on their relevance to the subject. Supporting healthcare professionals with making 'medication review(s)' is the main functionality of STRIPA. These reviews are focussed on patients who chronically take more than five medicines, called 'polypharmacy'. STRIPA makes these medication reviews based on the STRIP method, whereby 'drug interaction' (how do drugs interact with each other), 'medicine criteria' (when to start and stop medication and dosage information), and 'medicine disease connection' (which medicines should be used for what disease) are of importance.

For each search key subject, there are five layers of identified search keys. In the first layer the 'general' search keys for that subject are identified. The second and third layer combine the subjects with the words 'app' and 'e-health' which is respectively short for application and electronic health. Since STRIPA is ought to be an app to help general practitioners and pharmacists conduct medication reviews, and therefore falls in the category of e-health, these search keys are included. The last two layers combine the subjects with the words 'general practitioner information system' and 'clinical decision support software'. STRIPA is a clinical decision support software which can be integrated with the general practitioner information system. By also including these four extra layers literature based on similar projects can be found.

For each search key the two databases were scanned. These search keys yielded a total of 3.005.135 results divided over PubMed (112.847) and Google Scholar (2.892.288).

- **Screening:** Of the roughly 3 million results, 4.759 records are screened. This is done by reading the titles and when deemed maybe relevant for this research, also the abstract. Per relevant keyword the first 100 results were searched trough for PubMed as well as Google Scholar. However, some keywords did not yield 100 results and other keywords were deemed not relevant enough. This was the case when after 40-60 results none of the screened titles were slightly interesting for this research, at which case the search stopped. Therefore divided over the 34 search key and two databases, 4.759 records are screened. Based on this title and abstract screening, 105 articles were deemed interesting for this research. 38 of which originated from PubMed and 67 of Google Scholar. Consequently, 4.654 screened articles were deemed not interesting enough and are dismissed.
- **Eligibility:** In order to access the eligibility of the remaining 105 articles, the full texts of these articles are sought. Of 34 articles no full text could be found and were thus excluded from further analysis. The full texts of the remaining 71 articles were further analysed on their added value to the literature review. The articles are divided into four different categories; (1) duplicates, (2) not applicable and (3) not relevant and (4) include in review. Duplicates are articles which came up in Google Scholar as well as PubMed and were in both cases included in the results. In total, there were eight duplicate articles. Seven articles were deemed not applicable. These articles were often solely based on countries apart from the Netherlands, and could not be generalized due to their research approach or because they were based on a device not used for STRIPA, for example PDAs. The last category in which articles are excluded are the articles which are deemed not relevant. An article is deemed not relevant when it is too specifically aimed at a disease, technique, target group etcetera. Also articles which were too general or did not fit in after all were deemed not relevant. In total 29 articles belong to this group.
- **Included:** A total of 27 articles were left to include in the literature review. When analyzing these articles some general concepts can be distinguished compliant with the search keys. In Appendix 2, Table 2, the references to the included literature as well as the concepts they entail, can be found. E-health is identified in most of the articles. A broad view of e-health is used, in which also apps, clinical decision support software and general practitioners information systems are reckoned in this term. The only search key missing in the concept table, is medication disease connection, because none of the selected articles are found based on this search key. In the next section, related literature, the result of the analysis of the included articles can be found.

1.2 Related Literature

There are a lot of papers reporting about side-effects of different drugs. Scanlin (2013) stated: "The more powerful a drug is, the more likely it is to have harmful side effects. The Institute of Medicine (IOM) of the National Academies estimates (...) that there are between 44.000 and 98.000 hospital deaths annually attributed to medical errors, more than 7.000 of which are due to medication errors." Based on this it can be stated that reviewing medication is important, the question remains whether or not it is also effective. Multiple studies have been conducted to measure the effectiveness of these medication reviews (Seidling et al., 2011; Krska et al., 2001; Vinks, Egberts, de Lange & de Koning, 2009; Bindoff, Tenni, Peterson,

Kang & Jackson, 2007). In a randomized control trial in the United Kingdom 332 records of patients aged 75 years and above or those who uses multiple medicines were reviewed. The results showed that all patients had at least two drug related problems (Krska et al., 2001). A study about the effectiveness of medication reviews in the Netherlands between GPs and pharmacists, showed a positive influence in reducing potential drug related problems for elderly. The results showed a significant reduction in the mean number of drug related problems per patient. The mean number of drugs per patient did not significantly reduce after the medication review (Vinks, Egbert, de Lange & de Koning, 2009). Another effectiveness study about medication reviews in Australia showed that not only significantly less drug related problems per patients were found, but also that a decision support system (DSS) for medication review found significantly more potential drug related problems than healthcare professionals without a DSS (Bindoff, Tenni, Peterson, Kang & Jackson, 2007). A medication review is an important measure to mitigate medication risks, but not all patients can be reviewed. A research in Sweden looked at the potential drug-drug interactions and concluded that there was a strong correlation between drug related problems and the use of multiple drugs. The pronounced increase in polypharmacy over time implies a growing reason for prescribers and pharmacists to be aware of drug-drug interactions (Åstrand, Åstrand, Antonov & Petersson, 2007).

A conclusion can be drawn that medication reviews can substantially improve patient safety and that information systems positively support the medication review process by reducing drug error rates (Kaushal, Shojania & Bates, 2003; Bates & Gawande, 2003; Miller, Gardner, Johnson & Hripcsak, 2005; Drenth-van Maanen, van Marum, Knol, van der Linden & Jansen, 2009; McInnes, Saltman & Kidd, 2006). The main driver behind implementing a computerized physician order entry for Dutch physicians is patient safety (Aarts & Koppel, 2009). Ko et al., (2007) stated "both prescribers and pharmacists indicated that the CPOE [Computerized Prescriber Order Entry] system had a neutral to positive impact on their jobs". Even though these information systems are useful, they are not always adopted by the users. A research about the implementation of DSS or CPOE systems in seven western countries showed that the United States and the Netherlands have the highest use rates. Healthcare professionals in the United States, the United Kingdom, Switzerland, and the Netherlands are most likely to have integrated decision support systems (Aarts & Koppel, 2009).

An American survey held under 1.745 pharmacists concluded that DSSs often fail to protect patients from harmful drug-drug interactions. The potential risk of drug-drug interaction depends on a number of drug and patient specific factors. DSSs should pay attention to these factors (Horn et al., 2013; Smithburger, Buckley, Bejian, Burenheide, Kane-Gill, 2011). In addition, "the effectiveness of computerized clinical decision support systems (CDSSs) depends on the quality of the knowledge they refer to" (Mille, Degoulet & Jaulent, 2007), it is important to include the knowledge of healthcare professionals in the system.

One major problem with the adoption of DSSs is that users ignore the alerts and advises the system produces. A lot of research has been done to study how many alerts and advises are ignored and more importantly why they are being ignored (Ko et al., 2007; Kmetik, Chung, Sims & Found, 2007; Ahearn & Kerr, 2003). Through a questionnaire the reasons for overriding drug alerts by GPs in the UK are examined. 236 GPs participated in this study. 22% of the GPs admitted they frequently to very frequently override drug alerts. The main reason for overriding drug alerts was that the drug alert was not relevant. 90% of the GPs stated that it should be harder to ignore drug alerts (Magnus, Rodgers & Avery, 2002). When tiering the alerts into categories based on severity, alerts were not as quickly ignored. In an American research, 71.350 alerts were investigated. A correlation was found between tiering the alerts and overriding the alerts. From the 31.876 tiered alerts, 100% of the most severe alerts were accepted against 34% of the total 39.474 non tiered alerts. Also moderate severe alerts had a higher acceptance rate with tiered alerts (29%) against non tiered alerts (10%) (Paterno et al., 2009). In Australia the opinion of 191 GPs and 138 Pharmacists about alerts were analysed. The vast majority of the respondents wanted to be able to differentiate the alerts by severity. The research states that it should also be harder for physicians to override alerts for severe interactions and that it should be mandatory to provide a reason when they do override an alert (Yu, Sweidan, Williamson & Fraser, 2011).

Another aspect that should be taken into consideration with these kinds of systems is that "Doctors may not be aware of all the drugs their older patients are taking. Frank and colleagues reported that in a study in Canada 37 per cent of patients were taking drugs without their doctors' knowledge, and 6 per cent of patients were not taking medications that were on their doctors' lists" (Duerden, Avery & Payne, 2013). STRIPA does not include these aspects in their assessment, while these aspects do influence the interaction with other drugs.

1.3 Lessons Learned

There are several lessons learned from the analysis depicted in the previous section. First of all, medical errors due to medication errors are not uncommon and polypharmacy is increasing in frequency. The dangers accompanying polypharmacy can be reduced due to medication reviews, since these reviews improve patients' safety substantially. This is proven to be true in multiple countries including the Netherlands. DDSs supporting these medication reviews can significantly increase the chance of identifying potentially harmful drug-drug interactions. These DDSs have a chance to do well in the Netherlands, since a study about the implementation of DSS or CPOE systems in seven western countries showed that the United States and the Netherlands have the highest use rates. However traditional DSSs (not aimed at medication reviews) often fail to protect patients from harmful drug-drug interactions, therefore future (versions of) DSSs should take the number of drugs, their interactions and specific patient factors into account. In addition, the effectiveness of a DSS also depends on the quality of knowledge they refer to. However one major problem with the use of DDS's by healthcare professionals is that they often ignore the alerts and advises the system produces. This can be overcome by tearing the alerts into categories based on severity, this leads to less ignored alerts and advices. Besides that, health care professionals are not yet aware of all the drugs patients are taking. This is because some drug can be bought without a prescription and patients are not sharing this with their doctors or because patients are without informing their doctors stop taking their prescribed medicine.

2. STRIPA

STRIPA is an online software service to support GPs when prescribing medication and to support GPs and pharmacists when reviewing the medication of their polypharmic patients. STRIPA is based on the STRIP method for medication review. The STRIPA tool can be used as an integrated software program which interacts with the information system of the GPs and the pharmacies. The medication reviews performed by the GPs and pharmacists are used to improve the generated advises when performing another medication review. Next to this integrated software, an app version is also available.

2.1 STRIPA Drivers

Polypharmacy is associated with several medical problems, like under-prescription, overtreatment, increasing risk of adverse effects, and decreased drug adherence. These medical problems are also described as drug related problems. When performing medication reviews, drug related problems can be identified. Previous research shows that out of 1.489 medication reviews, an average 3.1 drug related problems were found per patient. In this research pharmacies did not include all problems they identified, therefore the actual amount of drug related problems per patient is even higher (Service Apotheek, 2012). KNMP (2013) investigated 507 medication reviews in the Netherlands and found an average of 3.6 drug related problems per patient. Identifying these problems and adjusting the medication leads to less medication usage and less hospital admissions. Booz (2012) stated that when 10-12% of the medicine usage can be stopped and 15-17% of the unplanned hospital admissions can be prevented when executing a medication review, the Netherlands can save up to €150-200 million. To support medication review a decision support system, STRIPA is designed. Meulendijk (2012) stated that after implementation of STRIPA in the Netherlands, mortality can be expected to decrease by 3 to 19 persons on a yearly basis, morbidity by 4 to 28 persons, and the financial cost by 10 to 45 million euros per year. To encourage GPs and pharmacists to perform medication reviews, in the Netherlands healthcare providers financially reward the GPs and pharmacists whenever they perform a medication review (Inspectie voor de Gezondheidszorg, 2009). They can send an invoice to the healthcare providers for each medication review they performed. Through a step-by-step method GPs are forced to critically review the medication use of their polypharmic patients. All the patient information is included in the STRIPA system.

2.2 STRIPA Functionalities

All screenshots mentioned in this section about STRIPA can be found in Appendix A, this includes an overall screenshot of STRIPA which is screenshot 1. In this case the dossier of Mrs. Kwarten is shown. Mrs. Kwarten is 93 years old and suffers from several conditions. These conditions are visualized at the top of the screen. She suffers from flatulence, nausea and an impaired kidney function. Next to the conditions, the lab results for several functions, like heart rate and blood pressure, are stated. These results help the GP with diagnosing the patient. The GP has previously diagnosed several diseases with Mrs. Kwarten. These diseases are shown at the left side of the system. The GP can enter more diseases if deemed necessary. For some of these diseases the GP has prescribed medication. The medicines are listed at the right side.

Besides an overview of the conditions, diseases and medication, STRIPA has several other functionalities. First, the STRIPA makes it possible to link the prescribed medication to the diseases. In screenshot 2, all the medicines are linked to the diseases. This process highlights redundant medication. STRIPA holds data about the lab results and conditions of the patients, with this information STIRPA gives advice about when a patient should start with a new medicine. Because Mrs. Kwarten suffers from atrial fibrillation, a new medicine should be prescribed. STRIPA suggests several medications which addresses the condition and do not interfere with other prescribed medication. The GP can choose whether to add the new medicine to the prescription or to ignore the condition. When the GP adds the medication, the medicine is automatically linked to the disease in the disease overview, this is shown in screenshot 3. In addition, STRIPA advises when a patient can stop using a certain medicine. In screenshot 4 STRIPA advises to stop using *acetylsalicylic acid* because the patient has no history of coronary, cerebral or peripheral vascular symptoms. Again, the GP can choose to accept or deny the given advice. The subscribed medication has several side effects. The GP can connect these side effects to the associated medication. Screenshot 5 shows the GP drag the side effect nausea to the medicine euthyrox tablet. Furthermore, STRIPA shows drug-drug interaction warnings. When prescribing multiple medicines to a patient, the consequences of using these medicines together are not always clear. Whenever a drug has an unwanted effect, a warning is given, shown in screenshot 6. In the example of Mrs. Kwarten four drug-drug interactions are identified. STRIPA advices another medication which has the same effect but has different constituents. The GP can choose whether to ignore the interaction warning and prescribe the drug he wanted, or listen to the advice of the STRIPA. Based on values like weight and age, medications have a certain dosage. After checking for drug-drug interactions, STRIPA checks for under- and overdosage. In the case of Mrs. Kwarten only two tablets of *calci chew d3* are allowed while two and a half tablets are prescribed (Screenshot 7). After these steps, the medication is optimized. STRIPA shows an overview, visualized in screenshot 8. On the left side all the medications are connected to the diseases and any potential side effects are listed with their associated medication. On the bottom the bin is located, which shows all the deleted or changed medication during the process. In this overview the GP can check if everything is correct and if necessary reverse the made changes or make new changes.

2.3 STRIPA Status

The STRIPA system is currently being tested through a clinical experiment and a one-year longitudinal study. The results of these tests will become available in 2015. Next to the STRIPA system, an app is being developed. This app currently is in the development phase. O'Hagan (2012) stated "there is tremendous potential for customized care through mobile devices". Therefore transforming the STRIPA system into an app can be an added value. The app will work on both Android and iOS and is developed for tablets (10 inch or bigger). The app is not designed for smartphone use. Because the app contains a lot of information, fitting it on a screen of 10 inch or smaller would indicate leaving out important information. All the information addressed in STRIPA is of equal importance therefore no information can be left out.

2.4 Preliminary Results

Looking at the lessons learned in section 2.3 and the information about STRIPA, some tentative conclusions can be formulated. Since there is an increasing amount of people with polypharmacy, and medication errors are not uncommon, there is a market for tools, software, procedures etc. which helps prevent these medication errors. Especially medication reviews are proven to reduce the amount of drug related errors. Therefore it can be stated that there is a use for an app that focuses on medication reviews, like STRIPA. DDS and CPOE systems have the highest use rates in the Netherlands, this positively influences the chances STRIPA is adopted by Dutch healthcare professionals. It is suggested that DDSs should take the number of drugs, their interactions and specific patients factors into account in order to protect the patient better from harmful drug-drug interactions. STRIPA considers both the drug and patient specific factors, and therefore addresses this issue. In addition, the effectiveness of a DSS also depends on the quality of knowledge they refer to. The generated advises by STRIPA are based on the medication reviews performed by the healthcare professionals themselves, this insures the quality of the advices are continuously improved and of a high standard. A problem with DDS's is that advises and alerts are often ignored by the healthcare professionals. This can be overcome by tearing the alerts into categories based on severity, which is proven lead to less ignored advices and alerts. STRIPA does not differentiate its alerts and advices by severity, it should be a good alternation in future releases.

3. RELATED APPS

There are two other apps for general practitioners and pharmacies to support the medication review process similar to STRIPA: Epocrates and MicroMedex Drug Information (O'Hagan, 2009), both apps focus on healthcare professionals. There are multiple other applications available but these applications only focus on one aspect of medication reviews.

Epocrates has a free version, which provides information on drugs, interactions and a pill ID function. According to the Epocrates website "nearly 1 in 2 U.S. physicians rely on Epocrates routinely to help inform their decisions in the moments of care, making it the #1 medical app available to U.S. physicians" (Epocrates, 2014). Epocrates also has a premium version with features like suggesting alternative drugs, insight in lab results and disease information. Epocrates is available for Android and iOS. For Android alone, Epocrates has been downloaded over a million times with an average rating of 4.3 out of 5 (n = 19.356) (Google Play, 2014a). For iOS the average rating is 3 out of 5 stars (n = 46.934) (iTunes, 2014a).

MicroMedex Drug Information is a subscription based app and costs $2.99 a year. It provides an extensive index of medications, which includes information about dosage, drug-drug interactions and a tool for pill identification (Micromedex, 2014). MicroMedex Drug Information has been downloaded over 5000 times for Android, with an average rating of 4.2 (n = 56) (Google Play, 2014b). For iOS the average rating is around 2 stars (n = 176) (iTunes, 2014b).

In Appendix 2, Table 3, the apps are compared according to functionality and rating. This table shows the uniqueness of STRIPA. STRIPA is the only app with information about the condition of the patient, start and stop criteria of the all medication and a functionality which links the medication to the diseases. Only pill identification, information about other healthcare professionals and general drug information is not included in STRIPA. However, these functionalities are not of importance when conducting a medication review.

4. EFFECTIVENESS STUDY DESIGN

An effectiveness study should be carried out in order to test whether or not STRIPA is deemed effective for GPs and pharmacists. This effectiveness study should focus on the perceived effectiveness of STRIPA. In order to test this, the development of STRIPA should be finished first.

There are several study designs possible for researching effectiveness (Baxter & Jack, 2008). Based on research of Stolberg, Norman and Trop (2004) and the University of Ottawa (uOttawa, n.d.) the randomized controlled trial (RCT) is chosen to test the effectiveness of STRIPA. In RCT a predefined group of subjects is randomly allocated to two or more groups. One group is assigned as a control group, while the other groups are assigned to an intervention. After the experiment, outcomes of both groups are measured and compared to each other. An advantage of RCT above other study designs, is that it is considered to be the most reliable form of scientific evidence (Akobeng, 2005), since it reduces bias (uOttawa, n.d.). This is the case because the subjects are randomly allocated to one of the two groups, which ensures homogeneity between the groups. Therefore, when comparing the groups to each other after the experiment, the difference between those groups must be caused by the experiment. There are also some disadvantages to RCTs. For example, the external validity can be not sufficient enough. This is for instance the case when only comparing groups in a specific country, then it may not be generalizable to another country with different norms, believes et cetera. In the STRIPA experiment however, this is not a threat. Since STRIPA is intended to use within specific boundaries (in the Netherlands, by GPs and pharmacists) these boundaries can serve as an input for the experiment. Another disadvantage of RCTs is the narrowness of the studied question. RCTs usually only inspect one variable or very few variables, rarely looking at the full picture (Singh, Kumar & Sarkar, 2011).

When conducting an experiment about STRIPA the subjects are GPs and pharmacists. It is estimated that in the Netherlands there are 9.115 GPs and 2.644 pharmacists (Nederlandse patiënten consumenten federatie, 2014). In order to have a representative sample for an effectiveness study, 396 GPs and 336 pharmacists should participate in the experiment. This is calculated based on the standard sample size formulae, with a confidence interval of 5 and a confidence level of 95%. The participating GPs and pharmacists should both be randomly divided in two equal sized groups. Both groups should be presented with the two cases discussed in the end-user survey section, since those cases are representative for the problems that can arise when conducting a medication review for polypharmic patients. The control group should perform the STRIP method with pen and paper and the intervention group by means of STRIPA. When comparing the results of the two groups there are several measurements that should be taken into account. The time to complete the two cases is of importance, since it is claimed that STRIPA reduces the time to complete a medication review. In addition, also the medication and dosage changes should be measured. In this way it can be measured if one of the groups makes significantly more errors than the other. An error is deemed as such, whenever a healthcare professional does not change a medication or medication dosage when it forms a threat to the patient or when a healthcare professional changes the medication or medication dosage into a (still) harmful combination. In addition, also not optimal changes in medication should be deemed as errors. Whether one of the groups makes significantly more errors than the others can be calculated with any statistical significance test.

The costs for such an experiment can be high, when paying all subjects their normal hourly wage in addition to the costs of the researchers themselves and the distribution of STRIPA to the intervention group. In order to suppress the costs, the healthcare professionals should be asked to participate voluntarily. Since only 4.05% of the GPs and 12.7% of the pharmacists in the Netherlands are needed to conduct this experiment, it should be possible to at least move a part of the healthcare professionals to participate voluntarily. This voluntary group will do so based on intrinsic drivers, like ‘the feeling to give something back’ and ‘the feeling of accomplishment’ (Noels, 2001), or the possibility to get to try STRIPA free-of-charge. When the intended amount of subjects is not reached, the healthcare professionals should be promised a financial compensation, in order to move them to participate.

5. DISCUSSION

In this research, there are some threats to the validity of the results. First of all, the SLR could be extended by including more specific search keys. However, the most crucial parts are already included in this version. Besides a more thorough SLR, from the 39 analyzed respondents, there were only four pharmacists. To give more accurate conclusions about the adoption of STRIPA, the overall sample size needs to be bigger and especially more pharmacists should be included. STRIPA is an extensive tool to support medication reviews. It gives alerts whenever a drug related problem is present. In STRIPA healthcare professionals can easily ignore these alerts. Previous research has shown that tiering these alerts into categories according to severity help healthcare professionals to make the right decisions. Alerts from the most severe category are rarely ignored. STRIPA should add these tiered alerts. Furthermore, STRIPA does not include information about the reliability of a patient. How frequent a patient takes their medication has great consequences on the overall treatment. Taking the patient reliability into account could be an added value for STRIPA since alternative treatments may be more effective.

In the experiment and end-user survey mentioned in this research two cases were used to test the potential usefulness of STRIPA. To measure the effectiveness, a RCT should be performed. The RCT clearly shows the effectiveness of STRIPA but it remains an experiment in a controlled environment. All the participants use the same cases to perform a medication review and the differences between the control group and the intervention group are measured. The narrowness of the studied question within a RCT does not examine the 'complete picture'. Therefore this type of experiment does not include how healthcare professionals use STRIPA in their daily life for real patients. This includes that in practice GPs and pharmacists will regularly work together whenever they perform a medication review. In the experiments mentioned before the differences between usage of GPs and pharmacists is measured but how these groups collaborate is not known.

6. CONCLUSION

Based on the research described in this paper, the main research question of this paper 'what is the potential usefulness of STRIPA for conducting medication reviews?' can be answered.

There are between 44.000 and 98.000 hospital deaths annually attributed to medical errors, of which more than 7.000 are due to medication errors. These medication errors cost a lot of money and certain errors can be avoided. A medication review is one way of reducing the amount of medication errors. Multiple studies have proven the effectiveness of medication reviews. To support GPs and pharmacists when performing medication reviews for polypharmic patients STRIPA offers a step by step-by-step method. STRIPA has several functionalities, including an overview of the patient with the condition and medication; linking the medication to the associated diseases; advice when to start and stop with a certain medication; linking side effects to the associated medications; drug-drug interactions; and advice about medication dosage. There are some similar apps to STRIPA, but none of these apps has all the functionalities combined in one app like STRIPA does. Furthermore, research showed that after implementation of STRIPA in the Netherlands, mortality can be expected to decrease by 3 to 19 people on a yearly basis, morbidity by 4 to 28 people, and the financial costs by 10 to 45 million euros a year. An effectiveness study should be carried out in order to test whether or not STRIPA is deemed effective for GPs and pharmacists.

7. FUTURE RESEARCH

Because the effectiveness study and the performed experiment mentioned in this research both include two predefined cases, an observational study would be a good addition. In the observation study the healthcare professionals use the system when conducting medication reviews on real patients. The RCT and observational study together give a complete picture how effective STRIPA is. This observation study includes studying the collaboration of GPs and pharmacists. To test if the functionalities of tiered alerts and taking the patient reliability into account is an added value for STRIPA, additional research should be performed. No research is available whether these two extra functionalities increase the value of STRIPA. This future research should examine how GPs and pharmacists handle the alerts when they are tiered into different categories according to severity. Moreover, not all patients take their medication according the prescription. This affects the overall treatment and should be included in the medication review. Missing in this research is the preference of healthcare professions for a mobile app. At the moment STRIPA is only available for a PC. In the future STRIPA will be available for tablets, both Android and iOS. However it can be the case that healthcare professionals do not see the need to use STRIPA on tablets, this need exists should be researched. In this research, as well as previous research about STRIPA, only GPs and pharmacists are included. STRIPA could be beneficial for a broader audience. STRIPA is mainly focused on elderly patients, therefore STRIPA could be useful for other healthcare professionals like homecare nurses and even for informal care providers and patients themselves. Certain STRIPA functionalities, like the drug-drug interaction feature, can support these other user types. For example, functionalities like start and stop criteria for medication are not applicable for all types of users. A study should be conducted in order to understand the needs of the different user types. This need can be translated into different user profiles with different STRIPA functionalities. A reason for healthcare professionals to use STRIPA may also rely on a CE certification, which can cost a lot of money and time. Before determining the most appropriate level of CE certification, research should be conducted to examine the extent to which the attitude towards STRIPA changes positively with a CE certification level. This should also include a cost benefit analysis. STRIPA is tentatively scheduled for public release on the Dutch market in 2015.

REFERENCES

Baxter, P., & Jack, S. (2008). Qualitative case study methodology: Study design and implementation for novice researchers. *Qualitative Report*, *13*(4), 544–559.

Booz & Company. (2012). *The potential for pharmaceutical quality services*. Available from: http://www.knmp.nl/downloads/nieuws/nieuws-2012/booz-co-the-potential-for-pharmaceutical-quality-services.pdf/view

Epocrates. (2014). *Epocrates overview*. Available from: http://www.epocrates.com/products

Google Play. (2014a). *Epocrates*. Available from: https://play.google.com/store/apps/details?id=com.epocrates&hl=en

Google Play. (2014b). *Micromedex Drug Interactions*. Available from: https://play.google.com/store/apps/details?id=com.thomson.druginteractions&hl=en

Inspectie voor de Gezondheidszorg. (2009). *Het resultaat telt*. Available from: http://www.farmaactueel.nl/beleidsstukken/IGZ2009.pdf

iTunes. (2014a). *Epocrates*. Available from: https://itunes.apple.com/en/app/epocrates/id281935788?mt=8

iTunes. (2014b). *Micromedex Drug Interactions*. Available from: https://itunes.apple.com/us/app/micromedex-drug-interactions/id391763035?mt=8

KNMP. (2013). *Medicatiebeoordeling*. Available from http://www.knmp.nl/downloads/organisatie-regelgeving/organisatie-regelgeving-normen-en-richtlijnen/richtlijnmedicatiebeoordeling.pdf

Krska, J., Cromarty, J. A., Arris, F., Jamieson, D., Hansford, D., Duffus, P. R., & Seymour, D. G. et al. (2001). Pharmacist-led medication review in patients over 65: A randomized, controlled trial in primary care. *Age and Ageing*, *30*(3), 205–211. doi:10.1093/ageing/30.3.205 PMID:11443021

Liberati, A., Altman, D. G., Tetzlaff, J., Mulrow, C., Gøtzsche, P. C., Ioannidis, J. P., & Moher, D. et al. (2009). The PRISMA statement for reporting systematic reviews and meta-analyses of studies that evaluate health care interventions: Explanation and elaboration. *Annals of Internal Medicine*, *151*(4), W-65. doi:10.7326/0003-4819-151-4-200908180-00136 PMID:19622512

Meulendijk, M. (2014). *STRIPA*. Unpublished Raw Data.

Nederlandse patiënten consumenten federatie. (2014). *Zorgkaart Nederland*. Available from: http://www.zorgkaartnederland.nl/apotheker

Nederlandse Service Apotheek Beheer B.V. (2012). *Analyse Medicatiebeoordeling t.b.v. de KNMP Richtlijn Medicatiebeoordeling*. Available from: www.serviceapotheek.nl

Noels, K.A. (2001). New orientations in language learning motivation: Towards a model of intrinsic, extrinsic, and integrative orientations and motivation. *Motivation and Second Language Acquisition, 23*, 43-68.

Stolberg, H. O., Norman, G., & Trop, I. (2004). Fundamentals of Clinical Research for Radiologists. *AJR*, *183*, 1539–1544. doi:10.2214/ajr.183.6.01831539 PMID:15547188

TPO. (2012). *Nieuwe multidisciplinaire richtlijn polyfarmacie ouderen. Tijdschrift voor praktijkondersteuning*. Available from http://www.tijdschriftpraktijkondersteuning.nl/archief/volledig/id689-nieuwe-multidisciplinaire-richtlijn-polyfarmacie-ouderen.html

uOttawa. (n.d.). *Study Designs*. Available from http://www.med.uottawa.ca/sim/data/Study_Designs_e.htm

Zorginstituut Nederland. (2013). *Fact sheet polyfarmacie*. Available from: https://www.zorginstituutnederland.nl/binaries/content/documents/zinl-www/kwaliteit/meerjarenagenda/agenda-2013/agenda-2013/agenda-2013/zinl%3Aparagraph%5B7%5D/zinl%3Adocuments%5B4%5D/1306-fact-sheet-polyfarmacie/Fact+sheet+polyfarmacie.pdf

APPENDIX 1: FIGURES

Figure 1. Systematic literature review process

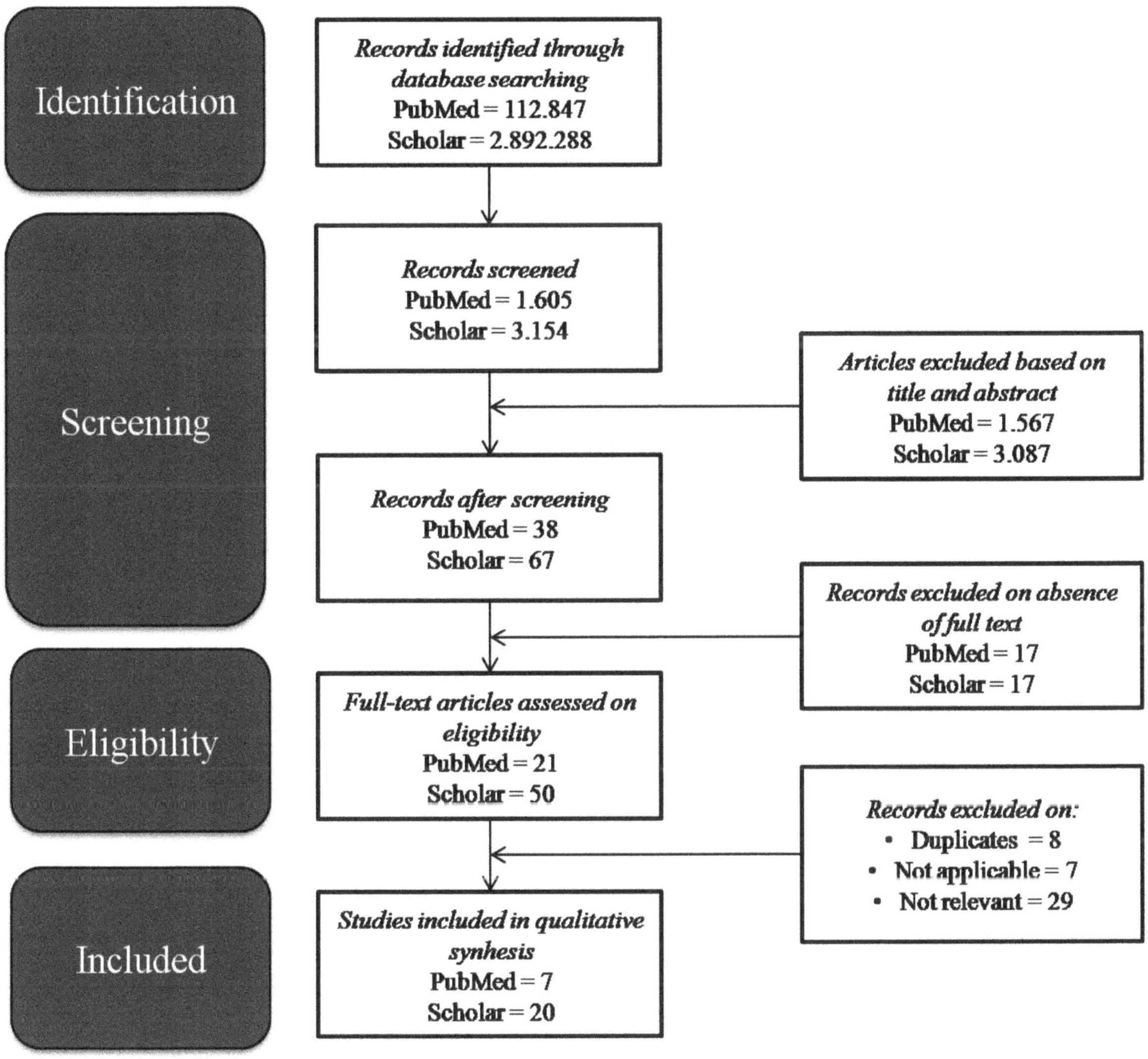

Figure 2. Screenshot 1: STRIPA

Figure 3. Screenshot 2: drug-medication combination

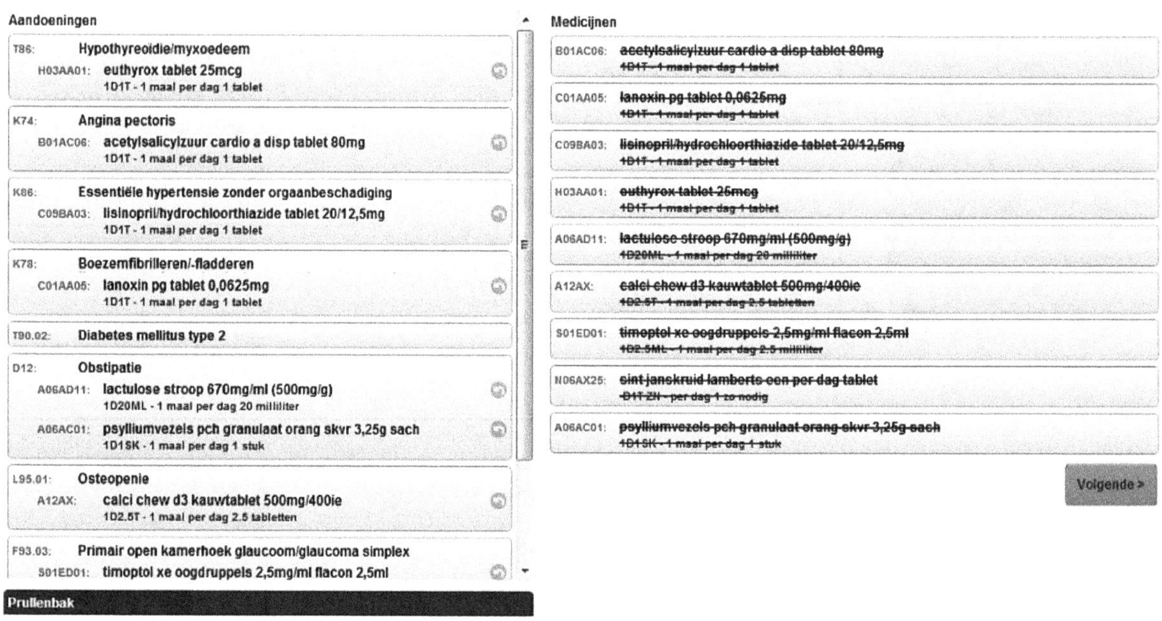

Figure 4. Screenshot 3: start criteria medication

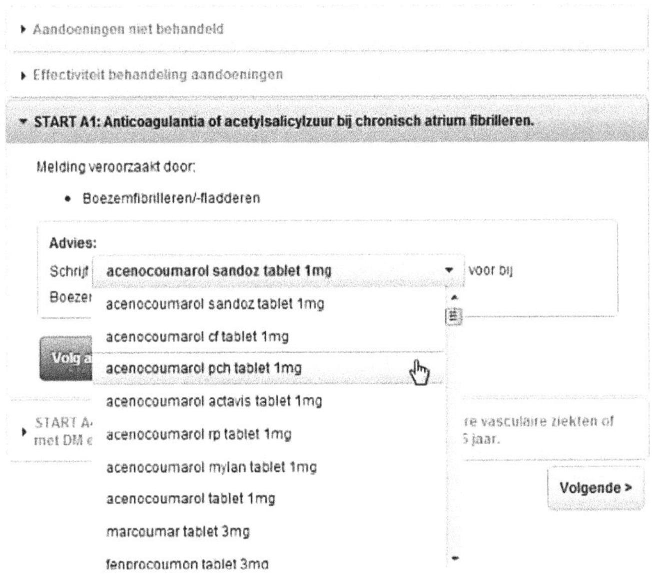

Figure 5. Screenshot 4: stop criteria medication

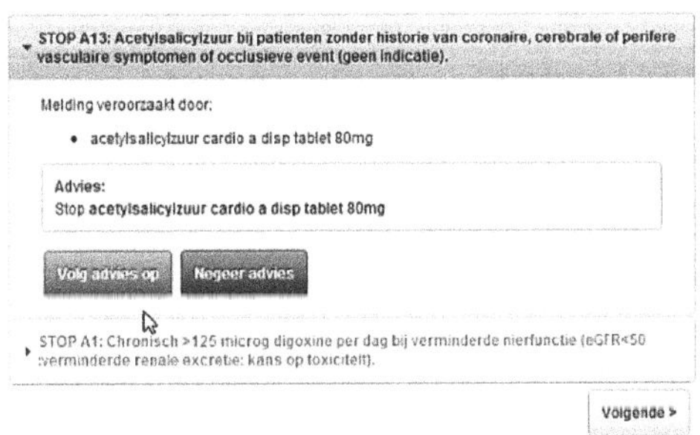

Figure 6. Screenshot 5: connection effect to medication

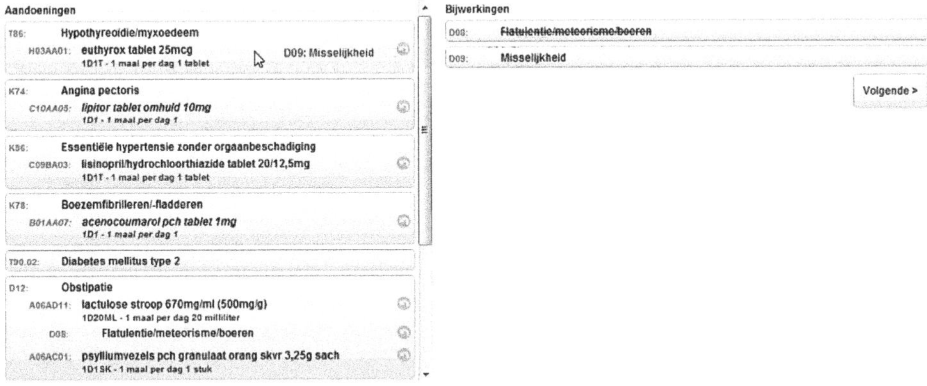

Figure 7. Screenshot 6: drug-drug interaction alert

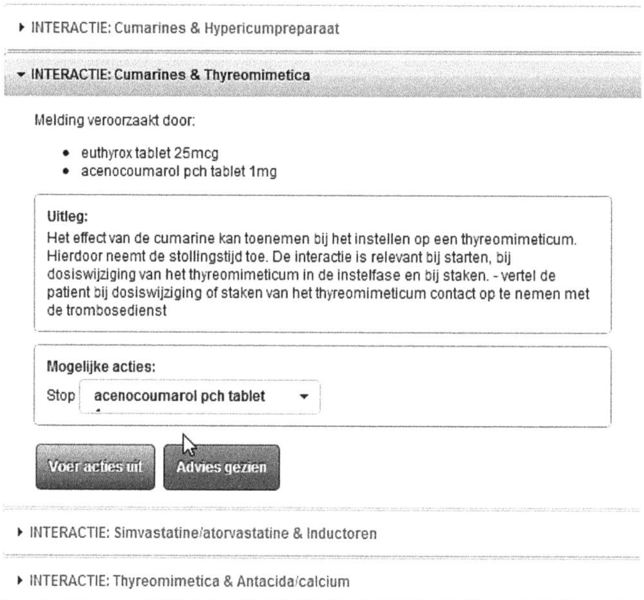

Figure 8. Screenshot 7: dosage medication alert

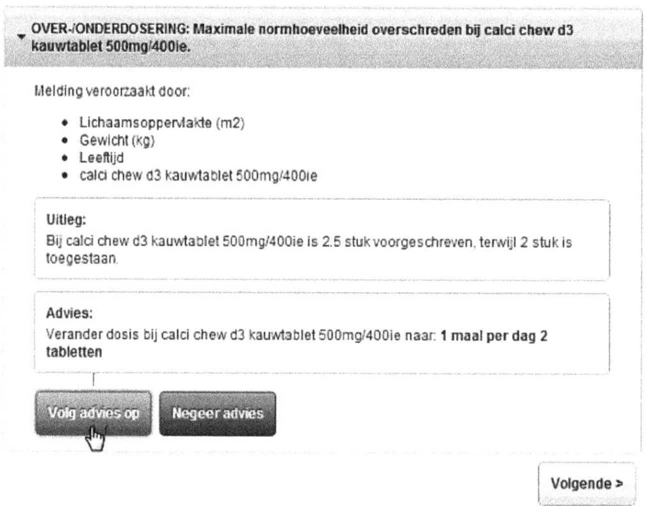

Figure 9. Screenshot 8: overview

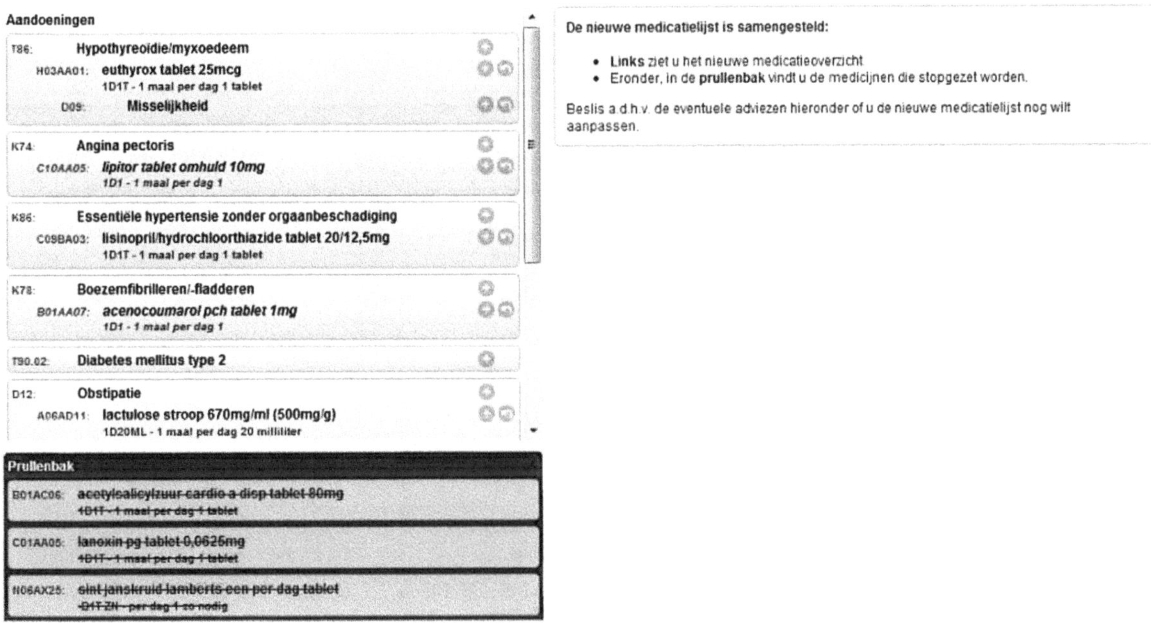

APPENDIX 2: TABLES

Table 1. Search keys

	Drug Interaction	Polypharmacy	Medicine Criteria	Medicine Decease Connection	Medication Review
General	Drug interaction	Polypharmacy	Medicine criteria	Medicine disease connection	Medication review
	Drug interactions	Polypharmacy elderly			
	Drug drug interaction	Polypharmacy APK			
	Drug drug interactions	Medicine multiplicity			
	Drug interaction study	Polypharmaxy optimization method			
		POM			
App	Drug interaction app	Polypharmacy app	Medicine criteria app	Medicine disease connection app	Medication review app
E-health	Drug interaction e-health	polypharmacy e-health	Medicine criteria e-health	Medicine disease connection e-health.	Medication review e-health
General practitioners information systems	General Practitioners information system drug interaction	General practitioners information system polypharmacy	General practitioners information system medicine criteria	General practitioners information system medicine disease connection	general practitioners information system medication review
Clinical decision support software	Clinical decision support software Drug interaction	Clinical decision support software Polypharmacy	Clinical decision support software medicine criteria	Clinical decision support software medicine decease connection	Clinical decision support software medication review

Table 2. Concept table

Reference	Drug-Drug Interaction	E-Health	Health Care Specialists	Polypharmacy	Medication Review
Magnus, D., Rodgers, S. and Avery, A. J. (2002). GPs' views on computerized drug interaction alerts: questionnaire survey. Journal of clinical pharmacy and therapeutics, 27 (5), 377-382.	X	X	X		
Kmetik, K.S., Chung, J., Sims, S. and Found, N.R. (2007). Reasons provided by prescribers when overriding drug-drug interaction alerts. Am J Manag Care, 13, 573-580.	X	X	X		
Paterno, M.D., Maviglia, S.M., Gorman, P.N., Seger, D.L., Yoshida, E., Seger, A.C., ... and Gandhi, T. K. (2009). Tiering drug–drug interaction alerts by severity increases compliance rates. Journal of the American Medical Informatics Association, 16(1), 40-46.	X	X			
O'Hagan, E. (2012). Getting started with medical apps: Apps you should know about. Journal of Hospital Librarianship, 12(2), 162-170.		X			
Yu, K. H., Sweidan, M., Williamson, M. and Fraser, A. (2011). Drug interaction alerts in software-what do general practitioners and pharmacists want?. The Medical journal of Australia, 195(11-12), 676-680.	X		X		
Åstrand, E., Åstrand, B., Antonov, K. and Petersson, G. (2007). Potential drug interactions during a three-decade study period: a cross-sectional study of a prescription register. European journal of clinical pharmacology, 63(9), 851-859.	X			X	
McInnes, D.K., Saltman, D.C. and Kidd, M.R. (2006). General practitioners' use of computers for prescribing and electronic health records: results from a national survey. Medical Journal of Australia, 185(2), 88.		X	X		
Bates, D.W. and Gawande, A.A. (2003). Improving safety with information technology. New England journal of medicine, 348(25), 2526-2534.		X			
Miller, R.A., Gardner, R.M., Johnson, K.B. and Hripcsak, G. (2005). Clinical Decision Support and Electronic Prescribing Systems A Time for Responsible Thought and Action. Journal of the American Medical Informatics Association,12(4), 403-409.		X			
Aarts, J. and Koppel, R. (2009). Implementation of computerized physician order entry in seven countries. Health Affairs, 28(2), 404-414.		X			
Ko, Y., Abarca, J., Malone, D.C., Dare, D.C., Geraets, D., Houranieh, A., ... and Wilhardt, M. (2007). Practitioners' views on computerized drug–drug interaction alerts in the VA system. Journal of the American Medical Informatics Association, 14(1), 56-64.	X	X	X		
Krska, J., Cromarty, J.A., Arris, F., Jamieson, D., Hansford, D., Duffus, P.R., ... and Seymour, D.G. (2001). Pharmacist-led medication review in patients over 65: a randomized, controlled trial in primary care. Age and Ageing, 30(3), 205-211.			X	X	X
Vinks, T.H., Egberts, T.C., de Lange, T.M. and de Koning, F.H. (2009). Pharmacist-based medication review reduces potential drug-related problems in the elderly. Drugs & aging, 26(2), 123-133.			X	X	X
Hajjar, E.R., Cafiero, A.C. and Hanlon, J.T. (2007). Polypharmacy in elderly patients. The American journal of geriatric pharmacotherapy, 5(4), 345-351.				X	
Scanlin, A. (2013). Reducing the Risks of Medication Errors. BioSupply Trends Quaterly, 26-29.	X				
Meulendijk, M. (2012). Development of a decision-support system in the primary care sector. In CAiSE (Doctoral Consortium).		X	X	X	X
Meulendijk, M., Spruit, M., Drenth-van Maanen, C., Numans, M., Brinkkemper, S. and Jansen, P. (2013). General practitioners' attitudes towards decision-supported prescribing: An analysis of the Dutch primary care sector. Health informatics journal, 19(4), 247-263.		X	X	X	X

continued on following page

Table 2. Continued

Reference	Drug-Drug Interaction	E-Health	Health Care Specialists	Polypharmacy	Medication Review
Horn, J.R., Gumpper, K.F., Hardy, J.C., McDonnell, P.J., Phansalkar, S. and Reilly, C. (2013). Clinical decision support for drug-drug interactions: Improvement needed. American journal of health-system pharmacy: AJHP: official journal of the American Society of Health-System Pharmacists, 70(10), 905-909.	X	X	X		
Mille, F., Degoulet, P. and Jaulent, M.C. (2007). Modeling and acquisition of drug-drug interaction knowledge. In Medinfo 2007: Proceedings of the 12th World Congress on Health (Medical) Informatics; Building Sustainable Health Systems (p. 900). IOS Press.		X			
Ahearn, M.D. and Kerr, S.J. (2003). General practitioners' perceptions of the pharmaceutical decision-support tools in their prescribing software. The Medical Journal of Australia, 179(1), 34-37.	X		X		
Smithburger, P.L., Buckley, M.S., Bejian, S., Burenheide, K. and Kane-Gill, S.L. (2011). A critical evaluation of clinical decision support for the detection of drug-drug interactions. Expert opinion on drug safety, 10(6), 871-882.	X	X			
Seidling, H.M., Phansalkar, S., Seger, D.L., Paterno, M.D., Shaykevich, S., Haefeli, W.E. and Bates, D.W. (2011). Factors influencing alert acceptance: a novel approach for predicting the success of clinical decision support. Journal of the American Medical Informatics Association, 18(4), 479-484.		X			
Drenth-van Maanen, A.C., van Marum, R.J., Knol, W., van der Linden, C.M. and Jansen, P.A. (2009). Prescribing optimization method for improving prescribing in elderly patients receiving polypharmacy. Drugs & aging, 26(8), 687-701.		X	X	X	
Duerden, M., Avery, T. and Payne, R. (2013). Polypharmacy and medicines optimisation.	X		X	X	X
Albrecht, U.V. (2012). Transparency of health-apps for trust and decision making. Journal of medical Internet research, 15(12), e277-e277.		X			
Kaushal, R., Shojania, K.G. and Bates, D.W. (2003). Effects of computerized physician order entry and clinical decision support systems on medication safety: a systematic review. Archives of internal medicine, 163(12), 1409-1416.		X			
Bindoff, I.K., Tenni, P.C., Peterson, G.M., Kang, B.H. and Jackson, S.L. (2007). Development of an intelligent decision support system for medication review. Journal of clinical pharmacy and therapeutics, 32(1), 81-88.		X	X		X
Meulendijks, E. A. (2013). Requirements Engineering for Medical Consumer Applications.	X	X	X	X	X

Table 3. App comparison

Features	STRIPA	Epocrates	MicroMedex Drug Information
Overview conditions patient	a		
Overview diseases patient	a	a *	
Overview lab results patient	a	a *	a
Overview medication patient	a	a	
Linking medication to the diseases	a		
Advice start new medicine	a		
Advice stop current medicine	a		
Drug-drug interactions	a	a *	a
Advice about medication dosage	a		a
Pill identification based on its imprint and physical characteristics		a	a
Information about other health care professionals		a	
General drug information		a	a
Costs	Unknown	Free version or a premium version for $159.99 a year	$2.99 a year
Rating	Unknown	Android: 4.3 (n = 19.356)	Android: 4.2 (n = 56)
		Apple: 3 (n = 46.934)	Apple: 2 (n = 176)

* This is a premium only feature.

Chapter 6
How to Identify Rheumatic Diseases by General Physicians

Eduardo C. Contreras
Autonomous University of Coahuila, Mexico

Gustavo J. Puente
Autonomous University of Coahuila, Mexico

ABSTRACT

A large part of the population in countries in process of development ignores what Rheumatic Diseases are, and general practitioners are in most cases unaware of enough information to identify them and the treatments to successfully control them. A proposal to help those general practitioners to detect if an articular condition belongs to a Rheumatic Disease case is to present them the clinical semiology that should lead them to redirect the given conditions to a specialist on the subject, a rheumatologist. The clinical semiology is presented by an automated algorithm inside a goal-based software agent, containing all the necessary information to identify the seven most common inflammatory Rheumatic Diseases, and fourteen of the non-inflammatory ones. The purpose of this tool is to provide the general practitioner with the correct information to redirect the patient with a rheumatologist, in order for it to receive the appropriate medication to be controlled.

INTRODUCTION

Project

This project is a field investigation, applied to medical area on the Rheumatology specialty related with Clinical Semiology and the Information Technology. It is in itself, an automated medical tool to identify joints affections, based on an Intelligent Agent from Artificial Intelligence specialty, based on the heuristic compilation of the tacit knowledge of rheumatologists. This study tries to associate the factors than relate the support process on Information Technology on a General Doctor, towards a modern tool as didactic software, which facilitates the most common rheumatic's affections identification.

DOI: 10.4018/978-1-5225-0248-7.ch006

This identify tool has a section of rheumatic inflammatory conditions and a section of rheumatic non-inflammatory conditions. Inflammatory conditions referred to joint. In the section of inflammatory conditions the user will find windows that include the different aspects of the joint disease that must be considered in diagnosis such as kind of onset, number and size of involved joints, topography of the arthropathy and symmetry of involvement. The section of non-inflammatory conditions considers only localized, regional and generalized involvement. Both sections follow a reasonable and logic algorism that allows reaching diagnosis of the rheumatic conditions included in the diagnostic tool.

OBJECTIVE

To value the utility of the algorism for the identification of Rheumatic Affections" with a general practitioner, so that he will be able to provide an accurate diagnosis of joint diseases and redirect the patient with a rheumatologist for its early evaluation.

The proposed clinic semiology is based upon the human experience that is used to build a software agent didactic tool related to the Information Technologies; the software agent is a tool that allows the doctor to offer an appropriate evaluation of the rheumatic diseases.

BACKGROUND

Rheumatic Diseases

The study and management of rheumatic diseases in a rigorously scientific way has been in charge of the medical specialty of Rheumatology, a branch of Internal Medicine, since the middle of the 19th century, making it an medical area relatively new and this explains why, even in the beginning of the 21st century, rheumatic diseases are largely unknown both by the population in general and by doctors, who are not dedicated to this area of health.

Rheumatic diseases are heterogeneous clinical entities which affect musculoskeletal (Schumacher et al, 1998) system and other systems and organs, whose diagnosis often comes late due to little knowledge of these diseases and scarcity of rheumatologists, among other reasons (Taranta & Markowitz, 1981). Delay in diagnosis of these diseases may lead to damage of joints and related tissues, organs and systems with subsequent functional disability and increase in morbidity and mortality.

A person who suffers from a rheumatic disease that has affected the hip joint is facing a severe pain when attempting to walk. The first three or five steps walks with great difficulty, as it decreases the pain, once that it lessens can walk with less pain and greater agility. When attempting to stand the pain is so acute that it does not balance and when sitting drops the body waiting for that pain to decreases, spending time it dulls, but knowing that when get up again will face that terrible acute pain. When the joints of the hand are concerned it is very difficult to remove the lid to a jar, as pain does not allow you to force. If the person suffers from a knee, usually the cartilage that covers the joint decreases its thickness, staying in touch with the bone part, decreasing joint mobility and increasing pain to move it. Usually the patient will live with the pain articulates for a lifetime, with the probability that the joint will be merged and significantly decrease their movement, in such degree that it turns into a person with a disability. Most of the people must be in constant motion, as it permits the disease, otherwise it would be in for life in a wheelchair or bed.

When a rheumatic disease affects joints, they are damaged in an irreversible process; therefore it is substantial to identify them as soon as possible. If general practitioners supply inappropriate medication, they will be at great risk of affecting the patient severely. The rheumatologist is the only appropriate person to administer the correct medication, due to the existence of the called "antirheumatics".

Distinctive features of this disease are pain, stiffness and articular inflammation that would only decrease after the consumption of the rightful medication.

Definition of the Common Ailments

- **Rheumatoid Arthritis:** A chronic inflammatory disease of unknown etiology which involves multiple diartrodial joints leading to hyperplasia of synovial tissue and destruction of cartilage and subchondral bone with subsequent deformity and functional limitation. (Stone, 2009)
- **Osteoarthritis or Degenerative Joint Disease:** A no inflammatory disorder of cartilage of the movable joints which suffers deterioration due to biochemical and structural changes in its matrix and in the subchondral bone.(Cañete,2008)
- **Gout:** A metabolic disease characterized by hyperuricemia and episodes of acute arthritis and progression to a generalized arthropaty with accumulation of monosodium urate (Wright & Harvey, 1988).
- **Spondiloarthritides:** A heterogeneous group of diseases characterized by inflammation of the joints of the spine, sacroiliac, limbs joints and in variable frequency affects also skin, mucosal of the digestive tract, urogenital tract and eyes.(Kats,1977)
- **Systemic Lupus Erythematous:** An autoimmune disease characterized for the appearance of multiple autoantibodies which participate in inflammatory disease of diverse organs and systems.
- **Juvenile Arthritis:** A term which refers to a group of inflammatory diseases of the joints in pediatric populations until the age of 16 years.
- **Infectious or Septic Arthritis:** A disease caused by invasion of the joints by bacteria and other microorganisms.
- **Fibromyalgia Syndrome:** The name of a condition characterized by chronic widespread musculoskeletal neuropathic pain in specific sites. (Clauw, 2005)
- **Carpal Tunnel Syndrome:** A clinical entity characterized by entrapment of the median nerve when it passes through a tunnel formed by the carpal bones and joints and the transverse carpal ligament.
- **Digital Stenosis Tenosynovitis:** (Of the fingers) is a constriction of the lumen of the fibrous sheath of flexor tendons of the fingers which impedes its free gliding.(Salter, 1998)
- **D'Quervain Tendinitis:** A condition that appears due to excessive friction between the tendon abductor policis longus and extensor pollicis brevis and its common fibrous sheath.
- **Epicondylitis or Tennis Elbow:** A degenerative tendon disease located in the origin of the forearm extensor muscles.
- **Rotator Cuff Tendinitis:** An inflammatory condition located at the insertion of the rotator muscles of the shoulders in the humeral head.
- **Dupuytren's Contracture:** A progressive retraction of the fibrous tissues of the palmar fascia.

Biological Therapies

Biological therapy (also called immunotherapy or biotherapy) often employs substances called biological response modifiers (BRMs). Biological therapy is thus any form of treatment that uses to stimulate or restore the ability of the immune (defense) system to fight infection and disease. Biologic therapy to block the action of a messenger of inflammation called tumor necrosis factor (TNF) is being used to treat conditions such as Crohn's disease and rheumatoid arthritis.

Biologics are genetically engineered proteins derived from human genes. They are designed to inhibit specific components of the immune system that play pivotal roles in fueling inflammation, which is a central feature of rheumatoid arthritis.

Antirheumatic drugs are drugs used to treat rheumatic diseases. The major classes of antirheumatic drugs include Nonsteroidal Anti-Inflammatory Drugs (NSAIDs), Corticosteroids, Disease Modifying Anti-Rheumatic Drugs (DMARDs), Slow-Acting Antirheumatic Drugs (SAARDs).

- **Nonsteroidal Anti-Inflammatory Drugs (NSAIDs):** Drugs belonging to this class bring symptomatic relief of both inflammation and pain, but have a limited effect on the progressive bone and cartilage loss associated with rheumatoid arthritis. They act by slowing the body's production of prostaglandins.
- **Corticosteroids:** These drugs are very powerful anti-inflammatory agents. They are the synthetic analogs of cortisone, produced by the body. Corticosteroids are used to reduce inflammation and suppress activity of the immune system.
- **Disease Modifying Anti-Rheumatic Drugs (DMARDs):** Influence the disease process itself and do not only treat symptoms, hence their name. DMARDs also have anti-inflammatory effects, and most were borrowed from the treatment of other diseases, such as cancer and malaria.
- **Slow-Acting Antirheumatic Drugs (SAARDs):** Are a special class of DMARDs and the effect of these drugs is slow acting and not so quickly apparent as that of the NSAIDs. (Medical Dictionary)

The treatment in a patient with anti-rheumatic drugs active ankilosing spondylitis (AS) and spondyloarthritides (PsA) whose condition is not sufficiently controlled with non-steroidal anti-inflammatory drugs (NSAIDs) in the case of axial disease, and sulphasalazine or methotrexate in the case of peripheral arthritis. Anti-tumor necrosis factor (Anti-TNF) therapy seems to be a powerful tool for the treatment of AS and other SpA.

Biological therapies for the treatment of rheumatoid arthritis, currently approved, are available: Etanercept, Infliximab, Adalimumab, Anakinra, Tocilizumab, Rituximab, Abatacept. Biologic agents most frequently used as first option are infliximab and etanercept. The progress in manufacturing biotechnology has contributed to the development of several other prospective agents that may form the basis for the therapy of rheumatoid arthritis in the near future. (Šenolta et al, 2009)

On a study in the hospital La Fé located in Valencia Spain, descripts the registry of adverse events of biological therapies in rheumatic diseases. It contains the description of: frequency of adverse events by groups, frequency of serious adverse events, frequency of fatal adverse events, incidence rate of adverse events and incidence rate of serious adverse events. (Research Unit of the Spanish Society of Rheumatology, 2013.)

The treatment in a patient with anti-rheumatic drugs active ankilosing spondylitis (AS) and spondyloarthritides (PsA) whose condition is not sufficiently controlled with non-steroidal anti-inflammatory drugs (NSAIDs) in the case of axial disease, and sulphasalazine or methotrexate in the case of peripheral arthritis. Anti-tumor necrosis factor (Anti-TNF) therapy seems to be a powerful tool for the treatment of AS and other SpA. (Braun & Sieper, 2004)

Biosimilars is the term used for generic versions of biologic agents, with identical primary sequences in which changes in the manufacturing processes have led to important differences in safety and efficacy. Introduction of biosimilars requires a specifically designed pharmacovigilance plan. (Kolkata, 2012)

Home Remedies for Arthritis and Joint Pain

In countries in process of developing most of the people relies on home remedies when they suffer a condition, given that they have little information, because of their culture, economic situation or the fact that they don't have access to specialized medical attention. In the case of suffering a joint pain without receiving a professional diagnosis, people often believe this pain is derived from a rheumatic origin, thus applying to themselves popular remedies such as: putting a potato, a magnet or a piece of copper or silver in their pockets; applying a poultice of ingredients like lettuce, smashed garlic, nutmeg, figs or cabbage; ingesting sage, leek or poppy seed infusions; and finally, drinking Turmeric and Ginger Tea, White Willow Tea, or Juniper Berry Tea.

Another home remedies are: Lubricate joints With Extra Virgin Olive Oil, Peppermint Eucalyptus Oil Blend, Blackstrap Molasses Drink and Pectin and Grape Juice.

These remedies don't change the course of the arthritis itself, but they do have an analgesic, or pain-relieving effect.

Clinical Semiology

In medicine, the clinical semiology is the group of knowledge that deals with the identification of the various pathological manifestations (signs and symptoms), from people and animals (in the veterinary field), in both cases with the objective of how to collect data.

Semiotics. With the objective of compile this data in syndromes and how to interpret them (clinical semiology). This discipline allows the physician not only directed at the diagnosis but have a prognostic assessment.

Unless other diseases, the traditional semiology teaching-learning process is focused on isolated concepts and practices. Rheumatic diseases, for example, need a systemic approach for their diagnosis.

This approach allows the study of the whole thing, it's not enough to study an only part or region of the human body because the rheumatic diseases come all from a mismatch on the immunological system. If a general practitioner applies immunosuppressers to a patient with a rheumatoid arthritis condition, with glaucoma, diabetes or an infectious condition, without considering the concomitant effect, this would affect the patient severely. For this reason, only a specialized practitioner, a rheumatologist, can prescribe the adequate medication without the risk of secondary effects.

For all the reason previously mentioned, the semiology must have a systematic approach (Bertalanffy, 2011). Semiology is the mainstay of clinical medicine. It is an art and a science. It presents a method of sorting knowledge (clinical method) and the goal is the diagnosis of health problems.

The methods applied on Semiology are complex and differ one from another, mainly because their use as a method and recognition as a science have stopped in the present time.

According to the systemic approach, questionnaire, observation, physical examination and lab exams are not enough, because it involves historical methods, sociology, genetics (biological inheritance), psychology, psychiatry, etc., among other sciences.

- Rheumatology Method,
- Questionnaire,
- Physical Exploration on Joints,
- Inspection,
- Palpation,
- Mobility,
- Arthrocentesis,
- Lab exams,
- Reactive C Protein,
- Rheumatoid factor,
- HLA-B27 HLA-DRB1 HLA- DR4,
- Erythrocyte sedimentation rate (ESR),
- Antinuclear antibodies,
- Radiological examination,
- Anteroposterior,
- Lateral,

Radiographic semiology to detect:

- Ankylosing hyperostosis,
- Ankylosing spondylitis,
- Rheumatoid Arthritis,
- Paget's Disease,
- Neoplasms.

Software Agent Definition

In artificial intelligence, a software agent (IA) is an autonomous entity which observes and acts upon an environment and directs its activity towards achieving goals. (Rich & Knight, 1991) An agent is everything that can be considered that perceives its atmosphere by means of sensors and that response or acts in such atmosphere by means of effectors. (Russell & Norvig, 2003) A sensor is a device that measures a physical quantity and converts it into a signal, which can be read by an observer or by an instrument. For example, voltmeter, thermometer, thermocouple, seismometer, parking sensor, carbon monoxide detector. Effectors are agents or structures that cause an activity, like a nerve causing a muscle to flex, or a cell triggering an immune response to a foreign substance.

In robotics, an end effector is a device or tool that's connected to the end of a robot arm where the hand would be. The end effector is the part of the robot that interacts with the environment. The structure of an end effector and the nature of the programming and hardware that drives it depend on the task the robot will be performing. (Resconi, 2004)

Intelligent agents are often described schematically as an abstract functional system similar to a computer program. On the Internet, an intelligent agent (or simply an *agent*) is a program that gathers information or performs some other service without your immediate presence and on some regular schedule.

Models to Construct Agent's Programs

- **Agents from Simple Reflex:** (Russell & Norvig, 2003) The rules condition-action allows the agent to establish the connection between perceptions and actions.
- **Agents Good Informed into Everything What Happens:** It updates the information based in its own internal state.
- **Agents Based on Goals:** It is a combination of the previous ones, with the information relative to the result than will produce the possible actions than will undertake. It handles the information that use to update the internal state and this way to select those actions that allow reaching the goal.
- **Agents Based on Utility:** It allows to differentiate a state from another and to take the decision that offers major utility to the agent.

Of these types we will be based only on Agents Based on Goals. Besides the states, agents need some kind of information about its goals. Goals will to detail the situations to that it is desired to arrive. The agent program could to combine goals with actions and in this way, to be able to choose those actions that allow reaching the goal.

Design of an Ideal Agent

To specify which kind of action should undertake an agent, as an answer to a certain sequence of perceptions constitutes the design of an ideal agent.

Agent Type: Medical Identification System

- **General Doctor's Perceptions:**
    - Symptoms, clinical signs, physical examination, heredity, personal data
- **Actions:**
    - Identifies: joint state, anatomical region, number of joints.
    - How the patient evolves.
    - Differentiate diseases.
- **Goals:**
    - Affection identification, time reduction, canalization of patient in case of such another disease.
    - Atmosphere.
    - Patient, general doctor.

Medical Identification System for Rheumatic Diseases (MISRheD)

The actions than this agent undertakes, are based exclusively in an integrated knowledge (provided by the rheumatologist), and so, its own perceptions are ignored (it does not count with sensors towards the environment), therefore the agent has not autonomy. Its conduct is defined by the human agent experience. It is reasonable to equip an agent of artificial intelligence with certain initial knowledge and capacity to learn (the rheumatology heuristic).

Similar Studies

Rejón established relationship between the characteristics of clinical judgment, medical semiology, and psychiatric semiology. Abductive inferences are especially useful to balance universal and singular information. (Rejón, 2013)

Noachtar and Peters mentioned that the Semiology has served for a long time as the major tool to classify epilepsy syndromes. They gives an introduction into the semiological seizure classification. (Noachtar & Peters, 2009)

Sadly most physicians depend on imaging services and laboratory tests far more frequently that a carefully executed history and physical examination for establishing a diagnosis or monitoring a therapeutic response. (Saravolats, 2007)

Monteagudo on the doctoral thesis presents didactic software for teaching the clinical semiology of the respiratory system. (Monteagudo, 2003)

In the game "Camino Cardíaco", it's goal was to promote a new educational resource in teaching the topic cardiovascular nursing assessment, the subject of semiotics, in addition to encouraging greater student interaction in the teaching-learning process. (Martínez, 2010)

De León proposes the use of hypertext in medical students for learning rheumatic diseases by using images to identify and interpret the signs and symptoms higher frequency. (De León, 2012)

Rheumatic Diseases: New Insights for the Healthcare Professional: 2013 Edition is a ScholarlyBrief that delivers timely, authoritative, comprehensive, and specialized information about Additional Research in a concise format.

A descriptive and prospective observational study was conducted "Ocular manifestations of rheumatic diseases in children" with the objective of characterizing the most frequently ocular manifestations in the course of the rheumatic diseases. (Parapar, 2007)

An Ontology-based intelligent agent for respiratory waveform classification. In this paper presents an ontology-based intelligent agent for respiratory waveform classification to help the medical staff with the judging the respiratory wave form from the ventilator.(Lee &Wang 2006)

Was as applied Utility-based Agents for Hospital Scheduling: A Conjoint-Analytical Approach for the scheduling of centralized operating theatres in large hospitals can be regarded as an archetypal cooperative decision problem.

The University of Aberdeen researchers are interested to collaborate with India on development of trusted mobile technologies for rural healthcare. The aim of the project is to explore the potential of trusted mobile technologies and to improve the management of chronic diseases in rural areas of the UK and India.

A Novel Approach for Cardiac Disease Prediction and Classification Using Intelligent Agents. In this paper the goal is to develop a novel approach for cardiac disease prediction and diagnosis using intelligent agents.

Exploiting social reasoning of open multi-agent systems to enhance cooperation in hospitals.

This article explains how agents can exploit their capacity in order to support the medical staff in a hospital complex specialized in the treatment of burn patients, either by cooperating with other agents, or with human actors that they represent in the virtual system.(Horn,1999)

MAIN FOCUS OF THE CHAPTER

Problem Definition

Problem Identification

Most of the Mexican population, including the medical people, does not have the necessary information on rheumatic diseases to enable them to understand the importance of them and their socio-economic and psychological impact.

The Rheumatology is a relatively new medical specialty. In Mexico it was until the 1960s when it began with the training of specialists in this area of health. Most of the schools and faculties of medicine, both the public and private sector lack, even today, of the field of Rheumatology as part of their curricula and there is a limited number of specialists in our country. According to data collected and provided by the National Institute of statistics and geography until the year 1995 there were 59 faculties of medicine in our country; and it was until 1990 when it included the field of Rheumatology in these powers, so that medical graduates prior to this date were not given adequate information on this area, and only obtained some knowledge about the pathophysiology Traumatology, orthopedics and dermatology; matters that are related to the Rheumatology but do not have full knowledge about the same. For this reason there are currently a very small number of specialists in this area in our country.

Even though the Mexican College of Rheumatology Association Civil, has carried out information campaigns through the specialists affiliated to it and has also established the *National Day of the Rheumatic Patient* and other institutions have been involved in campaigns such as the Ministry of public health and private health institutions, as well as schools and faculties of medicine in our country have not achieved the objectives to inform the population in general on this condition.

All of these factors are contributors to the very low and erroneous information on rheumatic diseases in our country. Despite efforts to disseminate information on rheumatic diseases, are still insufficient the established programs for this purpose until the moment.

The population taken care by general doctors, or those from the population that has no access to the health services, due to its low economical condition or extreme poverty, this due to in Mexico is an indispensable requirement to access this health service system to be employed by some company, institution or government. Most of the population in Mexico is served by the main clinics hospitals of Social Security (Social Security Mexican Institute, Social Security Institute to Serve the State Workers, and specifically in Saltillo's Municipality by Magisterial Clinic Hospital) the rightful claimants must request its appointment in first instance with the general doctor, and this as well depending on the disease and also with some luck, it will alternate to the specialist doctor: however, the patient waits till three months to obtain an appointment with the specialist doctor. This without a doubt has consequences that it can worsen the rheumatic disease of patients due to they are not served by a specialist doctor with the knowledge to identify the affection and to treat it suitably.

Medical specialists in the field estimated that 10% of the population develops over the course of his life any rheumatic disease and because general practitioners not diagnosed in time and correctly the disease is progressive and irreversible can subsequently cause a disability which may have been avoided if they had received the proper attention of a specialist.

Problem Quantification

In Other Countries

Rheumatic diseases in Spain according from data obtained from the National Institute of Social Security were in 2007 spending of 1.678 billion euros in temporary work disability. It is estimated that in Spain, approximately ten million people suffer from a rheumatic disease, 22.6% of population and cause between 10 and 15% of the primary care.

According to Statistics in USA: nearly 70 million Americans suffer from some form of arthritis or symptoms of joints chronic, rheumatic diseases are the leading cause of disability in people of 65 years of age and older, rheumatoid arthritis, how more disabling of arthritis affects approximately 2,100,000 Americans and attacks on women are two to three times more than males. These data come from Centers for the Disease Control and Prevention, National Institute of Arthritis and Musculoskeletal and Skin Diseases, which are part of the National Institutes of Health and Arthritis Foundation.

According to the statistics managed by the National Center for rheumatic diseases (CNER), in Venezuela, 20 of every 100 Venezuelans have been diagnosed with a disease of this kind.

In Our Country

In Mexico is not satisfied with the recommendation of the World Health Organization (WHO), that there is a rheumatologist per hundred thousand inhabitants, because there are only 560 Rheumatologist certificates, this is, a specialist for every 200 thousand inhabitants, it is not enough.

Rheumatic diseases are a major cause of morbidity and mortality. The available information indicates that the Systemic Lupus Erythematous has a mortality rate of 10% for the five years of its inception and 20% after 10 years; the rheumatoid arthritis reduces the survival in a range of 3 to 15 years compared with the general population and are diseases that on course have significant medical complications. Both diseases often affect people in productive age (18 to 60 years).

Statistics of the Mexican Institute of Social Security (IMSS) in our country report that until 1974 25% of disability pensions were granted by rheumatic ailments, and in 1978, 34% of the dictums of disability were granted for patients with diseases of the muscular-skeletal system. Rheumatic diseases are among the top ten causes of demand for health care in the family doctor system.

In Our Region

A number of questionnaires applied in the city of, Saltillo, Coahuila. To a based sample of 190 people older than 13 years (except for medical specialists in rheumatology) gave the following results:

From the sample of the population of Saltillo the 18.42% have information about rheumatic diseases, the 62,63% they have never been to talks or known information about these diseases, the 50.52% don't know that there is a medical specialty dealing with rheumatic diseases, the 27.36% believe that rheumatic diseases affect only to people of the third age, the 55.78% believes that rheumatic diseases affect little in economic or social, to the 61.57% almost aren't interested in further information on rheumatic diseases..

In the city of Saltillo with a population of 578 046 inhabitants only has 4 medical specialists in the medical industry.

Development Method of the Software Agents

The proposal to solve the problem is to identify the type of software agents (intelligent agent) suitable, according to the following elements:

Human agent: to identify a rheumatic disease, knowledge and heuristics of the specialist physician in rheumatology, which will provide the basis of knowledge, rules, diagnostic approach in rheumatology, interpretation of radiological studies, laboratory and physical examination is necessary. The experience of the physician rheumatologist for issuing a diagnosis relies 70% on interrogation, 20% in physical exam and 10% in laboratory and radiological tests.

- **Environment:** Specify the environment in which the agent will run, in this case the main actors are: the general practitioner and the patient.
- **Percepts:** Determine perceptions, entry of data to the agent, identify the types of sensors that the agent will communicate with the environment; in this case the sensors are replaced by the perceptions of the general practitioner.
- **Effectors:** Determine the effectors, mechanisms of response of the agent, in our case is the information displayed on the monitor.

When the above elements have been defined it identifies the type of appropriate agent that will better solve the problem on the basis of the needs and resources available.

Agent Type will be based only on Agents based on goals. Besides the states, agents need some kind of information about its goals. The goals will detail the situations which it is desired to arrive. The agent program could combine goals with actions and in this way, to be able to choose those actions that allow reaching the goal.

The actions than this agent undertakes, are based exclusively in an integrated knowledge (provided by the rheumatologist), and so, its own perceptions are ignored (it does not count with sensors towards the environment), therefore the agent has not autonomy. Its conduct is defined by the human agent experience. It is reasonable to equip an agent of artificial intelligence with certain initial knowledge and capacity to learn (the rheumatology heuristic).

Identification System of Rheumatic Diseases

Agent Type: Medical Identification System of Rheumatic Diseases

- **Percepts:**
    - General Doctor's perceptions.
    - Symptoms, clinical signs, physical examination, heredity, personal data.
- **Actions:**
    - Identifies: joint state, anatomic location, number of joints, joints size.
    - How the patient evolves.
    - Differentiate diseases.
    - Onset type.
    - Inflammatory.
- **Goals:**
    - Affection identification, time reduction, canalization of patient in case of another disease diagnosis.
- **Environmental:**
    - Patient.
    - General Doctor.

Unlike Russel and Norving (1996), in his model *an agent with explicit goals* applied in robotics, software agents type medical diagnosis system, instead of using sensors, software agents are given as percepts (agent receives as perceptions) of input data provided by the general practitioner (symptoms, clinical signs, physical examination, heredity, personal data, evidences and the patient answers) to report the current state of the environment (patient), this information is established a current state on the agent, the agent in this time meets the age, sex, type of disease, whether it is inflammatory or not inflammatory, Identifies: joint state, anatomical region, number of joints, joints size. Which will later be modified to know how will evolve the patient. Immediately questioned actions to take based on a table of terms, when he has taken an action check out the new State in which the patient has evolved to differentiate diseases and turns to question what happens if you take a different action, with this action obtains a goal (identifies the disease). The effectors are the responses of the agent to suggest a group of diseases in hierarchical order which shows the interaction of the user.

Requirements for Software Agents Development

Requirements

It's required, (Pressman, 2002) provide a rheumatic identification by means of an software agents which contains a knowledge base, and the needed algorithms, to make a search process that allows reaching a defined goal.

It's required, create a data bank which contains the information of general data of the patient, age, sex, anatomic location, startup type, inflammatory and so on), clinical signs, symptoms, laboratory analysis.

The human agent (rheumatologist) requirements are: provide the information to elaborate the structured questions, diagnostic approach in rheumatology, the initial evidences, rules and goals to that it is desired to arrive.

Table 1. Rheumatic diseases that cause inflammation in the joints

1	Gout
2	Infectious arthritis
3	Juvenile arthritis
4	Spondiloarthritis
5	Systemic Lupus Erythematosus
6	Rheumatoid arthritis
7	Degenerative joint disease

The requirements for the computer program are: a user interaction design that allows it to capture the needed information on a structured form, and that shows the same way an answer by means of a text file and images.

Data Analysis

It is necessary to identify and structure the information to facilitate the development of the agent, according to the heuristic of the physician rheumatologist, is described as follows:

The agent identifies a total of 41 diseases, same that are separated into two categories: inflammatory and non inflammatory. Referred to joint.

Diseases that Identifies Software

Inflammatory. See in Table 1, rheumatic diseases that can identify the software.

Description of the Knowledge Base

To search based on goals, we need the Knowledge Base, with well-defined information, established the same, with characteristics of diseases as well as by their categories.

For inflammatory joint disease characteristics are as follows:

1. **Sex:** Sex may refer to:
   a. Male.
   b. Female.
2. **Age:** Inflammatory joints diseases that will be taken into account for the agent are distinguished by a range of age that is presented; these ranges include the following way:
   a. 0 - 12 months or newly born.
   b. 1 - 16.
   c. 17 - 30.
   d. 31-50 years.
   e. 51 years old, or later.

Non inflammatory. See in Table 2, rheumatic diseases that can identify the software.

Table 2. Rheumatic diseases non inflammatory in the joints

1	Deltoid Bursitis	18	Cervical root compression syndrome
2	Meniscus injuries	19	Thoracic outlet syndrome
3	Knee ligaments injuries	20	Carpal tunnel syndrome
4	Chondromalacia patellae	21	Lumbar root compression syndrome
5	Osgood-Schlatter disease	22	sciatic neuritis
6	Prepatellar bursitis	23	Epicondylitis or tennis elbow
7	Anserine bursitis	24	Peroneal syndrome
8	Ankle valgus	25	Growing pains
9	Plantar Fasciitis	26	Fibromyalgia Syndrome
10	Flat feet	27	Multiple myeloma
11	Hallux valgus	28	Febrile diseases
12	rotator cuff Tendinitis	29	Epitrocleitis
13	Back sprains	30	D'Quervain tendinitis
14	Scoliosis	31	Digital stenosis tenosynovitis
15	Spondylosis	32	Dupuytren's Contracture
16	Lumbar disc herniation	33	subtrochanteric Bursitis
17	Vertebral fracture	34	Disorders of the spinal alignment

3. **Onset Type:** The onset type or in other words the way in which the disease occurs in the person concerned may be:
    a. **Acute:** The disease occurs suddenly, in a way not very recurrent.
    b. **Insidious:** The disease occurs so insistent.
4. **Number of Joints:** The number of joints refers to the total number of joints that are being affected by the disease in the patient taking the following classification.
    a. **Monoarticular:** Disease found only in a joint. I.e. only affects a part of the patient's body.
    b. **Oligoarticular:** Disease is two to four joints. I.e. affects two to four parts of the body of the patient.
    c. **Polyarticular:** The disease is found in more than five joints. I.e. affects more than five parts of the body of the patient.
5. **Affected Region:** The region affected by the patient will be taken in three ways:
    a. Upper limbs.
    b. Lower limbs.
    c. Spine.

In the case of inflammatory diseases they are not given the joint, only the involved region, upper limbs may be the hand without specifying (ulna, radio, Carpus and Metacarpus), illness can involve multiple regions such as it can affect upper and lower limbs or spinal column.

6. **Size of the Joints:** The size of the joints in inflammation may refer to:

   a. **Small:** Metacarpophalangeal, interphalangeal, metatarsophalangeal.
   b. **Great:** Femur, knee, elbows, ankles, shoulders, and wrists.

Here we define the basic features that will be taken into account to enable the agent to the search for inflammatory diseases.

For "non" inflammatory disease characteristics are as follows:

1. **Similar to the Inflammatory:** sex, age, type of home, number of affected joints.
2. **Affected Region:** The region in the patient shall be taken in three ways:
   a. Upper limbs.
   b. Lower limbs.
   c. Spine.
3. **Location:** In the case of "non" inflammatory diseases wanted the part of the body that is affected by the disease, categorized in the following way:
   a. **Upper Limbs:**
      i. Shoulder(s).
      ii. Elbow(s).
      iii. Neck.
      iv. Hand(s).
      v. Wrist(s).
      vi. Finger(s).
   b. **Lower Limbs:**
      i. Foot(s).
      ii. Knee(s).
      iii. Leg(s).
      iv. Hip.
      v. Ankle(s).
      vi. Thigh(s).
   c. **Spine:**
      i. Lumbar.
      ii. Dorsal.
      iii. Cervical.

Some medical symptoms and clinical symptoms are similar in different rheumatic diseases. For this reason, software agents need take several actions to reach a goal.

4. **Clinical Signs:**
   a. Conjunctivitis.
   b. Seizures.
   c. Deformity.
   d. Volume increase of joints.
   e. Digital clubbing.
   f. Lymphadenopathy.
   g. Hair loss.

   h. Cyanosis (hand) or blue hand.
   i. Convulsions (seizure).
   j. Boutonnière Deformity of Finger.
   k. Varus or valgus knee deformity.
   l. Varus or valgus ankle deformity.
   m. Ulnar deviation or ulnar drift.
   n. Cubitus varus (gun stock deformity).
5. **Medical Symptoms:**
   a. Cough.
   b. Fatigue physical.
   c. Chest pain.
   d. Morning stiffness.
   e. Joint tenderness.
   f. Xerostomia(dry mounth).
   g. Dry eyes.
   h. Pyrexia(fever).
   i. Blurred vision.

Design Software Agents

Agent Attributes and Rules of Behavior

Listed in Table 3 are a group of conditions that will allow software agents to take action, which is not enough, therefore need a group of conditions that meet to take the appropriate action is what happens if I take action A? What happens if I take action B? What happens if I take action N…?

With the action taken and to meet the rules set out in the knowledge base to identify a disease, the effectors are triggered, showing the interaction of user, the name of the disease, its features and images. For the general practitioner according to its criteria can give its opinion.

In this type of disease clinical signs, symptoms, and affected region, may be similar, which must have clearly defined rules, which are software agents that help identify correctly the condition. Then we'll show some rules to identify the rheumatoid arthritis.

Table 3. Identifying inflammatory diseases

Condition	Action
If female	To identify sex
If the age is 7 years	Set to range between 1-16
If the disease is insistent	Sets onset type insidious
If you have an affected joint	Monoarticular
If you have from two to four	Oligoarticular
If you have more than five	Polyarticular
If it occurs in the shoulder	Identifies region affected upper limbs
If it occurs in the knee	Identifies region affected lower limbs

R1: Starts with polyarticular pain.
R2: Volume increase of joints.
R3: It presents between 30 and 50 years old.
R4: It affects the wrists, metacarpophalangeal, interphalangeal joints.
R5: 20% of the cases have extrarticular disease.
R6: It affects shoulders and elbows in more than 50% of cases.
R7: The rheumatoid factor is positive in 70% to 80%.
R8: LC-reactive protein is positive.
R9: X-rays show ankylosis joints.
R10: Its elevated erythrocyte sedimentation and C-reactive protein is positive.

Table 4.

Located	Action	Goal
Upper limbs	Shoulder	Deltoid bursitis
		Rotator cuff Tendinitis
	Elbow	Epicondylitis
		Epitrocleitis
	Hand	D'Quervain tendinitis
		Digital stenosis tenosynovitis
		Dupuytren Contracture
Lower limbs	Hip	Subtrochanteric bursitis
	Knee	Injury of meniscus
		Ligaments injury
		Chondromalacia patellar
		Osgood-Schlatter disease
		prepatellar Bursitis
		Anserine bursitis
	Feet	Ankle valgus
		Plantar Fasciitis
		Flat feet
		Hallux valgus
Back	Spine	Sprain
		Scoliosis
		Spondylosis
		Hernia of intervertebral disc
		Vertebral fracture
General	--	Fibromyalgia Syndrome
		Multiple myeloma
		Febrile diseases

For Non Inflammatory Diseases

The main condition is that the disease is present without joins inflammation, the action taken by the agent is to locate the anatomical region to go to a possible goal, which is simpler to identify because they are very specific conditions of such diseases.

Table 4 shows the agent identifying a condition and taking action to reach a possible goal.

Algorithm

To classify diseases in inflammatory and "not" inflammatory, the search for Software Agents is much simpler, because that some diseases are discarded by groups thus are.

1. **Start:** First the user is asked to identify if the disease is inflammatory.
   a. If it is inflammatory
      i. Go to table of inflammatory diseases.
   b. else
      i. Go to table of Non inflammatory diseases.
2. **Input Percepts:** Calls for the data depending on whether the disease is inflammatory or not
   a. If is inflammatory then:
      i. {Age and sex.
      ii. Onset Type.
      iii. Size of joints.
      iv. Number of joints.
      v. Region affected.
      vi. Signs.
      vii. Symptoms}.
   b. Else "non" inflammatory:
      i. {Age and sex.
      ii. Onset type.
      iii. Region affected.
      iv. Part (s) concerned (s)}.
3. **State Update for Inflammatory Diseases:**
   a. To have entered the age and sex, is looked for in the Data Base diseases that match the age and sex that the user entered and stored. Result1 = Percepts1 ∩ **A,** where are the diseases recorded in the table knowledge base, percepts1 data entered above.
   b. To have entered the start type is looked for in the database diseases to coincide with the start type that the user entered and stored. Result2 = percept2 ∩ **A**.
   c. To have entered the size of joints diseases that matches the size of joints that the user entered is sought in the database and is stored. Result3 = percept3 ∩ **A**.
   d. To have entered the number of joints diseases that matches the number of joints that the user entered is sought in the database and is stored. Result4 = percept4 ∩ **A**.
   e. To have entered the affected region seeks a disease that matches the affected region that the user entered in the database and saved. Result5 = percept5 ∩ **A**.
   f. To have admitted the sign or signs is sought in the database diseases matching the sign or signs that the user entered and stored. Result6 = percept6 ∩ **A**.

g. To have admitted the symptom or symptoms database is search diseases that match the symptom or the symptoms that the user entered and stored. Result7 = percept7 ∩ **A**.

Thus the sum of partial results is: Partial Result (PR)

PR = Result1 ∩ Result2 ∩ Result3 Result4 ∩ Result5 ∩ Result7 Result6

4. **Formulate Goal:** With the partial result, it obtained a first goal, performs a preliminary identification, at the moment the software agents known to the goal to be achieved, in other words, so far has identified a group of diseases that share common characteristics.
5. **Formulate Problem:** The problem to be solved is to select only one of the Groups of diseases, whereas the rules established for each of the diseases, we find differences that allow us to establish a new state that leads to a new target.
6. **Plan:** We will establish a plan to select one of the diseases of the identified group. What happens if I take action A? Or, if I take action B?

If taken to action leads to a goal to which we do not know if it is correct. This forces us to create a search-space. (Nilson, 2004)

7. **Search:**

```
while (plan not empty) do {
if (agree state == true) then{
If there are similarities with the provided data (state) establishes an action
leading to a recommendation, by selecting an option in the space of search
(search-space), considering the rules that are specific to each disease (these
are those that allow us to do a proper identification). Emits a response
through the effectors (user interaction) to the environment (general doctor,
patient) environment
If one accepts the answer ends.
}
Else {
If does not return to do a new search
It establishes a new plan calling for new data (creates a new state) to have
options to generate new actions that would lead us to a new target from the
selected groups of diseases (remainder), in every new search, it will decrease
the (remainder) until exhausted. }
Output Action
If there are no more actions to take, displays message that does not have suf-
ficient information to identify a disease of this kind.
end
```

Description of Medical Identification System of Rheumatic Diseases (MISRheD)

Clinical Method

In order MISRheD can correctly locate a disease, it is necessary that the information is provided by a general doctor, due to the account with the basic knowledge of medicine. Another person without this knowledge can hardly reach the desired goal.

First there is an interaction between the doctor and the patient to collect information to determine the clinical semiology considering the following factors:

- **Background Inherited Family:** Diseases of the connective tissue, urinary lithiasis, hyperuricemia, psoriasis, uveitis.
- **Particular Personal History:** Urinary lithiasis, hyperuricemia, adverse obstetric events, thrombotic, skin diseases, thyroid disease, uveitis events.
- **Evolution:** Episodic, intermittent, continuous, all types.
- **Joints Physical Examination:** Complete, orderly, symmetric, comparative.
- **Physical Examination:** Deformity, increased volume, functional limitation, snapping, painful points.

Below are the basic information needed from MISRheD to identify the following diseases.

Procedure for Identification of: Rheumatoid Arthritis

- **Female sex:**
  - Age 42 years
  - Inflammatory classification
  - Onset type: insidious
  - Number of joints 5
  - Region affected upper limb
  - Size of small joints
  - Radiological study erosion articulate with decrease of space joint
  - Study laboratory Hematology thrombocytosis, rheumatoid factor immunological studies 75%
  - Procedure for identification: spondiloarthritides
- **Male:**
  - 17 Years old
  - Inflammatory classification
  - Onset type: insidious
  - Number of joints 2
  - Region affected lower limb
  - Size of large joints
  - Radiological study of pelvis: sacroiliitis
  - Laboratory study: histocompatibility HLA-B27 antigen
  - Uveitis

Procedure for identification of: Gout

- **Male:**
    - Age 47 years
    - Inflammatory classification
    - onset type: Acute
    - Number of joints 1
    - Region affected lower limb
    - Size of small joints
    - Radiological study: without bone injury
    - Laboratory study: hyperuricemia
    - Previous urinary lithiasis

User Interaction

Description of the Templates for The capture of Data Description Templates

The template in Figure 1 shows the capture of data to select age and sex of the patient. The Figure 2 template captures data on the number of damaged joints. The template in Figure 3 captures clinical signs of the patient. The template in Figure 4 captures symptoms of the patient. Finally, in the template shown in Figure 5, the general practitioner can select the affected body region.

Figure 1. Templates to select age and sex of the patient

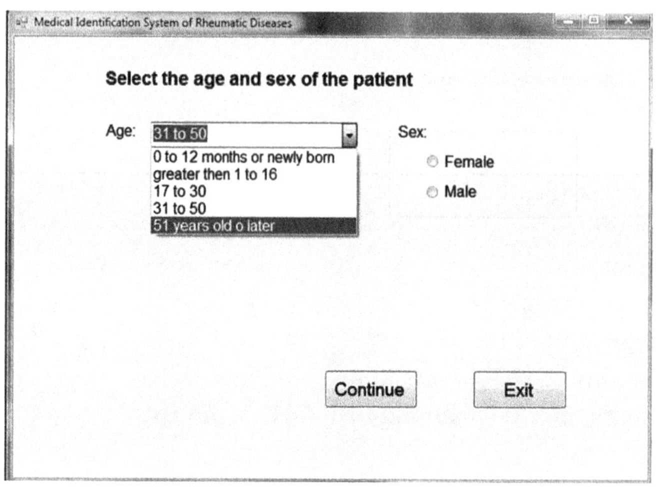

Figure 2. Templates to select the number of joints

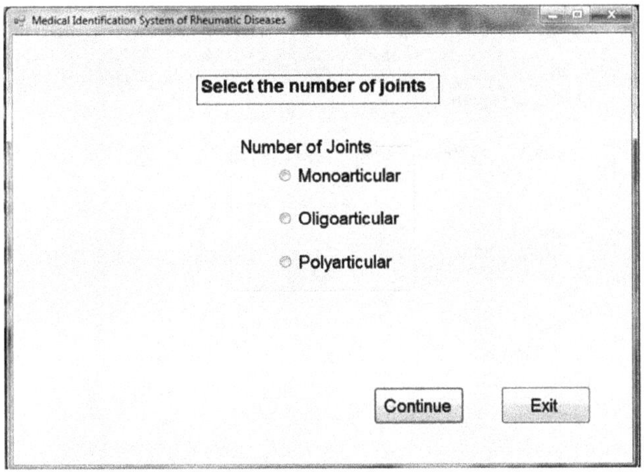

Figure 3. Templates to select clinical signs

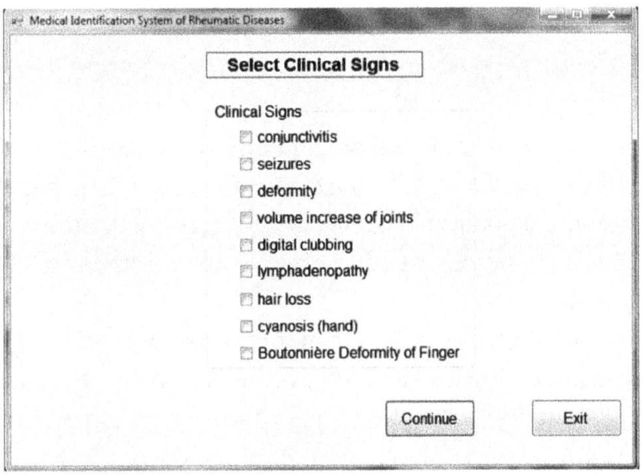

Figure 4. Templates to select symptoms

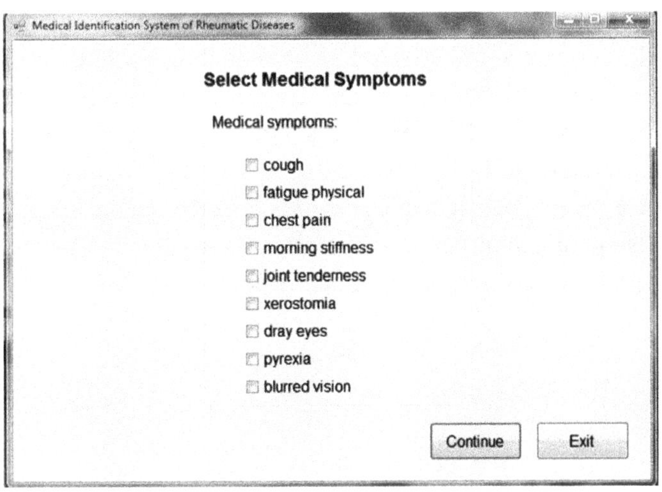

Figure 5. Templates to select region of the body

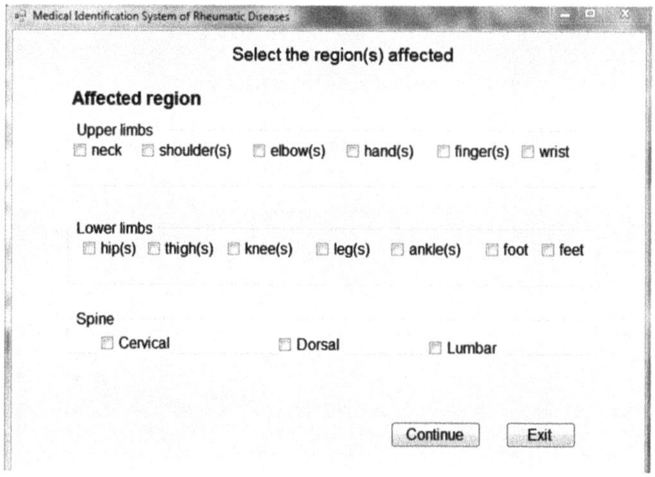

SOLUTIONS AND RECOMMENDATIONS

Information to the Population and General Practitioners

Information campaigns and efforts from Government employ an appropriate strategy to the culture of the population. For example in our country, Mexico, the majority of the population is not accustomed to read texts nor to attend talks on cultural themes being these media the principal for disseminating information, in addition there are no records of formal studies of the results and impact obtained by these means of dissemination.

General practitioners who have years on practicing commonly received little information on rheumatic diseases, there is an especial need to offer them courses through the medical institution to which they belong. In this way it can improve its approach to identify a rheumatic condition to ensure that the patient receives timely specialized medical attention. As mentioned above the general medical depends on the detection and prompt attention, which has become a vital part to address the problem in time.

Modifying the culture of the population in terms of popular beliefs of the rheumatic diseases are cured with home remedies, also such diseases are not curable, only controllable. To alleviate ailments people fall into apathy and let the time pass to receive adequate professional attention.

On the economic side, companies are affected by issued labor disability and pensions granted to its staff, and that is why they should collaborate with the public health institutions requesting that the doctors could detect in time and control these diseases, to avoid large amount of costs that may be unnecessary.

In the social aspect personal disability causing these diseases, becomes a serious public problem that has been increasing in recent years, when a person is affected during his productive age affects also greatly to the family. Give people in productive age alternative effective for the control of the disease, so that in this way they prolong their lives economically active.

In Educational Institutions

Medical schools promote among their students the training of medical specialists in rheumatology to meet the demand of the population and to comply with the ratio of inhabitant/doctor (50/1) that in our country is not enough, in relation to Latin American countries such as Cuba (170/1).

The medical students in their courses receive information and expertise in the specialty of Rheumatology. Encourage the use of new technologies as alternative didactics, in our country this is an educational problem for the lack of resources and training to the medical teaching in the application of teaching tools.

In computer schools that offer courses on information technology, computer and related, promote the development of applications that focus to the medical area to address some of the demanding problems in our country in students and teachers. Projects are being implemented in conjunction with specialists from the sector health.

Physician Rheumatologist

Our country usually the physician rheumatologist provides medical care in his Office and very few are involved in research projects.

It is recommended that they are supported by government agencies or of private initiative for research activities. As mentioned in previous subjects nowadays still much is unknown with what respect to rheumatic diseases. On the other hand medical specialists are encouraged to increase the participation in events that allow the publication of information in books of texts, journals, magazines, proceedings, etc.

FUTURE RESEARCH DIRECTIONS

Perspective

In the future the new trends are to deploy applications using new information technologies in the medical sector to enable decision-making based on the knowledge of specialists and not only information, but complementing information and knowledge to establish a process of reasoning that leads to a certain diagnosis, we know that it is a major challenge for developing countries, because general practitioners do not have the culture of the use of tools of this type and on the other hand the conjunction medical specialist and ICT sector is difficult, because the physician has no researcher training. In the future, the narrow health gap and ICT should be closed.

The social aspect has lagged behind in the development of the technology because it has focused on productivity, development of quality and economic improvements with cost-cutting in enterprises and the labor and productive sector.

From the technological aspect, it is not enough to develop applications to solve economic or business issues, the social aspect is very important, we know that public demand increasingly focuses on a better quality of life and better health services. In this chapter, we propose a way of decreasing the costs involved in attending to an employee with a rheumatic disease, (payments by inability, absenteeism, and lower productivity) that may temporarily or permanently disabled or improve care to an older adult, and as mentioned that these diseases affect a large population of this type. Using MISRA we propose new solutions based on technologies that will improve social and health services to treat a patient in time and take effective control of their disease.

This is neither the only solution nor the only tool, but new applications must be developed where The ICT sector is highly present in this context.

Viability

The viability of this proposal is feasible from the economic point of view, considering that the majority of medical institutions have computer equipment, for the implementation of MISRA requirements is minimal. The technical aspect was discussed in the model and how much the attitude of the physician to accept this proposal was discussed in the paradigm.

Model

On the way in which we are proposing the use of the model goal-based agents related to the medical area, can be very useful for students of computer science with application in artificial intelligence that helps to understand the algorithm and the structure of software agents and allows them later to develop such applications. For the medical students it can show them how the ICT can help them in their professional development. Use of ICT in their professional development, as well as software developers are to be considered an alternative resource for medical researchers.

Paradigm

In our country along the lines of others, to break paradigms and traditions is very difficult, the predominant practice of medicine used to make diagnoses in the presence of the patient, through questioning and exploration complemented by studies of laboratory and Cabinet, using clinical semiology traditional. Not relies on the computing resource, whether an expert system, computerized analysis of images, etc. because there isn't this resource implemented in our public health system. Public health institutions do not have this medium. The technological advancement of today's world requires that all areas of human activity are updated and this includes the medical work. It is desirable that the health institutions have computer hardware in their areas of medical care and that also to have computer tools to facilitate the diagnosis of common and frequent diseases in the various areas of medicine.

Some public institutions of health and in the majority of private institutions electronic formats are used to take control of the record of the patient, consultation to pharmaceutical dictionaries, existence of medicines at the pharmacy of the institution itself, but not on computer auxiliary diagnosis processes. This tells us that in the future the general practitioner can accept computer programs to improve their job performance and to enable them to make a reliable diagnosis with high expectations of accepting a new paradigm.

Future Research

A proposal for future research opportunities using software agent in medicine is the application:

Dermatology

The proposed algorithm can be applied wit the next similarity; it's required to create a data bank, which contains the information of general data of the patient (age, gender), visual data (rash, grain, color, anatomic location, size, shape and so on), clinical signs, symptoms, and descriptive images of several sufferings. The human agent (dermatologist) requirements are: provide the information to elaborate the structured questions, the initial evidences, rules and goals to that it is desired to arrive.

The software agent is required of a tables group, as they are enlisted next:

- Table of questions,
- Table of initial evidences,
- Table of conditions,
- Table of goals,
- Table of rules,
- Table of actions,
- Table that shows the pre-diagnosis.

The difference for these software agents is the use of a Bank of images for general practitioner so they can compare the injury of the patient with the disease that is shown in the image and software agents compare these features to perform the search.

Physical Medicine

The goal of software agents is to suggest to the general practitioner when he should send the patient for attention by the specialist in physical medicine and to receive physical rehabilitation therapy. Physical rehabilitation is considered as alternative medicine, for this reason not all doctors resort to it. The areas of medicine that frequently require management by physical medicine as support are Traumatology and orthopedics, Neurology and Rheumatology, however other areas can benefit from its use. An algorithm with the characteristics of the disease that afflicts the patient can achieve a timely delivery to a rehabilitative treatment to enable the recovery of a patient.

Similar to the previous proposal, apply the same algorithm with appropriate amendments, in this case the treatment received by the patient (cold, ultrasound, heat, infrared, hydrotherapy, physical exercise) that the patient needs is going according to the conditions of the joints or the affected part of the body.

The development of a medical diagnosis requires the mental processing of data provided by the patient, the exploration the doctor will take and support studies. For this reason any area of medicine could benefit from the creation of tools similar to which is intending in this chapter.

CONCLUSION

The use of CIT as a tool to solve social and health problems, leads to the following:

1. Create multidisciplinary Workgroups. General physician, Rheumatologist, CIT's engineering, students that still are in the formative stages of engineering.

2. The project development and its completion through MISRheD, did not require many financial resources, most valuable resource was the human resource.
3. Takes time to develop such tools, for approximately two years we collected information from the experience of the rheumatologist, with approximately thirty years of professional experience.
4. The shortage of rheumatologists at our country indicates that few people have the knowledge and experience in this specialty, moreover, is a medical area where still much is unknown, MISRheD is a way of preserving part of this experience and transmitting it to the public health community.
5. There is a large backlog in the use of CIT in the institutions of public health in our country; there is the possibility that MISRheD will support telemedicine to apply it in places where there are no rheumatologists.
   The algorithm may be useful for general practitioners, professors, and computer science students, as an assistance that would help them to solve similar problems
6. This tool is an effort between researchers, doctors and students to improve the patients in health care, which are invited to participate in a future in collective projects to solve society's problems.
7. Antirheumatic drugs are needed to control them, such as: Nonsteroidal Anti-Inflammatory Drugs, Corticosteroids, Disease Modifying Anti-Rheumatic Drugs, Slow-Acting Antirheumatic Drugs and Biological Therapy.

REFERENCES

Bertalanffy, L. (2011). Teoría General de los Sistemas. Ed. Fondo de Cultura y Económica.

Braun, J., & Sieper, J. (2004). Biological therapies in the spondyloarthritidesthe current state. *Oxford Journals Medicine & Health Rheumatology, 43*(9). Retrieved august 31, 2014, from http://rheumatology.oxfordjournals.org/content/43/9/1072.full.pdf+html

Cañete, J. (2008). *Manual S.E.R. de las Enfermedades Reumáticas. 3a edición. Sociedad Española de Reumatología*. Editorial Médica Panamericana.

Carbonel, J. (2006). *Semiología de las enfermedades reumáticas*. Madrid: Editorial Médica Panamericana.

Clauw Daniel, J. (2005). *Fibromyalgia and Other Central Pain Syndromes* (1st ed.). Lippincott Williams and Wilkins.

De León, M. (2012). El aprendizaje de las enfermedades reumáticas desde una perspectiva tecnológica. *Revista cubana de Reumatología, 15*(19).

Horn, W. (1999). *Artificial intelligence in medicine*. Joint European Conference on Artificial Intelligence in Medicine and Medical Decision Making, AIMDM'99, Aalborg, Denmark.

Kats, W. A. (1977). *Rheumatic Diseases: diagnosis and management*. Lippincott.

Kolkata, A. (2012). *Biologics in Rheumatology 2012. Medicine Update 2012 Vol. 22*. Retrieved august 30, 2014, from http://www.apiindia.org/pdf/medicine_update_2012/rheumatology_05.pdf

Lee, C., & Wang, M. (2006). *An Ontology Based Intelligent Agent for Respiratory Waveform Classification.* Advances in applied artificial intelligence: 19th international conference on industrial and engineering applications of artificial intelligence and expert systems, IEA/AIE, Annecy, France. doi:10.1007/11779568_131

Martins, V. (2010). *Validation of an educational game for teaching cardiovascular assessment.* Investigación y Educación en Enfermería Colombia.

Medical Dictionary. (n.d.). Retrieved august 12, 2014, from http://medicaldictionary.thefreedictionary.com/Antirheumatic+Drugs

Monteagudo, P. (2003). *Software educativo para la enseñanza de la semiología clínica del sistema respiratorio.* (Master's Thesis). Ciudad de La Habana: ISCM de La HabanaCentro de Cibernética Aplicada a la Medicina (CECAM).

Nilson Nils, J. (2004). *Inteligencia artificial una nueva síntesis*. Mc GrawHill.

Noachtar, S., & Peters, A. (2009). *Semiology of epileptic seizures: A critical review*. University on Munich Germany.

Parapar. (2007). Manifestaciones oculares de algunas enfermedades reumáticas en el niño. *Hosp. Pedro Borras*. Retrieved April 12, 2014, from http://www.portalesmedicos.com/publicaciones/articles/634/1/Manifestacionesocularesdealgunasenfermedadesreumaticasenelnino.html

Pressman, R. S. (2002). *Ingeniería del Software. Un enfoque práctico*. Madrid: McGrawHill.

Rejón, C. (2013). *Logic structure of clinical judgment and its relation to medical and psychiatric semiology.* U.S.A. Pub med / U.S. National Library of Medicine.

Resconi, G., & Jain, L. (2004). *Intelligent agents Theory and applications*. Springer Verlag.

Research Unit of the Spanish Society of Rheumatology. (2013). Spanish registry of adverse events of biological therapies in rheumatic diseases. *Biobadaser*. Retrieved August 12, 2014, from http://biobadaser.ser.es/biobadaser/eng/docs/SER_informe_web2012_ENG.pdf

Rich, E., & Knight, K. (1991). *Artificial Intelligence*. Mc GrawHill.

Russell, S., & Norving, P. (1996). *Inteligencia artificial un enfoque moderno*. México: Prentice Hall Hispanoamericana, S.A.

Salter, B. (n.d.). *Nonarticular Rheumatism in Textbook of Disorders and Injuries of the Musculoskeletal System* (3rd ed.). Lippincott Williams.

Saravolats, L. (2007). *The role of Medical Semiology in clinical infectious diseases: Back to the basic*. Lippincott Williams & Wilkins, Inc.

Schumacher Ralph, H., Klippel John, H., & Robinson Dwight, R. (1988). *Principios de las Enfermedades Reumáticas (12th ed.)*. Arthritis Foundation.

Šenolta, L., Vencovskýa, J., Pavelkaa, K., Ospeltb, C., & Gayb, S. (2009). Prospective new biological therapies for rheumatoid arthritis. *Autoimmunity Reviews, 9*(2).

Stone, J. (2009). *Pearls and Myths in Rheumatology*. London: Springer Dordrecht Heidelberg.

Taranta, A., & Markowitz, M. (1981). *Rheumatic fever: a guide to its recognition, prevention, and cure*. MTP Press.

Wright, V., & Harvey, A. (1988). *Ilustraciones diagnósticas en reumatología: Pruebas de autoevaluación*. Interamericana McGraw Hill.

ADDITIONAL READING

Bichindaritz, I. (2010). *Computational Intelligence in Healthcare*. Springer Verlag. doi:10.1007/978-3-642-14464-6

Brahnam, S., & Jain, L. (2001). *Advanced computational intelligence paradigms in healthcare. Virtual reality in psychotherapy, rehabilitation, and assessment*. Berlin, Heidelberg: Springer.

Brooks, F. P. (1995). *The Mythical manmonth, Essays on Software Engineering*. AddisonWesley Pub Co.

Combi, C., Shahar, Y., & AbuHanna, A. (2009). *Artificial intelligence in medicine*: 12th Conference on Artificial Intelligence in Medicine, AIME 2009, Verona, Italy, July 1822, 2009: proceedings. Berlin; New York: Springer.

Dar E. and Ghani M. (2010). Utility Based Agent for Test Paper Generation. International Journal of Control, Automation, and Systems Vol. 1, No. 2, June 2003 ACM Press.

Englebardt, S., & Nelson, R. (2002). *Health care informatics: an interdisciplinary approach*. USA: Mosby.

Håkansson, A. (2009). *Agent and multiagent systems: technologies and applications*: third KES international symposium, KESAMSTA 2009, Uppsala, Sweden, June 35, 2009: proceedings. Berlin: Springer.

Hakim, L. (2007). *Web mobilebased applications for healthcare management*. Hershey, Pa., USA: IGI Global. doi:10.4018/978-1-59140-658-7

Kim D. (2003). *Modeling and Design of Intelligent Agent System*. International Journal of Control, Automation, and Systems Vol. 1, No. 2, June 2003.

Lim, C. T., & Goh, J. C. (2009). 13th International Conference on Biomedical Engineering: ICBME 2008 36 December 2008, Singapore. Berlin: Springer.

McCarty D. & Koopman W. (1993). Arthritis and allied conditions: a textbook of rheumatology Lea & Febiger.

Moreno, A., & Nealon, J. L. (2003). *Applications of software agent technology in the health care domain*. Basel; Boston USA: Birkhauser Verlag. doi:10.1007/978-3-0348-7976-7

Müller, J., Singh, M., & Wooldrige, M. (2000). *Intelligent agents: agent theories, architectures, and languages*. Berlin: Springer.

Padgham, L., & Winikoff, M. (2004). *Developing intelligent agent systems: a practical guide*. Chichester: Wiley. doi:10.1002/0470861223

Resconi, G., & Jain, L. (2004). *Intelligent agents: theory and applications Berlin*. Springer. doi:10.1007/978-3-540-44401-5

Riaño, D. (2008). *Knowledge management for health care procedures: from knowledge to global care, AIME 2007 Workshop K4CARE 2007, Amsterdam, The Netherlands, July 7, 2007*. Berlin, New York: Springer. doi:10.1007/978-3-540-78624-5

Salter, R. B. (1999). *Textbook of disorders and injuries of the musculoskeletal system: an introduction to orthopaedics, fractures and joint injuries, rheumatology, metabolic bone disease, and rehabilitation*. Philadelphia: Lippincott Williams & Wilkins.

Sarker, R., & Ray, T. (2010). *Agentbased evolutionary search*. Berlin: SpringerVerlag. doi:10.1007/978-3-642-13425-8

Silverman, B. G. (2005). *Intelligent paradigms for healthcare enterprises: systems thinking*. Berlin, New York: Springer. doi:10.1007/b99809

Sordo, M. (2010). *Advanced computational intelligence paradigms in healthcare*. Berlin, Heidelberg: Springer.

Tan, J. (2005). *Ehealth care information systems: an introduction for students and professionals*. San Francisco: JosseyBass.

Tsihrintzis, G. A. (2010). *Intelligent interactive multimedia systems and services*. Berlin, Heidelberg: Springer. doi:10.1007/978-3-642-14619-0

Wickramasinghe, N., Gupta, J., & Sharma, S. (2005). *Creating knowledgebased healthcare organizations*. Hershey Penns, USA: Idea Group Pub. doi:10.4018/978-1-59140-459-0

Yosida, H. (2007). *Advanced computational intelligence paradigms in healthcare. BerlinNew York*. Springer. doi:10.1007/978-3-540-47527-9

Yuan, S., & Yokoo, M. (2001). *Intelligent agents: specification, modeling, and applications*. Berlin: Springer. doi:10.1007/3-540-44637-0

KEY TERMS AND DEFINITIONS

Agents Based on Goals: The goals are to itemize the situations to that it is desired to arrive. The agent program can combine goals with actions and in this way, be able to choose those actions that will allow reaching the goal.

Antirheumatic Drugs: Antirheumatic drugs are drugs used to treat rheumatic diseases. The major classes of antirheumatic drugs include Nonsteroidal Anti-Inflammatory Drugs (NSAIDs), Corticosteroids, Disease Modifying Anti-Rheumatic Drugs (DMARDs), Slow-Acting Antirheumatic Drugs (SAARDs).

Biological Therapy: Treatment to stimulate or restore the ability of the immune (defense) system to fight infection and disease.

Clinical Semiology: In medicine, the clinical semiology is the body of knowledge that deals with the identification of the various pathological manifestations (symptoms and signs) and compiling them into syndromes and how to interpret.

Disease Modifying Anti-Rheumatic Drugs (DMARDs): DMARDs influence the disease process itself and do not only treat symptoms, hence their name. DMARDs also have anti-inflammatory effects, and most were borrowed from the treatment of other diseases, such as cancer and malaria.

Formulate Goal: To set a model for an intelligent agent due to define to achieved the goal, generating new states and testing them against a goal.

Inflammatory: Term to differentiate the types of rheumatic diseases. This is an inflammatory process that occurs in joints. Regardless of nerves or ligaments as prepatellar bursitis, it is the inflammation of the bursa that is positioned in front of the kneecap (patella).

Medical Signs: Term to identify the types of rheumatic diseases. The physician makes his judgments from the information that he gathers during a physical examination of a patient (conjunctivitis, seizures, deformity, volume increase of joints, digital clubbing, among others).

Medical Symptoms: Term to identify the types of rheumatic diseases. They are abnormal perception of a patient (cough, fatigue physical), chest pain, morning stiffness, joint tenderness, etc.).

Nonsteroidal Anti-Inflammatory Drugs (NSAIDs): Drugs belonging to this class bring symptomatic relief of both inflammation and pain, but have a limited effect on the progressive bone and cartilage loss associated with rheumatoid arthritis. They act by slowing the body's production of prostaglandins.

Recommendation: Filtered information is displayed to the user in order make decisions. Software agents suggest rheumatic disease.

Remainder: Selected information group, which matches the search. Rule out disease that does not match your search.

Rheumatic Diseases: They are systemic diseases of unknown origin that happen when the immune system goes haywire. The immune system attacks the linings of joints, ligaments, tendons, muscles, skin, eyes, cause inflammation, swelling, and pain. There are more than 100 rheumatic diseases.

Search Space: The group of records in a database that were selected according to the search that matches the goal that you want to reach (identify rheumatic diseases).

Slow-Acting Antirheumatic Drugs (SAARDs): A special class of DMARDs and the effect of these drugs is slow acting and not so quickly apparent as that of the NSAIDs.

Chapter 7
Improving Pharmaceutical Care through the Use of Intelligent Pharmacoinformatics

Tagelsir Mohamed Gasmelseid
International University of Africa, Sudan

ABSTRACT

The expansion of drug-related problems urged healthcare organizations to adopt Pharmacoinformatics to signal, analyze and report Adverse Drug Reactions (ADRs). Data for this study have been compiled from local and international sources such as WHO. The study resulted into the development of an intelligent multi-agent decision support system including a process model, a multi-agent architecture and an integrated data processing model with clear description of agent functionalities. The model reflects three main modules: a data capture and update module, diagnosis module and a pharmaceutical care and drug monitoring module. The study also reflected on the practical and managerial environment of the model and the basic considerations to be taken into account. The study also provided some important recommendations.

INTRODUCTION

The use of information systems for the improvement of healthcare is gaining paramount importance. They increase information accessibility to healthcare providers (Makus, 2001) and increased convenience for patients (Fitzgerald, Piris & Serrano, 2008; Glaser & Foley, 2008). Their deployment by pharmaceutical care assists in the assessment and management of therapeutic outcomes in patients as well as in detecting, signaling, evaluating, and solving potential and actual drug-related problems (including adverse drug events or drug interactions). Within the context of a wider hospital management system, the use of information systems to assist in pharmacy-related decision making is known as "pharmacoinformatics". In addition to databases, such systems make use of different technological settings (including informatics, the internet, and interactive technologies) to assist in improving pharmaceutical care, patient safety and enhancing hospital management processes. They aim at improving the capacity of clinical practitioners

DOI: 10.4018/978-1-5225-0248-7.ch007

to efficiently acquire and develop new treatment strategies. The use of such systems by practitioners tends to focus on three main areas: facilitating information exchange, detecting and managing adverse drug events and enabling the supply chain management process.

However, despite the potential outstanding benefits to be gained by hospitals as a result of the deployment of powerful pharmacoinformatics applications, the realization of such applications is still limited. Currently, pharmacoinformatics applications are used as sub-modules of the corporate information system. Their roles tend to be limited to stock control, monitoring drug availability and issuance at outpatient and ward pharmacies. Even for stock control, there seems to be no emphasis on the use of electronic ordering and procurement processes for which no standard operating procedures exist. The analysis of drug therapies and management of prescription inconsistencies are not supported by the current pharmacoinformatics applications in use. Moreover, there is a considerable lack of emphasis on developing and deploying information systems for signaling and detecting adverse drug events manifested in patients and recorded by healthcare professional. While the role of Pharmacovigilance centers in analyzing and managing adverse drug reactions is fundamental, there is a considerable lack in using information systems to establish relationships between such centers and hospitals through which adverse events can be reported and tracked. As a result, hospitals tend to miss the technical support provided by national Pharmacovigilance centers.

Therefore, the aim of this project is to develop a reference multi-agent pharmacoinformatics model to assist in detecting adverse drug events and drug interactions, as well as improving hospital management. The model will be based on the use of multi-agent technologies which has proven to be suitable for tackling large, real-world problems (Gasmelseid, 2006).

1. PROJECT'S OBJECTIVES

This project highlights the different Pharmacoinformatics channels that have been used (by both healthcare professionals and patients) in hospital management and the provision of pharmaceutical care which involves identifying, solving, and preventing potential or actual drug-related problems (DRPs) with regards to a patient's drug therapy.

Despite the fact that there are a number of different approaches to the development of Pharmacoinformatics applications used for decision support, there is a general agreement that decision support systems are developed most successfully by an interactive, prototyping approach. The suitability of this approach for this project stems from the fact that complete information requirements of stakeholders are difficult to identify in advance and may change significantly over time. To achieve the aims of this project emphasis will be made on:

1. The articulation of the issues which can potentially affect hospital management processes and the quality of pharmaceutical care in public hospitals in pursuit of providing insights into how Pharmacoinformatics can potentially affect the future of healthcare.
2. The development of a process mode (including all decision models and processing functions relevant to the organization as a base for the transformation of data into information) and
3. The development of a data processing model (establishing relationships between the system's structure and dynamics.

2. PROJECT'S SIGNIFICANCE

The significance of the study originates from the following considerations:

1. This study deals with the context of healthcare with emphasis on the improvement of pharmaceutical care processes as basic elements in the delivery of improved healthcare services.
2. The growing importance of the deployment of advanced information systems in response to the fundamental changes exhibited in the healthcare system and its subsystems. The growing incidence of medical errors and adverse drug events, the complications of communications among medical professionals in many hospitals (mainly physicians and pharmacists), and the escalation of drug procurement costs, are signs that motivate the emphasis on pharmacoinformatics.
3. The emergence of new business models in response to technological advancements, increasing use of medical informatics and communication systems and technologies.

3. METHODOLOGY

The methodology for this study is generally a descriptive analytical survey with both inductive and deductive methods applied. Due to the nature of the study a number of other methods and empirically driven theories capable of dealing with the topic in question, both quantitatively and qualitatively, could also be used. The use of several methods and instruments of research is to allow some flexibility in order to capture the specific relations between the main variables and the factors affecting them. To account for the diversity of information across the different managerial landscapes and to ensure the validity of the instruments of research, Anthony's taxonomy of managerial levels is adopted. The project also made use of the concepts of "information modeling" and agent oriented software engineering methodologies and other "process-centered" and "resource-oriented" approaches for the articulation of variables and the development of the entire model and its corresponding modules.

The majority of the project's data is collected from electronic sources of many healthcare organizations and regulatory agencies such as the National Pharmacovigilance Center at the Food and Drugs Authority (FDA) at the Ministry of health. The study also used a variety of tools and methods of analysis of data provided by international organizations such as WHO. The tools include data matrix, tables, diagrams, models and output of computer programs. To achieve the aims and objectives of this study, data has been compiled from the use of questionnaires and interviews with key personnel in different healthcare organizations across the kingdom. In order to reduce the variance of estimators and gain sampling precision, the technique of optimum allocation has been used.Despite the fact that a wide number of approaches can be used for the development of agent-oriented information systems, the multi agent distributed reference framework proposed in this study follows TROPOS agent oriented software engineering methodology. To improve the face validity of the methods of research as well as the outcomes of the study, close consultation and review for the whole project is done by different experts in many international universities.

4. REPORT STRUCTURE

This report falls in nine main parts. Part one through part five provide a basic introduction about the main thread of the project and the context of problems to be investigated, project objectives, significance of the project, methodological framework and report structure. Part six is devoted for the development of a conceptual framework guided by a synthesis of related work. While part seven shows the description of the proposed pharmacoinformatics reference model, part eight and nine provide some discussion and recommendations.

5. SOFTWARE AGENTS

Software agents' technology is witnessing an increasing deployment in different sectors such as healthcare, e-business, environmental management, entertainment, telecommunications, e-banking and e-learning, among others. There is a general consensus that software agents interact autonomously and share resources to achieve objectives assigned to them by their users and/or other agents, they also operate individually to achieve personal objectives. Resource sharing and objective realization usually take place within the domain of a specific organizational structure. Because there is no agreement about what constitutes "agency", the conceptualization of agent-based applications and modeling tend to be done by focusing on the qualities possessed by software agents such as autonomy, learning, intelligence, proactivity, reactivity and social ability, among others. Moreover, such qualities-based characterization of agents has been used to develop a typology for agent types as mobile, interface, information, broker, recommendation agents, among others. Software agents' technology has been widely used to perform different tasks and automate different processes in e-commerce, environment, e-banking, airlines, communication and industry.

Software agents are software entities that perform some tasks on behalf of their users, other agents or programs with some degree of autonomy using the appropriate information and communication platforms. Their roles include task delegation, users training, event monitoring, information search, matchmaking and filtering (Gasmelseid, 2009a). They possess important properties such as autonomy, social ability, reactivity and pro-activeness, learning, mobility, benevolence, rationality, independence, cooperation, reasoning, intelligence and adaptivity (Lisa et al, 2001; Persson et al, 2001; Bonarini & Trianni, 2001; Hu & Weliman, 2001). According to Wooldridge and Jennings (1995), autonomy, which represents the main quality, reflects the capacity of agents *to* operate without the direct intervention of humans or others, and have some kind of control over their actions and internal state. Social ability denotes the need that agents should be able to interact with other agents (and possibly humans) via some kind of agent-communication language. Reactivity reflects the ability of agents to perceive their environment and respond in a timely fashion to changes that occur in it. This environment may be the physical world, a user via a graphical user interface, a collection of other agents, the Internet, or perhaps all of these combined. Pro-activeness denotes the ability of agents to act in response to their environment and exhibit goal-directed behaviour by taking the initiative.

Different ways have been used to classify software agents in accordance with their agency attributes and functions. According to Ponnurangam (2005), software agents can be broadly classified into two categories: user agents (direct task-oriented) and service agents (implicit task assistance). Alternatively, agents may be classified according to the tasks they perform, their control architecture, the range and effectiveness of their actions, the degree of sophistication of their internal states, or the environment in which agents find themselves (Anumba & Newnham, 2000).

Especially for complex systems, software agents tend to be deployed in the form of a multi-agent organization in which multiple agents interact and exchange information in a decentralized and "social" manner to solve larger and complex problems. Applications such as electronic commerce, traffic control, health care provisioning, portfolio management and telecommunications, prove to be suitable for complex, distributed problems involving a multiplicity of interconnected processes whose solutions demand the allocation of fusion of information and expertise from demographically distributed sources (Sycara 1998). A multi-agent system can be defined as a collection of possibly heterogeneous, computational entities, having their own problem-solving capabilities, and which are able to interact in order to reach an overall goal (Oliveira, et al, 1999). According to Ferber (1999), a multi-agent system is composed of an environment (a space), a set of situated objects (agents) with the possibility of being associated with a position in the environment). It also includes an assembly of agents representing the active entities, an assembly of relations that link objects together. The assembly of operations it includes allows agents to perceive, produce, transform, and manipulate objects in the set of objects and operators with the task of representing the application of these operations and the reaction of the world.

According to Anumba et al (2003), there are several common ways of structuring the agent organizations:

1. **Organizational Structuring:** In which all agents share the same explicit end goal. Due to the fact that only one agent possess a global view of the full task, it is very difficult for peer agents with different goals to resolve their difficulties.
2. **Contracting:** Where a manager agent breaks the entire problem into component problems and then announces each task for contracting agents to make their bids. The manager then reviews the bids and awards the contract. Since contract nets are best used when the problem can be broken down via a well-defined hierarchical nature into a set of tasks, planning has to be done centrally.
3. **Multi-Agent Planning:** Which involves a central arbiter who reviews all potential plans of individual agents and resolve conflict when they exist.
4. **Peer-to-Peer Negotiation:** Where agents communicate directly with each other in the pursuit of achieving both individual and collective goals.

Multi-agent concepts incorporate technological, organizational and decision making issues that support the realization of the objectives of collaborative work in a wider socio-organizational context. In such a context, the objectives of the collaborating parties and their tasks are represented in terms of agents. Such representation goes hand-in-hand with the measures that govern the organizational structure, division of labor and hierarchy of responsibilities at different decision making and organizational levels. Therefore, the agent organization reflects the organization of task assignments together with their related hierarchy of command and goal congruence mechanisms. The diversity of organizational settings, therefore is reflected in the type and functionalities and privileges of the agents interacting within the entire multi-agent organization. The potential and applicability of multi-agent paradigm to support collaborative work is driven by the following considerations. Especially in complex situations, multi-agent configurations facilitate collaboration and manage tasks through the shift of some collaboration tasks to agents and allowing some space for wider interactions and the production of user-created content. Throughout the collaboration process, the collaborating parties can mainstream their tasks in a functional way by using functional and task-oriented agents such as the information and interface agents without breaking organizational chains of command. They also contribute to the negotiation of tasks and the use of cooperative distributed applications.

Different software engineering methodologies have been used for the design and development of multi-agent systems such as Gaia, TROPOS, MESSAGE and Multi-agent Systems Engineering methodology (MaSE), among others (Gasmelseid, 2013). While some of the available AOSE methodologies represent an extension to and/or application of existing "conventional" software engineering methodologies (especially object oriented) to agent oriented applications, others have focused on defining a number of models that use the basic guidelines of agent theory as a guide for analyzing and designing agent oriented applications and systems (Iglesias et al, 1998; Wooldridge, et al, 2000; Odell et al, 2000; Kendall, 1999; Lind, 1999). Gaia methodology deals with both the societal (macro) and agent (micro) level aspects of information system design. The main concepts in Gaia are divided into two categories: "abstract" and "concrete". At the high level of conceptualization, "abstract" concepts involve the definition of roles including protocols, permissions, activities and responsibilities. On the other hand, "concrete" concepts put more emphasis on the articulation of agent types, services and acquaintances. The outcomes of the analysis phase are used as inputs for design phase which focuses on creating an agent-based model.

TROPOS methodology pays more attention to understanding the organizational context of the multi-agent information system by focusing on the very early phase of software development in which the system as well as it's context are studied as a larger social-technical system (Bresciani, et al, 2002; Perini, et al, 2001; Giunchiglia et al, 2001; Giunchiglia et al, 2002). It includes five phases of software development: early requirements analysis, late requirements analysis, architectural design, detailed design and implementation. Being an organizationally-oriented methodology, TROPOS tries to improve the ability of designers to justify the entire multi-agent system as well as its "traceability" dimensions by enabling artifacts produced during later phases to be clearly referred back to artifacts or requirements produced earlier. MESSAGE methodology is more oriented towards knowledge representation. Most of the knowledge level entity concepts of the MESSAGE methodology fall into three main categories: "Concrete-Entity", "Activity", and "Mental-State-Entity" (Caire, et al, 2001). The main types of Concrete-Entity are: agents, organizations, roles, and resources. The main types of Activity are: tasks, interaction and interaction protocols. The basic type of Mental-State-Entity is the goal. A goal (whether intrinsic or transient) associates an agent with a situation in terms of a rule-based utility function. An analysis model in MESSAGE is contingent upon a number of views describing organization, goal-task, agent-role, interaction, and domain. The Prometheus methodology consists of three phases: system specification, architectural design and detailed design (Padgham & Winiko, 2002a; 2002b). The system specification phase, involves two activities: determining the system's internal and external environment (in terms of percepts denoting incoming information from the environment and actions by which an agent affects its environment), and determining the goals and functionality of the system. The architectural design phase involves three activities: defining agent types, designing the overall system structure, and defining the interactions between agents. The internals of each agent and how it will accomplish its tasks within the overall system are addressed in the detailed design phase which focuses on defining capabilities, internal events, plans and detailed data structure for each agent type identified in the previous step. Multi-agent Systems Engineering methodology (MaSE) is similar to Gaia with respect to generality and the application domain supported, but in addition MaSE goes further to support for automatic code creation through the MaSE tool (DeLoach, 1999; Wood & DeLoach, 2000). It is divided into seven phases: goal capture, application of use cases, goal refinement, agent classes' creation, conversations construction, agent class assembly and system design.

Some of these methodologies have been criticized for their limited deployment due to the lack of maturity and their failure to capture the autonomous and proactive behavior of agents, as well as the richness of the interactions (Zambonelli, 2001). Current agent-oriented methodologies focus mainly on multi-agent systems analysis and design, but without providing straightforward connections to the implementation of such systems (Wooldridge, et al 2000; DeLoach, et al 2001). Some of the existing software development techniques are characterized with a fundamental mismatch between the concepts used by object-oriented developers and the agent-oriented view (Wooldridge, 1997). In particular, extant approaches fail to adequately capture an agent's flexible, autonomous problem-solving behavior, the richness of an agent's interactions, and the complexity of an agent system's organizational structures. Most of these methods feature a technology-driven, model-oriented and sequential approach for the development of multi-agent applications that may not be always adequate for the development of agent-oriented systems. Most of them assume (in advance) the suitability of multi-agent technology for the entire problem domain. While the model orientations of methodologies are obvious, the process of model coupling and integration process does not explicitly reflect the links between models (Lind, 1999). Besides as the main issues to be addressed by agent oriented software engineering methodologies (such as autonomy, reactivity, proactive-ness and social ability), the concern for mobility have been growing over time (White, 1997). In spite of the growing diffusion of mobile agent technology, little research has been done to settle "design" directions to be followed in order to determine when mobile agents are convenient to be used or not. However, the current agent oriented software engineering methodologies used for developing multi agent systems do not provide methods to determine in which cases mobile agents should be used. Many of the existing methodologies intentionally do not support intelligent agents; rather, they aim for generality and treat agents as black boxes.

Except for the case of TROPOS and to some extent Gaia, AOSE methodologies tend to focus on system design (basically definition of agents, capabilities and communication) without providing a thorough understanding of the "actual problems" facing decision makers, the causes and consequences of these problems and whether the use of multi agent systems can contribute to the solution space envisioned. Because these methodology are characterized by the limited communication between the "decision maker" and the "designer" and the fact that agent technology has already been decided in advance to be the appropriate paradigm for solving problems, the chance of developing systems on the basis of "imagined" potentials of designers is highly probable.

The use of MESSAGE methodology results into a large amount of data and demands additional processing resources (time, cost and skills) in a way that complicates the process of system design because it is necessary to undertake resource-consuming investigations of the interactions among different "views" and associating them with corporate objectives. Such a consideration may result into complexities with regards to information update and system maintenance. Because of the complex view-coupling process any mistake that exists at later stages of the system development process demands intensive rework that delays design. Although the methodology does not provide in depth emphasis on "problem analysis" and the early phases of decision making, the process of coupling problem dynamics with agent capabilities will be cumbersome. While the process of system analysis and design require extensive user involvement to enrich and enable the articulation of system linkages and agent functionalities, they validity of such methodology will be seriously threatened if such involvement is not maintained appropriately.

Prometheus methodology does not provide emphasis on the early phases of decision making (decision intelligence and problem-objective orchestration) therefore, the resulting system specifications are not directly related to the basic "information requirements" of the users who "delegate" some of their authorities to "agents. As it is the same limitation in conventional system development and engineering methodologies, the entire system dynamic will be based on the designer's perception of "what will be his information requirements if he were the decision maker for whom the entire agent module or organization provides support" rendering the entire agent-oriented applications to be "operationally" out right failures although "technically sound" and reduce their applicability with regards to the provision of appropriate decision support. Although Prometheus is regarded as supporting the development of intelligent agents in accordance with goals, beliefs, plans, and events (Padgham & Winikoff, 2002), many other methodologies treat agents as "simple software processes that interact with each other to meet an overall system goal" (DeLoach, 2001). Like other methodologies, Prometheus is envisioning implementation through the emphasis on the "automation" of the stages of the methodology rather than the "implementation" of the target system.

Although Gaia is viewed as being neutral with respect to both the target domain and the agent architecture, it does not present a procedure to guide roles and protocols identification when building agent-based systems when it is regarded as an organizational design process. While the analysis models used in Gaia methodology are based on well-defined concepts, they only represent a subset of the concepts required for agent oriented analysis with the design models remain not clearly explained. Although it has been proven as a good approach for developing closed domain agent-systems, the above mentioned restrictions delimit its applicability to open and unpredictable domain of internet applications. The methodology has been extended by adding some organizational concepts such as organizational rules, structures and patterns (Zambonelli et al, 2001).

Because the MaSE methodology is similar to Gaia with respect to generality and application domains supported, it inherits the same limitations. The "process" of using the MaSE methodology tends to be oriented towards the "tool itself" by focusing on the use of diagrams rather than focusing on the orientation of the whole process directed towards the investigation and solution of decision-making problems. On the other hand, MaSE methodology has been characterized by a long system development process that complicates the process of integration in addition to the fact that it does not support the initial phases of decision making.

Multi-agent systems have been widely used in the healthcare sector because they possess a higher capacity to address complex and distributed problems involving a multiplicity of interconnected processes, users, information sources, and locations. The development of agent organizations in the healthcare sector tend to be structured in accordance to organizational structures characterized in functional terms in which superior agents are supported by subordinate ones (e.g., information, interface, task, etc). Superior agents have the capacity to act as "task mangers" to ensure the effectiveness of the entire agent organization in the achievement of corporate objectives based on the functionalities of their corresponding subordinate agents responsible for task performance. While in some cases pharmaceutical care applications are addressed within the context of a wider hospital information system (as a sub-module), there has been a tendency to address pharmacy-related issues in separate applications, such as the use of multi-agent systems for monitoring and reporting of adverse drug events, electronic prescriptions, managing drug therapies and mainstreaming pharmaceutical procurement activities, among others.

The use of software agents in healthcare has been widely cited in e-health, electronic medical records as well as functional applications supporting medical and paramedical business units (e.g., pediatrics, internal medicine, geriatrics, pharmacy, nursing, intensive care and emergency, obstetrics, etc.) at the level of hospitals. The use of multi-agent systems in healthcare allows for the realistic modeling of complex, heterogeneous and distributed medical and healthcare systems and environments. They are used to improve the performance of a computerized system in terms of interoperability, scalability and configurability (David, David, & Antonio, 2010), information integration (Yifeng & Bhattacharyya, 2010; Kostkova, et al 2003; Bodenreider, 2004), task automation (such as scheduling of meetings, negotiation and learning (Bireshwar & Mishra, 2010; Hokyin, Minhong, & Huaiqing, 2011).

Godo et al (2003) used the multi-agent systems to monitor the prescription of restricted use antibiotics within the context of an electronic Institution (i.e., hospital) incorporating agents, roles, dialogic framework, scenes, and performative structure. The architecture included six types of (functional) agents: patient (a.k.a guardian angel), physician secretary, laboratory manger, pharmacy expert, and nurse agents. Different scenes were used to address communication among the agents such as: Patients Room, Physicians Room, Laboratory, Pharmacy, Dialog scene, Waiting Room and halls. Based on the patient information acquired from the electronic medical record, the entire agent structure focuses on three main functions: antibiogram authorization, antibiogram results and modification of the entire electronic medical record. However, despite its advantages, the architecture has a limited scope to provide comprehensive pharmacoinformatics assistance.

The use of multi-agent systems for the analysis and signaling of adverse drug events originates from the need for collaboration and exchange of complementary skills from different experts (e.g., epidemiologists, biostatisticians, pharmacists and physicians) for the analysis and interpretation of reports, collecting additional relevant information, and drawing reliable conclusions (Yanqing et al, 2005). Yanqing et al, (2007); (2010) used intelligent agents with a fuzzy recognition-primed decision model to develop a distributed adverse drug reaction detection system. They proposed a multi-agent framework for early detection of adverse drug reactions (ADRs) by utilizing electronic patient data distributed across many different sources and locations. In this framework, intelligent agents assist a team of experts based on a well-known human decision-making model called Recognition-Primed Decision (RPD). They generalized the RPD model to a fuzzy RPD model and utilize fuzzy logic technology to not only represent, interpret, and compute imprecise and subjective cues that were commonly encountered in ADRs but also to retrieve prior experiences reported by patients, physician and hospitals by evaluating the extent of matching between the current situation and a past experience.

Multi-agent systems have also been widely used for the management of supply chains using different technologies (Pathak et al. 2000; Fu & Piplani, 2000; Frey et al., 2003; Davidsson & Wernstedt, 2004; Walsh & Wellman, 2000; Lu & Wang 2008). Govindu and Chinnam (2007) proposed a generic process-cantered methodological multi-agent supply chain management framework to simplify multi-agent systems (MAS) development for supply chain applications. Shirazi and Soroor (2007) presented architecture for strategic information systems. Turcu et al. (2008) discussed the use of multi-agent systems and radio frequency identification (RFID) technologies in healthcare to track pharmaceuticals from the manufacturer, distributor, and pharmacy to the point of administering medications to the patient. Trappey et al. (2009) presented a multi-agents system collaborative production system to support the collaborative and autonomous mold manufacturing outsourcing processes. Seyed et al (2010) focused on merging remote sensing data and population surveys in large, empirical multi-agent models. Gottfried et al, (2011) combine a multi-agent framework with ontology-driven solutions to support and automate the procurement process. Gaurav Jetly et al (2009) developed a multi agent system to simulate the supply chain of the pharmaceutical industry.

Figure 1. Stakeholders representation and involvement

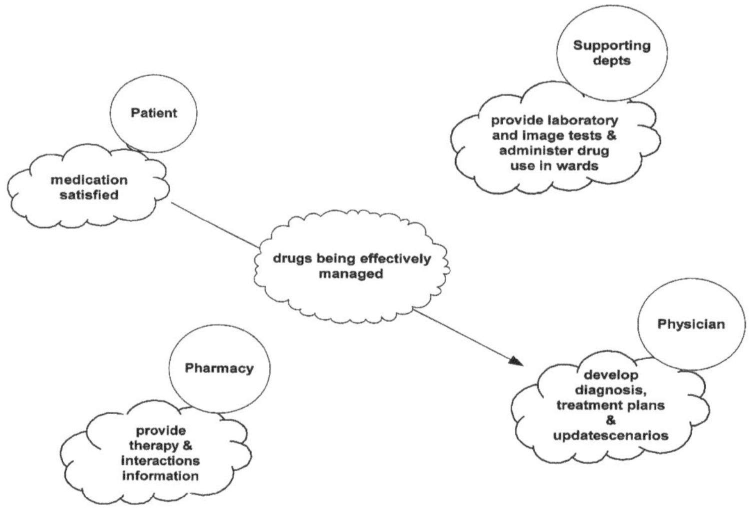

6. MODEL DESCRIPTION

Process Modeling

The use of multi-agent systems is guided by a wide range of agent oriented software engineering methodologies. Using TROPOS, Figure 1 shows the context of interaction among the stakeholders involved (directly or indirectly) in the process of pharmaceutical decision making. Patients, physicians, pharmacists and other supporting departments constitute the stakeholders whose interactions are reflected in the functionality of the multi-agent organization and its model.

Following a process-cantered approach, the flow of analysis depicted in Figure 2 reflects the set of flows that take place to facilitate communication between patients, physicians, pharmacists and other departments. The flow also depicts the movement of requests and orders in order to ensure drug availability for treatment. As shown in Figure 2, the central entity in the relational conceptualization is the drug order initiated and itemized by the physician (in the form of a prescription), hospital, outpatient or ward pharmacies. It represents an authorization to provide a specified type and quantity of a specified drug from a specific drug source to a specific user. Drug inventories exist at different locations such as hospital, outpatient and ward pharmacies and are capable of satisfying all drug orders using different drug types. The entity representing transactions that moves drugs in, out and through the entire pharmacy network is a movement or a flow. When drugs (flow) are provided against a drug order (prescription), a flow (movement) is recorded by the drug provider from a particular location (e.g., outpatient pharmacy). Should there be an excess flow (drugs) provided by mistake to a patient, it should be monitored and addressed by the "providing" location or other entities in the hospital during the medication course. This may entail another kind of movement transaction to record it. When "stock control" is carried out (with adjustments increasing or reducing the stock balances at different hospitals), an increase can be recorded as a movement to the "inventory" with no "from source of" being indicated and a decrease recorded as a movement from the "inventory" with no "to destination of" being identified.

Figure 2. Flow analysis in pharmaceutical care systems

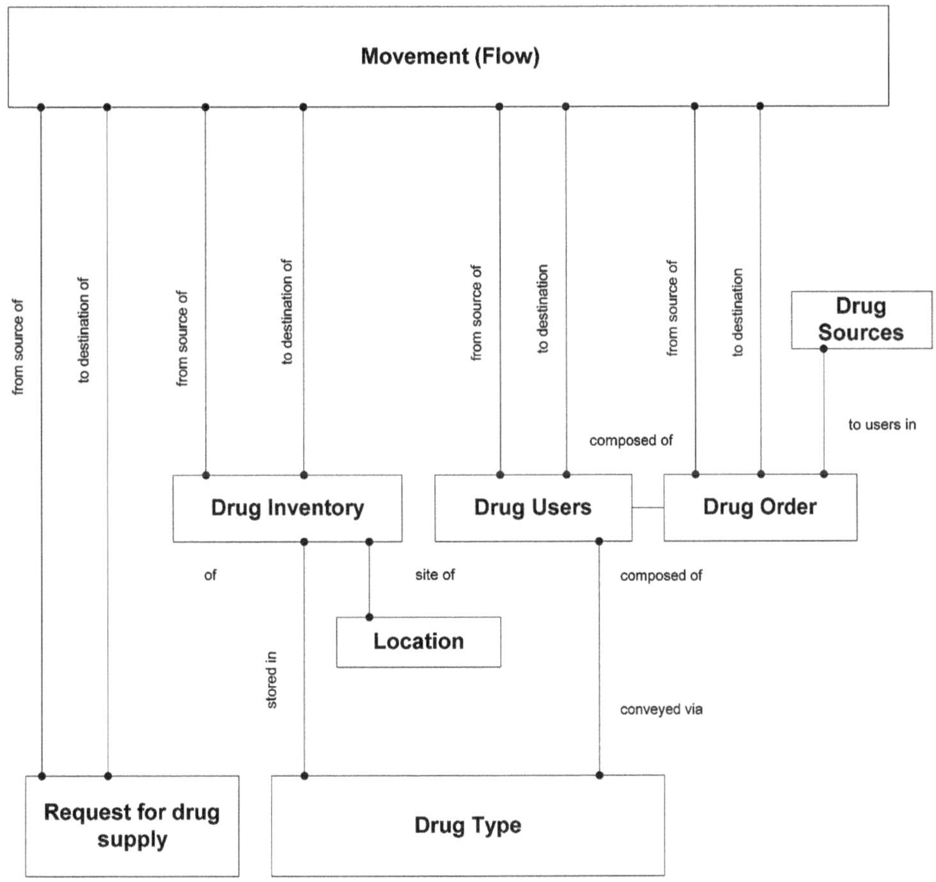

Multi Agent Architecture

The proposed multi-agent architecture follows the superior-subordinate structure that incorporates superior and subordinate functional agents. It includes interface agents (providing access to databases and knowledge-based systems), information agents (enabling communication and information exchange among health care practitioners), and functional agents (human agent assistants such as physician and pharmacist). In their functional capacity, software agents can assist in the procedure of revising a therapy (getting clinical information, deciding alternative therapies, etc.),

As shown in Figure 3, the proposed multi-agent reference pharmacoinformatics model includes two types of agents:

1. Superior agents representing health organizations (such as hospitals, Pharmacovigilance centres and Food and Drug Authorities (FDAs).. Such agents act on behalf of their organizations as well as their supporting subordinate agents assisting in the implementation of hospital-wide functions organized in business units such as surgery, paediatrics, etc. The communication process is essential for the assistance in the management of the entire health organization (such as a hospital) or coordination of health organization efforts such as Pharmacovigilance centres and FDAs. In addi-

Figure 3. Multi-agent architecture

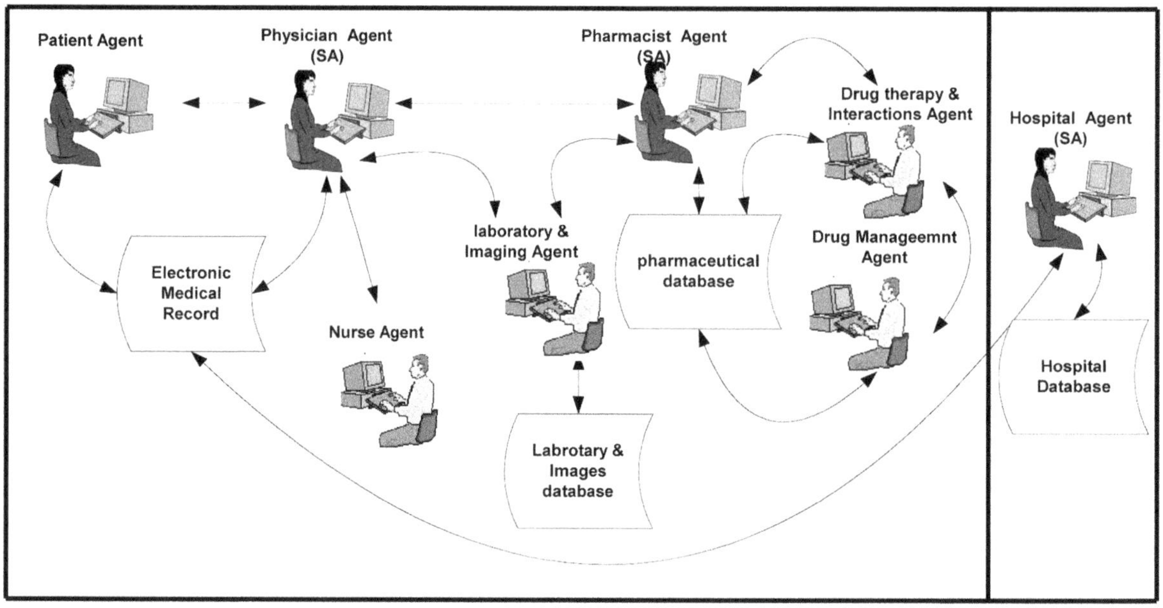

tion to the facilitation and management of internal and external communication, superior agents are responsible for ensuring that organization-wide databases, such as electronic medical records (EMRs), are managed and appropriately accessed.

2. Subordinate agents providing support to their corresponding superior agents as "functional", "information" or "interface" agents. Agents that support departments, physicians, pharmacists and nurses, among others, are examples.

The functionality description of each agent is provided below.

1. **The Hospital Agent (HA):** This is the superior agent responsible for the corporate management of the entire hospital. It works with other supporting subordinate agents who act in the capacity of superior agents for their corresponding subordinate agents. In its capacity as a main superior agent, the hospital agent directs all other agents towards the realization of corporate objectives and represents a link between the agents interacting inside the hospital and other agents or entities in the external environment. To do so, it should be able to access all functional databases and investigate the role of agents and their operating procedures in facilitating joint functionality.
2. **The Patient's Agent (PA):** This is the patient's main agent responsible for exchanging information between the patient and other agents or personnel in the entire healthcare organization, including physicians, medical records, the pharmacy and laboratory. In addition, it allows an interface exchange between the patient, the pharmacist and the physician in case of problems with regards to drug management, after-discharge complexities, and reactions and alerts in relation to appointments, drug usages and confirmations. It also helps in reporting and signaling adverse drug reactions.
3. **The Physician Agent (PhA):** This is the agent that supports physicians with regards to their processes, functions, schedules and communication. It allows an interface between the physician,

patients, pharmacists, nurses and other laboratory experts dealing with microbiological analyses, pharmaceutical screening and drugs ordering, modification and administration. Exchanging alerts among physicians and other experts, such as pharmacists and patients, are also addressed. The agent also interacts with the patient's agent with regards to appointments, diagnosis, reporting of adverse drug reactions (especially post-discharge) and the confirmation and modification of drugs.

4. **The Laboratory and Images Agent (L&IA):** This is the agent that supports data acquisition, processing and communication to support laboratory and imaging professionals in the hospital. It takes care of the laboratory data (laboratory test requests and results), as well as imaging requests and results. The L&IA updates of the entire laboratory and imaging database and improves the availability and accessibility of information in a comprehensive hospital information system. It interacts with other agents such as the PA, PhA and PrA with regards to the requests and results of laboratory test requests.
5. **The Pharmacy Agent (PrA):** This is the main agent in the pharmaceutical setting because of its functional expertise. In addition to its role in enabling the implementation of pharmaceutical functions, it plays a significant role in facilitating information acquisition and communication, analysis, reporting and recommendation of different alternative medication scenarios using technical expertise and patient's data. It liaises between the hospital agent and its subordinate agents with regards to drug therapies, pharmaceutical and medical analyses, as well as signaling, analyzing and reporting of adverse drug reactions. Since it is responsible for the management of pharmaceutical care at the level of the entire health organization, the pharmacy agent gets support from the following subordinate task agents:
    a. **Drug Therapy and Interactions Agent (DT&I):** The drug therapy and interactions agent works as a subordinate agent for the pharmacy agent. It uses information from the patient's EMR, medical and laboratory test results, and prescription records to assist in managing drug therapies. In addition to reporting drug therapies and confirming them, this agent has the responsibility to recommend changes in medications and drugs based on a data-mining algorithm. It is responsible for screening drug therapies and signalling adverse drug reactions occurring for hospitalized and discharged patients throughout the medication process, in collaboration with the PA and PhA. Upon verification, this agent informs the pharmacy agent to take necessary actions.
    b. **Drug Management Agent (DMA):** The drug management agent acts as a subordinate agent for the pharmacy agent and is responsible for drug management at the level of the hospital, outpatient and ward pharmacies, where it manages drug availability, procurement, ordering and removal of near-expiry drugs. Pharmacy data recorded by the DMA at the level of the hospital pharmacy include name of drug, drug's serial number, drug code, drug components, drug close substitute, dose, quantity available in pharmacy, producing company (with updateable contact address), drug importing company (with updateable contact address), drug's date of production, expiry date, drug food interactions, drug-drug interactions, pharmacist on duty, etc.
6. **The Nurse Agent:** This is the subordinate agent responsible for collecting medication instructions from the physician agent and managing drugs by monitoring drug availability in wards in accordance with the medical order forms forwarded to it. This agent uses a standard operating procedure to monitor and request routine drugs and ad-hoc requests in collaboration with the DMA.

Figure 4. Data processing model

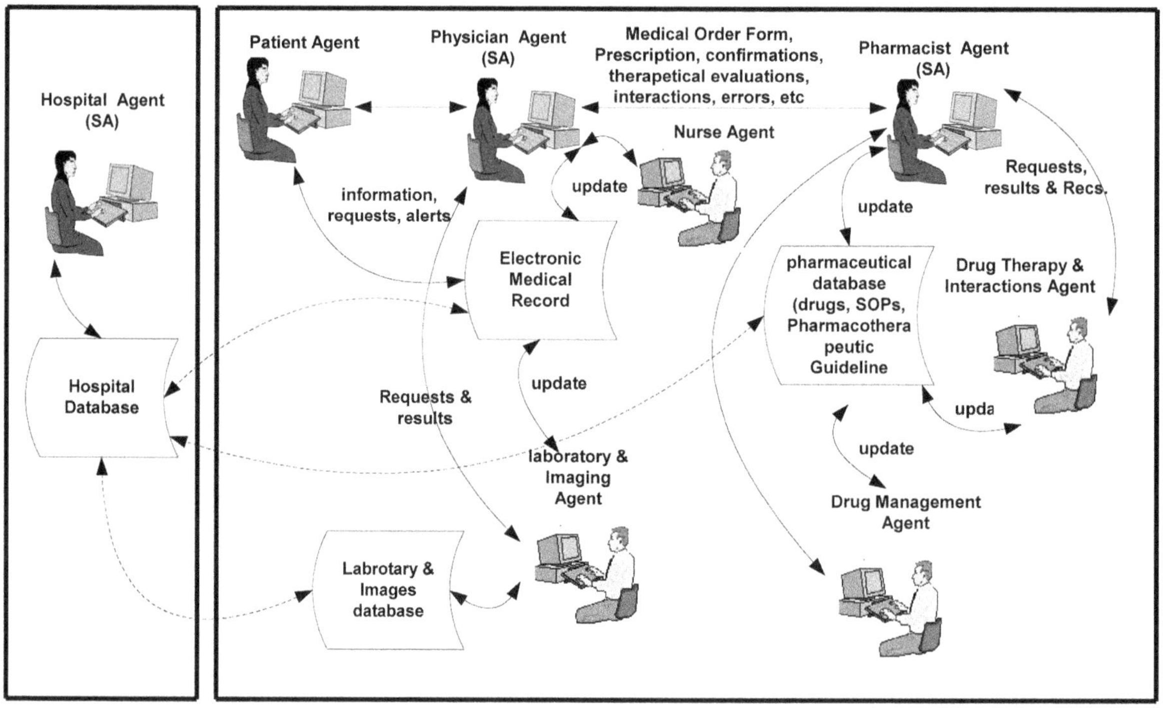

Multi-Agent Pharmacoinformatics Data Processing Model

The reference model is implemented through three main modules:

1. Data capture and update (implemented through the functionalities of the agents representing the patient, physician, laboratory and images, and nurses),
2. Diagnosis (implemented through the agents representing physicians and pharmacists), and
3. Drug monitoring (therapies, interactions analyses and requests for supply implemented through the agents representing physicians and pharmacists). The interactions shaping the functionality of these modules are shown in Figure 4. In addition to using different colors to differentiate among superior and subordinate agents, superior agents are labeled with (SA).

Data Capture and Update Module

The process starts with the use of the patient agent to make appointments at the hospital and provide basic information to establish the electronic medical record, which consists of standard set of data about the patients, their diseases and diagnoses, and therapies. For purposes of patient identification, this module captures the name, age, addresses of the patient and his/her closest relative(s), the department, room number, room phone, equipments used for the patient in the room (if any), medical records and reports, medical images and the treating physician. Diagnosis data include the patient's medical history, current disease diagnosis and its causes. Current and previous therapies include the current drug(s) used by the patient, dosages, frequencies, types of administration, and start and dates of treatment. Some of these data elements are generated after the consultation by the physician agent as a part of EMR updates. Any change on the patient's electronic medical record done by other staff (such as nurses and image specialists) and their corresponding agents takes the form of data input (e.g. imaging requests or medication instructions) in response to requests made by the physician agent. Updates originating from the pharmacist and his/her corresponding agents concerning any comments about drugs and drug therapies that delay its issuance are a result of the physician agent's approval and authorization. The delay in drug issuance then leads to an automatic update of the electronic medical record that will not require the approval of the physician anymore.

Diagnosis Module

Upon the completion of diagnosis, a prescription is made by the physician and communicated to the patient with a copy to the pharmacy. The issuance of the prescription automatically updates the electronic medical record and informs both the patient agent and the pharmacy agent.

Pharmaceutical Care Analysis and Drug Monitoring Module

This is the module managed through the pharmacy agent and its subordinate agents. It is the core module of the entire multi-agent pharmacoinformatics module as it includes the analysis of drug therapies, detection and signaling of drug interactions and adverse drug reactions, and maintaining drug availability at the different levels of the pharmacy (hospital, outpatient and ward). Upon the receipt of a copy of the prescription by the pharmacy agent, the following supporting agents are activated: the drug allergy and interactions agent (to examine drug therapies) and the drug management agent (to examine drug availability). This process is complex and requires the integration between the pharmaceutical database, electronic medical record and other agents such as the physician agent and the laboratory and images agent for further examination and confirmation.

The pharmaceutical database constitutes information about the:

1. Standard operating procedures that govern the examination of medical documents and requests coming to the pharmacy department, such as medical order forms and requests for drug supplies placed by different units in the hospital. These procedures also govern the interaction between the pharmacy department and other units in the form of document exchanges and requests placed to other internal units (e.g. laboratory) and external organizations (e.g. suppliers, ministry of health and Food and Drug Authority (FDA)). While such procedures are generally related to the functionality of the pharmacy department, they remain an essential part of the hospital standard operating procedures.
2. Hospital Therapeutic Guide, which includes information about all routine drugs and medicines used in the hospital, as well as guidelines for acquiring and managing ad-hoc drugs. It also includes detailed recording of drug attributes such as correctness of drug types, dosage, route of administration, frequency, length of use etc.
3. Allergy and microbiological information necessary for making associations between prescribed drugs and diseases, for example, isolated pathogen micro-organisms, and their corresponding sensitivities.

4. Drug groupings created in accordance with diseases and the data individual patients. Each group includes all drugs related to a particular disease (in accordance to drug pharmaceutical attributes and laboratory analyses) for patients with certain allergies and symptoms.

Upon the receipt of the order by the PrA, it examines its "procedural context" and its compliance with the standard operating procedures. Orders in compliance with such procedures are communicated for the corresponding subordinate agents. Non-complying prescriptions are returned by the PhA for further analysis. The DTIA access all prescriptions approved by the PrA. It consults the pharmacy database, HTG, and the patient's electronic medical record to screen order attributes such as appropriateness of drug types, dosages, routes of administration, frequencies and lengths of use; and reports any inconsistencies to the pharmacy agent. These inconsistencies are then communicated to the PhA for further examination. Once the drugs are considered suitable, the DTIA recommends the issuance of these drugs. The DMA examines the availability of drugs by screening and updating the HTG, and drug inventory of routine and ad-hoc drugs. Based on the laboratory and/or diagnostic analysis results posted in the electronic medical record, the PrA can now provide some recommendations to the PhA, such as dosage adjustments, changes in prescribed drugs, and use of medicines. Upon the confirmation of the PhA, the PrA approves the order and interacts with the PhA to update the electronic medical record and the PA for drug administration.

In case of suspected adverse dug events, the DTIA examines the results of general lab analysis of the patient by accessing his/her electronic medical record and informs the PrA to warn the PhA in case of orders prescribed without general lab analysis. The process is also applicable for the detection and signaling of adverse drug reactions reported to the PrA through the PhA in accordance with a signal originating from either the PA or other medical staff involved in medical activities in wards. The (DAIA) examines the event and gives an initial assessment whether it is Type (A) or (B) or (C). In case of type (A), the recommendation posted to the PrA focuses on improving dosage characteristics, since a large proportion of adverse drug reactions are dosage-based and related to the pharmacokinetic properties of drugs. In case of Type (B) detected adverse drug reactions, the DAIA recommends the PrA to consult with both the PhA, and L&IA for further shared discussions because Type (B) events are generally related to the patient's reactions and tend to be allergic, idiosyncratic, immunological, or non-immunological (Meyboom et al, 2000). The same consultation process is required to address Type (C) events since they tend to be serious and may result in significant implications on public health. They originate from the impacts of drugs used for improving the quality of life of patients with serious chronic diseases. The PrA also informs the HA about all reported adverse drug reactions and interactions so as to communicate them to the National Pharmacovigilance Centre.

Drug Management Module

This is the module managed by the Drug Management Agent (DMA) in collaboration with the Nurse Agent (NA), which gets information from the physician agent. The basic aim of this module is to ensure the availability of suitable drugs included in the HTG. The nurse agent is concerned with monitoring drugs for patients in wards and ordering them on regular and ad-hoc bases. In the case of regular drugs (i.e. drugs included in the hospital's drug list and HTG, the nurse agent (NA) examines the ward repository for the drugs included in the prescription made by the physician and prepares it for the patient's treatment (if available in stock). If not, it posts a request to the Drug Management Agent (DMA) through the pharmacy agent to screen for drug availability and prepare the medicine for delivery to the requesting ward or post a request (through the hospital agent) for the FDA (in case of public hospitals) or suppliers (in case of private ones) and deliver it to the ward upon arrival. In case of ad-hoc requests, the NA (in consultation with the physician agent) sends the prescription to the pharmacy agent for examination and ordering. In case of detected inconsistencies, if any, the physician agent is informed with recommendations. Upon the receipt of confirmation from the physician agent, the pharmacy agent follows the standard operating procedures to acquire the drug and deliver it to the requesting wards. The DMA monitors drug availability and controls stock at the level of ward repositories and outpatient pharmacies. Such systems provide information about drug orders for in-patients, availability statuses of the drugs through alerts such as 'No quantity', 'Near expiry', 'Re-order', 'restrictions' and requests for authorization and Verification.

7. DISCUSSION

The area of focus of the project is a new research domain where over years, the question of pharmacoinformatics is being misinterpreted and misunderstood by dealing at the process of pharmaceutical care as a function in the entire hospital management process. The study has cleared some of the methodological doubts and gaps with regards to the crystallization and conceptualization of the term by relating it to process of decision making rather than the use of computers in drug production.

The study focused on the development of a multi agent reference pharmacoinformatics model in pursuit of improving hospital management through improved The use of intelligent agents for the improvement of pharmaceutical care processes at the hospital level is gaining importance due to several reasons. The outstanding capabilities shown by intelligent agents in addressing complex problems provide significant assistance to the healthcare in general and pharmacy information systems in particular. Being used as stand-alone applications or within the context of a wider hospital information systems, the use of multi agent pharmacy-related information systems are gaining paramount importance due to their growing roles in improving the acquisition and exchange of drug-related information, examining drug therapies and interactions and integrating information.

The practical managerial context of the model is shown in its potential benefits in improving communication between physicians and other professionals (such as nurses and laboratory) on the one hand and between the physician and the pharmacist, a relation that tend to be complicated in real life. The outcome of such improved communication should be reflected in the quality of medical services and pharmaceutical care as it creates a conducive environment for negotiation, accountability and coordination. The prescription of (suitable) drugs will guided by a pharmaceutical cross check that investigates and alters the compliance with procedures (standard operating procedures) and requirements (such as microbiological analysis) in a way that minimizes medical errors and the occurrence of adverse drug reactions. This will eventually improves signaling and analyzing adverse drug and exchanging information with national Pharmacovigilance centers and attracting technical support.

However, many issues need to be addressed. The process of agent interactions, consultation and information exchange calls for a clear role description especially for agent-assigned tasks and user-based intervention in accordance with the degree of autonomy allowed for the entire software agent. Semantics, ontologies and knowledge representation are also important considerations that need to be addresses in the light of the problem domain and solution spaces envisioned. Moreover, the preparedness of the medical staff to use intelligent tools and interact in an agent-based organization, willingness to provide inputs and comply with continuous improvement and security requirements are also important issues.

8. RECOMMENDATIONS

In the light of the work done and its outcomes, the following recommendations can be made to help healthcare organizations (hospitals, FDAs, Pharmacovigilance Centres and ministry of health) improve their efficiency in providing high quality pharmaceutical care services:

1. Restructuring hospital-wide processes to reflect the central role of pharmaceutical functions and incorporating clear standard operating procedures to ensure the incorporation of suitable tasks and activities.
2. Adopting an integrated approach for the crystallization of pharmacoinformatics very early in the medication process to ensure its involvement in the analysis of drug therapies, cross-checking prescriptions, alerting physicians and other medical professionals about the non-existence of drug prescription pre-requisites such as general lab analysis and recommending change of drug regimens and the use of free-use drugs wherever applies.
3. Making pharmaceutical care processes more patient oriented to optimize the utility matrix of the stakeholders in the entire healthcare system. As a result, the pharmacoinformatics can be of high importance in improving the knowledge of patients about drugs, the dysfunctional consequences of adverse drug events and their capacity and willingness to report such events.
4. Developing clear hospital-wide information system strategies and information security policies to facilitate effective communication among medical professionals, interactions with patients and, most importantly, interactions with the National Pharmacovigilance Centre of the Federal Food and Drug Authority at the Ministry of Health.

REFERENCES

Anumba, C., Ren, A., Thorpe, O., Ugwu, O., & Newnham, L. (2003). Negotiation within a multi-agent system for the collaborative design of light industrial buildings. *Advances in Engineering Software*, *34*(7), 389–401. doi:10.1016/S0965-9978(03)00038-3

Anumba, C. J., & Newnham, L. N. (2000). Computer-based collaborative building design: Conceptual model. *Int J Construct Inform Technol*, *8*(1), 1–14.

Bonarini, A., & Trianni, V. (2001). Learning fuzzy classifier systems for multi-agent coordination. *Information Sciences*, *136*(1–4), 215–239. doi:10.1016/S0020-0255(01)00149-9

Caire, G., Chainho, P., Evans, R., Garijo, F., Gomez Sanz, J., Kearney, P., & Stark, J. (2002). Agent Oriented Anlysis Using MESSAGE/UML. In Agent-Oriented Software Engineering II, (pp. 119-135). Springer.

Davidsson, P., & Wernstedt, F. (2004). *A framework for evaluation of multi-agent system approaches to logistics network management. In Multi-Agent Systems: An Application Science*. The Netherlands: Kluwer.

DeLoach, S. A. (1999). Systems engineering: A methodology and language for designing agent systems. In *Proceedings of Agent Oriented Information Systems* (pp. 45–57). Academic Press.

DeLoach, S. A., Wood, M. F., & Sparkman, C. H. (2001). Multiagent system engineering. *International Journal of Software Engineering and Knowledge Engineering, 11*(3), 231–258. doi:10.1142/S0218194001000542

Ferber, J. (1999). *Multi-Agent Systems: An Introduction to Distributed Artificial Intelligence*. Reading, MA: Addison-Wesley.

Frey, D., Stockheim, T., Woelk, P., & Zimmermann, R. (2003). Integrated Multi-agent-based Supply Chain Management. In *Proceedings of 1st International Workshop on Agent-based Computing for Enterprise Collaboration*.

Fu, Y., & Piplani, R. (2000). Multi-agent enabled modelling and simulation towards collaborative inventory management in supply chains. In *Proceedings of the 2000 Winter Simulation Conference.*

Gasmelseid, T. (2006). Multi Agent Negotiation Framework in Resource Bounded Environments. In *Proceedings of the Information and Communication Technologies, 2006. ICTTA '06.* (pp. 465 – 470). Retrieved from http://ieeexplore.ieee.org/xpl/freeabs_all.jsp?tp=&arnumber=1684414&isnumber=35470

Gasmelseid, T. (2009). Improving clinical practice through mobile medical informatics. In U. Bhuvan (Ed.), *Handbook of Research on Mobile Business: Technical, Methodological and Social perspective, Project HRMB2* (pp. 604–614). IGI Publishing. doi:10.4018/978-1-60566-156-8.ch056

Gasmelseid, T. (2009a). On the design of multi-agent, context aware and mobile systems. In Handbook on Modern System Analysis and Design Applications and Technologies. Idea Group Publishing.

Gasmelseid, T. (2014). Managing Stakeholder Concerns in Large-Scale Multi-Agent Information Systems. *International Journal of Agent Technologies and Systems, 5*(4), 68-86.

Gaurav, J., Christian, R., & Robert, H. (2009). *A Multi-Agent Simulation (MAS) of the Pharmaceutical Supply Chain (PSC).* POMS 20th Annual Conference, Orlando, FL.

Giunchiglia, F., Perini, A., & Mylopoulus, J. (2002). The Tropos software development methodology: Processes, models and diagrams. In *Proceedings of the First International Joint Conference on Autonomous Agents and Multi agent Systems* (pp. 63-74). Bologna, Italy: ACM Press. doi:10.1145/544741.544748

Giunchiglia, F., Perini, A., & Sannicol, F. (2001). Knowledge level software engineering. In *Intelligent Agents VIII (LNCS)* (Vol. 2333, pp. 6–20). Seattle, WA: Springer-Verlag. doi:10.1007/3-540-45448-9_2

Godo, L., Puyol-Gruart, J., Sabater, J., Torra, V., Barrufet, P., & Fàbregas, F. (2003). A multi agent approach for monitoring the prescription of restricted use antibiotics. *Artificial Intelligence in Medicine, 27*(3), 259–282. doi:10.1016/S0933-3657(03)00006-X PMID:12667739

Govindu, R., & Chinnam, R. (2007). MASCF: A generic process-cantered methodological framework for analysis and design of multi-agent supply chain systems. *Computers & Industrial Engineering, 53*(4), 584–609. doi:10.1016/j.cie.2007.06.003

Hu, J., & Weliman, M. (2001). Learning about other agents in a dynamic multi-agent system. *Cognitive Systems Research*, *2*(1), 67–79. doi:10.1016/S1389-0417(01)00016-X

Iglesias, C., Garrijo, M., & Gonzalez, J. (1998). A survey of agent-oriented methodologies. In *Intelligent Agents V – Proceedings of the 1998 Workshop on Agent Theories, Architectures and Languages*. Academic Press. doi:10.1007/3-540-49057-4_21

Kendall, E. (1999). Role modeling for agent system analysis, design, and implementation. In *Proceedings of Third International Symposium on Mobile Agents* (MA'99). Palm Springs, FL: Academic Press. doi:10.1109/ASAMA.1999.805405

Koppensteiner, G., Merdan, M., Lepuschitz, W., Moser, T., & Reinprecht, C. (2011). *Multi Agent Systems combined with Semantic Technologies for Automated Negotiation in Virtual Enterprises. In Multi-Agent Systems - Modelling, Control, Programming, Simulations and Applications* (pp. 221–242). InTech.

Lind, J. (1999). *A review of multiagent systems development methods: Technical report*. Martlesham Heath, UK: British Telecom, Adastral Park Labs.

Lisa, M., Hogg, L., & Jennings, N. (2001). Socially intelligent reasoning for autonomous agents. *IEEE Transactions on Systems, Man, and Cybernetics. Part A, Systems and Humans*, *31*(5), 381–393. doi:10.1109/3468.952713

Lu, L., & Wang, G. (2008). A study on multi-agent supply chain framework based on network economy. *Computers & Industrial Engineering*, *54*(2), 288–300. doi:10.1016/j.cie.2007.07.010

Odell, J., Parunak, V. D., & Bauer, B. (2000). Extending UML for agents. In *Proc. of the Agent-Oriented Information Systems Workshop at the 17th National Conference on Artificial Intelligence*. Academic Press.

Padgham, L., & Winiko, M. (2002a). Prometheus: A pragmatic methodology for engineering intelligent agents. In *Proceedings of the OOPSLA 2002 Workshop on Agent-Oriented Methodologies* (pp.97-108). Seattle, WA: OOPSLA. doi:10.1145/544741.544749

Padgham, L., & Winiko, M. (2002b). *Prometheus: Engineering intelligent agents*. Tutorial notes. Unpublished.

Paolo, B., Anna, P., Paolo, G., Fausto, G., & John, M. (2004). TROPOS: An agent-oriented software development methodology. *Journal of Autonomous Agents and Multi-Agent Systems*, *8*(3), 203–236. doi:10.1023/B:AGNT.0000018806.20944.ef

Pathak, D., Nordstrom, G., & Kurokawa, S. (2000). Modelling of supply chain: a multi-agent approach. *IEEE International Conference on Systems, Man, and Cybernetics*.

Perini, A., Bresciani, P., Giunchiglia, F., Giorgini, P., & Mylopoulos, J. (2001). A knowledge level software engineering methodology for agent oriented programming. *Proceedings of the fifth international conference on Autonomous agents* (pp. 648-655).

Persson, P., Laaksolahti, J., & Lönnqvist, P. (2001). Understanding socially intelligent agents—a multilayered phenomenon. *IEEE Transactions on Systems, Man, and Cybernetics. Part A, Systems and Humans*, *31*(5), 349–360. doi:10.1109/3468.952710

Ponnurangam, D., & Uma, G. (2005). Fuzzy complexity assessment model for resource negotiation. *Expert Systems with Applications*, *29*(1), 105–119. doi:10.1016/j.eswa.2005.01.008

Ronald, M., Marie, L., & Antonie, E. (2000). An ABC of drug-Related problems. *Drug Safety*, *22*(6), 415–423. doi:10.2165/00002018-200022060-00001 PMID:10877036

Seyed, M., Rizi, M., Maciej, M., Latek, M., & Armando, G. (2010). *Merging Remote Sensing Data and Population Surveys in Large, Empirical Multi-agent Models: The Case of the Afghan Drug Industry*. Presented during the Third World Social Simulation Congress, Kassel, Germany. Retrieved April, 30, 2011 from: http://www.css.gmu.edu/projects/irregularwarfare/remotesensing.pdf

Shirazi, M., & Soroor, J. (2007). An intelligent agent-based architecture for strategic information system applications. *Knowledge-Based Systems*, *20*(8), 726–735. doi:10.1016/j.knosys.2006.10.004

Sycara, K. (1998). Multi-agent Systems. *AI Magazine*, *19*(2), 79–92.

Trappey, A., Lu, T., & Fu, D. (2009). Development of an intelligent agent system for collaborative mold production with RFID technology. *Robotics and Computer-integrated Manufacturing*, *25*(1), 42–56. doi:10.1016/j.rcim.2007.06.002

Turcu, C., Turcu, Popa, V., & Gaitan, V. (2008). Identification and Monitoring of Patients Using RFID and Agent Technologies: Synergy and Issues. *Electronics and Electrical Engineering, 6*(86), 17-22.

Walsh, W., Wellman, A., Walsh, W., & Wellman, M. (2000). Modelling supply chain formation in multi-agent systems. In M. Alexandros, Y. Fredrik, & S. Carles (Eds.), Agent Mediated Electronic Commerce II (LNCS), (vol. 1788, pp. 94–101). Springer-Verlag. doi:10.1007/10720026_5

White, J. (1997). Mobile agents. In *Software agents*. AAAI Press.

Wooldridge, M. (1997). Agent based software engineering. *IEEE Proceedings of Software Engineering, 144*(1), 26–37.

Wooldridge, M., Jennings, N. R., & Kinny, D. (2000). The gaia methodology for agent-oriented analysis and design. *Journal of Autonomous Agents and Multi-Agent Systems*, *3*(3), 285–312. doi:10.1023/A:1010071910869

Wooldridge, M. J., & Jennings, N. (1995). Agent theories, architectures and languages: A survey. *The Knowledge Engineering Review*, *10*(2), 115–152. doi:10.1017/S0269888900008122

Yanqing, J., Hao, M., Farber, S., John, Y., Peter, D., Richard, M., & Michael, M. (2010). A Distributed, Collaborative Intelligent Agent System Approach for Proactive Post marketing Drug Safety Surveillance. *IEEE Transactions on Information Technology in Biomedicine*, *14*(3), 826–837. doi:10.1109/TITB.2009.2037007 PMID:20007038

Yanqing, J., Hao, Y., John, Y., Shizhuo, Z., Daniel, C., Barth, J., & Michael, M. et al. (2007). A distributed adverse drug reaction detection system using intelligent agents with a fuzzy recognition-primed decision model. *International Journal of Intelligent Systems*, *22*(8), 827–845. doi:10.1002/int.20230

Yanqing, J., Hao, Y., Yen, J., Shizhuo, Z., Massanari, M., & Barth, J. (2005). Team-based multi-agent system for early detection of adverse drug reactions in post marketing surveillance. In *Proc Proceedings of the 24th North American Fuzzy Information Processing Society*.

Zambonelli, F., Jennings, N. R., & Wooldridge, M. (2001). Organizational rules as an abstraction for the analysis and design of multi agent systems. *International Journal of Software Engineering and Knowledge Engineering*, *11*(3), 303–308. doi:10.1142/S0218194001000505

KEY TERMS AND DEFINITIONS

Adverse Drug Reactions: Complications that occur as a result of drug-related or patient-related factors.

Autonomy: The capacity of agents *to* operate without the direct intervention of humans or others, and have some kind of control over their actions and internal state.

Electronic Medical Record (EMR): The record created by healthcare providers to include all patient-related and medication-specific information that can be used for medical examination, prescription and investigation. Usually it is referenced using the patient's name or number as a key filed.

Healthcare Business Units: All departments and sections that are specialized in the provision of healthcare service at the level of service provision such as specialized departments in hospitals. They constitute the base of the management of pharmaceutical services and the identification of agent-based functionalities.

Information Systems: The collection of physical resources (hardware, software, telecommunication and database) and personnel (such as analysts, designers, developers and programmers) involved in the acquisition, storage, processing, dissemination and provision of information to enable decision makers achieve organizational objectives effectively and efficiently.

Multi-Agent System: A collection of, possibly heterogeneous, computational entities, having their own problem-solving capabilities and which are able to interact in order to reach an overall goal.

Pharmaceutical Care: The set of activities, processes and procedures carried out at different levels of the health care chain in pursuit of improving the quality of pharmaceutical interventions and decisions made.

Pharmacoinformatics: The use of computer based applications for the provision of information and implementation of tasks to assist medical policy makers and professionals in healthcare organizations to control and monitor drug therapies, prescription, use, signaling and analyzing of adverse drug events and stock control.

Reactivity: The ability of agents to perceive their environment and respond in a timely fashion to changes that occur in it.

Software Agents: Computational entities that interact and cooperate together and use resources collectively in order to achieve their own objectives or the objectives of their owners or other agents.

Superior Agents: High rank software agents operating within the context of the entire software organization. Their basic aim is to ensure that all subordinate agents contribute to the realization of the corporate objective to be met by the multi-agent organization. They exercise control and direction upon their subordinate agents and communicate on behalf of them.

Chapter 8
Cluster Origin of Solvation Features of C-Nanostructures in Organic Solvents

Francisco Torrens
Universitat de València, Spain

Gloria Castellano
Universidad Católica de Valencia, Spain

ABSTRACT

The existence of fullerenes, Single-Wall Carbon Nanocones (SWNCs), especially Nanohorns (SWNHs), Single-Wall Carbon Nanotube (SWNT) (CNT) (NT), NT-Fullerene Bud (NT-BUD), Nanographene (GR) and GR-Fullerene Bud (GR-BUD) in cluster form is discussed in organic solvents. Theories are developed based on columnlet, bundlet and droplet models describing size-distribution functions. The phenomena present a unified explanation in the columnlet model in which free energy of cluster-involved GR comes from its volume, proportional to number of molecules n in cluster. Columnlet model enables describing distribution function of GR stacks by size. From geometrical considerations, columnlet (GR/GR-BUD), bundlet (SWNT/NT-BUD) and droplet (fullerene) models predict dissimilar behaviours. Interaction-energy parameters are derived from C_{60}. An NT-BUD behaviour or further is expected. Solubility decays with temperature result smaller for GR/GR-BUD than SWNT/NT-BUD than C_{60} in agreement with lesser numbers of units in clusters. Discrepancy between experimental data of the heat of solution of fullerenes, CNT/NT-BUDs and GR/GR-BUDs is ascribed to the sharp concentration dependence of the heat of solution. Diffusion coefficient drops with temperature result greater for GR/GR-BUD than SWNT/NT-BUD than C_{60} corresponding to lesser number of units in clusters. Aggregates $(C_{60})_{13}$, SWNT/NT-BUD$_7$ and GR/GR-BUD$_3$ are representative of droplet, bundlet and columnlet models.

DOI: 10.4018/978-1-5225-0248-7.ch008

INTRODUCTION

Interest in nanoparticles (NPs) arises from the shape-dependent physical properties of materials at the nanoscale (Faraday, 1857; Murphy *et al.*, 2010). Occurrence of single-wall carbon nanocones (SWNCs) was used to investigate nucleation and growth of curved C-nanostructures (NSs) suggesting pentagon role. When a pentagon is introduced into a graphitic sheet nanographene (GR) (Figure 1d) *via* extraction of a 60° sector from the sheet one forms a cone leaf. Pentagons presence in an SWNC apex is analogue of their occurrence in single-wall C-nanotube (NT) (CNT) (SWNT) tip topology (*cf.* Figure 1b). Terminations of SWNTs attracted interest once Tamura & Tsukada (1995) theoretically predicted peculiar electronic states related to GR topological defects. Kim *et al.* (1999) observed resonant peaks in density of states (DOS) in SWNTs and Carroll *et al.* (1997), in multiple-wall (MNTs) C-nanotubes (MWNTs).

The SWNCs with discrete opening angles (apices, θ) of 19°, 39°, 60°, 85° and 113° of cone (*cf.* Figure 2) were observed in a C-sample generated by hydrocarbon (HC) pyrolysis (Krishnan *et al.*, 1997), which was explained by a cone-wall model composed of wrapped GR sheets where geometrical requirement for seamless connection accounted for semidiscrete character and absolute values of cone angle. Total

Figure 1. Arrangement of C-nanostructures: (a) C_{60}; (b) SWNT; (c) NT-BUD; (d) GR; (e) GR-BUD

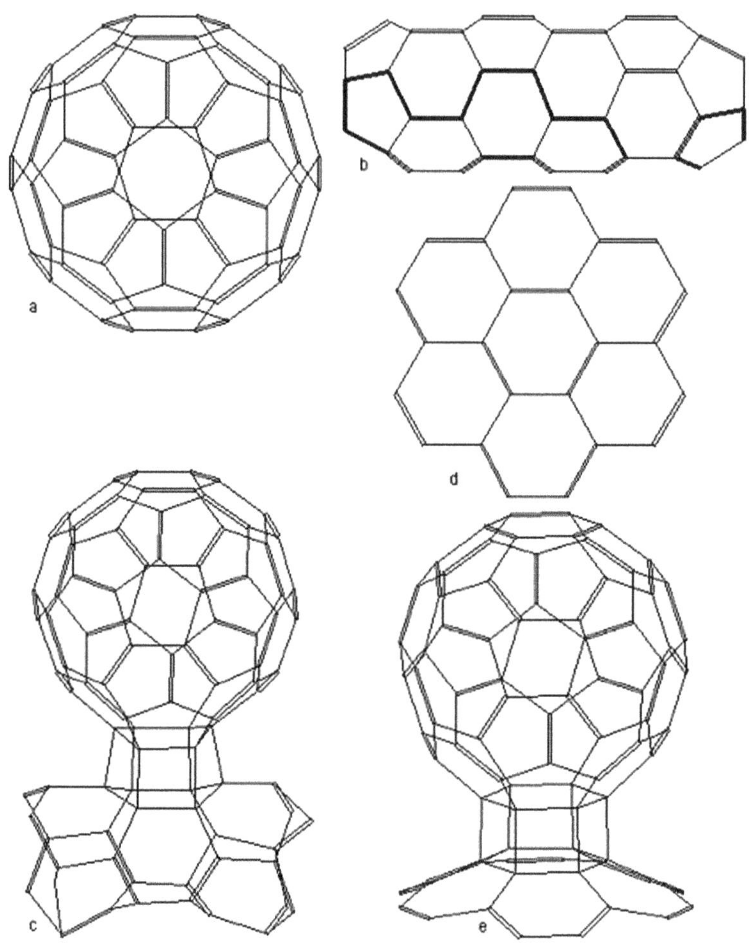

disclinations of all conic GR microstructures are multiples of 60º corresponding to the presence of a given number ($P \geq 0$) of pentagons in SWNC apices. Considering GR sheet symmetry and Euler theorem, five types of SWNCs (corresponding to angles) are obtained from a continuous GR sheet matching to $P = 1$–5. Cone angle (θ) is given by $\sin(\theta/2) = 1 - P/6$ leading to SWNC angles where flat discs and caped SWNTs correspond to $P = 0$ and 6, respectively. The SWNC with $P = 5$ pentagons ($\theta = 19$º) is named nanohorn (SWNH). Several configurations exist for an SWNC angle depending on the form in which pentagons are arranged in conic tips. According to *isolated pentagon rule* (IPR) derived from fullerenes (Figure 1a) (Kroto, 1987) configurations containing isolated pentagons lead to isomers that are more stable than those with grouped pentagons for NSs. Additional rules were derived from *ab initio* calculations (Jan & Jaffe, 1998) performed to evaluate stability of NSs containing isolated and grouped pentagons. Consideration of a curvature-producing pentagon as a defect, in a planar net of hexagons, results in that two-pentagon arrangement in a hexagonal lattice is specified by a hexagonal co-ordinate (*a*,*b*) with a pentagon in (*a*,*b*) and another in (0,0). Nearest-neighbouring pentagons are (1,1)-co-ordinated in C_{60} (all pentagons connected by a C–C bond) and are (1,1)/(2,0)-co-ordinated in C_{70} (the latter corresponds to two pentagons separated by one hexagon). In accordance with density-functional-theory (DFT) calculations pentagons (1,1) lead to more stable SWNC tip structures than those of pentagons (2,0), which is attributed to lower *stress* induced by each pair (1,1) in relation to pairs (2,0). The SWNCs present geometry asymmetry and are semiconductors. Insolubility of NSs in all solvents and tendency to agglomerate should be overcome before applications. Tagmatarchis *et al.* (2006) reported SWNC covalent functionalization with NH_4^+ to improve solubility but still leaving aggregation behind. Progress was made to SWNC solubilization, which was achieved by covalent functionalization of its skeleton (Cioffi *et al.*, 2006, 2007; Pagona *et al.*, 2007a)/highly strained cone-ends (Pagona *et al.*, 2006a) and supramolecular π–π stacking interactions (Pagona *et al.*, 2006b, 2007b; Zhu *et al.*, 2003) with pyrenes (Py's)/porphyrins.

Li *et al.* (2003), Nasibulin *et al.* (2007ab, 2008) and Anisimov *et al.* (2010) synthesized C-NanoBuds™ (NT-BUDs, Figure 1c): fullerene-functionalized SWNTs, which combine CNTs and spheroidal fullerenes. In NT-BUDs, fullerenes are covalently bonded to the outer sidewalls of underlying CNT: they exhibit CNT and fullerene properties; *e.g.*, mechanical features and electrical conductivity are similar to corresponding CNTs. However, because of attached-fullerene reactivity, NT-BUDs were functionalized *via* fullerene chemistry. Attached fullerenes were used as molecular anchors to prevent CNTs slipping in several nanocomposites (NCs) modifying mechanical properties. All NT-BUDs are calculated semiconductors (Fürst *et al.*, 2009; Meng *et al.*, 2008; Wu & Zeng, 2008) and some, magnetic (Wang &

Figure 2. (a) Extraction of a section $D_\theta = 60^o$ from GR $\theta \approx 113^o$; (b) incorporation of one pentagon P = 1 into GR forming a cone; (c) HR e^- microscopy image of an SWNC with P = 3 in apex $D_\theta = 180^o$, $\theta \approx 60^o$

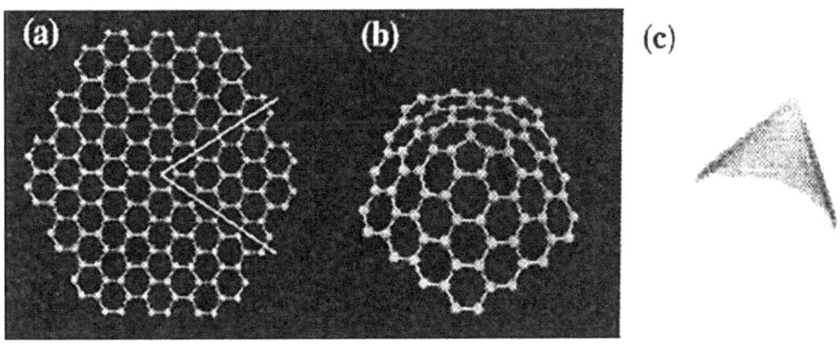

Li, 2011; Zhu & Su, 2009). More non/C-NSs exist (*e.g.*, NT-based borides, oxides of several elements, *etc.*). The GR, a two-dimensional (2D) membrane one-atom thick, emerged as material, which consists of hexagonal arrangement of C-atoms in 2D honeycomb crystals (Novoselov *et al.*, 2004). Geim & Novoselov (2007) called it *mother of all graphitic forms* because it is wrapped into fullerenes, rolled into CNTs and stacked into $C_{graphite}$. It is building block for nanomaterials: if it is wrapped up into a ball a 0D fullerene is obtained, when rolled, a 1D CNT and if stacked, a 3D $C_{graphite}$. It differs from 3D materials. Its atoms exhibit sp^2 hybridization that, together with atomic thickness, makes it unique: extremely high electron (e^-)/hole (h^+) mobilities even at room temperature (RT), high thermal conductivity at RT, 2.3% of light absorbance over a wide visible (VIS), mechanical strength and impermeability to gases. Basic GR is a semimetal and zero-gap semiconductor. *Zigzag*-edges nature imposes e^--density localization with maximum at border C-atoms leading to flat conduction and valence bands formation near Fermi level. Localization states are spin polarized and in case of ordering e^- spin along *zigzag* edges, GR is established in anti/ferromagnetic phase; the former breaks GR sublatice symmetry that changes its band structure and opens a gap. The GR sparked potential to be ingredient of devices (*e.g.*, single molecule gas sensors, ballistic transistors, spintronics). Hernandez *et al.* (2008), Khan *et al.* (2010) and Zhang *et al.* (2010) analyzed solvent selection. In this laboratory Bosch-Navarro *et al.* (2012, 2013ab) examined multifunctional hybrid NCs based on CNTs and chemically modified GR. Gadet *et al.* (1992), Fang *et al.* (2005), Novoselov *et al.* (2005), Coronado *et al.* (2006, 2010, 2011) and Abellán *et al.* (2013) analyzed other 2D materials. Wu and Zeng (2009) calculated magnetic some GR-fullerene nanobuds (GR-BUDs, Figure 1e). In earlier publications SWNT periodic classification was discussed (Torrens, 2004a, 2005a). A program based on model AQUAFAC allowed calculating SWNT aqueous coefficients (Torrens, 2006a). A molecular modelling analysis of SWNT solvents classified them into *best*, *good* and *bad* ones (Torrens, 2005bcde, 2006b; Torrens & Castellano, 2011a, in press). Packing effect was analyzed (Torrens & Castellano, 2007b). Asymptotic analysis of cluster coagulation–fragmentation equations was performed (Torrens & Castellano, 2007c). Similarity laws determining thermodynamic characteristics of fullerite crystals were analyzed (Torrens & Castellano, 2008a, 2010a). *Bundlet* model of SWNT (Torrens & Castellano, 2005, 2007abcd, 2009a), SWNH (Torrens & Castellano, 2010b, 2012, 2013ab, 2014ab) and NT-BUD (Torrens & Castellano, 2004d) clusters was presented. *Columnlet* model of GR clusters (Torrens & Castellano, 2013cd, 2014c) was informed. The aim of the present report is to perform a comparative analysis of C_{60}, SWNT and SWNC properties. A class of phenomena accompanying SWNT-, SWNC- and SWNH-solution behaviour is examined from a unique viewpoint taking into account cluster-formation tendency. Based on droplet model, bundlet and columnlet ones are proposed representing more complex systems. The next section describes the methods. Following that, several sections review recent outcomes. Next, the calculation results are illustrated and discussed. The last section summarizes our conclusions.

COMPUTATIONAL METHODS

Solubility of C-Nanostructures

A solubility mechanism is based on NS-cluster formation in solution. Aggregation changes NS thermodynamic parameters, which displays phase equilibrium and changes solubility. Droplet, *bundlet* and *columnlet* models are valid when NS characteristic number in cluster $n >> 1$ (*cf.* Figure 3).

Figure 3. Scheme of the studied models: (a) droplet, (b) bundlet and (c) columnlet

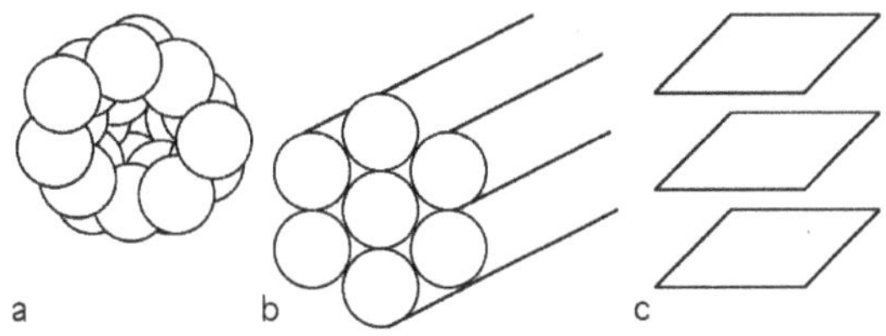

In a saturated solution, chemical potentials per NS for dissolved substance and for a crystal are equal; the equality is valid for NS clusters. Cluster free energy is made up of two parts: the volume one proportional to the number of NSs n in cluster and surface one, to $n^{1/2}$ (Bezmel'nitsyn *et al.*, 1994, 1995, 1998). Clusters of $n >> 1$ parallel NS present cylindrical bundlet shape. Gibbs energy G_n for an n-sized cluster is:

$$\begin{aligned} G_n &= G_1 n - G_2 n^{2/3} \quad (droplet) \\ G_n &= G_1 n - G_2 n^{1/2} \quad (bundlet) \\ G_n &= G_1 n - G_2 \quad (columnlet) \end{aligned} \tag{1}$$

where $G_{1/2}$ are responsible for contribution to Gibbs energy of NSs placed inside volume and on surface of cluster. Chemical potential μ_n of an n-sized cluster is:

$$\mu_n = G_n + k_B T \ln C_n \tag{2}$$

where k_B is Boltzmann constant and T, absolute temperature. With Equation (1) it results:

$$\begin{aligned} \mu_n &= G_1 n - G_2 n^{2/3} + k_B T \ln C_n \quad (dr.) \\ \mu_n &= G_1 n - G_2 n^{1/2} + k_B T \ln C_n \quad (bn.) \\ \mu_n &= G_1 n - G_2 + k_B T \ln C_n \quad (cl.) \end{aligned} \tag{3}$$

where $G_{1/2}$ are in joules. In a saturated NS solution, cluster-size distribution function is determined by equilibrium condition linking aggregates of specified size with solid phase, which corresponds to equality between chemical potentials for NSs incorporated into clusters and crystal, resulting in cluster-size distribution function:

$$\begin{aligned} f(n) &= g_n \exp\left(\frac{-An + Bn^{2/3}}{k_B T}\right) \quad (d) \\ f(n) &= g_n \exp\left(\frac{-An + Bn^{1/2}}{k_B T}\right) \quad (b) \\ f(n) &= g_n \exp\left(\frac{-An + B}{k_B T}\right) \quad (c) \end{aligned} \tag{4}$$

where A is the difference between interaction energies of an NS with its surrounding in solid and cluster volume, B, that for NSs located on aggregate surface and g_n, the statistical weight of a cluster of size n. Normalization for distribution function (4):

$$\sum_{n=1}^{\infty} f(n)n = C \tag{5}$$

requires $A > 0$, and C is solubility in relative units. As $n >> 1$, normalization (5) results:

$$\begin{aligned} C &= \bar{g}_n \int_{n=1}^{\infty} n \exp\left(\frac{-An + Bn^{2/3}}{k_B T}\right) dn = C_0 \int_{n=1}^{\infty} n \exp\left(\frac{-An + Bn^{2/3}}{k_B T}\right) dn \quad (droplet) \\ C &= \bar{g}_n \int_{n=1}^{\infty} n \exp\left(\frac{-An + Bn^{1/2}}{k_B T}\right) dn = C_0 \int_{n=1}^{\infty} n \exp\left(\frac{-An + Bn^{1/2}}{k_B T}\right) dn \quad (bundlet) \\ C &= \bar{g}_n \int_{n=1}^{\infty} n \exp\left(\frac{-An + B}{k_B T}\right) dn = C_0 \int_{n=1}^{\infty} n \exp\left(\frac{-An + B}{k_B T}\right) dn \quad (columnlet) \end{aligned} \tag{6}$$

where $\bar{g}_n$ is the statistical weight of cluster averaged over n, and C_0, NS molar fraction. The $A = 320$K, $B = 970$K and $C_0 = 5 \cdot 10^{-8}$ were taken from C_{60} in hexane, toluene and CS_2. A correction comes from different packing efficiencies between C_{60}, SWNT/SWNC/NT-BUDs and GR/GR-BUDs:

$$\begin{aligned} A' &= \frac{\eta_{cyl}}{\eta_{sph}} A \quad \& \quad B' = \frac{\eta_{cyl}}{\eta_{sph}} B \quad (clndr.) \\ A' &= \frac{\eta_{con}}{\eta_{sph}} A \quad \& \quad B' = \frac{\eta_{con}}{\eta_{sph}} B \quad (cone) \\ A' &= \frac{A}{\eta_{sph}} \quad \& \quad B' = \frac{B}{\eta_{sph}} \quad (stack) \end{aligned} \tag{7}$$

where $\eta_{cyl} = \pi/2(3)^{1/2}$, $\eta_{sph} = \pi/3(2)^{1/2}$ and $\eta_{con} = (1-1/\pi)^{1/2}$ are packing efficiencies of cylinders, spheres (face-centred cubic, FCC) and cones. Dependences of cluster-size distribution function on concentration and temperature lead to those of thermodynamic and kinetic parameters characterizing NS behaviour. For an unsaturated solution, distribution function is determined by cluster equilibrium condition. From Equation (3) one obtains distribution function *vs.* concentration:

$$\begin{aligned} f_n(C) &= \lambda^n \exp\left(\frac{-An + Bn^{2/3}}{k_B T}\right) \quad (dr.) \\ f_n(C) &= \lambda^n \exp\left(\frac{-An + Bn^{1/2}}{k_B T}\right) \quad (b.) \\ f_n(C) &= \lambda^n \exp\left(\frac{-An + B}{k_B T}\right) \quad (c.) \end{aligned} \tag{8}$$

where parameter λ depends on concentration; it is determined by normalization condition:

$$C = C_0 \int_{n=1}^{\infty} n\lambda^n \exp\left(\frac{-An + Bn^{2/3}}{k_B T}\right) dn \quad (droplet)$$
$$C = C_0 \int_{n=1}^{\infty} n\lambda^n \exp\left(\frac{-An + Bn^{1/2}}{k_B T}\right) dn \quad (bundlet) \qquad (9)$$
$$C = C_0 \int_{n=1}^{\infty} n\lambda^n \exp\left(\frac{-An + B}{k_B T}\right) dn \quad (columnlet)$$

where C_0 defines absolute concentration; $C_0 = 10^{-4}$mol·L^{-1} is found requiring saturation in Equation (9). Formation energy of an n-sized cluster is:

$$E_n = n\left(An - Bn^{2/3}\right) \quad (droplet)$$
$$E_n = n\left(An - Bn^{1/2}\right) \quad (bundlet) \qquad (10)$$
$$E_n = n\left(An - B\right) \quad (columnlet)$$

Via cluster-size distribution function one obtains a formula for the thermal effect of an NS solution per mole of dissolved substance:

$$H = \frac{\sum_{n=1}^{\infty} E_n f_n(C)}{\sum_{n=1}^{\infty} n f_n(C)} N_A = \frac{\sum_{n=1}^{\infty} n\left(An - Bn^{2/3}\right)\lambda^n \exp\left[\left(-An + Bn^{2/3}\right)/k_B T\right]}{\sum_{n=1}^{\infty} n\lambda^n \exp\left[\left(-An + Bn^{2/3}\right)/k_B T\right]} N_A \quad (droplet)$$
$$H = \frac{\sum_{n=1}^{\infty} E_n f_n(C)}{\sum_{n=1}^{\infty} n f_n(C)} N_A = \frac{\sum_{n=1}^{\infty} n\left(An - Bn^{1/2}\right)\lambda^n \exp\left[\left(-An + Bn^{1/2}\right)/k_B T\right]}{\sum_{n=1}^{\infty} n\lambda^n \exp\left[\left(-An + Bn^{1/2}\right)/k_B T\right]} N_A \quad (bundlet) \qquad (11)$$
$$H = \frac{\sum_{n=1}^{\infty} E_n f_n(C)}{\sum_{n=1}^{\infty} n f_n(C)} N_A = \frac{\sum_{n=1}^{\infty} n\left(An - B\right)\lambda^n \exp\left[\left(-An + B\right)/k_B T\right]}{\sum_{n=1}^{\infty} n\lambda^n \exp\left[\left(-An + B\right)/k_B T\right]} N_A \quad (columnlet)$$

where N_A is Avogadro number and λ depends on the total concentration by normalization condition (9).

Transfer Phenomena in C-Nanostructures in Solutions

Diffusion coefficient characterizes C_{60} behaviour in solution, governs crystallization, separation and purification conditions, and is estimated by Stokes' formula for diffusion of a spherical particle in a viscous fluid:

$$D = \frac{k_B T}{6\pi\eta r_s} \tag{12}$$

where η is the dynamic viscosity coefficient of the liquid and r_s, the particle radius. Equation (12) is valid for low Reynolds number for a diffusing particle:

$$\mathrm{Re} = \frac{\overline{r}_s \overline{v} \rho}{\eta} << 1 \text{ or } r_s >> \frac{T\rho^2}{\eta^2 \rho_p} \tag{13}$$

where $\overline{v} \approx \left(T/m\right)^{1/2}$ is the particle characteristic velocity, m, mass, ρ, solvent mass density and ρ_p, particle mass density. For viscosity coefficients of typical organic solvents $\eta \sim (1–3)\cdot 10^{-3}\mathrm{N\cdot s\cdot m^{-2}}$, limitation (13) reduces to $r_s >> 10^{-12}$m. Radii r_s determined by Equation (12) with C_{60} in several solvents exceed that of the molecule because of aggregation. Cluster existence suggests dependence of diffusion coefficient on concentration (Bezmel'nitsyn, 1994). For low concentration almost no clusters are formed and diffusion coefficient is close to a molecule. As C_{60} concentration rises, cluster size increases and diffusion coefficient decays [Equation (12)]. For NSs, cluster-size distribution function for non/saturation is expressed by Equations (4) and (8). Diffusion coefficient D comes from:

$$J = -D\nabla C \tag{14}$$

where J is matter flux under concentration-gradient action. From cluster origin of NS solubility, Equation (14) results:

$$J = \sum_n J_n = -\sum_n D_n \nabla C_n \tag{15}$$

where J_n, D_n and C_n are partial values of flux, diffusion coefficient and concentration of n-sized cluster. We derive the relation between diffusion coefficient D_n and radius r_n of n-sized cluster based on models, Stokes' Equation (12) and relations:

$$\begin{aligned} r_n &= \left(\frac{3Mn}{4\pi\rho}\right)^{1/3} \quad (droplet) \\ r_n &\propto n^{1/2} \quad (bundlet) \\ r_n &= constant \quad (columnlet) \end{aligned} \tag{16}$$

where M is C_{60} molecular mass and ρ, cluster density. Combining Equations (14)–(16) and cluster-size distribution function (8), NS diffusion coefficient results:

$$D = D_0 \frac{\int_{n=1}^{\infty} n^{5/3} \lambda^{n-1} \exp\left[\left(-An + Bn^{2/3}\right) / k_B T\right] dn}{\int_{n=1}^{\infty} n^2 \lambda^{n-1} \exp\left[\left(-An + Bn^{2/3}\right) / k_B T\right] dn} \quad (dr.)$$

$$D = D_0 \frac{\int_{n=1}^{\infty} n^{3/2} \lambda^{n-1} \exp\left[\left(-An + Bn^{1/2}\right) / k_B T\right] dn}{\int_{n=1}^{\infty} n^2 \lambda^{n-1} \exp\left[\left(-An + Bn^{1/2}\right) / k_B T\right] dn} \quad (bn.) \qquad (17)$$

$$D = D_0 \frac{\int_{n=1}^{\infty} n \lambda^{n-1} \exp\left[\left(-An + B\right) / k_B T\right] dn}{\int_{n=1}^{\infty} n^2 \lambda^{n-1} \exp\left[\left(-An + B\right) / k_B T\right] dn} \quad (col.)$$

where D_0 is the diffusion coefficient of an NS unit, which was taken equal to that for C_{60} in toluene $D_0 = 10^{-9} m^2 \cdot s^{-1}$. Concentration dependence of cluster-size distribution function causes concentration dependence of NS diffusion coefficient. If the solution contains a mixture of different NS sorts, diffusion of a given one is determined by cluster formation. The NS minor impurity does not form clusters, and shows higher diffusion coefficient than that whose concentration is close to saturated and that is present as large clusters. Source of NSs is provided by a plane layer of solid material constituting mixed NSs of two sorts in which one predominates and the other results a minor impurity (Bezmel'nitsyn, 1996a), which is characterized by diffusion coefficient D_0 of a molecule. Diffusion coefficient of predominating-sort NSs depends on concentration and because of cluster formation it is lesser than isolated NSs. Diffusion equations for predominating sort (concentration C_1) and minor impurity (C_2) present the form:

$$\frac{d}{dx} D_1\left(C_1\right) \frac{dC_1}{dx} + \frac{dC_1}{dt} = 0 \qquad (18)$$

$$D_2 \frac{\partial^2 C_2}{\partial x^2} + \frac{\partial C_2}{\partial t} = 0 \qquad (19)$$

where $D_{1/2}$ denote diffusion coefficients for the first and second components. Equations (18) and (19) present automodelling solutions dependent on $x/t^{1/2}$; however, for concentration dependence of diffusion coefficient the solution calls for numerical calculations. Equation (18) was solved with initial conditions:

$$\begin{aligned} &C_1\left(x = 0, t = 0\right) = C_1^* \\ &C_1\left(t = 0\right) = 0 \\ &C_1\left(x = \infty\right) = 0 \end{aligned} \qquad (20)$$

which match to 1D diffusion from instantaneously actuated plane source where C_1* is saturated concentration. Equation (19) solution with initial conditions:

$$\begin{aligned} &C_2\left(x = 0, t = 0\right) = C_2^0 \\ &C_2\left(t = 0\right) = 0 \\ &C_2\left(x = \infty\right) = 0 \end{aligned} \qquad (21)$$

is known for $C_2^0 << C_1^*$:

$$C_2 = \frac{K}{\left(4\pi Dt\right)^{1/2}} \exp\left(-\frac{x^2}{4Dt}\right) \tag{22}$$

where K is a normalization factor. Equations (18) and (19) solutions were reported as spatial dependences of NS enrichment factor η_e:

$$\eta_e = \frac{C_2\left(x,t\right)C_1\left(x=0,t=0\right)}{C_1\left(x,t\right)C_2\left(x=0,t=0\right)} \tag{23}$$

Difference between diffusion coefficients of isolated NSs of different sorts is neglected. Enrichment factor of NSs some time-dependent distance x^* away from source assumes maximum η_m. Because of automodelling character of Equations (18) and (19) solutions, η_m is time independent and ≈ 20. Elementary separation cell consists of two volumes divided by a porous partition: the initial solution containing two-sort NSs is slowly pumped *via* one part of the cell; pure solvent is pumped in the opposite direction *via* the other. Because of diffusion *via* the porous partition the solution in the second part is enriched with the minor impurity. Maximum enrichment factor $\eta_0 \approx 1.3$ corresponds to the ratio between diffusion coefficients for the two components. In a multistage system, enrichment factor η_f *vs.* number of stages m results: $\eta_f = \eta_0^m$. Temperature–concentration dependences of cluster-size distribution function show possibility of mechanism of NS thermal diffusion. Thermal diffusion coefficient D_T of NS is defined by relation between thermal diffusion flux J_T and temperature gradient (Bezmel'nitsyn, 1996b,c):

$$J_T = -C\frac{D_T}{T}\nabla T \tag{24}$$

Time required for equilibration of cluster-size distribution function [Equations (4) and (8)] is lesser than that required for smoothing spatial temperature nonuniformities. By Equations (4) and (8), temperature gradient causes gradients in cluster partial concentrations, which produces diffusion flows proportional to the temperature gradient. Partial diffusion flux of n-sized clusters is:

$$\begin{aligned} J_n &= -D_n\nabla C_n = -\frac{\nabla T}{T}D_n\left(\frac{-An+Bn^{2/3}}{k_BT}\right)f\left(n\right) \quad (drplt.) \\ J_n &= -D_n\nabla C_n = -\frac{\nabla T}{T}D_n\left(\frac{-An+Bn^{1/2}}{k_BT}\right)f\left(n\right) \quad (bndl.) \\ J_n &= -D_n\nabla C_n = -\frac{\nabla T}{T}D_n\left(\frac{-An+B}{k_BT}\right)f\left(n\right) \quad (col.) \end{aligned} \tag{25}$$

where cluster-size distribution function $f(n)$ [Equations (4) and (8)] shows temperature dependence in the exponential factor. Net diffusion flux is calculated by Equation (25) integration over n, which allows Equation (24) to determine the thermal diffusion coefficient. Diffusion coefficient D_n of n-sized clusters is determined *via* Stokes' Equation (12). Thermal diffusion coefficient of NSs results:

$$D_T = D_0 \int_1^{\infty} \frac{-An + Bn^{2/3}}{k_B T} \frac{f(n)}{n^{1/3}} dn \quad (droplet)$$
$$D_T = D_0 \int_1^{\infty} \frac{-An + Bn^{1/2}}{k_B T} \frac{f(n)}{n^{1/2}} dn \quad (bundlet) \tag{26}$$
$$D_T = D_0 \int_1^{\infty} \frac{-An + B}{k_B T} \frac{f(n)}{n^{1/2}} dn \quad (columnlet)$$

Another mechanism of NS thermal diffusion is caused by larger size of solute molecule as compared to the solvent one. In a temperature gradient, C_{60} molecule is subjected to a force, which is proportional to the difference of pressures acting from the side of fluid on the two opposing hemispheres of the unit, which causes a drift of molecules whose velocity w is estimated by Stokes' formula:

$$w = \frac{k_B \nabla T}{4\pi\eta r} \tag{27}$$

where r is the molecule radius, which results in the estimation of the thermal diffusion coefficient:

$$D_T \approx \frac{k_B T}{4\pi\eta r} \tag{28}$$

Under cluster formation the thermal diffusion mechanism is more efficient than the general one.

Fractal Analysis of C-Nanostructures in Solutions

A C_{60}-solution in benzene at concentration $1 g \cdot L^{-1}$ was studied at RT with static (SLS) and dynamic light scattering (DLS): SLS correlates between relative variation of radiation intensity scattered at given angle and particle average mass; DLS measures spectral line width of scattered radiation because of particles Brownian motion (BM). As characteristic velocity of particle BM is inversely proportional to mean particle radius it allows information on dissolved-particle dimensions. Combining SLS–DLS one determines cluster ratio mass/size. In benzene, C_{60} forms fractal aggregates with dimension 2.1. Cluster growth was observed over 100 days. Structures are destroyed by light shaking after which fractal structure formation and growth are restarted. Growth dynamics of fractal structures gave cluster hydrodynamic radius R_h *vs.* time. If solution was prepared in $N_{2(g)}$ the hydrodynamic radius was 20% higher than in air. Average radius of fractal cluster is 170nm. From relation between cluster fractal dimension D, radius R and number of particles n:

$$n = \left(\frac{R}{r_0}\right)^D \tag{29}$$

where r_0 is C_{60} radius, maximum number of particles during observation time of $4{\cdot}10^6$s is 10^5. Consider coalescence of two particles under Re << 1 [Equation (13)] (Eletskii *et al.*, 1997). The BM follows Stokes–Einstein–Smoluchowski approach. Rate constant *k* for particle aggregation is defined by diffusion:

$$k = 4\pi\left(D_1 + D_2\right)\left(r_1 + r_2\right) \tag{30}$$

where $r_{1/2}$ are the particle radii and $D_{1/2}$, the diffusion coefficients. From Stokes' Equation (12) one derives the rate constant of particles coalescence:

$$k = \frac{8k_B T}{3\eta} F\left(r_1, r_2\right) \tag{31}$$

where function:

$$F\left(r_1, r_2\right) = \frac{\left(r_1 + r_2\right)^2}{4r_1 r_2} \tag{32}$$

is $F \approx 1$ for $r_1 \approx r_2$ and $F \approx 0.25r_1/r_2$ for $r_1 >> r_2$. Typical value for NS saturated concentration in used solvents is $N_0 \sim 10^{18}$cm^{-3}. Characteristic dynamic viscosity coefficient is $\eta \sim 0.01$P. Rate constant for coalescence of two NS clusters of comparable sizes is 10^{-12}cm$^3{\cdot}$s^{-1}, which corresponds to attachment time under diffusion approach $\tau \sim (N_0k)^{-1} \sim 10^{-6}$s. Time for equilibrium-size distribution function of small clusters is of the same order. Real time of fractal-cluster growth 10^6s exceeds estimation by 12 orders of magnitude. Simplest model of fractal cluster growth is diffusion-limited cluster aggregation (DLCA) in which cluster aggregation results of attachment of comparable-size clusters. Rate constant is determined from relations (30)–(32) and is independent of cluster sizes. Growth kinetics of fractal aggregates with average number of particles *n* results:

$$\frac{dn}{dt} = N_0 k' \tag{33}$$

Equation (33) right side is independent of *n* because concentration of *n*-sized clusters is N_0/n while attachment of an *n*-sized cluster to a given one increases its size by *n*. Cluster growth rate is proportional to their product and results N_0k'. From Equation (29), DLCA equation of growth kinetics of fractal cluster of average size *n* results:

$$R = r_0\left(N_0 k' t\right)^{1/D} \tag{34}$$

Time to increase fractal cluster radius by a factor of 500 is 1s, which differs from experiments by six orders of magnitude; DLCA does not apply to experiments. Another model is based on diffusion-limited aggregation (DLA) in which cluster growth results of attachment to a given aggregate of individual particles. If initial number density N_0 of NSs and average concentration N_c of growing clusters are time independent one derives equation describing time variation of average cluster size n:

$$\frac{dn}{dt} = \left(N_0 - nN_c\right)k' \tag{35}$$

In accordance with relations (29–32) one obtains:

$$k' = n^{1/D}\frac{2k_BT}{3\eta} = k_0'n^{1/D} \tag{36}$$

Equation (35) is independent of the size of a small cluster attaching to a large n-sized cluster. Let number of NSs in a small cluster be equal to n_s and concentration of clusters of this size, N_s. Growth rate of large clusters from attachment of n_s-sized small ones results:

$$\left(\frac{dn}{dt}\right)_s = k'N_sn_s \tag{37}$$

Equation (37) summation over all values $s << n$ in view of normalization condition:

$$N_cn + \sum n_sN_s = N_0 \tag{38}$$

provides Equation (33). Growth rate of large fractal clusters does not depend on the shape of size distribution function of small clusters. The feature is caused by the form of cluster size dependence on attachment rate constant (32), which in the limiting case of differing-size clusters does not depend on smaller-cluster size. Equation (33) solution with initial condition $n(t = 0) = 1$ results:

$$t = \frac{1}{k_0'N_c\left(1-1/D\right)}\int_1^n \frac{dn^{1-1/D}}{\overline{n}-n} \tag{39}$$

where $\overline{n} = N_0/N_c$ is the maximum number of particles. Expression (39) for $D = 2$ results:

$$\left(\frac{n}{\overline{n}}\right)^{1/2} = \frac{R}{R_m} = \frac{\exp\left\{t/\left[\tau\left(\overline{n}\right)^{1/2}\right]\right\}-1}{\exp\left\{t/\left[\tau\left(\overline{n}\right)^{1/2}\right]\right\}+1} \tag{40}$$

where $R_m = \left(\bar{n}\right)^{1/2} r_0$ is maximum cluster radius and $\tau = (N_0 k'_0)^{-1}$. In accordance with Equation (40), characteristic time of cluster growth is $\tau\left(\bar{n}\right)^{1/2}$. The conclusion does not correspond to experiments. Because dependence $R(t)$ is close to saturation at last growth stage one assumes that $R_m \approx 200$nm. The $\bar{n} \approx \left(R_m / r_0\right)^2 \approx 3 \cdot 10^5$ and characteristic time of cluster growth $\tau\left(\bar{n}\right)^{1/2} \approx 10^{-3}$ s. Measured time exceeds the estimation by nine orders of magnitude so DLA is unsuitable. Another model is based on reaction-limited cluster aggregation (RLCA) in which cluster growth results of add-on of different-size clusters, with attachment probability of approaching clusters $\gamma << 1$ so that for a pair of clusters to fasten they should undergo a large number of collisions. Equation describing cluster growth kinetics results:

$$\frac{dn}{dt} = \gamma N_0 \left(\frac{k_B T}{2\pi\mu}\right)^{1/2} 4\pi\left(R_1 + R_2\right)^2 \tag{41}$$

where $R_{1/2}$ are the radii of approaching clusters and μ, the reduced mass. *Via* Equation (29) and averaging Equation (41) over cluster-size distribution function one derives:

$$\frac{dn}{dt} = J\gamma N_0 \left(\frac{k_B T}{2\pi m_0}\right)^{1/2} 4\pi r_0^2 n^{2/D - 1/2} \tag{42}$$

where r_0 is C_{60} molecular radius and m_0, mass. Dimensionless coefficient J depends on cluster-size distribution function and fractal dimension D. The $J = 6.8$ for $D = 2$ and simplest form of function:

$$f \approx \exp\left(-\frac{n}{n_0}\right) \tag{43}$$

where n_0 is the average number of particles in the cluster. Integration of (42) results:

$$R = r_0 \left[8\pi\gamma N_0 J \left(\frac{3}{2} - \frac{2}{D}\right)\left(\frac{k_B T}{2\pi m_0}\right)^{1/2} r_0^2 t\right]^{2/(3D-4)} \tag{44}$$

The RLCA leads to unlimited growth of cluster radius with time. Because $D \approx 2$, Equation (44) is almost linear, which differs from experiments, so RLCA is not applicable to fractal-cluster growth. Agreement between calculated and measured evolution of fractal-cluster growth is reached after RLCA modification; let us assume that cluster attachment probability γ depend on cluster size as:

$$\gamma = \gamma_0 \left(\frac{r_0}{R}\right)^{\alpha} \tag{45}$$

which results:

$$R = r_0 \left[4\pi\gamma_0 N_0 \left(\frac{3}{2} + \frac{\alpha}{2} - \frac{2}{D} \right) \left(\frac{k_B T}{2\pi m_0} \right)^{1/2} r_0^2 t \right]^{4/(6D+2\alpha D-8)} \quad (46)$$

Equation (46) calculated for $D = 2.08$, $\alpha = 2$ and $\gamma_0 = 10^{-7}$ showed that the dependence agrees with experiments and computations in the simplified model with $D = 2$.

Thermodynamic Parameters of Fullerite

From Equations (3) and (4), C_{60} solubility is expressed by the difference between chemical potential of a molecule in solution and solid. Change in C_{60} crystal structure in equilibrium with solution is reflected on solution properties (Smirnov, 1992; Vorob'ev & Eletskii, 1995, 1996). To the class of structures of inert-gas crystals belongs fullerite at a temperature exceeding phase-transition point accompanied with molecular-orientation disordering and rotation defreezing. Crystal structure of inert gases is FCC. All thermodynamic parameters in dimensionless form coincide because of low sensitivity of thermodynamic characteristics to small differences in potential $V(R)$ for atoms of several kinds; to determine the substance thermodynamic characteristics two parameters are needed: well depth D and equilibrium distance R_0. The D–R_0 determine a dimensionless variable set that generates thermodynamic-function similarity. Temperature dependence of saturated vapour pressure $p(T)$ for inert gases is:

$$\frac{p}{p_0} = \phi\left(\frac{T}{D} \right) \quad (47)$$

where $p_0 = D/R_0^2$ is pressure natural unit. As density natural unit $\rho_0 = \mu/R_0^3$ is used where μ is atomic mass. System critical parameters are expressed by inherent natural units:

$$p_c = (0.131 \pm 0.001)\, p_0 \quad (48)$$

$$\rho_c = (0.301 \pm 0.001)\, \rho_0 \quad (49)$$

$$T_c = (1.04 \pm 0.02)\, D \quad (50)$$

Numerical factors in Equations (48)–(50) were obtained from experiment statistical averaging for inert gases. Other thermodynamic parameters are determined by system natural units. Taking into account Expressions (48)–(50), Equation (47) results:

$$\frac{p}{p_c} = f\left(\frac{T_c}{T}\right) \tag{51}$$

The relation is universal; C_{60} in solid is related to the class of systems of spherically symmetrical close-packed and short-range interacting particles. In C_{60} and inert-gases crystals, similarity allows estimating C_{60} thermodynamic quantities, which fit experiments. Comparison of estimates to numerical calculations, and with approximations of intermolecular interaction potential obtained from measurements of saturated vapour pressure, permits parameter refinement and determination of sensitivity of C_{60} thermodynamic characteristics to details in intermolecular interaction potential. Let us consider C_{60}-crystal at temperatures exceeding critical point for an orientational disorder phase transition, which is accompanied by defreezing free thermal rotation of C_{60} molecules in crystal about their axes. Molecules of C_{60}-crystal are spherically symmetric with a pair intermolecular interaction potential. Binding energy of C-atoms in a C_{60} molecule is more than that of C_{60} units in a crystal. In systems with short-range interaction and FCC, R_0 determines nearest-neighbour distance. For C_{60}-crystal at $T = 260$K, $R_0 = 1.006$nm (Eletskii & Smirnov, 1991), which provides crystal density $\rho = 1.69$g·cm^{-3} with uncertainty. Calculated–experimental D values result in $D = (0.044 \pm 0.306)$eV, which statistical spread hinders D use for estimation of fullerite thermodynamic magnitudes. One reason is in obtaining D from fitting measured temperature dependences of saturated vapour by Arrhenius equation. Since experiments are performed in limited temperature range whereas temperature dependence $p(T)$ differs from Arrhenius exponent, experiments in dissimilar temperature ranges vary from each other. Parameter accurate estimation is obtained from similarity relations (47–51); to use the dependence, similarity for close-packed systems with short-range interaction is appropriate. Measured temperature dependences of saturated vapour pressure for C_{60} are expressed in dimensionless form (51). The D is determined by fitting function $p(T_c/T)$ for C_{60} with the inert-gases function. Dependences of reduced saturated vapour p/p_c on T_c/T for {Ne, Ar, Kr, Xe} are calculated. Similarity law (51) is fulfilled. Quantum effects cause data deviation for Ne in low-temperature (LT) region. Dependence $p(T)$ presents a characteristic bend at triple point. Data are measured and calculated for saturated vapour pressure of C_{60} (Abrefah *et al.*, 1992; Baba *et al.*, 1994; Chen *et al.*, 1992; Korobov & Sidorov, 1994; Mathews *et al.*, 1992; Richter *et al.*, 1994; Tokmakoff *et al.*, 1991) processed with similarity laws (47–51), which critical temperatures–pressures were calculated *via* Equations (48) and (50) for fixed D. Both are proportional to D. Dependences $p(T)$ do not agree with similarity law (51) for inert gases. Best fitting of dimensionless temperature dependences of saturated vapour pressure for C_{60} with inert gases is reached for $D = 0.257$eV, which lies in D range. Temperature dependence of saturated vapour pressure for C_{60} determined with D is calculated, which agrees with universal dependence for inert gases. Allowable D range is 0.013 ± 0.257eV, which is narrower. Assuming similarity laws (47)–(49) and $D = 0.257$eV one obtains C_{60} critical parameters:

$$\begin{aligned} T_c &= 3100K \\ p_c &= 53.6atm \\ \rho_c &= 0.5g \cdot cm^{-3} \end{aligned} \tag{52}$$

Compressibility factor at the critical point is:

$$Z_c = \frac{\mu p_c}{RT_c\rho_c} \tag{53}$$

where μ is the molecular mass and R, gas constant. Dimensionless heat of evaporation q at $T = 0$ is:

$$q = \frac{q_0\rho_c}{p_c} \tag{54}$$

where q_0 is specific heat of evaporation at $T = 0$. Data related to isothermal compressibility modulus K_T are evaluated, which for C_{60} were taken from Hebard (1993). Similarity laws (47)–(51) cause constant ratio K_T/p_c. Law fulfilment is poor because of experiment limited accuracy. Energy at Debye temperature is proportional to characteristic energy of crystal-lattice vibration $\hbar\omega$. From D, R_0 and μ, a quantity with frequency dimension is built:

$$\omega_D \sim \frac{\left(D/\mu\right)^{1/2}}{R_0} \tag{55}$$

Debye temperature is proportional to frequency, which is inherent natural unit. Its dimensionless value is:

$$\theta_0 = \frac{\omega_D R_0}{\left(D/\mu\right)^{1/2}} \tag{56}$$

In accordance with similarity laws, dimensionless parameter for all crystals in the same class is close. Values of θ_0 are determined by Equation (56) from experiments and estimations. Similarity relation (56) is fulfilled for inert-gas crystals. Relation violation for Ne is attributed to LT quantum effects. Value of θ_0 for fullerite is twice that for inert-gas crystals, which is not surprising since over a temperature range near Debye temperature (74K) fullerite crystal is orientationally ordered so that its structure is dissimilar to FCC. A C_{60} molecule whose thermal rotation is frozen is not spherical. Similarity-law fulfilment, which is valuable for particles with spherically symmetric interaction potential, is not expected. Contribution to fullerite Debye spectrum is made by intramolecular vibrations, which are not considered in similarity laws. If intra and intermolecular vibration frequencies in Debye temperature range are of the same order this is reflected in Debye frequency value. For similarity law (56), fullerite Debye temperature is set at $\omega_D =$ 40K. Can C_{60} exist in a liquid state (Ashcroft, 1993; Cheng *et al.*, 1993; Hagen *et al.*, 1993)? Calculations with a model intermolecular interaction potential give positive and negative answers (Girifalco, 1991, 1992). The latter are explained by difference of intermolecular interaction potential from Lennard–Jones (LJ) one. An insignificant deviation of C_{60} thermodynamic parameters from similarity laws shows a weak sensitivity of parameters to specific form of intermolecular interaction potential, which permits expecting that fullerite, as other systems of close-packed spherically symmetric molecules with a short-range interaction, exist in liquid state in some temperature region. At atmospheric pressure, a narrow region ranges from melting temperature T_m = 1729K to boiling temperature T_b = 1820K. Obstacle is thermal decomposition of C_{60} molecules; however, calculations (Zhang *et al.*, 1993) and experiments (Kolodney *et al.*, 1994) show C_{60} thermal stability at atmospheric pressure up to 2000K.

Real-Space Imaging of Nucleation and Growth in Colloidal Crystallization

In colloidal crystallization, surface–bulk energy competition is reflected in free energy for a spherical crystallite:

$$\Delta G = 4\pi R^2 \gamma - \frac{4\pi}{3} R^3 \Delta\mu N \quad (57)$$

where R is the radius, γ, surface tension, $\Delta\mu$, difference between liquid and solid chemical potentials and N, number density of particles in crystallite (Gasser *et al.*, 2001). Critical-nucleus size is $R_c = 2\gamma/(\Delta\mu N)$ corresponding to maximum ΔG [Equation (57)]. Crystallite radius of gyration R_g is related to number of particles n in every crystallite as $n(R_g) \propto R_g^{\,D}$, with fractal dimension $D = 2.35 \pm 0.15$ for all values of packing volume fraction ϕ; fractal behaviour reflects surface roughness. Crystal/fluid interfacial tension is a key parameter that controls nucleation yet γ is difficult to calculate and measure experimentally but one measures γ examining smallest-nuclei statistics. For $R << R_c$, surface term in Equation (57) dominates crystallite free energy and one expects number of crystallites to be $n_{cry}(A) \propto \exp[-A\gamma/(k_BT)]$ where A is surface area approximated by an ellipsoid. The $\gamma \approx 0.027 k_BT/r_0^2$ (r_0 is particle radius =1.26μm) and decays slightly with rising ϕ; it agrees with DFT for hard-spheres and LJ systems. Low-γ measurement is consistent with observed rough surfaces of crystallites, which reflects softer-potential effects because of particles weak charges. Approximating critical nucleus as an ellipsoid with $n_c \approx 110$ one obtains A_c = 880μm^2, $\Delta\mu \approx 0.13 k_BT$ and $\Delta G(A_c) \approx 7.4 k_BT$.

Dimensional Analysis for the Early and Later Stages of Fusion-Site Expansion

Both stages of cluster fusion, a fast early and a slower later stage, were detected in vesicle fusion. During the former, fusion site opened rapidly: expansion velocity of site rim was ≈4cm·s^{-1}. Fusion pore opens up to micrometres in a hundred microseconds. One relates time τ_{early} to fast relaxation of membrane tension. Cluster tension achieved before fusion was in membrane stretching regime. The τ_{early} should be governed by membrane stretching relaxation. Viscous dissipation is associated with two contributions: in-plane dilatational shear as fusion site expands and intermonolayer slip among multiplayer-membrane leaflets in fusion-site zone; the latter is negligible for fusion-site diameter $L > 0.5$μm. The $\tau_{\text{early}} \sim \eta_s/\sigma$ where η_s is surface dilatational viscosity coefficient of membrane ≈0.35μN·s·m^{-1} with units [bulk viscosity coefficient]·[membrane thickness] and σ, membrane tension (Haluska *et al.*, 2006). For $\sigma \approx$ 5mN·m^{-1} close to rupture (≈7mN·m^{-1}) one obtains $\tau_{\text{early}} \sim$ 100μs in agreement with experiments (≈300μs). During later fusion stage, site expansion velocity slowed down by two orders of magnitude. Dynamics was governed by fluid volume displacement ΔV around fusion site between fused clusters. Restoring force was related to membrane bending elasticity. Decay time $\tau_{\text{late}} \sim \eta\Delta V/\kappa$ where η is bulk viscosity coefficient of solvent, $\Delta V \sim R^3$ and κ, bending elasticity modulus of membrane (≈10^{-19}J). For a cluster size of R = 20μm one obtains $\tau_{\text{late}} \sim$ 100s, which is time scale measured for complete fusion-site opening. When two clusters fuse at contact points and form fusion sites their coalescence leads to small, contact-zone clusters. Consider three fusion sites, which expand and touch each other in such a way that they enclose a roughly triangular segment of contact zone. If all sites are circular and grew up to a diameter L_1, enclosed contact-zone segment will form a contact-zone cluster of radius $R_{czc} = [1/3^{1/2}–1/2]L_1 \approx 0.08L_1$. Sites coalescence leads to small contact-zone clusters encapsulating solvent. Clusters are interconnected by thin tethers because pinching membrane off requires energy. Fusion-induced cluster formation resembles membrane processes during cell division when one looks at them in a time-reversed manner. During initial division stages, cell accumulates membrane in small-vesicle form, which defines division plane and transforms into two adjacent cell membranes. From dimensional analysis an appropriate time scale τ for the later stage of fusion-site expansion is found. The expansion driving force is provided by membrane tension σ whereas hydrodynamic-Stokes friction is governed by solution viscosity coefficient η. The system is characterized by two length scales: membrane thickness l and a typical cluster size R. It is chosen $R = (R_1+R_2)/2$ where $R_{1/2}$ are clusters radii before contact. Time scale, which one can obtain from a combination of the four variables σ, η, l and R, results $\tau = (\eta R/\sigma)f(l/R)$ with dimensionless function $f(l/R)$. Because $l << R$ one can replace $f(l/R)$ by $f(0)$ and ignore corrections of order (l/R). Let v (in m·s^{-1}) be average site expansion velocity for a single site. The same order of magnitude for average expansion velocity is deduced assuming that fusion process starts with $N > 1$ fusion sites. Fusion sites grow until they start to touch and coalesce; they then create a coalesced site of diameter L if each site grew up to $L/N^{1/2}$, which implies an average expansion velocity $\approx v/N^{1/2}$ of the same order of magnitude even if N were as large as 10.

Description of the Asymptotic Analysis of Coagulation–Fragmentation Equations

Approximating coagulation–fragmentation equation behaviour is challenging; it is presented by asymptotic analysis. Results were checked *vs.* numerical solutions dealing with Becker–Döring equations. Binding-energy models of an n-sized cluster for spherical and rod/column-like aggregates are:

$$\begin{aligned} \varepsilon_n &\sim (n-1)\alpha_{BD}k_BT - \frac{3}{2}\sigma_{BD}n^{2/3} \quad (spherical\ aggregts.) \\ \varepsilon_n &= (n-1)\alpha_{BD}k_BT \quad (rod\ /\ column-like\ aggrgts.) \end{aligned} \tag{58}$$

where $\alpha_{BD}k_BT$ is monomer–monomer binding energy, $\sigma_{BD} = 2\gamma(4\pi v^2/3)^{1/3}$, γ, surface tension and $v = V/M$, molecular volume [$A \sim (n–1)\alpha_{BD}$, $B \sim 3\sigma_{BD}/2 = 3\gamma(4\pi v^2/3)^{1/3}$] (Cañizo, 2005; Cañizo *et al.*, 2004; Neu *et al.*, 2002). In Becker–Döring model, reactions between monomers and other clusters are taken into account. Equations (58) are suitable for aggregates of certain kinds of lipids forming rod-like clusters (Notman *et al.*, 2006). Lipid molecules present a hydrophilic head and hydrophobic tail so in aqueous solution, they spontaneously arrange themselves so that tails are away from surrounding water and heads in contact with it. Depending on particular-molecule shape they form spherical aggregates with tails pointing inwards and heads outwards, or form lipid bilayers where lipid molecules form a double layer with heads on the outside and tails inside. Lipid clusters in aqueous solution are called *micelles* and formation is named *micellization*; to determine time scale one needs a measure of kinetic coefficient of *d* decay reaction, which was set equal to one. A relation is an equation, which in dimensional units is $<n> \approx (d\pi t)^{1/2}$. In case self-similar size distribution is not reached during intermediate phase, another way to determine *d* is to study equilibration era and compare experimental size distribution with model numerical solution. Combining $\tau_{\text{early}} \sim \eta_s/\sigma$ with $<n> \approx (d\pi t)^{1/2}$ it is obtained $<n> \approx (d\pi\eta_s/\sigma)^{1/2}$.

Leak-In Kinetics of Pre-Equilibrium Membrane Pores: Mono/Multi-Exponential

Heimburg (2010) reviewed lipid ion channels. Amphitropic proteins of family B-cell lymphoma type 2 (Bcl-2) regulate apoptosis controlling permeability of outer mitochondrial membrane, which is because of Bcl-2-associated X (Bax) pores. Nature, mechanism and properties remain elusive because of lack of structural information about Bax-membrane complexes; however, Fuertes *et al.* (2010) showed that Bax main α-structure allowed designing minimal active versions. Peptides encompassing sequences of amphipathic helices from central hairpin of Bax, Bcl-2 homology domain-3 interacting domain death (Bid) and Bcl-extralarge (xL) exhibited membrane binding and permeabilization, which resemble complete proteins. Peptide fragments of Bax cause membrane poration. An action-mechanism study (*e.g.*, modelling, quantitative analysis of vesicle-leakage kinetics) permitted counting/sizing pores, and studying formation, equilibration and dynamics. Pores formed by pro-apoptotic Bax fragment α5 relaxed to a smaller size and were kept at equilibrium. Fuertes *et al.* (2011) published a lipocentric view of peptide-induced pores. Time constant results:

$$\tau_{flux} = \frac{V}{A_0 P} = \frac{V}{A_0 D/m} \tag{59}$$

and allows calculating pore area A_0 corresponding to *initial* (short-term) pores where *V* is vesicle volume, *D*, *diffusion* coefficient, *m*, effective pore length and $P = D/m$, pore *permeability* coefficient. Pore-size change was modelled as an exponential decay from a state corresponding to an initially *large* area A_0 into other of a *smaller* one A_∞ with relaxation time τ_{relax}:

$$A(t) = A_\infty + (A_0 - A_\infty)e^{-t/\tau_{relax}} \tag{60}$$

meaning that characteristic flux time of the dye leak-in is time dependent and Equation (59) converts into:

$$\tau_{flux}(t) = \frac{V}{A(t) D/m} \tag{61}$$

Drug–Membrane Interactions: Analysis, Drug Distribution, Delivery, and Modelling

Seydel and Wiese (2002) reviewed drug–membrane interactions: analysis, drug distribution and modelling. Drug design interest focussed on ligand–protein interaction in specific receptor and enzyme form. Target proteins are embedded in membranes and it is assumed that ligand bioactivity arises from binding to membrane-embedded proteins. Lipid environment plays a more passive role. Ligand–membrane interaction influence on drug activity and selectivity was underestimated. *Nonspecific* drug–membrane interaction involves dealings with specific phospholipid structures. Although lipid layer is a dynamic fluid it is organized. Membranes do not consist of only lipid but possess polarized phosphate groups, and neutral or cationic/anionic head groups, and they are structured and chiral. Interaction with the structures presents a role in membrane drug partitioning, orientation, conformation and physicochemical properties/functioning. Drug–membrane interactions play a position in drug transport, distribution, accumulation, efficacy and resistance. Molecular classification and periodic table of local anaesthetics (procaine analogues) were reported (Castellano & Torrens, 2009; Torrens & Castellano, 2006, 2011b). Iordache (2011, 2012, 2014) reviewed local anaesthetic effects on membranes in general polystochastic model (PSM). Effect of chemical inactivation of membrane functions by drugs and chemicals consists of modifying conformations of active membrane components. Anaesthetics act on excitable membrane in a charged form inside nerve axon (Strichartz, 1987). Exponential correlations were proposed for amplitude relaxation of compound action potential. Frangopol & Morariu (1988) and Iordache *et al.* (1988) applied multi-scale models to experiments describing action potential decay by procaine. Data were action potential relative amplitude of procaine 10mM *vs.* time step n. Value of α correlates to mean residence time $\bar{n} = 1/(1-\alpha)$. Taking $\bar{n} = 6.25$ it results $\alpha = 0.84$ (time step 5min). Objective function $S(M)$ for different numbers of scales M allows selecting a model, which is a truncation number M. The $M = 1$ provides a picture for which coefficients are $q_0 = 0.878$ and $q_1 = 0.081$. Perturbed ($M = 1$) and ideal ($M = 0$) models were compared. If flow and kinetics statistics and coefficients q_m are known, multi-scale modelling represents a progress towards an *a priori* design of anaesthetic treatments. More process scales suggest that excitation and inhibition of a given one is efficient for drug kinetics control; to take into account stochastic character at different scales, variable confidence intervals are added. Local anaesthetic effects on multi-compartmental membranes were studied. Stochastic compartmental representations are widely used in modelling, *e.g.*, drug kinetics (Jacquez, 1985; Yu & Wehrly, 2004). In biochemical experiments, ligand population decay is observed because of capture by appropriate compartment systems; *e.g.*, drug–membrane interaction. Drug classes (*e.g.*, protein-linking aldehydes, local anaesthetics) present relaxation effects on functions of membrane ionic channels and pumps. Multi-compartment case was studied. A compartment is defined indicating not only location but also change of a material-object state. Object transport includes position and nature changes. Model prediction was discussed with respect to effect of cross-linking glutaraldehyde (GA), 0.25% on amplitude of compound action potential of frog sciatic nerve. Experimental results are height of compound action potential *vs.* time. Objective function $S(M)$ for multicompartmental models was calculated. Relaxation rate a resulted $a = (k+1)/\bar{\tau}$ with $\bar{\tau} = 70.275$ min mean relaxation time of studied system and $k + 1$ compartments. Two compartments and four scales ($M = 3$) provide a relaxation picture as objective function $S(M)$ stays unchanged for $M \geq 3$. It resulted $q_0 = 0.98$, $q_1 = -0.055$, $q_2 = -0.226$ and $q_3 = -0.133$. Results are interpreted as a *label* of a specific interaction where coefficients q_m depend on contribution of relaxation m-th scale. The two compartments obtained in correlation are interpreted in terms of *two-membrane theory*, which considers skin as two functional membranes in series: apical and latero-basal membranes. Spectrum q_m is similar for several cross-linking aldehydes, which supports an only physical mechanism of membrane interactions. The theory is applied to a broad class of biophenomena. Applications include membrane function inactivation, bacteria killing by disinfectants, ion-channel kinetics, *etc.* Case studies pertain to bio/ecosystems. Entropy production for a non-equilibrium system originally initiated by biophysical and chemical problems will be considered as a model-categorization example; to account for membrane-transport selectivity, membrane-pore concept was developed as a transport channel for different ion and molecule types. The pore is a 1D array of k stable sites for ion/molecule kinds. Particle state is determined at time n *via* row vector $p(n) = [p_1(n),\ldots p_k(n)]$, where $p_i(n)$ is the probability that the particle be in i-th stable site after n transitions. Denote by p_{ij} the one-step probability of transition from i to j and by $\boldsymbol{P} = (p_{ij})$, $(1 \le i, j \le k)$, the corresponding transition matrix. Stochastic model results:

$$p(n+1) = p(n)\boldsymbol{P} \tag{62}$$

The model presents applications in drug kinetics, pharmacology, intracellular transport, metabolic systems, ecosystems, *etc*. A two-state model is Adam's co-operative one for nerve excitation (Schnakenberg, 1977) in which axon membrane contains an irregular lattice of active centres, every one binding monovalent K^+ (excited state 1) and bivalent Ca^{2+} (ground state 2). Denote by $p_1(n)$ the probability for an active centre to be in excited state and by $p_2(n) = 1 - p_1(n)$ in ground one. From Equation (62), difference equation results:

$$p_1(n+1) - p_1(n) = -p_{12}p_1(n) + p_{21}p_2(n) \tag{63}$$

In order to take into account different scales of time it is necessary to translate by model categorization discrete time n to expansion N:

$$N = n + \varepsilon w_1 n + \ldots + \varepsilon^M w_M n \tag{64}$$

with w_m, $m = 1,\ldots M$, constants. Probabilities $p_i(n)$, $i = 1, 2$ are translated to:

$$P_i(N) = p_i^0 + \varepsilon p_i^1(n) + \ldots + \varepsilon^M p_i^M(n) \tag{65}$$

Transition probabilities p_{ij} are translated by model categorization to:

$$P_{ij}(N) = p_{ij}^0 + \varepsilon p_{ij}^1(n) + \ldots + \varepsilon^M p_{ij}^M(n) \tag{66}$$

Difference Equation (63) becomes *via* model categorization, non-Archimedean (NA) equation:

$$\frac{P_1(N) - P_1(N_1)}{(N - N_1)} = -P_{12}(N_1)P_1(N_1) + P_{21}(N_1)P_2(N_1) \tag{67}$$

where $N \ne N_1$. Consider:

$$N = n + \varepsilon n + \ldots + \varepsilon^M n,\ N_1 = n - 1 + \varepsilon(n-1) + \ldots + \varepsilon^M(n-1) \tag{68}$$

In far-from-equilibrium phenomena [ion transport, nerve excitation, adenosine triphosphate (ATP) synthesis, *etc*.], linearity appears unrealistic and transition probabilities p_{ij} depend on $p_i(n)$, $1 \leq i \leq k$. Stochastic chain is non-Markovian. In co-operative model for nerve excitation it was considered that transition probabilities or rate constants are functions of Ca^{2+}/K^{+} concentrations on the two membrane sides. Co-operative mechanism assumes that transition probabilities, for a certain active centre to change from ground to excited state or inversely, depend on states of neighbouring active centres at transition time. The Ca^{2+} (K^{+}) bound to a certain centre in ground (excited) state receives an additional binding energy for each neighbouring centre that be in ground (excited) one, according to which it is supposed:

$$p_{12}\left(n\right) = k_1\left[1 - k_2 p_1\left(n\right)\right],\ p_{21}\left(n\right) = k_3\left[1 - k_4 p_2\left(n\right)\right] \tag{69}$$

where $0 < k_i < 1$, $i = 1, 2, 3, 4$ are constants. In NA frame, transition probabilities p_{ij} result translating $p_i(n)$ to $P_i(N)$:

$$P_{12}\left(N\right) = k_1\left[1 - k_2 P_1\left(N\right)\right] \tag{70}$$

$$P_{21}\left(N\right) = k_3\left[1 - k_4 P_2\left(N\right)\right] \tag{71}$$

Information entropy associated to a stochastic model as Equation (62) results:

$$S\left(N\right) = -\sum_i P_i\left(N\right)\ln P_i\left(N\right) \tag{72}$$

By model categorization, NA discrete entropy is defined by $S(N)$ expansion:

$$S\left(N\right) = s_0\left(n\right) + \varepsilon s_1\left(n\right) + \ldots + \varepsilon^M s_M\left(n\right) \tag{73}$$

In case of a two-state model as Adam's co-operative one of nerve excitation and two timescales it results from Equation (72), equating coefficients of different ε, powers of:

$$s_0\left(n\right) = -\left(p_1^0 \ln p_1^0 + p_2^0 \ln p_2^0\right) \tag{74}$$

$$s_1\left(n\right) = -\left(p_1^1 \ln p_1^1 + p_2^1 \ln p_2^1\right) - \left(p_1^1 + p_2^1\right) \tag{75}$$

It results:

$$p_1^0(n) = p_1^0(n-1) - k_1\left[1 - k_2 p_1^0(n-1)\right] p_1^0(n-1) + k_3\left[1 - k_4 p_2^0(n-1)\right] p_2^0(n-1) \quad (76)$$

$$\begin{aligned} p_1^1(n) &= p_1^1(n-1) - k_1 k_2 p_1^1(n-1)\, p_1^0(n-1) - k_3 k_4 p_2^1(n-1)\, p_2^0(n-1) \\ &+ k_1\left[1 - k_2 p_1^0(n-1)\right]\left[p_1^1(n-1) + p_1^0(n-1)\right] + k_3\left[1 - k_4 p_2^0(n-1)\right]\left[p_2^1(n-1) + p_2^0(n-1)\right] \end{aligned} \quad (77)$$

Variation of NA entropy $S(N)$ in NA time is:

$$\frac{S(N) - S(N_1)}{N - N_1} = \left\{s_0(n) - s_0(n-1), \ldots \left[s_M(n) - s_M(n-1)\right] - \left[s_{M-1}(n) - s_{M-1}(n-1)\right]\right\} \quad (78)$$

where N and N_1 are given by Equation (68). Zero-th-order level $M = 0$ is not necessarily an equilibrium one. Numerical simulations outlined that after a period $s_0(n)$, variation becomes extinguished. Valence topological charge-transfer indices were reported for dipole-moment calculation of transdermal-delivery drug models (percutaneous absorption enhancers) of cancer chemotherapy agent 5-fluorouracil (5-FU), phenylalcohols (Torrens, 2003), 4-alkylanilines (Torrens, 2004b) and aliphatic amines (Torrens & Castellano, 2009b). Fractal analysis of transdermal-delivery drug models was published: phenylalcohols (Torrens, 2004c)/4-alkylanilines (Torrens & Castellano, 2008b).

Modelling Mass Transfer Rate: Biocoagulation–Flocculation

Jimoda *et al.* (2013) developed a model for destabilized particle transfer during biocoagulation-flocculation of turbid drinking water, from relative motion of undermined particles to the reference one. Mass conservation of destabilized particles in the system was accomplished identifying assumptions defining appropriate initial and boundary conditions. Suppositions follow: homogeneous system is presumed, 1D mass (particle) transfer occurs, process is isothermal, initial suspended particles (turbidity) concentration in the system is uniform and external resistance to particle transport is neglected. Consider a set of BM spherical particles with radius R_j diffusing in the vicinity of a target sphere with radius R_i centred at origin. Frequency of *j*-spheres collisions on the target sphere is found *via* a stationary diffusion model: every *j*-sphere that hits the target one is removed from the bulk while simultaneously a new *j*-sphere is added to the bulk. Continuity equation for *j*-spheres number concentration C_j is:

$$\begin{aligned} &\textit{Time rate of change of reference particle} \\ &= \textit{influx of mass into the reference sphere particle} \\ &- \textit{outflux of mass from the reference sphere particle} \end{aligned} \quad (79)$$

$$\textit{Time rate of change of mass of the floc} = \frac{\partial(C)4\pi r^2}{\partial t} \quad (80)$$

$$\textit{Influx of mass into the floc} = J4\pi r^2 \tag{81}$$

$$\textit{Outflux of mass from the floc} = \left(J + \frac{\partial J}{\partial r}\right)4\pi r^2 \tag{82}$$

where J = flux of particle across unit area per time. Combining Equations (79)–(82) gives:

$$\frac{\partial\left(C\right)4\pi r^2}{\partial t} = J4\pi r^2 - \left(J + \frac{\partial J}{\partial r}\right)4\pi r^2 \tag{83}$$

Simplifying Equation (83):

$$\frac{\partial\left(C\right)4\pi r^2}{\partial t} = -\frac{\partial J}{\partial r}4\pi r^2 \tag{84}$$

Dividing Equation (84) by 4π gives:

$$\frac{r^2\partial C_j}{\partial t} = -r^2\frac{\partial J}{\partial r} \tag{85}$$

where C_j = concentration of spherical particle *j*. Recalling first Fick's law:

$$J = -D\frac{\partial C_j}{\partial r} \tag{86}$$

where D = particle diffusion coefficient. Substituting Equation (86) into Equation (85) gives:

$$\begin{aligned} \frac{r^2\partial C_j}{\partial t} &= r^2\frac{\partial}{\partial r}D\frac{\partial C_j}{\partial r} \\ \frac{\partial c_j}{\partial t} &= \frac{\partial}{\partial r}D\frac{\partial C_j}{\partial r} \end{aligned} \tag{87}$$

Equation (87) is the partial differential equation (PDE) that describes rate of mass (destabilized particle) transfer during biocoagulation/flocculation of coal-rich wastewater; however,

$$D = \frac{K_R}{8\pi R_p} \tag{88}$$

Menkiti & Onukwuli (2010) obtained Equation (88) from von Smoluchowski rate constant expression where $R_p = 2a$ = particle diameter, a = particle radius and K_R = von Smoluchowski rate constant for coagulation/flocculation. Initial condition is:

$$C_j\left(t=0,r\right)=C_{j0} \tag{89}$$

Boundary conditions are:

$$C_j\left(r=0,t\right)=C_{e0} \tag{90}$$

$$C_j\left(r=a,t\right)=C_{e0} \tag{91}$$

where C_{e0} = concentration of particles in the system in equilibrium with boundaries surface concentration.

$$C_0 = initial\ suspended\ particle\ concentration\ value\ of\ coal-rich\ effluent \tag{92}$$

From Equation (87) let $C_j r^2 = U$; Expression (87) becomes:

$$\frac{\partial C_j}{\partial t}=\frac{\partial}{\partial r}D\frac{\partial C_j}{\partial r}=\frac{\partial U}{\partial t}=\frac{\partial}{\partial r}D\frac{\partial U}{\partial r} \tag{93}$$

Let us define the following dimensionless parameters. Dimensionless space variable:

$$\eta=\frac{r}{R}$$

and dimensionless time variable:

$$\tau=\frac{Dt}{R^2}$$

Introducing dimensionless concentration function:

$$U^*=\frac{U_t-U_e}{U_0-U_e} \tag{94}$$

Dimensionless function changes initial and boundary conditions to:

$$U^*\left(\tau=0,\eta\right)=1 \tag{95}$$

$$U^*(\tau,0) = 0 \tag{96}$$

$$U^*(\tau,1) = 1 \tag{97}$$

Substituting dimensionless variables into Equation (93) gives:

$$\frac{\partial U^*}{\partial \tau} = \frac{\partial^2 U^*}{\partial \eta^2} \tag{98}$$

Dimensionless Equation (98) predicts the rate of suspended particle transfer at different conditions.

Chelate and Co-operativity Effects in Metal– and Protein–Ligand Bindings

Thermodynamics of complexes formed by metal cation and ligands parallel protein–ligands binding. Chelate and co-operativity-effect concepts are useful to analyze relations between the two fields. Chelate effect, which is apparent in complex-increased stability, occurs when two or more co-ordinating donor atoms belong to the same ligand molecule (Schwarzenbach, 1952), whereas co-operativity one occurs when both or more donor atoms belong to different molecules (Weber, 1975). Homotropic (the same donor atoms) and heterotropic (different donor atoms) chelate effects are observed as homo and heterotropic co-operativity ones exist. The analogy leads to choice of standard states, equilibrium constants and energy scale by which both are measured. Equilibrium constants suitable to establish common scale are calculated as ratio of operational equilibrium constants, every one is expressed in homogeneous reciprocal concentration units. Chelate effect comprehends co-operativity one. Two chelation equilibrium constants K_ε and K_η ($K_{\varepsilon'}$, $K_{\eta'}$ for heterotropic chelate effect) are proposed to evaluate chelate effect and one equilibrium constant K_γ ($K_{\gamma'}$) to evaluate co-operativity effect. Relations $K_\eta = K_\varepsilon K_\gamma$ and $K_{\eta'} = K_{\varepsilon'} K_{\gamma'}$ hold. Consistent energy scale results:

$$\begin{aligned} \Delta\mu_\gamma &= -RT\ln K_\gamma \\ \Delta\mu_{\gamma'} &= -RT\ln K_{\gamma'} \end{aligned}, \text{ etc.} \tag{99}$$

where $\Delta\mu$ indicates chemical-potential change (Schellman, 1975). On energy scale, co-operativity effect results $8 > \Delta\mu_\gamma > -6\text{kJ}\cdot\text{mol}^{-1}$ for metal– and macromolecule–ligand bindings; chelate one amounts to $-29 > \Delta\mu_\eta > -151\text{kJ}\cdot\text{mol}^{-1}$ for Cu^{II} complexes bound by polyamines.

Protein Adsorption: Area and Adsorbate Clustering Effects

Chatelier & Minton (1996) and Minton (2000, 2001) reported effects of excluded surface area and adsorbate clustering on protein surface adsorption. Adsorbate-cluster growth is assumed by reversible addition of ad-*monomer* single molecules to pre-existing clusters, which diffusional mobility in surface plane decays with rising size. Pre-existing-clusters annealing to form larger ones contributes negligibly to overall adsorption kinetics. Formation of $i + 1$-mer from i-mer and monomer in solution proceeds *via* two pathways. (1) Soluble-monomer adsorption to vacant surface (elementary process 1), followed by diffusion of adsorbed monomer to the edge of cluster species *i* and subsequent accretion (elementary process 2), which is referred to as *direct deposition + accretion pathway.* (2) Soluble-monomer deposition on to and insertion into cluster species *i* (elementary process 3), which is mentioned as *piggyback deposition pathway.* Combining the two pathways leads to rate equations set:

$$\frac{d\rho_1}{dt} = \left[k_1 c + \sum_{i=1} k_{-2}^{(i)}\rho_{i+1} + \left(k_{-2}^{(1)} + k_{-3}^{(1)}\right)\rho_2\right] - \left[k_{-1} + \sum_{i=1} k_2^{(i)}\rho_i + k_2^{(1)}\rho_1 + k_3^{(1)}c\right]\rho_1 \tag{100a}$$

where ρ_i denotes surface density of oligomeric adsorbate species *i*, *t*, time, k_j/k_{-j}, forward/backward rate constants of process *j*, and *c*, concentration of monomeric adsorbate in solution. For $i > 1$:

$$\frac{d\rho_i}{dt} = \left(k_2^{(i-1)}\rho_1 + k_3^{(i-1)}c\right)\rho_{i-1} + \left(k_{-2}^{(i)} + k_{-3}^{(i)}\right)\rho_{i+1} - \left[k_{-2}^{(i-1)} + k_{-3}^{(i-1)} + \left(k_2^{(i)}\rho_1 + k_3^{(i)}c\right)\right]\rho_i \tag{100b}$$

Since numerical calculations are performed for a finite set of clusters containing a maximum of imax promoters it follows that for *i* = imax, Equation (100b) reduces to:

$$\frac{d\rho_{i\max}}{dt} = \left(k_2^{(i\max-1)}\rho_1 + k_3^{(i\max-1)}c\right)\rho_{i\max-1} - \left(k_{-2}^{(i\max-1)} + k_{-3}^{(i\max-1)}\right)\rho_{i\max} \tag{100c}$$

Software used in the study is available from the author and is free for academics (torrens@uv.es).

Graphene: The First Two-Dimensional Material – Epitaxial Graphene

The GR is a $C_{graphite}$ single layer with amazing mechanical, electrical and optical properties, which enable devices (Geim, 2009). Meyer *et al.* (2007) reported structure of suspended GR, which is an unprecedented material. González *et al.* (1992) reported GR–C-material relation and GR special properties, and Levy *et al.* (2010), strains and gauge fields. The GR is a conductor and charge carriers within it scatter in unusual ways, which are revealed *via* quantum mechanical behaviour at LT (Tikhonenko *et al.*, 2008). Kim *et al.* (2010) reviewed production routes to exfoliated $C_{graphite}$ especially top-down strategies starting from $C_{graphite}$ oxide. Rieger & Müllen (2010) reviewed electronic structure–optical spectra relation of polycyclic aromatic HCs (PAHs), noting versatile properties from stable fully-benzenoids (high optical gaps) to reactive acenes (low gaps). Nair *et al.* (2010) reported a GR stoichiometric derivative with an F-atom attached to every C-atom: fluorographene (FG). Abe *et at.* (2010) studied surface chemistry involved in GR epitaxy sublimating Si atoms from surface of epitaxial 3C-SiC(111) thin films on Si(111). Dresselhaus *et al.* (2010) showed Raman scattering (RS) to provide a tool to differentiate between two distinct sp^2 C-NSs (NTs, GR), which present properties in common and others that differ. Zhang *et al.* (2010) prepared chemically modified GR/polyaniline (PANI) nanofibre (NF) NCs by *in situ* polymerization of aniline monomer in GR oxide (GO) presence under acidic conditions.

Boron Chemistry, Nanohorns, and Swnts (Optical Nanowaveguides)

Hawthorne *et al.* (2004), Oliva *et al.* (2005, 2009ab), Serrano-Andrés & Oliva (2006) and Serrano-Andrés *et al.* (2008) reported reaction mechanisms, and properties of mono/dimeric and (hetero)borane clusters in ground and excited electronic states for property prediction of B-based molecular higher architectural constructs. While researchers discuss differences among NTs Manuel (2010) identified their topological and structural similarities. Yudasaka *et al.* (2008), Pagona *et al.* (2009) and Guldi & Martin (2010) reviewed SWNH physicochemical applications. Zhu & Xu (2010) outlined SWNH research (*e.g.*, properties, functionalization, applications, outlook). Wei & Wang (2004) and Khosravi & Moradi (2007) solved Maxwell and hydrodynamic equations propagation of electromagnetic waves in SWNTs, showing that dispersion behaviours of plasma waves with modes TM and TE are similar. Moradi (2010) analyzed propagation of surface plasmon waves in metallic SWNTs in classical electrodynamics.

Tuning Plasmonic Properties of Au Nanorods and Quantum Dots

Au nanorods (NRs) exhibit optical properties, which are tunable with their shape leading to sensing, imaging and biomedical therapeutic applications (Faraday, 1857). Colloidal preparations of Au NRs impart surfactants or others on NR surfaces; a preparation leads to a surfactant bilayer on surface (Murphy *et al.*, 2010). Position of plasmon absorption band depends not only on metal nature but also on parameters (*e.g.*, particle size/shape, environmental dielectric properties) (Guerrero-Martínez *et al.*, 2010). An area when Au metal NPs (MNPs) are used to build macroscopic devices is plasmonics, because they are synthesized with tailored optical properties and assembled into optoelectronic devices to manipulate NP light transfer (Guerrero-Martínez *et al.*, 2009). Fan & Govorov (2010) achieved directed assembly and alignment control of Au NRs *via* incorporation on supramolecular chiral fibres, which permits tuning of circular dichroism (CiD) of Au NRs, which they compared to predictions by theories, and discussed parameters. Quantum dots (QDs) are fluorescent (FL) semiconductor particles (Esteve-Turrillas, personal communication). Fluorophores property is that an excitation is transferred from one fluorophore to another placed in close proximity *via* dipole–dipole interaction (Förster resonance energy transfer, FRET) in which the former acts as donor as it transfers a certain energy to the latter (acceptor) and emission of the former is partially quenched (Lakowicz, 1999) [*e.g.*, an assembly realized packing together two sets of QDs characterized by two different sizes, comparing photoluminescence (PL) emission from assembly with those of each set (Kagan *et al.*, 1996)]. Medintz *et al.* (2003), Clapp *et al.* (2006), Algar & Krull (2007) and Chong *et al.* (2007) exploited QD and FRET in bioapplication sensing. Xing *et al.* (2011) developed a versatile QD label modifying double-stranded deoxyribonucleic acid (DNA) (dsDNA) with biotin [vitamin (Vit)-B_7]/thiol –SH groups at opposite ends and attaching it to QDs *via* metal–SH bond. Mirkin *et al.* (1996) described a method for assembling colloidal Au NPs reversibly into macroscopic aggregates. Willets & Duyne (2007) reviewed spectroscopic studies, which reveal relations governing localized surface plasmon resonance (LSPR) spectral location and sensitivity to local environment (*e.g.*, NP shape, size). Sepúlveda *et al.* (2009) revised basis behind LSPR sensing and NS fabrication techniques/biosensing applications. Saleh *et al.* (2011) tested Zn/Ag-doped mullite ceramic discs as potentially resistant materials *vs.* bacterial adhesion and biofilm formation. Barnes *et al.* (2003), Zayat & Smolyaninov (2003) and Schuller *et al.* (2010) experimental and theoretically researched MNPs optical properties. Localized surface plasmons (LSPs) are responsible for molecular labelling (Bachelot *et al.*, 2003), NP optical field enhancement (Ibn El Ahrach, 2007), PL (Bouhlier *et al.*, 2005), solar energy harvesting (Atwater & Polman, 2010), photothermal tumour ablation (Haes *et al.*, 2004), light generation (Noginov *et al.*, 2009), *etc.* Deeb *et al.* (2010) reported an approach for quantifying depth and strength of optical near field of a single colloidal MNP associated with LSPs. Corma & Garcia (2008) reviewed supported Au NPs as heterogenous catalysts for organic reactions. McGilvray *et al.* (2006) and Marin *et al.* (2008) developed MNPs, *e.g.*, Au NPs, *via* fundamentals of photochemistry/physical organic chemistry and produced unprotected, *naked* Au NPs in aqueous solutions. The ZnO is present in different photovoltaic (PV) devices: as e^- transport material in dye-sensitized (DySCs)/hybrid solar cells (HSCs), antireflection coating in inorganic solar cells or optical spacer in polymer (HSCs, inverted organic) solar cells (Gonzalez-Valls & Lira-Canut, 2009, 2010; Law *et al.*, 2005). Bioactive materials formed by deposition of metal-oxide NSs on organic and polymer surfaces reveal an ability to depress/kill pathogenic bacteria, and their introduction to organisms prevents viruses/bacteria *fixing* to cell walls improving antibacterial/viral immunities (Brown, 1992; Burrell, 1997; Chule *et al.*, 2006). The ZnO NCs as coatings and NC films, with antitumour activity, were obtained depositing ZnO nanofilms on surfaces of ethyl ether salicylidene DL-tyrosine and its Cu^{II} chelate by magnetron sputtering of Zn target (Arakelova *et al.*, 2010). Advances in catalytic nanomaterial design [*e.g.*, NPs supported on porous materials (Pignataro, 2010b; White *et al.*, 2009), NTs, NFs (Gonzalez-Arellano *et al.*, 2009, 2010a)] showed that materials are prepared *via* different environment-benign methods (*e.g.*, microwaves, ultrasonication, milling) and possess catalytic activities. Balu *et al.* (2010) and Gonzalez-Arellano *et al.* (2010b) described synthesis of Fe-oxide NPs supported on porous materials, *e.g.*, MCM-41, SBA-15, *etc.*, and their catalytic applications in processes (*e.g.*, oxidation of alcohols/alkenes/sulphurs, *N*-alkylation of benzylic alcohols). Nacci & Cioffi (2011) edited a special issue on nanocatalysts for sustainable organic synthesis. Morant-Miñana (personal communication) functionalized syndiotactic 1,2-polybutadiene NPs in aqueous dispersion by free-radical mercaptan addition. Jongh & Adelhelm (2010) discussed confinement impact on metal hydride dihydrogen (H_2)-sorption.

Biomimetic Molecular Switches and Motors

Inspired by discovery of biomolecular machines crucial in living-cell function, synthetic molecular devices are a field in chemistry, physics and molecular biology intersection (Balzani *et al.*, 2008). Based on theoretical calculations Sampedro *et al.* (2004) designed the chemical structure that potentially showed best photochemical properties. Rivado-Casas *et al.* (2009) reported compounds that allow profit of solar light to perform photoisomerization, which results an advantage for applications. Photochromic entities undergo a light-induced reversible transformation between two isomers presenting distinct structural and electronic features (Jacquemin *et al.*, 2010a). If the properties of the two forms are different, *e.g.*, diarylethenes (DAs), photochromes act as building blocks in *on*/*off* nanodevices (Jacquemin *et al.*, 2010b). If one designs a single molecule encompassing several photochromes one stores more information than a simple bit 0/1 pattern, *e.g.*, a DA trimer keeps one byte (Jacquemin *et al.*, 2010c). Jacquemin *et al.* (2010de) converged to design of optimal coupled switches providing efficient-molecule suggestions. Jacquemin *et al.* (2010f) synthesized molecules incorporating more than one single photoswitchable unit, which underwent limitations. Difficulty is because photochromic excited state is snatched by closed DAs (Perrier *et al.*, 2011).

Electrochemistry and Properties of Single Nanoparticles

Despite properties and contrary to other conducting polymers (CPs), *e.g.*, polypyridine (PPy), poly(3,4-ethylendioxythiophene, EDOT) (PEDOT) use for enzyme immobilization matrix and biosensor preparation was restricted because water is not adequate for EDOT polymerization, owing to poor monomer solubility and molecules interaction with polymerization intermediate (Wen *et al.*, 2009). Turkarslan *et al.* (2010) used ionic liquids (ILs) as growth medium for PEDOT electrosyntheses to minimize the problem leading to improved electrochemical (EC) properties. Bard *et al.* (2010) described observation of single-NP collisions with an electrode *via* current amplification by electrocatalysis. Baba *et al.* (2010) investigated *in situ* EC and dielectric field properties of monomer co-electropolymerization with a π-conjugated polymer network (CPN) precursor. Park *et al.* (2011) synthesized D-(+)-galactose-conjugated SWNTs for use as biosensors to detect cancer marker galectin-3; to investigate galectin-3 binding to D-(+)-galactose-conjugated SWNTs they fabricated an EC biosensor *via* Mo electrodes. Plasmon-band position depends on Au NP size/form, medium in which they are placed and organic layer that covers them (Kreibig & Vollner, 1995; Werner *et al.*, 2010). Cartoixà & Rurali (2010) reviewed theoretical modelling of structural, electronic and transport properties of Si nanowires (NWs) (SiNWs) showcasing improved ways of passivating these, bringing up failure of conventional Si dopants to act as donors/acceptors in thin SiNWs, proposing workarounds for failure and studying effect that impurities cause in transport. Müller-Meskamp *et al.* (2010) studied electrical properties of α, ω-mercaptoalkyl ferrocenes FcC_n (n = 3, 5, 11), embedded in a self-assembled host matrix of alkanethiols on Au(111), by scanning-tunnelling microscopy (STM) and spectroscopy (STS). Electronic coupling of FcC_n to Au(111) substrate depends on length of alkane spacer chain (Müller-Meskamp *et al.*, 2009). Biology made significant advances in understanding molecular mechanisms governing life's processes (Aimé, 2010). Rohrschneider (1966) studied solvent properties. Snyder (1974) analyzed solvent properties, which are important in liquid chromatography (LC) separations especially reversed-phase (RP). In high-performance (HP) LC (HPLC) development Snyder (1997) found necessary to vary sample relative retention selectivity α changing experimental variables, *e.g.*, solvent type, pH, *etc.*

Nanoparticles Based On Fe^{II} Complexes with Spin Transition

In this laboratory Coronado *et al.* (2007) reported bistable spin-transition [ST, spin-crossover (SCO) or spin-equilibrium] NPs showing magnetic thermal hysteresis at RT. Galán-Mascarós *et al.* (2010) informed tuning size and thermal histeresis in bistable ST NPs. Volatron *et al.* (2008) analyzed ST co-ordination NPs. Boldog *et al.* (2008) examined ST nanocrystals with magnetic, optical and structral bistability near RT. Agustí *et al.* (2008) investigated thermal and light-induced STs in new 3D Hofman-like microporous metal–organic frameworks (MOFs) produced as bulk materials and nanopatterned thin films. Martínez *et al.* (2010) inspected ST in nanocrystals and NPs of $[Fe(3\text{-}Fpy)_2M(CN)_4]$ (M^{II} = Ni, Pd, Pt) 2D co-ordination polymers. Forestier *et al.* (2008) probed Fe^{II} ST NPs. Prins *et al.* (2011) published RT electrical addressing of a bistable ST molecular system.

Nanomaterials for Biomedical Applications

Tan *et al.* (2010) developed organic–metal–semiconductor-hybrid NC NPs with dimensions <100nm for biomedical applications. Nanotechnology fundamental component in medical imaging is NPs and microparticles (MPs) (Tan *et al.*, 2009). A liposome is a spherical vesicle composed of uni and multi-lammelar phospholipid–cholesterol membrane (Xu *et al.*, 1999). Emulsions are oil-in-water (OW)-type mixtures and stabilized with surfactants to maintain size/shape (Wickline *et al.*, 2006); they are readily produced, inexpensive and show no signs of toxicity *in vivo* (Guccione *et al.*, 2004; Hawker & Wooley, 2005). Microbubbles are made from albumin/charged lipids and function by resonating in an ultrasound beam rapidly contracting/expanding in response to pressure changes of a sound wave (Blomley *et al.*, 2001). Therapeutic agents carried by modified liposomes, emulsions, polymeric spheres and micro-bubbles provide decayed systemic toxicity, a more selectively targeted delivery to specific tissues and higher doses brought to aim (Rubesova *et al.*, 2002). In different imaging modalities, NPs designed for medical imaging were given several names; *e.g.*, Au NPs remain unoxidized to produce good contrast for X-ray-based computerized tomography (CoT) imaging (RSNA, 2005), QDs manifest stable FL at several wavelengths and offer multi-colour optical coding in gene expression/*in vivo* optical imaging (Stinaff *et al.*, 2006), and *metal nanoshells* consist of spherical dielectric NPs surrounded by an ultra-thin/conductive/metallic layer and are used as a near infrared (IR) (NIR) contrast agent for optical imaging (Hirsch *et al.*, 2006). Son *et al.* (2007), Deerink (2008) and Gilmore *et al.* (2008) (pre-)clinically tested biomedical NPs for targeted drug/gene delivery and to enhance diagnostic imaging output, *e.g.*, magnetic resonance imaging (MRI). Dendrimers were tested in pre-clinical models for drug/gene delivery and as imaging agents tailoring binding properties to several requirements, *e.g.*, facilitation of cellular uptake of (cancer) drugs (Boyd, 2008; Duncan, 2007; Duncan & Spreafico, 1994). Superparamagnetic NPs (SPIONs) are inorganic-based NPs presenting a Fe-oxide core coated by inorganic [silica (SiO_2), Au] or organic materials (phospholipids, fatty acids, polysaccharides, peptides, surfactants/polymers) (Babiç *et al.*, 2008; Euliss *et al.*, 2003; Gupta & Curtis, 2004). The characteristic makes applications: separation techniques and contrast enhancing effects for MRI to drug delivery systems (DDSs), magnetic hyperthermia (tumour therapy) and magnetically assisted transfection of cells (Gupta & Gupta, 2005; Horák, 2005; Jordan *et al.*, 2001; Neuberger *et al.*, 2005); because of specific adsorption they help physicians to identify dangerous arteriosclerotic plaques by MRI (Muhlen *et al.*, 2007; Smith *et al.*, 2007). Research is directed towards monitoring events on physiological and molecular levels, so that inflammatory diseases or tumours are detected *via* SPION accumulation and markers expressed on cell surface (Thorek *et al.*, 2006). Mailänder and Landfester (2009) reviewed harnessing interactions of polymer NPs synthesized by miniemulsion with different cell types. A nanodrug is an important product of developing biomedical nanotechnologies (Balogh 2009). Liu *et al.* (2010) addressed nanodrug capability on targeting to cells, penetrating *via* epicyte, controlled release and security resulting from using. Gai *et al.* (2010) reported synthesis, characterization and application in controlled drug release for monodisperse core–shell-structured $Fe_3O_4@n SiO_2@m SiO_2@NaYF_4$: Yb^{3+}, Er^{3+}/Tm^{3+} NCs with mesoporous, up-conversion luminescent and magnetic properties. Nanobiotechnology (Drexler, 1986; Feynman, 1960; Ozin & Arsenault, 2005; Taniguchi, 1974) is split into two areas: nanotechnological device use to understand biosystems and material exploitation in nanodevice fabrication (Pignataro, 2010). In the latter a biomolecule set was used [*e.g.*, nucleic acids (Mirkin *et al.*, 1996; Niemeyer, 2001; Niemeyer & Mirkin, 2004; Seeman, 2005), crystalline bacterial cell-surface layers, whole organisms (prokaryotic, eukaryotic cells) (Flenniken *et al.*, 2004; Kowshik *et al.*, 2002; Peelle *et al.*, 2005; Sweeney *et al.*, 2004), proteins (ferritin) (Douglas & Stark, 2000; Kramer *et al.*, 2004; Li *et al.*, 2003; Meldrum *et al.*, 1992), heat-shock proteins (HSPs) (Flenniken *et al.*, 2003, 2005; Klem *et al.*, 2005), viral capsids]. The latter were used as scaffolds for (in)organic material assembly, selective attachment, and presentation of chemo/biomoieties and building blocks for 1/2/3D-array construction (Arora & Kirshenbaum, 2004;

Bittner, 2005; Douglas & Young, 2006; Flynn *et al.*, 2003; Singh *et al.*, 2006). Efforts were devoted to structures design, preparation/characterization on sub-100nm scale and their use as novel functional devices (Whitesides, 2005). The NTs are building blocks for applications (*e.g.*, catalysis, DDS, optics, electronics, chemotherapy, transmembrane transport) because physical and chemical properties are tunable *via* size and shape control (Martin & Kohli, 2003). Although advances were made in covalently bonded NSs (Bong *et al.*, 2001) noncovalently bonded NTs offer significant advantages (*e.g.*, synthetic convergence, built-in error correction, control *via* unit design, self-organization) (Science, 2002). Self-assembling peptide NTs (SPNs), formed by stacking cyclic peptides stabilized by H-bonds, attracted attention because of probable ease with which they are endowed with structural and functional properties (Schwarz *et al.*, 2004). Seker & Demir (2011) reviewed material binding peptides for nanotechnology.

TiO_2 Photocatalyst, Dyes, and Applications

Sheetal & Pragati (2010) presented a systematic investigation of heterogeneous photocatalytic degradation of methylene blue (MB) in aqueous titania (TiO_2) suspension, *via* 8W low-pressure $Hg_{(v)}$ lamp, studying effect of parameters (*e.g.*, initial MB, TiO_2 concentrations) and addition of e^- scavenger H_2O_2 to complete MB degradation; they used a pseudo-first-order kinetics. Longfeng *et al.* (2010) synthesized nanocrystal TiO_2-anatase NPs by LT crystallization with homogeneous $Ti(OBu)_4$ hydrolysis, providing hydrolyzing water by acetic acid (HAc) + ethanol (EtOH) esterification. Erickson *et al.* (2010) reviewed historical foundation for NP materials in chemistry and environmental applications (*e.g.*, NP solvents, destructive sorbents, nanocatalysts, photocatalytic nanomaterials). Ali *et al.* (2010) compared adsorption characteristics of adsorbents [MWNTs, powdered (PAC), granular activated carbon (GAC)] in terms of dissolved organic-C (DOC) removal from synthetic wastewater. Liu & Giessen (2010) revised coupling in optical metamaterials, which became a hot field of photonics since seminal work of Pendry *et al.* (1996) on negative refractive index, invisibility cloacking and perfect lensing. Lanthanide-doped up-conversion (UC) NPs showed promise in biolabelling, imaging and therapeutics (Wang *et al.*, 2010). The MNPs of unique electronic/catalytic properties or supramolecular nucleic acid-based NSs provided materials for electroanalytical applications (Willner *et al.*, 2011). Moniruzzaman & Winey (2006) reviewed polymer NCs in which fillers are SWNTs and MWNTs. A universally accepted model of starch structure is lacking (Buléon *et al.*, 1998). Le Corre *et al.* (2010) revised starch NP preparation, characterization, properties and applications. Lee *et al.* (2010) discussed manufactured nanomaterials (MNMs) applications, which improved conventional construction materials, suggested likely environmental release scenarios and summarized potential adverse biotoxicological effects/mitigation. Haag (2011) presented a method to calculate supply air volume flow–airborne particle concentration relation. Abbasalipourkabir *et al.* (2011) proposed solid lipid NPs (SLNs) as an alternative colloidal DDS for water-soluble drugs. Jones (2011) discussed techniques and described how to combine them complementarily is beneficial whilst being aware of different results.

Charge Transport of Deoxyribonucleic Acid

Ordejón *et al.* (1996) and Sánchez-Portal *et al.* (1997) performed DNA feasibility tests in early stages of project SIESTA, relaxing a dry B-form poly(dC)-poly(dG) structure with a minimal basis set. In realistic-calculation preparation Machado *et al.* (submitted) studied 30 nucleic acid pairs addressing precision of approximations and double-ζ, singly polarized (DZP) bases, and accuracy of generalized gradient approximation (GGA) functional (Perdew *et al.*, 1996) obtaining results even for H-bonds. Based on preliminary tests and with confidence from results of base-pair calculations, Artacho *et al.* (2003) carried out a serious study of dry A-DNA with structure full relaxation and electronic-characteristic analysis. Artacho *et al.* (1999) performed linear-scaling *ab-initio* calculations for large and complex systems. Brandbyge *et al.* (2002) developed method TranSIESTA to describe electronic transport phenomena in nanometric devices off equilibrium, from first principles + DFT for non-equilibrium e^- transport. De Pablo (2000) calculated and measured absence of direct current (DC)-conductivity in λ-DNA. Molecular wires show promise in nanoelectronics but synthesis of uniform, long conductive molecules is a challenge. Precise-length DNA is synthesized but its conductivity over distances required for nanodevices was not explored. Slinker *et al.* (2011) showed DNA charge transport (CT) over 34nm in 100-mer monolayers on Au. Multiplexed Au electrodes modified with 100-mer DNA yielded EC signals from a distal, covalent Nile Blue redox probe. Signal attenuation upon incorporation of a single base-pair mismatch shows DNA-mediated CT. Efficient cleavage of 100-mers by a restriction enzyme indicates that DNA adopts a native conformation accessible to protein binding. Similar e^--transfer rates measured *via* 100/17-mer monolayers are consistent with rate-limiting e^- tunnelling *via* a saturated C-linker. The DNA-mediated CT distance of 34nm surpasses molecular wires.

Enumeration of Heterofullerenes: A Survey

Chemical-compounds enumeration was accomplished *via* Pólya-Redfield theorem for combinatorial details of graphs, polyhedra, chemical compounds, *etc.* Heterofullerenes are enumerated as fullerenes in which one or more C-atoms are replaced by heteroatoms (*e.g.*, B, N). *Via* Pólya's theorem, Ghorbani (2012) computed number of permutational isomers of fullerene graphs.

Advanced Magnetic Force Microscopy: Graphite and Graphene

Leite *et al.* (2012) reviewed theoretical models for surface forces/adhesion and measurement *via* atomic force microscopy (AFM). Horsell *et al.* (2011) reported GR mechanical measurements by AFM. Scanning force microscopy (SFM) allowed detecting short and long-range interactions; *e.g.*, magnetic force microscopy (MFM) characterized domain configuration in ferromagnetic materials (thin films grown by physical techniques, ferromagnetic NSs). Topography and magnetic signals were separated scanning at 25–50nm lift distance such that long-range tip–sample interactions dominated. The MFM permitted detecting magnetic fields (MFs) from low-dimensional complex systems, *e.g.*, organic nanomagnets, superparamagnetic NPs, C-based materials, *etc.* Magnetic nanocomponents and supporting substrate presented different electronic behaviour, *i.e.*, surface-potential differences causing heterogeneous electrostatic tip–sample interaction, which is interpreted as magnetic one; to distinguish tip–sample-forces origin, Jaafar *et al.* (2011) proposed Kelvin probe force microscopy (KPFM)/MFM combination. The KPFM allows compensating in real time (RTi) electrostatic tip–sample forces minimizing electrostatic contribution to frequency shift signal, which is a challenge in samples with low magnetic moment; they studied Co-NSs array, which exhibited an electrostatic interaction with MFM tip; with KPFM/MFM they separated electric from magnetic tip–sample interactions: it is challenging to understand ferromagnetic mechanism in C-based materials, which contain *s*/*p* e^- in contrast to traditional ferromagnets based on 3*d*/4*f* e^-. Çervenka *et al.* (2009) showed ferromagnetic order locally at defect structures in highly oriented pyrolytic graphite (HOPG) *via* MFM and bulk magnetization measurements at RT. Magnetic impurities were excluded as signal origin. Ferromagnetic effect originated from localized e^- states at HOPG grain boundaries forming 2D arrays of point defects. Theoretical magnetic ordering temperature based on weak interlayer coupling and magnetic anisotropy was comparable to experiments. Chemical environment of defects bonded in $C_{graphite}$ nets revealed *s*/*p* e^- role creating routes for spin transport in C-based materials. Martínez-Martín *et al.* (2010) studied possible ferromagnetic order on $C_{graphite}$ surface by MFM; they showed that tip–sample interaction along steps is external-MF independent. Combining KPFM/MFM they separated electrostatic from magnetic interactions along steps obtaining upper bound for magnetic force gradient: $16\mu N \cdot m^{-1}$; they showed ferromagnetic-signal absence in $C_{graphite}$ at RT. Jaafar *et al.* (2012) introduced drive-amplitude modulation (DAM) AFM as dynamic mode with performance in all environments. As with frequency modulation (FM), DAM followed feedback scheme with nested loops: the first kept cantilever-oscillation amplitude constant regulating driving force and the second used this as a topography feedback variable. A phase-locked loop was used as parallel feedback allowing non/conservative-interaction separation; they described DAM basis and exemplified performance in different environments. The DAM is stable, intuitive and easy-to-use, which is free of feedback instabilities associated with noncontact-to-contact transition, which occurs in FM. Jaafar *et al.* (2009) introduced a variable-external-field MFM microscope, which allowed stable images under variable external MF, which was applied in- and out-of-plane directions; they illustrated microscope performances for samples: HOPG, longitudinal magnetic storage media, FePt thin films with in-plane anisotropy and Ni NWs with axial easy axis embedded in a ceramic matrix; they used variable-field microscope as magnetic writing and reading. Tiberj *et al.* (2011) performed micro-RS and -transmission imaging experiments on epitaxial GR grown on C/Si-faces of on-axis 6H-SiC substrates. On C-face they showed that SiC sublimation resulted in growth of long and isolated GR ribbons ($\leq 600\mu m$), which were strain-relaxed and *p*-type doped, in which combining results of micro-RS with -transmission, they ascertained that uniform monolayer ribbons were grown and found Bernal stacked/misoriented bilayer ribbons. Full GR coverage of SiC surface was achieved but anisotropic growth occurred because of step-bunched surface reconstruction. While in the middle of reconstructed terraces thin GR stacks ($\leq$5 layers) were grown thicker GR stripes appeared at step edges; in both, GR layers–SiC substrate interaction induced compressive thermal strain and *n*-type doping. Escrig *et al.* (2007) studied effect of macroscopic size of Ni NW arrays on their remanence state; they developed a phenomenological magnetic model to obtain remanence *vs.* magnetostatic interactions in an array; they observed that because of long-range dipolar interactions between wires, sample size affected array remanence; they studied NWs magnetic state by variable-field MFM for remanent states; they deduced NWs distribution with magnetization in up/down directions and subsequent remanent magnetization from magnetic images. Two short-range magnetic orderings with similar energies explain typical labyrinth pattern observed in MFM images of NW arrays. Imaging of MFM allows locally studying NSs magnetic state. Asenjo *et al.* (2006) used MFM to characterize an ordered array of Ni NWs embedded in a porous membrane. Because of large aspect ratio (AR) of wires (30nm diameter:1000nm length) these presented an easy axis; considering NWs as single-domain structures and calculating amount of wires pointing to each direction they obtained average magnetization; they introduced a method to analyze MFM data considering distribution functions of magnetic contrast, with which they studied NW-array magnetization and compared results with major/minor hysteresis loops, measured by supercomputing quantum interface device (SQUID) magnetometer. Knorr and Vinzelberg (2012) reported charge writing and detection by electrostatic force microscopy (EFM) and KPFM scanning probe microscopy (SPM).

Photocatalytic Degradation of Organic Pollutants in Dispersion

Oppenländer (2003) evaluated water and air photochemical purification. Anpo and Kamat (2010) examined environmentally benign photocatalysts with applications of TiO_2-based materials. Legrini *et al.* (1993) reviewed photochemical processes for water treatment. Hoffmann *et al.* (1995) revised environmental applications of semiconductor photocatalysis. Kuznetsov and Serpone (2006) analyzed heat and photoinduced absorption spectra of TiO_2-Degussa-P25/polymer NCs; they described spectra as sum of overlapping absorption bands (ABs) with maxima at 2.90eV (427nm, AB_1), 2.55eV (486nm, AB_2) and 2.05eV (604nm, AB_3). Spectra correlated with experiments after TiO_2 reduction; they studied VIS spectra of TiO_2 photocatalysts reported in the literature. Relative narrow spectra were similar and photocatalyst-preparation independent. Average spectrum was described by $AB_{1/2}$ sum; VIS activation of TiO_2 specimens (anion doped, otherwise) implicated defects associated with O-vacancies, which started colour centres displaying ABs and not narrowing original band gap of TiO_2-anatase (E_{bg} = 3.2eV) *via* dopant and O-states mixing as suggested in the bibliography. Second-generation $TiO_{2-x}D_x$ photocatalysts doped with cations and anions (N, C, S) showed red-shifted absorption edge enhancing photonic efficiencies of photoassited surface redox reactions. Researchers proposed that the shift is caused by narrowing band gap of pristine TiO_2, *e.g.*, anatase absorption edge 387nm, while others suggested intragap localized states of dopants; they showed that commonality in all doped TiO_2 rested with O-vacancies formation and advent of colour centres (*e.g.*, **F**, $\mathbf{F}^+$, $\mathbf{F}^{2+}$, Ti^{3+}), which absorbed VIS. Serpone (2006) showed that absorption-edge shift was caused by colour centres formation and, while band-gap narrowing is not unknown in semiconductor physics, it needs heavy doping of metal-oxide semiconductor producing materials that present different chemical compositions from TiO_2 with different band-gap electronic structures. Kuznetsov and Serpone (2007) examined TiO_2/polymer-NCs photocolouration and colour-centres photobleaching at selected irradiation wavelengths from ultraviolet (UV) to NIR regions; they analyzed centres photoactivation irradiating into AB_{1-3}; they observed photostimulated absorbance changes: (1) rises; (2) decays. The latter is a direct experimental manifestation of photobleaching of coloured TiO_2/polymer NCs, which showed presence and photoinduced disappearance/destruction of TiO_2 colour centres. Photobleaching of coloured TiO_2/polymer NCs originated from *intrinsic* light absorption by TiO_2 ($h\nu > 3.2$eV) and *extrinsic* light absorption, by colour centres at wavelengths corresponding to absorption spectral bands (*i.e.*, $h\nu < 3.2$eV), which were active in photodestruction centres; they proposed a photobleaching photochemical mechanism involving O-assisted annihilation of O-vacancies; they showed that TiO_2 VIS absorption originated from colour centres and *not* band-gap narrowing of pristine TiO_2. However, valence and conduction bands, which some researchers suggested that are involved in observed red shifts of VIS absorption edges of doped TiO_2 because of apparent narrowing of TiO_2 band gap, are not photodestroyed; they modelled competitive photoinduced formation and destruction of colour centres, which resulted in qualitative agreement with experiments. D'Oliveira *et al.* (1993) studied photocatalytic degradation of six dichlorophenols (DCPs) and three trichlorophenols (TCPs) in TiO_2 aqueous suspensions; they correlated apparent first-order rate constants k_{app} of both and monochlorophenols (MCPs) disappearance with Hammett constants σ/1-octanol–water partition coefficients k_{ow}: k_{aw} (h^{-1}) = $-10\,\sigma + 5.2\log k_{ow} - 7.5$ (correlation coefficient $R = 0.987$, omitting 2,4,6-TCP). Aromatic intermediates, identified by UV (HPLC separation) and mass spectra [gas chromatography (GC) one], corresponded to aromatic-nucleus hydroxylation with reactivity: $p > o > m$; they detected *p*-benzoquinones with 0–2 Cl/0–1 OH and intermediates with two aromatic/quinone rings; they determined intermediates temporal variations. All compounds in which subsisted C6 ring were unstable, their maximum concentration was low compared to initial chlorophenol (CP). Release of Cl^- and CO_2 evolution followed apparent first-order kinetics in degradation's first part. Effects of Cl atoms number and position on kinetics were discussed. Wastewaters contained substituted phenols: Ksibi *et al.* (2003) photocatalytically degraded two hydroxyphenols [hydroquinone (hydro), resorcinol], 4-nitrophenol (4-NP), 2,4-dinitrophenol (2,4-DNP) and 2,4,6-trinitrophenol (2,4,6-TNP) in TiO_2 aqueous suspension. Substituent-nature effect on photodegradation was studied comparing initial degradation rates (v_0 relative to phenol) to substituted phenols, which reactivity followed Hammett law. Plot of v_0 *vs.* σ poorly correlated considering all substituted phenols; when nitrophenols (NO_2) were evaluated separately and hydroxyphenol (OH) was rejected correlations were observed. A p*H* effect upon kinetics of chemical oxygen demand (COD) disappearance was observed. Acidic p*H* was preferred over COD removal for phenolics. Photocatalysis transformed NO_2 in nitrophenols into NO_3^-/NH_4^+ *via* NO_2^- intermediate. Amount of NO_3^- depended on NO_2 number in nitrophenol. Selectivity in NO_3^- resulted 80, 56 and 66% for 4-NP, 2,4-DNP and 2,4,6-TNP, respectively; they found that in all phototreated solutions, 5-day biological oxygen demand ratios (BOD_5)/COD showed values higher than initial solutions, which indicated a positive effect.

Heterogeneous Photocatalysis: Emerging Technology for Water

Herrmann *et at.* (1993) surveyed heterogeneous photocatalysis principle and literature. Based on studies on substituted-aromatics degradation in UV-irradiated TiO_2 aqueous suspensions they indicated: kinetic characteristics, intermediates nature, degradation pathways/p*H* effect, common ions, *etc.*; they described photocatalytic recovery of noble metals and deteoxification of water containing inorganic ions; they discussed commercialized-photocatalysis advantages and disadvantages. Photocatalysis was based on photocatalyst aptitude to adsorb simultaneously reactants and efficient photons. Herrmann (1999) described fundamental principles and effects of kinetic parameters (catalyst mass, wavelength, initial concentration, temperature, radiant flux). Besides selective mild oxidation of organics performed in gas and organic liquid, UV-irradiated TiO_2 became total oxidation catalyst in water because of $OH^•$ photogeneration by neutralization of OH^- surface groups by photo-h^+. Organics were degraded and mineralized into CO_2 and harmless inorganic anions. Improvement in TiO_2 photoactivity by noble-metal deposition and ion-doping was detrimental. Toxic heavy-metal ions (Hg^{2+}, Ag^+, noble metals) were removed from water by TiO_2 photodeposition. Water-detoxification photocatalytic devices were commercialized. Solar platforms are working with large-scale pilot photoreactors, in which pollutants are degraded with quantum yields comparable to those determined with artificial light in a laboratory. Herrmann (2005) explained photocatalysis fundamentals with parameters, which govern kinetics: (1) catalyst mass; (2) wavelength; (3) reactants partial pressure and concentrations; (4) temperature; (5) radiant flux. Photocatalytic-reaction types concern:

1. Selective mild oxidation of HCs;
2. H_2 production;
3. Total oxidation reactions of organics in water presence.

The last constitute ensemble of recent photocatalysis developments. Most organic contaminants, *e.g.*, dangerous pesticides, are easily, completely degraded and mineralized. Dyes are not only decolourized but also mineralized in coloured aqueous effluents. Most abundant (azo) dyes present groups –N=N–

decomposed into $N_{2(g)}$, which represents an ideal decontamination. Photocatalytic engineering is under development *via* deposited TiO_2 in fixed beds. Solar photocatalytic pilot reactor and prototypes were described. Solar-energy used as a source of activating UV–A irradiation is called *helio-photocatalysis*. Catalyst TiO_2 adsorbs:

1. Reactants;
2. Efficient photons, which create e^-/h^+ responsible for redox reactions.

Photocatalysis is a complex polyphasic system involving solid, gas and liquid + UV-irradiation (*electromagnetic* phase). Action of physicochemical parameters, which control kinetics, was described. In dry media, photocatalysis is used for selective mild oxidation reactions; $OH^\bullet$ generation results in mineralization to $CO_2 + H_2O$ of pollutants, pesticides and dyes in water, and pollutants degradation in (humid) air. Herrmann (2006a) reported examples. A trend is use of fixed beds of TiO_2 fotocatalysts for air/water purification. Photocatalytic engineering developes *via* pilot plants, *e.g*., helio-photocatalysis. *Catalysis by metals* was transposed to *bifunctional photocatalysis* based on noble metals deposited on TiO_2 (Herrmann, 2006b). Participations of

1. Photo-active oxidic support and
2. Deposited metal were analyzed, delimited and associated; for H-involving reactions [cyclopentane-D isotope exchange (CDIE), ROH dehydrogenation], he evidenced e^-/h^+ roles.

Photogenerated e^- were spontaneously transferred to MNPs because of Fermi-levels alignment of the two solid phases; he neutralized H^+ and D^+ into H/D before H_2/HD recombination and evolution. The h^+ neutralized anions in oxidations responsible for organics activation. Anions were OD^- surface groups in CDIE and RO^- in ROH dehydrogenation; both reactions exhibited a stoichiometric threshold, which was exceeded hundreds of times defining catalysis. Noble metals deposited on TiO_2 appeared as auxiliaries and co-catalysts working independently of UV-irradiation for making reaction run catalytically. Bifunctional-photocatalysis accounted for interphasic photocatalyses, *e.g*., H/e^--transfer, photosensitization in VIS *via* CdS addition. Photocatalysis became a major discipline owning to:

1. Intuition of 20th-Century pioneers;
2. Mutual enrichment of scientists from different fields (photochemistry, EC, analytical chemistry, radiochemistry, materials chemistry, surface science, electronics, catalysis).

However, heterogeneous photocatalysis belongs to catalysis meaning that the bases are respected and it became imperative to refocus the frame to avoid misfits/conceptions:

1. Reaction-rate proportionality to catalyst mass (below plateau because of photon full absorption);
2. Implication of Langmuir–Hinshelwood mechanism of kinetics with initial rate proportional to coverages θ in reactants;
3. Obtaining conversions beyond *stoichiometric threshold*: number of potential active sites initially present at surface.

Photonics should be respected with photocatalytic activity being:

1. Parallel to photocatalyst absorbance;
2. Proportional to radiant flux Φ enabling to determine quantum yield, defined as ratio of reaction rate r (in molecules converted per second) to efficient photonic flux (in photons per second) received by a solid.

Photocatalytic normalized tests should be established to prove catalytic activity of irradiated solids independent of non-catalytic side-reactions; *e.g.*, dye decolourization is a misleading test, which provides dye *visible* and apparent *disappearance*, photochemical but not photocatalytic. Thermodynamics is respected: photon energy decay to VIS is thermodynamically detrimental for generating active species, *e.g.*, $OH^{\bullet}$. In solid-state chemistry, cationic doping is harmful to photocatalysis; anionic doping is rapidly clarified and abandoned. Recommendations are addressed and experiments operated in suitable conditions before claiming that one deals with a true photocatalytic reaction, whose veracity is proved following the correct protocol (Herrmann, 2010). Pino and Encinas (2012) compared photodegradation kinetics UV/TiO_2-mediated of 4-CP/2,6-DCP mixtures under the same experimental conditions with individual CPs; their aim was to approach critical processes involved in photoinduced heterogeneous catalysis of systems containing two contaminants in competition and evaluate particle-size effect *via* commercial TiO_2-325mesh/P25 as catalysts. Determination of equilibrium adsorption constants in dark showed that 2,6-DCP adsorption is higher than 4-CP in both particles but relation decays in mixture. Parameters affecting reaction rate were studied (initial phenol concentration, catalyst loading, p*H* change, particle-size distribution). Phenols degradation kinetics at low conversion showed similar reactivity towards photogenerating $OH^{\bullet}$; in agreement with reduced adsorption in mixture their degradation rate was lower in mix than individual. Photodegradation rate was higher on TiO_2-325mesh than -P25, which is opposite to particles surface area, indicating that other surface properties, *e.g.*, pore size, are important. Photoinduced CP degradation showed two intermediates, which correspond to each-CP degradation.

Serum Proteins on Intracellular Uptake: Phenol Red Cytotoxicity

In order to explore effects of C-NPs (CNPs) novel properties on cytotoxicity, Zhu *et al.* (2009) investigated serum proteins adsorption in cell culture medium on MWNTs and three kinds of C-blacks (CBs); they measured quantitatively CNPs uptake by Henrietta Lacks (HeLa) cells, *via* ^{99m}Tc radionuclide labelling/tracing techniques, and examined CNPs-uptake dependence on serum proteins; they assayed CNPs cytotoxicity in a medium with/out serum by method [3-(4,5-dimethylthiazol-2-yl)-2,5-diphenyltetrazolium bromide] (MTT). Cellular uptake was higher in CNP-exposed cells in serum-free than serum culture medium. Serum proteins adsorbed on CNPs attenuated inherent cytotoxicity and decreased rise with increasing serum proteins adsorbed on CNPs; they indicated reasons for serum-protein influence on cytotoxicity; to explore CNP properties in nanotoxicity assays Zhu *et al.* (2012) studied adsorption of phenol red (PR), a p*H* indicator for culture medium, by MWNTs and three kinds of nanosized CBs, and cytotoxic effects. The PR adsorbed and delivered into cells by CBs was responsible for toxicity to HeLa in the serum-free medium. Cellular uptake of PR was verified *via* ^{125}I-labelling. Size-dependent cytotoxicity of CBs correlates closely to PR adsorption, cellular uptake of PR–CB complexes and PR amount delivered into cells by CBs. Although CBs were non or slightly toxic as PR vehicles they played a role in PR-induced cytotoxicity; however, MWNTs showed intrinsic cytotoxicity independent of PR.

Quantitative Structure: Activity Relationship/Toxicity Prediction

Berhanu *et al.* (2012) reviewed applications of quantitative structure–activity/property relationships (QSARs/QSPRs) in nanotechnology toxicity prediction. Products based on nanotechnology entered the market and NP amounts are produced (Barnard, 2009; Chatterjee, 2008). Unintended exposure of humans and ecological receptors to nanomaterials results in adverse effects, which differ from the bulk (Dreher, 2004; Som *et al.*, 2010). Environmental protection plans, associated with nanomaterial manufacture and use, require understanding nanobiointerface interactions that govern nanomaterial bioactivity and potential toxicity (Rallo *et al.*, 2011). Studies of QSAR involved nanomaterials. Evaluation of desired and undesired bioeffects caused by manufactured NPs (MaNPs) is important for nanotechnology. Toxicological studies are time consuming, costly and impractical calling for development of efficient computational approaches capable of predicting MaNP bioeffects. Puzyn *et al.* (2011) investigated cytotoxicity of 17 metal-oxide NPs towards *Escherichia coli* bacteria; based on toxicity and computed structural descriptors they built a model to predict nanomaterials cytotoxicity. Fourches *et al.* (2010) developed a quantitative NS–activity relationship (QNAR) *via* machine learning approaches, *e.g.*, support vector machine (SVM)-based classification and *k*-nearest neighbours (kNNs)-based regression.External prediction power resulted 73% for classification with $R^2 = 0.72$ for regression. Studies of QNAR are employed for predicting nanomaterials bioactivity profiles and prioritizing design/manufacture to better/safer products. Measurement distinguished NP-toxicity changes (Dean, 2012).

Co-Operative-Effects Origin in ST: Supramolecular Intereactions

Dîrtu *et al.* (2010) tracked thermally induced hysteric ST, which occurs in polymeric chain compound $[Fe(NH_2trz)_3](NO_3)_2$ above RT ($T_c{\uparrow} = 347K$, $T_c{\downarrow} = 314K$) by ^{57}Fe Mössbauer spectroscopy, SQUID, differential scanning calorimetry (DSC) and X-ray powder diffraction (XRPD) at variable temperatures; from XRPD pattern indexation they observed orthorhombic primitive cell with parameters: $a = 11.83(2)$ Å, $b = 9.72(1)$Å, $c = 6.361(9)$Å at 298K [low-spin (LS)] and $a = 14.37(2)$Å, $b = 9.61(4)$Å, $c = 6.76(4)$ Å at 380K [high-spin (HS)]. They evaluated enthalpy and entropy variations associated to ST by DSC: $\Delta H = 23(1)\text{kJ·mol}^{-1}$ and $\Delta S = 69.6(1)\text{J·mol}^{-1}\text{·K}^{-1}$; they used thermodynamic data with two-level Ising-like model for statistical analysis of first-order reversal curve (FORC) diagram, which was recorded in cooling mode; they witnessed intramolecular co-operative effects by derived interaction parameter $J = 496K$; they obtained $[Cu(NH_2trz)_3](NO_3)_2{\cdot}H_2O$ crystal structure thanks to high-quality (HQ) crystals prepared by slow evaporation after hydrothermal pretreatment. Catena poly[μ-tris(4-amino-1,2,4-triazole-*N1,N2*)$Cu^{II}](NO_3)_2{\cdot}H_2O$ crystallizes in monoclinic space group *C2/c*: $a = 16.635(6)$Å, $b = 13.223(4)$Å, $c = 7.805(3)$Å, $\beta = 102.56(3)°$, $Z = 4$. The complex is a 1D infinite chain containing triple *N1,N2*-1,2,4-triazole bridges with intrachain distance of Cu...Cu = 3.903(1)Å; they observed dense H-bonding net with NO_3^- counterion involved in intra/interchain interactions. The supramolecular net causes large hysteresis loop ($\Delta T \sim 33K$) result of efficient propagation of elastic interactions. The hypothesis is strengthened by crystal structure and absence of crystallographic phase transition over whole temperature range as shown by XRPD. Compounds of $3d^n$ ($4 \leq n \leq 7$) transition metal exhibit LS/HS co-operative transition, which is abrupt and occurs with thermal hysteresis, which confers system memory effect (Kahn and Martinez, 1998). Intersite interactions and co-operativity are magnified in polymer compounds, *e.g.*, $[Fe(Rtrz)_3]A_2{\cdot}nH_2O$ in which Fe^{2+} are triply bridged by 4-R-substituted-1,2,4-triazole molecules. The ST is accompanied by change of colour violet/white (LS/HS). Materials transition temperatures were tuned *via* an approach based on a molecular alloy; *e.g.*, it is possible to design compounds for which RT falls in the middle of a thermal hysteresis loop. The materials present applications (*e.g.*, temperature sensors, active elements of several display types, information storage/retrieval). Roubeau *et al.* (2001) synthesized a family of polymer 1D chains of Fe^{II} species showing ST *via* 4-*n*-alkyl-1,2,4-triazoles as bridging ligands; they studied effect of alkyl-tails length on triazole ligands on ST showing steepness degrading with increasing length; they used a set of four counterions to access wider range of ST temperatures; they detected large hysteresis loops with small tails mainly for methyl- and ethyl-substituted products. Longer tails weaken co-operativity and hysteresis gradually decays to zero; however, they showed that with anions, hysteresis remains even with long tails. Co-operativity weakening arises from diminution of polymer-chain length with rising tails. The effect is anion dependent. Interactions along polymer chains are confirmed.

Inorganic Fullerene-Like Nanoparticles and Inorganic Nanotubes

After C_{60} discovery and fullerenes chemistry, attention was directed to $C_{graphite}$ cylindrical and polyhedral forms; however, observations of such closed structures were limited to C-system. Adini *et al.* (2011) reported formation of equivalent stable structures in layered semiconductor WS_2. After heating thin W films in $H_2S_{(g)}$, transmission electron microscopy (TEM) revealed concentric polyhedral and cylindrical structures growing from an amorphous matrix; they verified structures closed nature by e^- diffraction and lattice imaging. As with C-system, complete closure of WS_2 layers requires structural-defect presence or atom arrangement in polyhedra other than planar hexagonal geometry; they informed nested fullerene-like structures. Layered-compound NPs (*e.g.*, MoS_2, WS_2) presenting hollow closed-cage structures and known as inorganic fullerene-like (IF) and NTs (INTs) are synthesized in macroscopic amounts; they show tribological properties and serve as solid-state additives to lubrication fluids. Metallic films incorporating IF NPs were prepared *via* wet and physical vapour deposition. Incorporation of NPs endows such coatings self-lubricating behaviour, *i.e.*, low friction and wear, which is desirable for applications variety; they reviewed IF/INT materials synthesis and medical devices/DDS applications. Inorganic-compound NPs with layered 2D structures (*e.g.*, $C_{graphite}$, MoS_2) are unstable in planar form and fold on themselves forming seamless hollow structures (*e.g.*, MNTs, IF NPs). Tenne & Redlich (2010) discussed developments in the field and nanophases applications, *e.g.*, solid lubricants, ultra-strong NC catalysts, *etc.* Properties of GR renewed interest in inorganic, 2D materials with unique electronical and optical attributes. Transition metal dichalcogenides (TMDCs) are layered materials with strong in-plane bonding and weak out-of-plane interactions, enabling exfoliation into 2D layers of single-unit-cell thickness. Advances in nanomaterials characterization and device fabrication opened opportunities for 2D layers of thin TMDCs in nano and optoelectronics. The TMDCs (*e.g.*, MoS_2, $MoSe_2$, WS_2, WSe_2) present sizeable bandgaps that change from indirect to direct in single layers allowing applications (*e.g.*, transistors, photodetectors, electroluminescent devices). Wang *et al.* (2012) reviewed TMDCs development, methods for preparing atomically thin layers, electronical/optical properties and advances. Binning *et al.* (1986) invented AFM based on ideas from Binning and Rohrer. With AFM one obtains images of physical objects and biological/chemical entities (viruses/bacteria, atoms/molecules). Settlement of AFMs achieve a share of nanometres. *Via* AFM one studies two-objects interaction: (1) to measure friction, elasticity and adhesion forces; (2) move individual atoms precipitate and remove them with surface (Hartmann, 1985). The AFM revealed DNA double helix (Bruker, 2013). Low-dimensional materials (LDMs), *e.g.*, GR, are composed of a single or few layers of atoms; to resolve LDMs structure requires an instrument with sub-Ångstrom resolution (AFM). Going beyond simple topography measurements a host of mechanical and electrical characterization techniques exists, which relies on AFM cantilever being able to *feel* material mechanical and electrical properties. Bertolazzi *et al.* (2013) described applications where AFM is used to probe LDM mechanical and electrical properties. Monolayers structurally similar to GR with different atoms behave as semiconductors (MoS_2) and insulators (BN). C-lattice in GR presents strong, highly directional bonds. High bond energy means that defects are rare and do not propagate *via* lattice as dislocations in metals. Modulus and strain-to-failure for GR are high. Monolayer MoS_2 presents a lower modulus comparable to stainless steel yet shows a breaking strength 30 times higher than steel. As with GR, monolayer-MoS_2 strength is close to the theoretical intrinsic one of its constituent chemical bonds indicating high molecular perfection. Directional bonds give LDMs layered nature with unique frictional properties on layers surface; they used force curves to test elasticity and strength of MoS_2 membranes stretched over circular holes in a SiO_2 support. Its mechanical properties make MoS_2 an excellent candidate material for incorporation into flexible and robust electronic devices (*e.g.*, molecular sensors, actuators) (Bertolazzi *et al.*, 2011).

Tuning Spin State in Fe^{II} Complexes: Pentadentate Ligand bztpen

Ortega-Villar *et al.* (2005) synthesized mononuclear diamagnetic compound $\{Fe(bztpen)[N(CN)_2]\}(PF_6)CH_3OH$ [bztpen = *N*-benzyl-*N,N',N'*-tris(2-pyridylmethyl)ethylenediamine] and studied crystal structure. The complex is precursor of dinuclear, –N≡C–N–C≡N-bridged Fe^{II} complexes with formula $\{[Fe(bztpen)]_2[\mu\text{-}N(CN)_2]\}(PF_6)_3{\cdot}nH_2O$ (n = 1 or 0) that they characterized in solid and solution. In all three complexes, Fe atoms present distorted $[FeN_6]$ octahedral (O_h) co-ordination defined by bztpen and terminal/bridging –N≡C–N–C≡N–. In solid, complexes (n = 0, 1) are molecular isomers, which differ by phenyl-ring position in *Z/E* $\{Fe(bztpen)[N(CN)_2]\}^+$ halves. Depending on texture, complex n = 1 exhibits paramagnetic behaviour or incomplete ST at 1 atm. Complex n = 0 undergoes gradual two-step ST with no hysteresis in solid. Both steps are 100K wide centred at 200/350K with a 80K plateau separating transitions; they determined crystal structure of complex n = 0 in 50K steps in 90–400K, which provides insight into structural behaviour and ST nature. Order/disorder transitions occur in bridge –N≡C–N–C≡N– and PF_6^- with ST suggesting that transitions trigger two-step character. In solution, complexes (n = 0, 1) display continuous STs; their EC showed voltammograms typical of dimeric systems with metals electronic coupling *via* –N≡C–N–C≡N–. In this laboratory Coronado *et al.* (2007) reported bistable ST NPs showing magnetic thermal hysteresis near RT. Prins *et al.* (2011) informed RT electrical addressing of bistable ST molecular system.

Organic Photochemistry to Make NPs: NPs to Direct Chemistry

Stamplecoskie and Scaiano (2010) discovered a method for preparing Ag NPs of various sizes and morphologies (*e.g.*, dodecahedra, NRs, nanoplates); they described photochemical synthesis of citrate-stabilized spherical Ag NPs, which are used to prepare multiple NSs with predictable and controllable size and morphology *via* irradiation with narrow-band light-emitting diodes (LEDs); they explained common mechanism for Ag NP-types formation. Variations in morphology resulted in spectroscopic changes. Shukla *et al.* (2005) reported chain-amplified photochemical reaction initiated by e^- transfer from excited sensitizer to *N*-methoxypyridinium (NMP) salts, which leads to N–O bond cleavage. Abstraction of H by methoxy radical $MeO^•$ from alcohol $HC(R_1R_2OH)$ yields α-hydroxy radical $^•C(R_1R_2OH)$, which reduces another NMP and propagates chain, which amplification is enhanced with water (Shukla *et al.*, 2006). Kinetic studies of 4-cyano-*N*-methoxypyridinium (CMP) salt reaction with benzhydrol (BH) showed that the rate constant for CMP reduction by diphenyl ketyl radical [•C(Ph$_2$OH), 1.1·10^{6}L·mol^{-1}s^{-1}] rises by more than one order of magnitude in water presence, which results of coupling e$^-$ to H transfer from •C(Ph$_2$OH) to water, which decreases endothermicity. Unfortunately rise in rate constant for one of two propagation steps is accompanied by increase in rate constants of competing termination reactions of •C(Ph$_2$OH). Enhancement in chain amplification results of rise in ratio propagation to termination rate constants of MeO• reactions. Main chain-terminating reactions of MeO• are D-abstraction from solvent, CD$_3$CN, and reaction with sensitizer, thioxanthone. Effect of rise in ratios of propagation rate constant of MeO• (H-abstraction from BH) to those of both termination reactions is larger than unfavourable water effect on •C(Ph$_2$OH) reactions. Chain-amplification rise depends on reactants concentration; at 0.037M of both, quantum yield increases from 16 to 45 in <1% water. 4-Phenyl-*N*-methoxypyridinium (PMP) reaction with 4-methoxybenzyl alcohol does not proceed *via* chain amplification because of endothermicity for e$^-$ transfer from •C(HArOH) to PMP; however, the amplification is induced by water addition where at 10% water quantum yield of 5 was obtained. Water-induced, H-coupled e$^-$ transfer increases rate constant for PMP reduction from negligible to dominant level. Scaiano *et al.* (2012) reported that Norrish type-I photocleavage is source of reducing free radicals, which are used to convert soluble metal cations into atomic state that proceed to form NPs. Proton-coupled e$^-$ transfer (PCeT) is a tool to interpret mechanism for metal-cation reduction, which involves multisite PCeT with H and e$^-$ having separate receiving substrates. Bueno-Alejo *et al.* (2012) informed nanosecond light amplification by stimulated emission of radiation (laser) ablation synthesis for spherical Au NPs of 4nm in 5s (532nm, 0.66J·cm^{-2}), where protecting agent is selected in the protocol, which avoids repeated sample irradiation and undesired exposure of capping agent during ablation. The method takes synthesis advantage of clean unprotected polymorph and disperse Au NSs *via* H_2O_2 as reducing agent. Laser drop technique allows delivering controlled doses to small drops, which undergo assisted fall into capping agent solution and suspension yielding monodisperse custom-derivatized composite materials *via* a simple technique. Au-NPs photocxcitation in plasmon transition 530nm allows carrying high-energy reactions at RT. For dicumyl peroxide (activation energy E_a = 34.3kcal·mol^{-1}) reaction occurs in <1min under 532nm laser. Fasciani *et al.* (2011) suggested that the peroxide is exposed to temperatures of 500°C for submicrosecond times and guides which organic-reaction type benefits from plasmon-mediated energy delivery. Plasmon excitation (532nm) of Au NPs in resazurin and hydroxylamine presence leads to its efficient photocatalytic reduction to resorufin. Under laser-drop conditions the process is complete following 8ns pulse at 532nm. Bueno Alejo *et al.* (2011) proved that excitation with LED sources at 530nm is a simple and cost-efficient way to promote plasmon-assisted reactions; they proposed that catalytic reaction is thermally activated by Au NPs and takes advantage of high temperatures (HTs) under plasmon excitation. Stamplecoskie *et al.* (2011) showed that in VIS exposure of Ag NP-containing films enhances field around Ag NPs in thin film containing azo free-radical initiator (2,2'-azo-bis-isobutyronitrile, AIBN) and triacrylate selectively cross-links triacrylate within plasmonic region around NPs. Cross-linked polymer is lesser soluble than precursor and behaves as solubility switch. After the film is developed with EtOH, polymer-encapsulated NPs are preserved on surface. The polymer structure of 8–10nm, which encapsulates NPs, maps and preserves plasmon-field morphology in Ag NP-controlled NSs. Stamplecoskie *et al.* (2012) obtained a hierarchy of lithographic-type imaging generating 3μm lines incorporating subdiffraction limit features *via* two-step reaction; they used photochemically generated ketyl radicals to make Ag NPs-defined lines; they used excitation of NP-surface plasmons to generate localized heat, which causes polymerization selectively on excited-NPs surfaces. Generated nylon-6 polymer serves as solubility switch used to retain features on substrate selectively; they used several imaging techniques to establish nylon-shells nature; they exploited heat generated by plasmon excitation to make negative-type lithographic features with dimensions below diffraction limit. Carl Zeiss Microscopy (2013) obtained an image of double bowtie plasmonic device made in Au on glass by high-resolution (HR) microscopy. Fujishima and Honda discovered (1972) water photocatalytic splitting on TiO_2 electrode under UV influence (Fujishima *et al.*, 2000). Efforts were devoted to TiO_2 material, which led to applications in fields ranging from photovoltaics and catalisys to photoelectronics and sensors, which are divided into *energy* and *sustainable* categories, in which different-modification TiO_2 are used (Gratzel, 2001). Photodegradation was used in waste/water treatment in all methods, *e.g.*, froth flotation coagulation, *etc.*; it offers prospects for overall treatment of dyestuff effluent. Shanthi & Priya (2012) performed photodegradation of aniline blue and crystal violet dyes from aqueous solutions of binary mixture *via* TiO_2 as photocatalyst. Performing photodegradation varying experimental parameters they found optimum conditions required for maximum degradation; they performed dye photodegradation *via* different energy sources [*e.g.*, solar, microwave (MW)]; they performed degradations at temperatures 25, 35 and 45°C to calculate rate constant and activation parameters. Both sources are equivalent in causing degradation except time and photocatalyst dose: time is lesser for MW than for solar radiation but MW photocatalyst dose is higher. Results help to design effluent treatment plants in industries.

SUPERCONDUCTOR VORTICES ON MOVE: MOLECULAR MOTOR LABORATORY

Abrikosov (1957) studied magnetic properties of bulk superconductors for which Ginzburg–Landau (GL) theory parameter $\chi > 1/2^{1/2}$ (superconductors type-2, SCT2); he explained SCT2-alloys behaviour in MF. Mkrtchyan & Shmidt (1972) calculated free energy of SCT2 vortex in superconductor interacting with hollow cylindrical channel of radius $r << \delta_o$ parallel to it (δ_o = penetration depth); capture of a single vortex by the channel is energetically favourable. Pinning force is computed: $f_p = H_{cm}{}^2\xi(T)/2$ where H_{cm} = critical thermodynamic field strength. Buzdin (1993) discussed formation possibility of multiple-quanta vortices on columnar defects (CDs) produced by ion irradiation. Upper-critical field for localized superconductivity near CDs depends nonmonotonously on the radius. Takezawa & Fukushima (1994) examined insulating-inclusion effectiveness in SCT2 as pinning centre. With regard to single-quantum vortex they investigated dependence of minimum pinning potential U_p on radius R of cylindrical insulating inclusion solving GL equations numerically. The U_p is defined by free-energy difference between vortex inside and infinitely distant. The $U_p < 0$ for all Rs and is shallow for $R < \xi(T)$ (coherence length). Depth of U_p rises rapidly with increasing R up to penetration depth $\lambda(T)$ and there U_p saturates. The U_p for $R \geq \lambda(T)$ is much deeper than for $R \leq \xi(T)$; *i.e.*, even for single vortex insulating inclusion with $R \geq \lambda(T)$ it gives stronger attraction with vortex and works as more effective pinning centre than with $R \leq \xi(T)$. Numerically solving 2D GL equations, Takezawa & Fukushima (1997) evaluated optimal size of insulating inclusion for pinning single-quantum vortex in SCT2. Although it was believed that optimal size of pinning centre is twice coherence length they found that in case of low MFs and pinning centre density, prismatic insulating inclusion with square cross-section of side length equal to penetration depth gives strongest pinning force to vortex. Teresawa *et al.* (1998) investigated experimentally effects of CDs with splayed configurations on flux pinning and creep; they introduced parallel and splayed CDs into $La_{2-x}Sr_xCuO_4$ specimens by high-energy heavy-ion irradiation; they observed enhancement of flux pinning and critical current density J_c because of splayed CDs: (1) at $T < 15K$ and $H = 0.1T$, effective pinning potential U_0 of CD-splayed specimen was larger than with parallel CDs; (2) at $15K \leq T \leq 30K$ and $H = 0.1T$, U_0 of CD-splayed specimen was smaller than with parallel CDs. Harada *et al.* (1996) investigated matching microscopic mechanism in SCT2, which manifested as peaks production in critical current at specific values of applied MF with Lorentz microscopy to allow vortices observation in Nb thin film presenting a regular array of artificial defects; they observed vortices forming regular lattices at matching MF, multiples and fractions. Dynamic observation revealed that vortices are most difficult to move at matching field whereas excess ones move easily. Linke (2002) edited a special issue on ratchets and Brownian motors. Linke *et al.* (2006) reported that liquids perform self-propelled motion when they are placed in contact with hot surfaces with asymmetric topology; they observed pumping effect when liquid is in Leidenfrost (film boiling) regime for many liquids over wide temperature range; they showed that liquid motion is driven by viscous force exerted by vapour flow between solid and liquid.

Molecular Motors in the Core of Molecular Biology Deoxyribonucleic-Acid Genes

Ternary complexes of DNA-dependent ribonucleic acid (RNA) polymerase (RNAP) with DNA template and nascent transcript are transcription intermediates (Uptain *et al.*, 1997). Biochemical reactions affect RNAP progression in ternary complexes *via* various transcription units, which are signalled intrinsically by nucleic-acid sequences/RNAP or extrinsically by protein/regulatory factors, which affect processes (*e.g.*, promoter proximal, distal pausing in prokaryotes, eukaryotes) and regulate gene-expression. In eukaryotes at least two factors are related to cellular transformation and cancers. Models for ternary-complexes structure and mechanism by which they move along DNA provide explanations for biochemical reactions, which predict that RNAP moves along DNA without dissociation and termination constant possibility; RNAP moves in discontinuous and inchworm-like manner. Direct predictions were confirmed; however, one feature of RNA chain elongation is that DNA sequence determines whether RNAP moves discontinuous or monotonically. In at least two cases, RNAP and DNA block to elongation encounter induces specifically discontinuous synthesis mode, which provides insights into RNA chain elongation and understanding bioregulatory systems at molecular level. Research advances in structure and function of RNAP-II elongation complex enlightened mechanisms governing elongation of eukaryotic messenger (m)RNA synthesis (Shilatifard *et al.*, 1997). Elongation regulation features by DNA and RNA binding transcriptional activators were illuminated; action mechanisms of elongation factors that suppress pausing/premature arrest transcribing RNAP-II were defined and elongation factors implicated in human disease were identified. Biochemical and genetic studies shed light on structure and function of RNAP-II elongation complex and transcription factors, which control it (Reines *et al.*, 1999). Elongation factors were identified and action mechanisms, characterized; insights into elongation-factor bioroles were gained from genetic studies of mRNA-synthesis regulation in yeast, and links between RNAP-II elongation machinery and DNA-repair/recombination pathways emerged. Chain elongation of RNA is processive and accurate, which is regulated by numerous intrinsic and extrinsic signals. Bar-Nahum *et al.* (2005) described the mechanism that governs RNAP movement and response to regulatory inputs (*e.g.*, pauses, terminators, elongation). The RNAP of *E. coli* moves by complex Brownian ratchet mechanism that acts before phosphodiester bond formation. Incoming substrate and flexible F-bridge domain of catalytic centre serve as separate ratchet devices, which function in concert to drive forward translocation. Adjacent G-loop domain controls F-bridge motion keeping balance between elongation-complex productive and inactive states, which is critical for cell viability since it determines transcription rate, processivity and fidelity.

FROM ATOMIC FORCE MICROSCOPY TO MAGNETIC FORCE MICROSCOPY

The MFM was invented by Martin and Wickramasinghe (1987) equipped with sharp magnetic tip, which is sensitive to spatial derivatives of stray fields from a sample, which depends on magnetization divergence. Scanning samples surface and measuring long-range magnetostatic force between ferromagnetic tip and sample *vs.* position, Rugar *et al.* (1990) observed domain-walls, Bloch-lines and ripple-structures image; to separate topography and magnetic signals, tip–sample separations are relatively large so that long-range magnetostatic forces dominate. Traditional MFM presents magnetic-domain resolution >50nm, which limits NS-characterization applications; HR MFM invented by Swiss Probe reaches resolution <10nm, which compares with scanning electron microscopy (SEM) with polarization analysis (SEMPA). Probing short-range magnetic exchange interactions magnetic exchange force microscopy (MExFM), it is possible to reach atomic resolution and detect spin configurations (Kaiser *et al.*, 2007).

Functional CdSe and ZnS NPs Capped with Thiols: Sensors

Murphy and Coffer (2002) published a QD primer. Wuister *et al.* (2004) prepared luminescent CdSe and CdTe QDs in hot solvent of capping molecules [trioctylphosphine (TOP), trioctylphosphine oxide (TOPO) and hexadecylamine (HDA) for CdSe, TOP and dodecylamine (DDA) for CdTe]; they investigated exchange influence of capping molecules with different thiol types [amino ethanethiol, (3-mercaptopropyl) trimethoxysilane, hexanethiol, 2-propenethiol, 4-mercaptophenol] for CdSe/CdTe QDs; they observed a difference: capping exchange with thiols results in increased luminescence efficiency for CdTe QDs but induces quenching of excitonic emission of CdSe QDs; they explained difference between the two types of II-VI QDs by difference in energy of valence-band top. Lower energetic position of valence band for CdSe results in photogenerated-h^+ trapping on thiol, quenching luminescence. For CdTe, valence band is situated at higher energies with respect to thiols redox level inhibiting h^+ trapping and maintaining luminescence efficiency. One method to render CdSe/ZnS core–shell QDs water soluble is to functionalize surface with carboxylate groups *via* heterobifunctional ligands, *e.g.*, 3-mercaptopropionic acid, where thiolic end binds on to outer ZnS shell; however, ligand exchanges starting with TOPO-capped QDs lead to quantum-yields loss and colloids poor stability in water. Pong *et al.* (2008) used calculations to understand binding nature between alkyl thiols and ZnS wurtzite surfaces; guided by computations they modified ligand exchange and increased 3-mercaptopropionic-acid reactivity towards ZnS surface in $CHCl_3$. Functionalization reaction required mild conditions and led to QD NPs, which were individually dispersed in water with colloidal stability. Photoluminescence performance of QDs was preserved. In this laboratory Aguilera-Sigalat *et al.* (2011) reported that while alkyl-thiols addition reduces FL of CdSe core QDs(-C) it enhances emission of FL amine-capped CdSe and ZnS core–shell QDs(-CS). Aguilera-Sigalat *et al.* (2012a) synthesized high-FL organic- and water-soluble CdSe/ZnS QD-CS with –SH ligands chemisorbed on QD surface by amine-ligands replacement by alkyl thiols under mild conditions. The QDs exhibited greater photostability than initial amine capped QD-CS. Aguilera-Sigalat *et al.* (2012b) prepared a supramolecular system based on ketoprofen-functionalized CdSe/ZnS NPs and Py-modified β-*cyclo*dextrin (CyD_7), and used it for molecular sensing of different analytes; they reported strategy for individual recovery of all components of sensing assay. Fluorophores of Py in Py-functionalized CdSe QD(@Py) and alkylpyrene (R–Py)/Py undergo fast degradation in aerated $CHCl_3$ under UV-A ($316 < \lambda < 400$nm) illumination. Aguilera-Sigalat *et al.* (2012c) showed steady-state FL of irradiated $CHCl_3$ solutions of QD@Py formation of bands red-shifted compared to Py. Similar behaviour is observed for Py/R–Py. Column chromatography of Py photolysate in $CHCl_3$ allowed isolating photoproducts arising from Py degradation and obtaining structural information of photoproducts responsible for emission bands. Predominant photoproducts originated from Py reaction with $^{\bullet}CHCl_2$. Phototransformation of QD@Py/R–Py involves R-detachment from aromatic ring induced by $^{\bullet}CHCl_2$; R-oxidation at benzylic position was detected. By contrast Py's showed photostability in aerated CH_2Cl_2. Transient absorption showed $^3Py/Py^{+\bullet}$ formation for all Py's in halogenated solvents. Yield of $Py^{+\bullet}$ for Py is higher than QD@ Py/R–Py. The $Py^{+\bullet}$ was longer-lived in CH_2Cl_2 than $CHCl_3$. Reason for Py photostabilty in CH_2Cl_2 is different $^{\bullet}CH_2Cl$ and $^{\bullet}CHCl_2$ reactivities towards Py and O_2. Use of CH_2Cl_2 is alternative to $CHCl_3$ when solubility properties of halogenated solvents are needed to dissolve FL-probe Py. Wadhavane *et al.* (2012) characterized FL organogels (QD–organogel) prepared combining pseudopeptidic macrocycle and different types of CdSe QDs *via* optical and microscopic techniques. Presence of QDs not only does not disrupt supramolecular organization of internal fibrillar net of organogel but also decreases gelator critical concentration needed to form stable and thermoreversible organogels; regarding QDs photophysical properties they observed different trends depending on presence of ZnS inorganic shell around CdSe core; while QDs-CS preserve photophysical properties in organogel they observed high or moderate rise of FL intensity and lifetime for QDs-C embedded in organogel. Luminescent organogels based on QDs present applications as hybrid materials. Kumar & Wei (2013) reviewed QDs for nano/bioapplications as technological platform. The QDs are versatile inorganic probes with unique photophysical properties (*e.g.*, narrow/size-dependent FL with broad absorption spectra, strong FL intensity, excellent antiphotobleaching). The QDs were used in diverse bioapplications in different domains (*e.g.*, cell labelling, genomic detection, optical sensors, nanosensors, quantum mechanics-based DDSs, biomedical imaging). Nanotechnology presents potential to revolutionize medicine, electronics, textiles and energy production.

NANOMATERIALS TOXICOLOGY: NANO-QSARS – ADVANCES AND CHALLENGES

Shevchenko *et al.* (2003) analyzed nanoworld structural diversity and emergence; they considered different problem aspects (*e.g.*, nonequilibrium, coherence, hierarchical structure, fragmentariness, generalized symmetry); they treated NP structure inhomogeneity and nanoworld structural diversity as self-organization result of nonequilibrium nonlinear multivariant system. Engineered-nanomaterials proliferation presented dilemma to regulators regarding hazard identification. International Life Sciences Institute Research Foundation and Risk Science Institute convened expert working group to develop screening strategy for hazard identification of engineered nanomaterials. Working group report presented *elements* of screening strategy rather than detailed testing protocol (Oberdörster *et al.*, 2005). Based on limited-data evaluation, the account presented a data gathering strategy applicable to hazard identification in risk-assessment development for nanomaterials; they included oral, dermal, inhalation and injection exposure routes recognizing that depending on use patterns nanomaterials contact occurs by any route. Toxicity-screening elements are: physicochemical characteristics and *in vitro/vivo* assays.

1. Likelihood exists that NPs bioactivity will depend on physicochemical parameters not considered in toxicity screening. Physicochemical properties in understanding toxic effects of test materials include particle size/size distribution, agglomeration state, shape, crystal structure, chemical composition, surface area, chemistry/charge and porosity.
2. *In vitro* techniques allow specific biomechanistic pathways to be isolated and tested under controlled conditions in ways that are not feasible in *in vivo* tests; they suggested tests for portal-of-entry toxicity for lungs, skin and mucosal membranes, and target organ toxicity for endothelium, blood, spleen, liver, nervous system, heart and kidney; they considered non-cellular assessment of NP durability, protein interactions, complement activation and pro-oxidant activity.
3. They proposed Tier-1 *in vivo* assays for pulmonary, oral, skin and injection exposures, and Tier-2 evaluations for pulmonary contacts. Tier-1 evaluations include inflammation, oxidative-stress and cell-proliferation markers in portal-of-entry and selected remote organs/tissues. Tier-2 evaluations for pulmonary exposures comprise:
    a. Deposition, translocation and toxicokinetics/biopersistence studies;
    b. Multiple-contact effects;
    c. Potential effects on reproductive system, placenta and foetus;
    d. Alternative animal models;
    e. Mechanistic studies.

Leszczynski & Shukla (2009) edited a monograph on practical computational chemistry. Puzyn *et al.* (2009) highlighted achievements and challenges relating to QSAR application in risk assessment of nanometre-sized materials; they discussed advances in *classical*-QSAR context; they reviewed ways for nanocompound structural characterization; they evaluated toxicological data applicability for developing QSARs; they presented models; they highlighted need to develop interpretative nanosystems descriptors; they suggested that because of molecular-structures variability and different toxicity mechanisms, NPs individual classes should be modelled separately. Research related to GR grew at spectacular pace in disciplines while more and more scientists are considering health and ecosystem risks. Work and challenges were reviewed from metals and small molecules to human health and ecosystem (Hu & Zhou, 2013). The goal is to reduce gaps between expanding material applications and studies on human health/ecosystem risks *via* correct assay methods, valid administration procedures, long-term tests and meaningful data. Nel *et al.* (2013) edited a special issue on environmental health and safety considerations for nanotechnology (nano-EHS).

Colloidal Methods for Supported Alloyed Metal Nanoparticles

Alcohols oxidation to aldehydes with $O_{2(g)}$ in place of stoichiometric O-donors is crucial for fine-chemicals synthesis; however, identified catalysts are inactive with primary alkyl alcohols. Enache *et al.* (2006) showed that Au/Pd-TiO_2 catalysts give turnover frequencies (up to 270 000 turnovers·hr^{-1}) for alcohols oxidation, *e.g.*, primary alkyl ones; addition of Au to Pd nanocrystals improved overall selectivity and, *via* scanning TEM (STEM) combined with X-ray photoelectron spectroscopy (XPS), they explained that Au-Pd nanocrystals were made up of an Au-rich core with a Pd-rich shell indicating that Au electronically influences Pd catalytic properties. Dimitratos *et al.* (2006) studied effect of metal particle size of mono/bimetallic-supported catalysts (Au, Pd, Au-Pd)/C *via* model reaction liquid phase oxidation of glycerol. Tuning particle size 2–16nm they observed an activity decay and rise in selectivity to Na glycerate; they studied temperature effect and found that rising temperature with large particles, formed glycerate was retained and not overoxidized to tartronate.

Inhomogeneous Electron Distribution in InN Nanowires: Optics

Segura-Ruiz *et al.* (2009) investigated vertically self-aligned InN nanocolumns (NCo's) *via* SEM, RS and PL spectroscopy; they studied different NCo morphologies in dissimilar growth conditions of molecular beam epitaxy (MBE). The RS revealed strain-free crystalline NCo's. Longitudinal optical modes un/coupled to e^- plasma co-exist in RS pointing to two regions in NCo: a surface layer of degenerated e^-'s and a nondegenerated inner core. Characteristics of LT PL (LTPL) and dependence on temperature/excitation power were explained by a model considering localized h^+'s, recombining with degenerated e^-'s close to nonpolar surface. Differences observed in optical response of different samples with similar crystalline quality were attributed to variation in e^- accumulation layer with growth conditions. Werner *et al.* (2009) investigated electrical properties of InN NWs in four-point probe measurements. Conductance dependence on NW diameter distinguishes *core* bulk (quadratic) from *shell* sheet (linear) contribution. They reported formation of a thin In_2O_3 layer at NW surface by XPS. Shell conductivity is ascribed to an e^- accumulation layer forming at radial InN/In_2O_3 interface. Although conductance *via* accumulation layer dominates for NWs below a critical diameter of 55nm, core channel is not neglected even for small NWs. Segura-Ruiz *et al.* (2010) measured PL excitation (PLE) spectra for a set of self-assembled InN NWs and crystalline InN layer grown *via* MBE; they reproduced PLE experimental lineshapes by an absorption self-consistent calculation (SCC) in a cylindrical InN NW; they accounted PLE-spectra differences for inhomogeneous e^- distribution within NWs, caused by a bulk donor concentration (N_D^+) and 2D density of ionized surface states (N_{ss}^+). For NW radii >30nm, N_D^+ and N_{ss}^+ modify absorption edge and lineshape, respectively, and are determined from comparison with experiments. Molina-Sánchez *et al.* (2012) studied theoretical and experimentally the effect of surface e^- accumulation on optical properties of InN NWs; they measured PLE spectra for self-assembled InN NWs grown under different conditions; they reproduced PLE experimental lineshapes by absorption SCC in a cylindrical InN NW. With the SCC model one explores how optical absorption depends on NW radius and doping concentration; the model solves Schrödinger equation for a cylindrical NW of infinite length assuming a parabolic conduction band. Geometry of NCo's introduces effects in e^- density and SCC conduction band profile with no equivalence in planar layer. Differences in PLE spectra for inhomogeneous e^- distribution inside NWs are produced by N_D^+/N_{ss}^+. For NW radii >30nm, N_D^+/N_{ss}^+ modify absorption edge/lineshape and are determined from experiments.

Ti-Mask Selective-Area Growth by RF-Plasma-Assisted MBE: GaN NCo

Kishino *et al.* (2009) performed Ti-mask selective-area growth (SAG) of GaN NCo's at temperature of 900ºC, while decaying supplied $N_{2(g)}$ flow rate Q_{N2} from 3.5 to 0.5 standard cubic centimetres per minute (sccm); they showed uniform arrays of GaN NCo's. At low Q_{N2}, Ga desorption and diffusion from and on nitrided Ti mask were accelerated, which suppressed crystal nucleation on Ti-mask surface, and GaN-NCo's SAG was achieved even when spacing between NCo's was several hundred nanometres. Ga-desorption enhancement with decaying Q_{N2} brought about a drop in growth rate of GaN NCo's from 1.05 to 0.15μm·h^{-1}. Lateral growth rate of GaN NCo's rapidly rose above critical Q_{N2} = 1.5sccm and became 45nm/h at Q_{N2} = 3.5sccm. For low Q_{N2} < 1.5sccm, lateral growth rate lowered, 8nm/h, which contributes to controlled GaN SAG where underlying nanomask patterns are traced.

Elemental Distribution of In-Rich $In_xGa_{1-x}N$ Single Nanowires

III nitrides were studied because of potential from high-efficiency solid-state lighting and PV to high-power and temperature electronics. Ternary InGaN alloy is of interest for solid-state lighting and PV because of tuning its direct bandgap from near-ultraviolet UV (NUV) to NIR. Attempting to synthesize InGaN, growth techniques were tried; however, difficulty remains in making HQ InGaN films and freestanding NWs with tunability across compositional range. Kuykendall *et al.* (2007) reported growth of single-crystalline $In_xGa_{1-x}N$ (x = 0–1) NWs; they synthesized NWs by LT halide chemical vapour deposition (CVD) and NWs presented tunable emission NUV–NIR; composition tunability is because of low process temperature and NW-morphology ability to accommodate strain-relaxed growth, which suppresses tendency towards phase separation, which plagues thin-film community. Vajpeyi *et al.* (2009) grew single-crystalline/phase $In_xGa_{1-x}N$ nanopillars (NPi's) spontaneously on Si(111) substrate by plasma-assisted MBE; they analyzed surface morphology, structural quality and optoelectronic properties of InGaN NPi's *via* SEM, energy-dispersive X-ray analysis (EDXA), HR X-ray diffraction (XRD) and RT/LTPL spectra. Results of EXDA showed that NPi's were composed of InGaN and In amount in $In_xGa_{1-x}N$ NPi's was controlled changing growth temperature. Spectra of RT/LTPL revealed that emission wavelength was tuned from blue to green luminescence depending on growth temperature. Wavelength tuning was attributed to a higher In amount incorporated at a lower growth temperature, which was consistent with EDXA/HR-XRD results. *Via* synchrotron radiation nanoprobe Segura-Ruiz *et al.* (2011) reported elemental distribution in single $In_xGa_{1-x}N$ NWs grown by MBE directly on Si(111) substrates; they characterized single NWs dispersed on Al-covered sapphire by nano-X-ray FL, RS and PL spectroscopy. Maps of Ga/In reveal an inhomogeneous axial distribution inside single NWs. Analysis of different-dimension NWs showed In-segregation decay with NW-diameter reduction while Ga distribution remained unaltered. They measured PL/RS on NW ensembles, which exhibit compositional-disorder signatures.

Non-Lithographic Fabrication of Ni–Se Heterojunction Nanowires

Kumari *et al.* (2014) fabricated Ni–Se heterojunction NWs *via* template-assisted electrodeposition; they characterized NWs by SEM, EDXA, XRD and electrical transport. The SEM reveals uniform and dense growth of Ni–Se NWs. Pattern of XRD shows crystalline nature of Ni–Se NWs. Spectrum of EDS shows higher Ni percentage compared to Se. Collective current–voltage (I–V) characteristic of heterojunction NWs shows resonant tunneling diodes (RTDs)-like behaviour with peak to valley current ratio 1.37 at RT.

Multiphoton Molecular Release: NIR-Triggered Phototherapy

In this laboratory Voliani *et al.* (2011) developed a modular nanosystem for controlled photorelease of molecular payloads induced by yellow-green laser. Peptide-coated 30nm Au nanospheres were covalently conjugated to payloads by standard click chemistry *via* a 1,2,3-triazole-ring (*cf.* Figure 4a) linker (Figure 4b). Payload photorelease was triggered in living cells by low-power continuous-wave (CW) 561nm laser within a standard confocal microscope. In agreement with calculations the process follows a nonlinear effect promoted by LSPR stimulation. Ability to perform controlled photocleavage of

UV-labile molecules with VIS/NIR opens up nanomedicine possibilities. Nanomaterials composition in multimodular systems in which every component works in synergysm results in a nanomedicine-tool class. Development of NSs able to release drugs directly in target after a stimulus improves therapeutic efficiency reducing side effects.

Poly(ethylene glycol) (PEG)-functionalized (pegylated) SWNTs (Liu *et al.*, 2007) and nano-GR oxide (NGO) (Sun *et al.*, 2008) display PL in UV–VIS/NIR; they showed that anticancer drug doxorubicin (*cf.* Figure 5) was physisorbed on to the pegylated NGO surface, which was functionalized with antibodies, which allowed for selective killing of cancer cells *in vitro*, which could be monitored *via* FL. Anthracycline antibiotic doxorubicin is a cancer-chemotherapy drug with adverse effect of life-threatening heart damage. Voliani *et al.* discussed therapeutic efficiency of Au-based nanosystems for exogenous-controlled intracellular release of doxorubicin exploiting multiphoton stimulation; to shift triggering irradiation to NIR they presented results on a hybrid system formed by Au nanospheres decorated with upconversion nanocrystals.

Perspective is to use other active agents: drugs, DNA/RNA and antioxidants, *e.g.*, flavan-3-ol (flavonol) catechins (*cf.* Figure 6) (Castellano & Torrens, 2015; Castellano *et al.*, 2012, 2013, 2014).

Inversion Symmetric Two-Level Systems: Glasses LT Universality

Amorphous solids, polymers and disordered lattices show qualitative and quantitative similarities (of, *e.g.*, specific heat, thermal conductivity, internal friction) at LT <3K, which suggests existence of a mechanism intrinsic to matter-disordered state, which dictates physical properties. Standard STM with which the problem is treated is that of tunnelling two-level systems (TLSs) introduced by Anderson *et al.* (1972) and Phillips (1972). Questions (*e.g.*, TLS nature, mechanism dictating universality, energy scale ordering universal-regime range) are misunderstood. In this laboratory Gaita-Ariño & Schechter (2011) proposed a model of two TLS types, (nearly) symmetric τ and asymmetric S with respect to inversion symmetry: τ-TLS interacts weakly with phonon field yet gapes S-TLS at low energies; the model explains puzzles above/others related to universality and proves useful in treating problems [*e.g.*, $1/f$ noise, glass ageing, superconducting quantum-bit (qubit) decoherence].

Anisotropic Absorption of CdSe/ZnS Quantum Rods in Polymer Film

Mukhina *et al.* (2013) achieved spatially homogeneous, ordered ensemble of semiconductor quantum rods (QRs) in a polymer film of poly(vinyl butyral). The CdSe/ZnS QRs are embedded in the film; they stretched up the film to four times its length. Concentration of QRs is $2{\cdot}10^{-5}$mol·L^{-1}. Absorption spectra in orthogonally polarized light confirmed spatial-ordering occurrence in QR ensemble (QRE). They examined optical-properties anisotropy in ordered QRE; the method is used as a low-cost solution for preparing nanomaterials with anisotropic properties and high nanocrystals concentration.

Photocontrol of Bistable Spin-Transition Systems: Dynamics

Ligand-driven light-induced spin change (LD-LISC) is a photomagnetic effect based on modulation of ligand-field strength of suitable ST complex *via* ligand photochemical reaction; it allows switching electronic spin state of metal ion *via* light over a broad range of temperatures, *e.g.*, RT. Among photochemical reactions capable of triggering ST reversibly Boillot *et al.* (2004) selected *cis–trans* isomerization; they

Figure 4. Molecular structures of (a) 1,2,3-triazole and (b) 1,2,3-triazole linkage

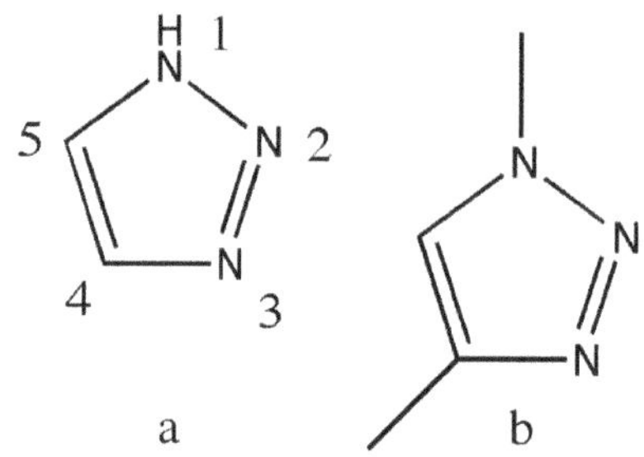

Figure 5. Molecular structure of anthracycline antibiotic doxorubicin

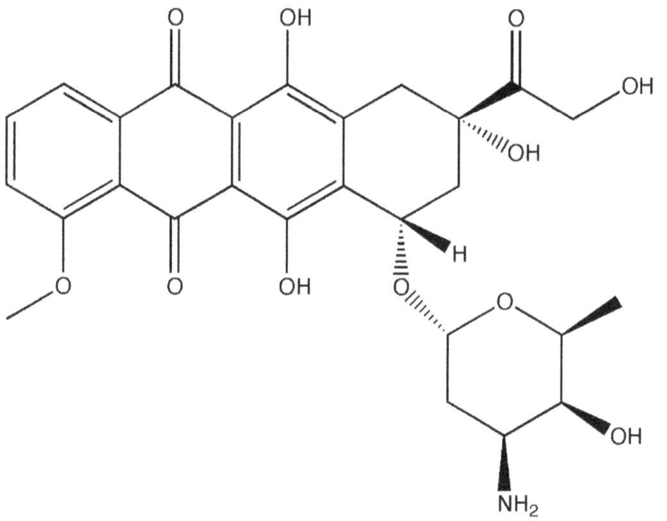

Figure 6. Molecular structure of antioxidant flavan-3-ol (flavonol) catechin

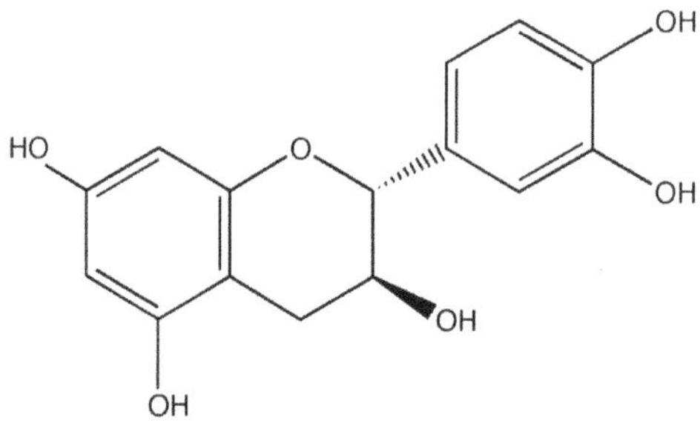

showed LD-LISC in $Fe^{II/III}$ complexes; varying molecular components they modulated working temperature and excitation wavelengths observing effect at RT upon sample VIS irradiation; they performed experiments on compounds in solution and included in polymeric matrices. Boillot *et al.* (2009) prepared two pseudo-O_h Fe^{II} complexes, $Fe(stpy)_4(NCSe)_2$, containing photoresponsive ligands (–CH=CH– *cis-trans* isomerization) with *trans*/*cis*-styrylpyridine (stpy) isomers; they showed that magnetic behaviour of polycrystalline solids depends on ligand-stpy configuration; they determined crystal X-ray structures at 293 and 104K for both isomers. All-*trans* and -*cis* compounds crystallize in orthorhombic ($Pna2_1$) and monoclinic space groups (*C*2/c), respectively. No symmetry change occurs upon cooling to 104K. Centres of Fe^{II} lie in axially compressed O_h with $NCSe^-$ anions in apical position and the four pyridine (pyr) Ns in southern plane. Variation of metal–ligand bond lengths *vs*. temperature reflects thermal S = 0 ↔ 2 ST of all-*trans* complexes and S = 2 ground state of all-*cis* ones. Unit-cell volumes per metal ion change accordingly, and relative variation because of ST compares to that associated with configuration formal alteration of the four stpy isomers; they investigated photomagnetic responses at 130K with doped polymer thin films containing all-*cis* (HS) or -*trans* (partly LS) species. Illumination at 130K of doped poly(methyl methacrylate) (PMMA) films leads to UV–VIS absorption typical for *cis*→*trans* photoisomerization of stilbenoid moiety. Direct magnetic measurements established LD-LISC. Excitation of 355nm of doped thin films produces long lifetime states, which are manifested by HS→LS (all-*cis* complex) and LS→HS (all-*trans* one) changes of Fe^{II} magnetic behaviour; the process is bi-directional; they proposed a structural analysis based on single-crystal XRD data to rationalize LD-LISC detected for doped PMMA thin films.

Nanostructuration and Guest Effects of ST {Fe(bpac)[Pt(CN)$_4$]}

Porous co-ordination polymers (PCPs) of {Fe(bpac)[M(CN)$_4$]}·guest (M = Pt, Pd) exhibit larger channels than 3D-Hofmann-like PCPs. The channels are partially occupied by unco-ordinated guest bis(4-pyridyl) acetylene (bpac) ligands and labile H_2O molecules. The PCPs exhibit scarce co-operative ST *ca*. RT with large hysteresis loop (up to 49K) and display sensitivity to humidity/guest molecules. Molecules-bpac inclusion in 3D net is avoided adding competitive volatile molecules during crystallization affording guest-free material. In this laboratory Bartual-Murgui, *et al*. (2011a) informed different guest and free-materials ST. Bartual-Murgui *et al*. (2012) reported synthesis and characterization of 3D-Hofmann-like clathrate porous MOFs {Fe(bpac)[M(CN)$_4$]} (M = Pt, Pd, Ni), which exhibit ST. Rigid bpac ligand is longer than 4,4'-azopyridine (azpy)/pyrazine (pz) and was selected with aim to improve PCP ST and porosity. Net of 3D is composed of successive {Fe[M(CN)$_4$]}$_n$ planar layers bridged by bis-monodentate bpac ligand linked in Fe apical-positions. Large void between layers, which represents 41.7% of unit cell, accommodates solvent molecules or free bpac; they used different syntheses to obtain STs with hysteresis loops *ca*. RT; they characterized samples by magnetic susceptibility, calorimetry, Mössbauer and RS spectroscopies; they revealed relation between quantity of included bpac molecules and ST completeness underlining role of π–π stacking interactions between host–guest bpac in net; although bpac inclusion increases quantity of active Fe centres they measured no ST-temperature variation; they investigated net ability to accommodate inclusion of molecules other than H_2O/bpac and studied host/guest interaction–ST synergy. Aromatic-molecule clathration revealed specific modifications of ST temperature. Temperature and completeness of ST are related to nature of metal associated with Fe centre and kind/quantity of lattice guest molecules. Bartual-Murgui *et al*. (2013) reported ST behaviour of 3D PCP {Fe(bpac)[Pt(CN)$_4$]} on adsorption of mono/polyhalobenzene guest molecules. Crystal structure of {Fe(bpac)[Pt(CN)$_4$]}·G (G = 1,2,4-trichlorobenzene) showed preferential guest sites establishing π···π stacking interactions with host framework, which explain relation between ST modification, chemical nature of guest molecule (electronic factors) and number of ad-molecules (*steric* factors). Agustí *et al*. (2008) synthesized 3D co-ordination polymers {Fe^{II}(azpy)[M^{II}(CN)$_4$]}·$n$$H_2O$ (M = Ni, Pd, Pt) as polycrystalline bulk materials and continuous/nanopatterned thin films on Au substrates. Single crystals of Pt derivative were isolated and XRD analysis at 140K indicated that it crystallizes in tetragonal *P4/mmm* space group [$a = b = 7.1670(5)$Å, $c = 13.0330(13)$Å, $V = 669.40(9)$Å^3, $Z = 1$]. Four square-planar $[Pt(CN)_4]^{2-}$ anions occupy equatorial positions of $[Fe^{II}N_6]$ O_h and four Fe^{II} atoms are co-ordinated to every $[Pt(CN)_4]^{2-}$ anion defining an infinite set of 2D planar layers pillared by azpy ligands, which bridge *via* axial-positions consecutive Fe^{II} atoms. Magnetic susceptibility, calorimetry and Mössbauer spectroscopy revealed thermal ST in powder samples, which depends on lattice H_2O-content in case of Pd/Pt derivatives. The ST is less co-operative with $T_c = 245$K in case of Ni but Pd (Pt) analogues display complete STs with a hysteresis loop centred at 292 (280) and 191 (183)K in hydrated and dehydrated forms. They grew thin films of the three compounds *via* sequential assembly *via* co-ordination reactions. The RS proved that thermal ST similar to bulk forms occurs in films. With exception of bulk Ni complex Bartual-Murgui *et al.* (2011b) detected complete light-induced excited spin-state trapping (LIESST) by RS in thin films and powders below 130K; they developed optimized layer-by-layer deposition of Hofmann clathrate-like co-ordination compound $\{Fe(pz)[Pt(CN)_4]\}$ as either continuous or nanopatterned thin films; they reported thickness and topography characterization of thin films by AFM and LSPR, which yielded layer's refractive index and losses; they found that films are of optical quality and AFM/LSPR results are in agreement with film-thickness predictions. Bartual-Murgui *et al.* (2011c) elaborated nanopatterned thin films of MOF $\{Fe(bpac)[Pt(CN)_4]\}$ by sequential assembly–lithography combination; they used RS to probe temperature dependence of Fe^{2+} spin state in films (40–90nm thickness) and revealed incomplete but co-operative ST comparable to bulk. Ad/desorption of pyr guest molecules influence thin-layer ST, which host/guest–ST interplay in thin films and nanopatterns shows material's potential use as microsensor.

Spin-Transfer Compounds for Technological Applications

Carbonera *et al.* (2006) investigated photomagnetic properties of Fe^{II} complexes: $[Fe(L^1)_2][BF_4]_2$, $[Fe(L^2)_2][BF_4]_2$, $[Fe(L^2)_2][ClO_4]_2$, $[Fe(L^3)_2][BF_4]_2$, $[Fe(L^3)_2][ClO_4]_2$ and $[Fe(L^4)_2][ClO_4]_2$ [L^1 = 2,6-di(pyrazol-1-yl)pyridine; L^2 = 2,6-di(pyrazol-1-yl)pyrazine; L^3 = 2,6-di(pyrazol-1-yl)-4-(hydroxymethyl)pyridine; L^4 = 2,6-di(4-methylpyrazol-1-yl)pyridine]. Compounds display a thermal ST centred into 200–300K and undergo LIESST at LT; they determined *T*(LIESST) relaxation temperature of photoinduced HS state for compounds. Sigmoidal-kinetics presence in HS→LS relaxation and observation of light-induced thermal hysteresis (LITH) loops under constant irradiation show co-operative nature of materials STs. All compounds follow linear relation between *T*(LIESST) and thermal ST temperatures $T_{1/2}$: $T(\text{LIESST}) = T_0 - 0.3T_{1/2}$. Compounds T_0 is identical to that found for another family of Fe^{II} complexes of related tridentate ligand; they described crystallographic characterization of HS/LS forms, light-induced HS and LS complex $[Fe(L^4)_2][BF_4]_2$. Bonhommeau *et al.* (2006) showed experimentally permittivity photoswitching for ST compound $[Fe(L)(CN)_2]\cdot H_2O$ and correlated it *via* DFT with the electronic-polarizability change that accompanies ST. Electrical detection of photoinduced ST promises opportunities for compounds use as switchable capacitors in electronic devices for information storage and processing. Research is necessary to enhance changes in ST-compound permittivity upon light irradiation. Permittivity tailoring is possible, *e.g.*, introducing polarizable atoms into materials. Relation between permittivity variation and structural/electronic changes accompanying ST should be investigated. Complex of $[Fe^{II}L(CN)_2]\cdot H_2O$, dicyano[2,13-dimethyl-6,9-dioxa-3,12,18-triazabicyclo[12.3.1]octadeca-1(18),2,12,14,16-pentaene]Fe^{II} monohydrate, exhibits thermally induced metal–ligand bond break reversible in solid state and associated to ST, which corresponds to structurally characterized modification of co-ordination metal environment hepta-co-ordinate HS → hexa-co-ordinate LS state (Guionneau *et al.*, 2007). Materials presenting stable and reversible switch of physical properties in solid state are of interest for fundamentation and applications. Design of metal complexes showing light-induced LS/HS ST leading to major change in magnetic and optical properties is a challenge. Materials of ST undergo reversible photoswitch between two magnetic states but photomagnetic-state lifetimes for compounds known so far are long enough at only LT, which prohibits applications. Létard *et al.* (2005) measured and collected temperatures above which photomagnetic effect disappears for >60 ST compounds. On database basis they drew correlation between nature of metal co-ordination sphere and photomagnetic lifetime, which allowed proposing general guideline for design of materials with long-lived photomagnetic lifetimes. Result opened way towards RT photonic materials based on ST, which will be of interest for communication devices. Sánchez Costa *et al.* (2007) investigated Fe^{II} complexes obtained from tetradentate, rigid, linear N_4 ligands to appraise steric effects and impact of *trans*-co-ordinated anions on ST. Well-designed ligands embrace metal centre resulting in O_h Fe^{II} complexes where basal plane is fully occupied by pyr/pz N_4 ligand, where anions/solvent molecules are axially co-ordinated; they used precursor complexes $[Fe(bpzbpy)(MeOH)_2](BF_4)_2$ and $[Fe(mbpzbpy)(MeOH)_2](BF_4)_2$ [bpzbpy = 6,6'-bis(*N*-pyrazolylmethyl)-2,2'-bipyridine; mbpzbpy = 6,6'-bis(3,5-dimethyl-*N*-pyrazolmethyl)-2,2'-bipyridine] for *in situ* preparation of structural analogues *via* exchange of weakly co-ordinated *trans*-methanol molecules by several anions, *e.g.*, thiocyanate, selenocyanate or dicyanamide; they investigated magnetic properties of Fe^{II} compounds: two Fe^{II} complexes $\{[Fe(bpzbpy)(NCS)_2], [Fe(bpzbpy)(NCSe)_2]\}$ exhibit gradual ST typical of isolated mononuclear species with weak co-operative interaction; they studied both materials by Mössbauer spectroscopy and LIESST revealing possibility to induce ST by temperature variation and light irradiation; they showed steric and anion effect–ST correlation. Bonnet *et al.* (2008) studied two-step ST of mononuclear Fe^{II} complex by magnetism, crystallography and calorimetry, revealing two successive first-order phase transitions and an ordered intermediate phase (IP) built by repetition of unprecedented [HS–LS–LS] motif. Bonnet *et al.* (2009) characterized three phases of mononuclear two-step ST compound $[Fe(bapbpy)(NCS)_2]$ [bapbpy = 6,6'-bis(2-aminopyridyl)-2,2'-bipyridine] by powder XRD, Mössbauer spectroscopy, RS, calorimetry and magnetic susceptibility. Mössbauer and RS spectroscopies confirmed results from single-crystal XRD showing that IP is characterized by 1:2 HS/LS ratio. Sample preparation and history studies showed that compound ST properties are robust, as steps and hysteresis cycles are present in powder/recrystallized samples and conserved when sample is cooled/heated several times. An *N,N*-dimethylformamide (DMF) solvate $[Fe(bapbpy)(NCS)_2]\cdot 2DMF$ showed ST but no co-operativity; they discussed the origin in the former compound; they applied it external perturbations: (1) light irradiation at LT promoted LS→HS ST with good optical conversions [~70%, *T*(LIESST) = 56K]; (2) hydrostatic-pressure application increased temperature and hysteresis width of higher-temperature (HT) ST (HS↔IP), while unexpectedly suppressing LT ST (IP↔LS). Tang *et al.* (2009) synthesized Fe^{III} co-ordination compound exhibiting a two-step ST with wide HS–LS plateau of 45K, from a hydrazino Schiff-base ligand with N,N,O donor set 2-methoxy-6-(pyridine-2-ylhydrazonomethyl)phenol (Hmph). Single-crystal X-ray structure of co-ordination compound $\{[Fe(mph)_2](ClO_4)(MeOH)_{0.5}(H_2O)_{0.5}\}_2$, determined at 150K, revealed presence of two different Fe^{III} centres in pseudo-O_h environments, generated by two deprotonated tridentate 2,6-di-tertbutyl-4-methylphenol (mph) ligands. Interactions of H-bonding instigated by well-designed ligand justified abrupt ST occurrence; they characterized the compound by temperature-dependent magnetic susceptibility, electron paramagnetic resonance (EPR) spectroscopy, DSC and ^{57}Fe Mössbauer spectroscopy, which confirm occurrence of two-step ST; they trapped Fe^{III} species in HS state and characterized by rapid cooling EPR. Complex $[Co(enbzpy)(dca)]_n(ClO_4)_n$ [enbzpy = *N,N'*-bis(2-pyridinylbenzylidene) ethane-1,2-diamine; dca = dicyanamide] is the first example of symmetry-breaking ST for Co^{II} ST polymer where rare abrupt HS↔LS ST exists; it is unprecedented X-ray crystallographically characterized structural phase transition accompanied with intermolecular-interaction reorganization. Spin-inactive components (*e.g.*, ligand, counterion) play role in ST *via* supramolecular contacts. Bhar *et al.* (2012) showed that transition between phases is crystallographically reversible. Modification in 1D-chain assemblage presents no repercussion in reversibility of symmetry-breaking abrupt ST. Calculations of DFT will allow understanding symmetry-breaking ST associated with co-operativity and hysteresis in electronic terms. Nanomolecular switch is used to store information in a single molecule. Although switching is detected electrically in form of change in molecule's conductance, adding spin functionality to molecular switches is key concept for realizing molecular spintronic devices. Miyamachi *et al.* (2012) showed that Fe-based ST molecules are individual and reproducibly switched between combined HS, high-conduction state and LS, low-conduction state, provided individual molecule is decoupled from metallic substrate by thin insulating layer; their results represented a step to achieving combined spin and conduction switching functionality on individual molecule level.

On/Off Photoswitch in Cyanide-Bridged Molecular Systems

Magnetic properties of many materials are controlled by external stimuli. Sato *et al.* (2007) reviewed thermal, photochemical, EC and chemical control of phase transitions, which involve changes in magnetization; they described molecular compounds ranging from metal complexes *via* pure organic compounds to composite materials; they were devoted to properties of valence-tautomeric compounds, molecular magnets and ST complexes, which find application in memory devices or optical switches. Sato *et al.* (1996) observed photoinduced magnetization in Prussian blue analogue $K_{0.2}Co_{1.4}[Fe(CN)_6]\cdot 6.9H_2O$; they detected rise in critical temperature from 16 to 19K caused by red illumination. Magnetization in ferrimagnetic region <16K was risen after illumination and was restored almost to its original level by thermal treatment, which effects are produced by internal photochemical redox reaction. Blue illumination was used to remove partly magnetization enhancement. Such control over magnetic properties by optical stimuli presents application in magneto-optical devices. Mondal *et al.* (2013) observed repeatable bidirectional paramagnetic↔diamagnetic photomagnetic effect in cyanide-bridged Fe–Co square complex $\{[Fe\{B(pz)_4\}(CN)_3]_2[Co(bik)_2]_2\}(ClO_4)_2\cdot 3H_2O$ [$B(pz)_4$ = tetrapyrazolylborate, bik = bis(1-methylimidazol-2-yl)ketone]. Magnetic measurements and LT single-crystal XRD experiments showed that complete e^- transfer, from diamagnetic Fe^{II}–Co^{III} to paramagnetic Fe^{III}–Co^{II} metastable state, is induced by 808nm laser irradiation whereas diamagnetic one is recovered in almost quantitative yield under irradiation at 532nm. Self-assembly of $[Fe^{III}(Tp)(CN)_3]^-$ (Tp = trispyrazolylborate) and $[Fe^{II}(bik)_2(S)_2]^{2+}$ affords cyanide-bridged mixed valence $\{Fe^{III}_2Fe^{II}_2\}^{2+}$ molecular square, which exhibits photomagnetic effect under laser irradiation at LT and shows thermal ST *ca.* RT (Mondal *et al.*, 2012).

Mesoporous TiO_2: Preparation, Doping, and Photocatalysis – Cotton

Mesoporous materials are used in solar cells, Li^+ batteries, sensors and supercapacitors. Mesoporous TiO_2 is active in photocatalysis owing to mesoporous-net benefits in promoting reactant/product diffusion and facilitating access to reactive sites on photocatalyst surface. Zhou & Fu (2013) reviewed development of mesoporous TiO_2, metal/nonmetal-doped mesoporous TiO_2 nets and mesoporous TiO_2 composites; they described mesoporous-TiO_2 design and synthesis (*e.g.*, formation of several mesostructured frameworks, pore-size control, large-pore-size fabrication, highly thermally stable mesoporous-net formation). From an application viewpoint, doped and multifunctional heterojunctions fabrication in mesoporous TiO_2 is necessary for improving its photocatalytic performance. Mesoporous-TiO_2 doping with non/metals extended absorption edge into VIS and improved separation efficiency of photogenerated e^-/h^+ pairs. Fabrication of mesoporous TiO_2 composites, especially heterojunctions, used each component characteristics for improving overall photocatalytic performance. Mesoporous TiO_2 possesses room for development in photocatalysis and is worthy. Afzal *et al.* (2012) formed self-assembled monolayers of *meso*-tetra(4-carboxyphenyl)porphyrin (TCPP) on TiO_2-anatase-coated cotton fabric to confer VIS-sensitized self-cleaning: complete degradation of methylene blue (MB) in 110 min and removal of coffee/red wine stains in 16h indicated that TiO_2/TCPP-coated cotton fabrics exhibit superior VIS self-cleaning, compared to bare TiO_2-coated cotton; they characterized the fabrics by UV–VIS, field emission scanning electron microscopy (FESEM), XRD and FL spectroscopy: VIS-induced self-cleaning cotton offers indoor self-cleaning applications. Afzal *et al.* (2013) formed thin films of Cu^{II} porphyrinato (CuTCPP) in conjunction with TiO_2 on cotton fabric; they investigated CuTCPP/TiO_2 self-cleaning properties *via* photocatalytic degradation of MB, coffee and wine stains under VIS. CuTCPP/TiO_2-coated cotton fabrics showed superior self-cleaning when compared to bare TiO_2-coated cotton; the fabrics presented photostability under VIS compared to free base TCPP/TiO_2-coated fabrics; they characterized the fabrics by FESEM, XRD and UV–VIS spectroscopy; they discussed an insight into mechanistic aspects of CuTCPP/TiO_2 photocatalysis. Self-cleaning cotton based on CuTCPP/TiO_2 catalyst driven by VIS exhibits stability and reproducibility for self-cleaning applications.

Compound Benefits in Computational Chemistry

In computational chemistry, computers play a role in substances development (Harris, 2014). Huskinson *et al.* (2014) described a battery technology based on a set of organic compounds: quinones (*cf.* Figure 7). Flow-battery system promises cheap, clean and reversible energy storage when it is needed usually seen as a missing piece in the quest for widespread renewable energy. However, a key part was selecting compounds to use in the first place. The type of interdisciplinary project spanning boundaries between laboratory and computer became important in academia and industry, especially pharmaceutical manufacturing. Computer-assisted screening of photocatalytic strategies showed that it is possible to couple different sunlight-fuelled photochemical reactions on different sides of open-ended inorganic NTs.

Figure 7. A quinone: 9,10-anthraquinone-2,7-disulphonic acid (AQDS) for metal-free flow battery

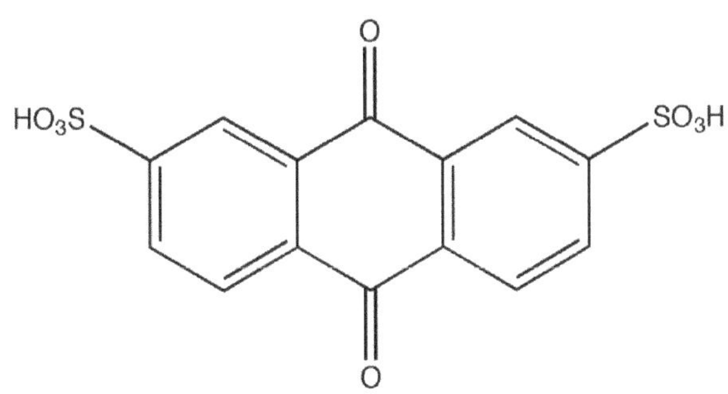

Procedure for Unsupervised Clustering: Solvent Classification

Torrens and Castellano (2011a) performed a molecular modelling analysis of SWNT solvents classifying them into *best*, *good* and *bad* ones. Artificial neural networks (ANNs) are a tool for solving classification problems. Difficulties are overcome for chemical-data application. Supervised-ANN use implies patterns initial distribution between pre-determined classes while objects attribution to the classes is uncertain. Unsupervised ANNs are free from the problem but do not always reveal data structure. Classification algorithms, which do not require *a priori* information about pattern distribution between the pre-determined classes and provide meaningful results, are of interest. Pushkarova & Kholin (2014) presented an approach based on Kohonen self-organizing map (SOM)–probabilistic ANNs combination, which enabled determination of number of classes and objects classification, which they illustrated for a 76-solvent set based on nine characteristics. The classification is chemically interpretable. The approach was applicable to examining C_{60} solubility. Solvents in the same group show similar abilities to dissolve C_{60}, which makes possible to estimate fullerene solubility in solvents for which there are no experimental data.

Ultrathin Films of NiFe–LDHs: C-Nanoform Hierarchical Synthesis

In this laboratory Abellán *et al.* (2010) reported the synthesis of a crystalline micrometric-sized hexagonal-shaped Ni^{2+}-Fe^{3+} layered double hydroxide (LDH) *via* modified homogeneous precipitation; material exfoliation in formamide leads to stable suspensions of hexagonal nanometric sheets, which they characterized; their data confirm that bulk intrinsic properties are retained by the segregated nanosheets, opening the door for use in development of layered multifunctional materials. Films of LDH were investigated because of applications (*e.g.*, catalysis, anti-corrosion coatings for metals, components in optical, electrical and magnetic devices). Guo *et al.* (2010) reviewed synthesis, properties and applications of functional LDH films offering perspectives for multifunctional-one design. For MgAl-LDHs *homogeneous precipitation* Okamoto *et al.* (2007) investigated factors influencing LDH-crystal size *via* two kinds of starting metal salts (Cl^-, NO_3^-) and two types of ammonia-releasing reagents (ARRs) [urea, hexamethylenetetramine (HMT)]; they conducted the reaction under a hydrothermal condition at 140ºC for 1 day; they varied ARR amounts while fixing Mg/Al ratio = 2.0 and total metal concentration [Mg + Al] = 0.15M; they revealed that crystal size ranged widest for NO_3^-/HMT in which HTM and anion amounts controlled size and state of resulting CO_3^{2-} LDH. For higher HTM concentration, LDH size was 1μm and reducing its concentration, crystal size rose to 5–6μm. With a further reduction in HTM concentration, rosette-like LDH aggregates as large as 20μm appeared; they investigated influences of other factors (*e.g.*, total metal concentration, reaction time); they assumed that LDH-crystal growth in a circle on Al hydroxide precursors to explain aggregation formation and LDH-crystal morphologies. Kobayashi *et al.* (2008) reported synthesis of crystal spinel platelets with exposed (111) faces, lateral dimensions in micrometre range and 20–50nm thicknesses, prepared by soft chemical dehydration of crystallized layered precursors. The method enables synthesis of metastable $NiCoAlO_4$, which cannot be prepared by conventional solid synthesis. Because of their nanothickness, mosaic structure, shape anisotropy and surface porosity, $NiCoAlO_4$ and $NiCo_2O_4$ platelets exhibit RT superparamagnetism (blocking temperature, T_B = 40 and 250K, respectively) despite fact that they present micrometre-size lateral dimensions. Accessibility of superparamagnetic spinel platelets (as opposed to cubes, spheres) is useful for studies in ferrofluids and related NC magnetic materials. Reports claiming RT spontaneous magnetization in LDHs were published; however, the materials concern as to whether co-operative magnetic behaviour comes from extrinsic sources, *e.g.*, spinel Fe-oxide NPs. Abellán *et al.* (2013) developed syntheses of crystalline Fe^{3+}-based LDHs with/out impurities highlighting care taken during synthesis to avoid magnetic-LDH misidentification. Abellán *et al.* (2014) synthesized alkoxide-intercalated CoFe-LDHs *via* non-aqueous methanolic route. According to XRPD and FESEM they exhibit a nanosized plate-like morphology with a basal space of 9.21Å; material hydrolysis leads to colloidal suspensions of nanosheets with 20nm lateral dimension and 4nm thickness as they showed by AFM and DLS. Atomic-resolution STEM combined with electron energy-loss spectroscopy (EELS) confirm crystal HQ and Co:Fe stoichiometry; they investigated CoFe-LDH magnetic properties *via* DC/alternate current (AC) magnetic susceptibility and isothermal magnetization, showing an LT magnetic ordering below 7K with a spin-glass-like behaviour and displaying hysteresis at 2K with a coercive field of 402G; they tested the sample as electrode material for supercapacitors in a three-electrode cell *via* cyclic voltammetry and galvanostatic dis/charge experiments, showing capacitances and stability; they explored electrocatalytic behaviour to water oxidation showing efficient and persisting performance under basic p*H*s with potential in energy-storage devices. Zhao *et al.* (2010) fabricated $NiFe_2O_4$ NPs (<10nm) embedded in a NiO matrix calcining corresponding $Ni^{II}Fe^{III}$-LDH precursors at HT (500°C). Compared to $NiFe_2O_4$/NiO NC obtained by calcination of a precursor prepared by a traditional chemical co-precipitation method, those derived from NiFe-LDH precursors show higher T_B~380K. Enhanced magnetic stability is ascribed to stronger interfacial interaction between $NiFe_2O_4$ and NiO phases, because of topotactic nature of LDH-precursor transformation to $NiFe_2O_4$/NiO NC. *Via* tuning Ni^{II}:Fe^{III} ratio of NiFe-LDH precursor, $NiFe_2O_4$ concentration is controlled and T_B/magnetic properties of the final material are regulated; their work represents an example of fabrication of ferro(ferri)magnetic (FeM)/antiferrimagnetic (AFeM) systems with magnetic stability from LDH precursors. The method is general and readily extended to other FeM/AFeM systems because of LDH-precursor range. Wu *et al.* (2005) presented a CNT solid synthesis from graphite islands [hexa-*peri*-hexabenzocoronenes (HBCs)] with a quantitative C-conversion yield. Obtained C-NSs (bamboo-shaped, straight CNTs) depend on employed heating. Similar metal–PAH (*e.g.*, graphite discs containing 96 C-atoms) complexes will be synthesized and thermolyzed to understand graphitic-PAH role in precursor complexes. Applications of CNT solid synthesis include NT selective growth by printing precursors on to certain surfaces followed by thermolysis or precursor pyrolysis in porous templates, SPM-tip modification and fabrication of CNTs/metal carbide heterostructures. Graphite-island nonvolatility in the precursor complexes and CNT-formation capability in solid allow *in situ* monitoring of CNT growth by TEM at HT without contamination. Abellán *et al.* (2012) described an LT chemical synthesis of C-nanoforms (*e.g.*, nano-onions, MWNTs).The method involves thermal decomposition of a sebacate-intercalated NiFe LDH at 400°C and benefits from catalytic activity of $FeNi_3$ NPs generated *in situ*. Natural resonance appears at 16GHz for (Fe, Ni)/C nanocapsules with (Fe, Ni) alloys as cores and graphite as shells. Liu *et al.* (2009) obtained reflection loss (RL) exceeding ~10dB in the whole Ku-band (12.4–18GHz) for a 2.0mm absorber thickness, while it exceeds ~20dB in range 13.6–16.6GHz. The bandwidth does not change for 1.87–2.10mm thicknesses for RL values exceeding ~10dB. The (Fe, Ni)/C nanocapsules with wide bandwidth absorption are used as electromagnetic-wave-absorption materials in the whole Ku-band. Combination of 1D and 2D building blocks leads to hierarchical NC formation, which take advantages of each material, which is effective for multifunctional-material preparation with properties. Among building blocks, nano-Cs (*e.g.*, CNTs, GR) and LDHs are two of the most powerful materials. Zhao *et al.* (2012) reviewed hierarchical NCs derived from Nano-Cs and LDHs; they described nano-C, LDH and combined NC properties; they presented efficient and effective fabrications for hierarchical NCs (*e.g.*, nano-C/LDH reassembly, LDH formation on nano-Cs, nano-C formation on LDHs). As-obtained NCs derived from nano-Cs and LDHs exhibited performance as multifunctional materials for applications in energy storage, NCs, catalysis, environmental protection and DDSs. Fabrication of LDH/C NCs provides a method for multifunctional-NC development based on existing nanomaterials. However, knowledge of assembly mechanism, robust/precise route for LDH/nano-C hybrid with designed structure and structure–property–application relation is inadequate. A multidisciplinary approach from scope of materials, physics, chemistry, engineering, *etc.* is required for development of the advanced functional NCs. Cassie and Baxter (1944) extended analysis of apparent contact angles for rough surfaces to porous ones particularly those encountered in natural and artificial clothing; they derived formulae for apparent contact angles and gave experimental data confirming them; they discussed water-repellent clothing structures *via* the analysis and showed that duck water-repellency is because of feathers structure. Lau *et al.* (2003) showed creation of a stable, superhydrophobic surface *via* nanoroughness inherent in a vertically aligned CNT forest together with a thin, conformal hydrophobic poly(tetrafluoroethylene) (PTFE) coating on CNT surface; they achieved superhydrophobicity down to the microscopic level where spherical, micrometre-sized water droplets are suspended on CNT-forest top.

Self-Assembled Monolayer over Permalloy

Little is known about mechanisms governing spin injection into organic molecules. Sanvito (2010) suggested that metal/organic interface paves the way for a field in which interfaces are designed for spin applications, which is *spinterface* science. Understanding CT of single molecules or a small collection of them sandwiched between electrodes is important for molecular electronics, which require device fabrication, which depends on factors (*e.g.*, testbed architectures, molecule number/defect density being tested, nature of molecule/electrode interface). Based on progresses achieved in experiments and theory Jia & Guo (2013) focused on insights into the influence of nature of the molecule/electrode interface, most critical issue hindering device development, on molecule conducting properties; they summarized strategies developed for controlling interfacial properties and how molecule–electrode coupling strength modulates device properties. The analyses are valuable for understanding relation between contact interface and CT mechanism, which is substantial for development of molecular, organic electronics, nanoelectronics and interface-related optoelectronic devices. Li *et al.* (2013) explored a strategy of using a phosphonic acid derivative as self-assembled monolayers (SAMs) on Si/SiO_2 surface and 6,13-pentacenequinone (PQ) as template layer to induce 6,13-diphenylpentacene (DPP) crystallization in vacuum deposited thin-film transistors, which showed field-effect mobility $8.3 \cdot 10^{-3} cm^2 V^{-1} s^{-1}$; they found that *n*-octadecylphosphonic acid (ODPA) SAMs modulate PQ morphology to form a flat layer, which is helpful for DPP crystallization, which indicated that ODPA bilayer-step surface controls PQ/DPP growth by vacuum thermal deposition. Analysis and confirmation of monolayer film thickness on metal-oxide surfaces is challenging. Quiñones *et al.* (2007) used XPS and AFM to investigate monolayer formation; however, the techniques are difficult to access and/or determine organic-molecule composition on surfaces; they showed ability of matrix-assisted laser desorption/ionization time-of-flight (MALDI–TOF) to charaterize long alkyl chain phosphonic acid molecules in thin films on Ti, Fe and stainless steel; the systems are stable, adhered films; they characterized the thin films by IR, AFM, contact angle measurements and confirmed the results by MALDI–TOF, which they used to differentiate between mono- and multilayers on planar surfaces. Applications of *in situ* techniques improved our understanding of self-organization of adsorbed molecular monolayers on solid surfaces (Schwartz, 2001). The process involves several steps starting with bulk solution transport and surface adsorption continuing with 2D organization on interest substrate; the later involves passage *via* one/more intermediate surface phases and is described *via* 2D nucleation/growth models. A picture emerged, which combines elements of interface surfactant-adsorption and epitaxial growth with complication of long-chain molecules with many degrees of freedom. Stoecklein *et al.* (1988) analyzed ferromagnetic resonance (FMR) spectra of Permalloy thin films exchange-coupled to Fe-Mn films; they made studies on bilayer, ferromagnetic-antiferromagnetic (FA) and trilayer (AFA) structures *vs.* *F*- and *A*-layer thicknesses ranged 20–800Å; they presented data at a frequency of 9.3GHz for in-plane/perpendicular directions of the applied field and at 34.1GHz, in plane. Data analysis enables magnetization, gyromagnetic ratio and exchange shift extraction because of spin-wave stiffness and perpendicular-surface anisotropy *vs.* layer thickness; they used azimutal dependence of in-plane FMR to determine exchange anisotropy magnitude (bias field). Magnetization and gyromagnetic ratio show little dependence on *F*/*A*-layer thickness down to 50Å implying that interfaces are sharp on a scale of a few lattice constants. In the interfacial region, magnetization is reduced because of interaction with the antiferromagnet; they showed that the perpendicular-surface anisotropy is created by exchange coupling to the antiferromagnet whose easy axes are not in the interface plane; they showed an exchange anisotropy model in which the antiferromagnetic domain pattern is not totally locked but adjusts in ferromagnetization response. The model qualitatively explains the bias field exerted by the antiferromagnetic layer deposited before the ferromagnet, field-training effect, FMR linewidth and bias-filed magnitude. Kataria *et al.* (2012) focused on substrate effect while characterizing different-hardness films *via* nanoindentation, which tests were performed on hard TiN and (Ti, Al)N coatings deposited on soft D9 steel substrates; they extracted coating hardness from distinct indentation depths for revealing substrate effect; for both coatings they found hardness to decline with rising indentation depths indicating substrate-effect severity at higher depths. However, decline in hardness was steeper in (Ti, Al)N than TiN film; they observed discontinuities (pop-ins), signature of film cracking, on load–displacement curves; *via* SEM they observed surface cracks near the indentation edge and inside indentation zone in both coatings; to understand the behaviour they used 2D finite element analysis to simulate indentation; they found plastic deformation in the substrate to take place in early stages in both cases resulting in an earlier hardness decline. However, the critical depth at which substrate started deforming plastically was 8% and 11% of the total film thickness for (Ti, Al)N and TiN films, respectively; they found pop-ins to appear on the load–displacement curves after substrate-yielding occurrence in both cases; they explained film cracking on the basis of stress distribution in the coatings during indentation; they concluded that substrate effect becomes more pronounced in harder films and manifest itself at indentation early stages. Sharifi *et al.* (2011) investigated the distribution of SO_3H-functional groups attached to the ordered inner pore walls of mesoporous Si-MCM-41 based on SO_2 by gas adsorption combined with *in situ* small-angle neutron scattering (SANS); they performed the functionalization by two methods: grafting and co-condensation. Adsorbates N_2 at 77K or mixed H_2O/D_2O (42:58) at 298K posses SANS length densities (SLDs) similar to SO_2 and quench diffraction signals of nonmodified SO_2. Measurements of SANS show that N_2 matches not only with pristine but also with Si-MCM-41-SO_3H functionalized by grafting; they proved adsorbate access in pores length. For analysis of functional-group distribution in the pores in dependence on functionalized method, however, more specific adsorbate H_2O/D_2O (42:58) is necessary because it reacts more sensitively to small changes in host-material SLD; for grafted MCM-41-SO_3H materials they observed an incomplete quenching indicating that the pore mouths were modified; for a sample functionalized by co-condensation they found almost no SANS quenching indicating a homogeneous distribution of the functional groups along the entire pores. Love *et al.* (2005) reviewed thiolate SAMs on metals as a form of nanotechnology. Petta *et al.* (2004) measured magnetic tunnel junctions made *via* a self-assembled-monolayer molecular barrier; they fabricated Ni-octanethiol-Ni samples in a nanopore geometry. The devices exhibit changes in resistance as they varied angle between magnetic moments in the two electrodes, showing that low-energy e^-'s traverse the molecular barrier while remaining spin polarized. Data voltage and temperature dependence analysis suggests that spin-polarized transport signals are degraded by localized states in the molecular barriers. Molecular electronic devices with spin-dependent tunnelling transport offer an enticing direction to spin electronics from fundamental and technological viewpoints. Wang & Richter (2006) provided evidence by inelastic e^- tunnelling spectroscopy of molecular species existence in the fabricated molecular magnetic tunnel devices; they used tunnelling spectroscopy to investigate spin-polarized inelastic e^- tunnelling in the molecular device; their results showed that inelastic scattering because of molecular vibrations causes the observed junction magnetoresistance bias dependence. Galbiati *et al.* (2012) unveiled self-assembled monolayers' potential for molecular spintronics and spin transport at high voltage; they observed an almost flat dependence of tunnel magnetoresistance (TMR) with bias voltage in the volt range. A question remains as to the physical origin of the unexpected behaviour.

Preparation of Ultrathin Films of Metal–Organic Frameworks

Rational assembly of MOF ultrathin films with controlled growth direction and film thickness is a critical and unrealized issue for enabling their use in nanotechnological devices (*e.g.*, sensors, catalysts, electrodes for fuel cells). Makiura *et al.* (2010) reported facile bottom-up fabrication at RT of such a perfect preferentially oriented MOF nanofilm on a solid surface (NAFS-1), consisting of metalloporphyrin building units; they achieved NAFS-1 construction by unconventional integration in a modular fashion of a layer-by-layer growth coupled with Langmuir–Blodgett (LB). The NAFS-1 is endowed with crystalline order in out-of- and in-plane orientations to the substrate as shown by synchrotron X-ray surface crystallography. The structural model incorporates metal-co-ordinated pyr molecules projected from 2D sheets, which allow each further layer to dock in an ordered interdigitated manner in NAFS-1 growth; they expected that versatility of the solution-based growth strategy allow fabricating ordered MOF nanofilms opening the way for applications.

Modulating Transition Temperature in ST MOFS

In this laboratory Coronado *et al.* (2013) presented an ST Fe^{II}-co-ordination polymer without permanent channels, which selectively sorbs CO_2 over N_2. Chains of 1D display internal voids of ~9Å diameter, every one capable of accepting one CO_2 molecule at 1 bar and 273K. The XRD provided structural evidence of gas-molecule location and revealed $O{=}C{=}O(\delta^-)\cdots\pi$-interaction. The physisorption modifies ST, increasing $T_{1/2}$ by 9K.

Design of Multifunctional Molecular Materials with Control

Crystal engineering (planning and construction of crystalline supramolecular architectures from modular building blocks) allows rational design of functional molecular materials with technologically useful behaviour [*e.g.*, super/conductivity, ferromagnetism, nonlinear optical (NLO) properties]. Because co-operative-properties presence in the same crystal lattice results in physical phenomena and applications, a goal is design of molecular materials with properties that are difficult or impossible to combine in a conventional inorganic solid with a continuous lattice. A strategy for creating the *bi-functionality* targets hybrid in/organic crystals comprising functional sub-lattices exhibiting distinct properties. In this laboratory Coronado *et al.* (2000) combined organic π-electron donor bis(ethylenedithio)tetrathiafulvalene (BEDT-TTF) and derivatives, which base molecular super/conductors with molecular magnetic anions, yielding not only semi/conducting materials but also superconducting, paramagnetic systems; however, interesting bulk magnetic properties fail to develop owing to discrete nature of inorganic anions. Another strategy for achieving co-operative magnetism involved insertion of functional bulky cations into a polymeric magnetic anion, *e.g.*, bimetallic oxalate complex $[Mn^{II}Cr^{III}(C_2O_4)_3]^-$, but they obtained insoluble powders; they reported synthesis of crystals formed by infinite sheets of the magnetic co-ordination polymer interleaved with layers of conducting BEDT-TTF cations and showed that the molecule-based compound displays ferromagnetism/metallic conductivity.

Graphene-Based Transparent Conductive Electrodes

Will tin-doped indium oxide (ITO) give way to GR for transparent conductive electrodes (TCEs)? Although it is impossible to predict, progress improved the figure-of-merit (FOM) of GR-based TCEs from <1 to >100, which was accomplished increasing grain size and crystallinity, reducing defects, appropriate doping and hybridizing with NWs/NTs. Research in GR sparks TCE improvement; *e.g.*, direct GR growth on dielectric substrates is a research area. Progress made catalyst-free growth possible. Plasma-enhanced CVD (PECVD) enables LT GR growth on SiO_2/Si (550–650ºC), which is compatible with microelectronics-industry infrastructure. Yu & Chen (2014) showed direct GR growth on dielectric substrates at 400ºC by hydrocarbon+H_2 plasma control to balance etching and nucleation; they anticipated that direct GR growth on plastic substrates will become a reality.

Large-Scale Synthesis of Au NRS: Continuous Secondary Growth

Au NRs exhibit a tunable LSPR, which depends on AR. Independently controlling Au NR AR and size remains challenging but it is important because scattering intensity depends on size. Kozek *et al.* (2013) reported a secondary (seeded) growth where continuous addition of ascorbic acid (Vit-C) to a stirring solution of Au NRs, stabilized by cetyltrimethylammonium bromide (CTAB) and synthesized by a common growth, deposits the remaining (70%) of Au precursor on to NRs; the growth phase of Au-NR synthesis is performed without stirring since agitating reduced the yield of rod-shaped NPs but they reported that whisking, coupled with continuous Vit-C addition during secondary growth, allows control over Au NR AR/size; after a common primary Au NR growth, LSPR is 820nm, which they tuned in 700–880nm during secondary growth adjusting Vit-C-addition rate or adding benzyldimethylhexadecylammonium chloride hydrate (BDAC); they used the approach for secondary growth with primary Au NRs of different ARs to achieve dissimilar LSPRs and extended it to diverse-shape and other-metal NPs.

Sustainable Chemistry: Learning from Nature

Learning from nature, one devises (photo)catalysis, nanotechnology, clean chemical and industrial processes. In clean chemical processes, sustainable chemistry allows minimizing residues and designing functional compounds and nanomaterials.

Lipid-Based Colloidal Carriers for Peptide and Protein Delivery

Martins *et al.* (2007) reviewed lipid-based colloidal carriers and pharmaceutical implications in peptide and protein delivery for oral and parenteral administration, *e.g.*, biomacromolecules used in therapeutics, which are delivered *via* liposomes and lipid NPs [*e.g.*, solid lipid NPs (SLNs), nanostructured lipid carriers (NLCs)]. Production is applied to achieve association efficiency between the bioactives and carrier depending on their physicochemical properties, and production procedure, which leads to improved bioavailability, or in case of oral administration a more consistent temporal profile of absorption from the gastrointestinal tract; they noted advantages and drawbacks of colloidal carriers; they described strategies used for peptide/protein formulation, methods for association-efficiency assessment and practical consideration regarding toxicological concerns.

Medicinal Applications of Fullerenes

Fullerenes attracted attention in different fields since their discovery in 1985. Investigations of physical, chemical and biological properties of fullerenes yielded information. Size, hydrophobicity, 3D and electronic configurations make them appealing in medicinal chemistry; their only C-cage structure coupled with derivatization scope make them a therapeutic agent. Bioapplication study attracted attention despite low C-sphere solubility in physiological media. Fullerene family, especially C_{60}, presents appealing photo, EC and physical properties, which are exploited in medical fields. Fullerene fits in the hydrophobic cavity of the human immunodeficiency virus (HIV) proteases, inhibiting substrate access to enzyme catalytic site. It is used as radical scavenger and antioxidant. If exposed to light, fullerene produces singlet oxygen in quantum yields which, together with direct e^- transfer from excited state of fullerene and DNA bases, is used to cleave DNA. Fullerenes were used as a carrier for gene and DDS; they are used for serum protein profiling as material-enhanced laser desorption/ionization (MELDI) for biomarker discovery. Bakry *et al.* (2007) reviewed fullerene medicinal applications. Fullerenes constitute C-allotropic variations, which called scientist attention. In nanotechnology, fullerenes presented applications in medicinal chemistry and pharmaceutical sciences, especially C_{60}-derivative design and bioevaluation for use as antimicrobial agents. Considering pathogen-resistance emergence to current antibiotics and challenge of finding more efficient antimicrobial chemotherapies, Cordeiro (personal communication) revised the last advances in the discovery of C_{60}-containing NSs as models of NP-based antibacterial agents; considering the importance and concern of NP-associated toxicity, they introduced a general multitasking model for quantitative structure–biological effect relationships (mtk-QSBERs); they built up the model from a large and heterogeneous dataset of compounds (>47200) for simultaneous prediction of multiple absorption, distribution, metabolism, excretion (ADME) and toxicity (ADMET) profiles. The mtk-QSBER classified more than 90% of the cases in training and prediction sets, being employed for screening of diverse ADMET properties of different molecular architectures containing C_{60}; results, which were in agreement with experiments, confirmed that the increment in the number of polar regions associated to C_{60} improves the safety profiles, which showed that their mtk-QSBER is used as tool for virtual assessment of different safety profiles of large compound libraries under different experimental conditions.

Side-Chain Control of Porosity Closure in Porous Materials

Porous materials are attractive for separation and catalysis. In MOFs, interactions are controlled *via* a flexible structural response to guest presence. In this laboratory Martí-Gastaldo *et al.* (2014) reported an MOF that consists of glycyl–serine (Gly–Ser) dipeptides co-ordinated to metal centres, and presents a structure that evolves from a solvated porous state to a desolvated non-porous one as a result of ordered co-operative, displacive and conformational changes of the peptide. The behaviour is driven by H-bonding that involves the side-chain hydroxyl groups OH of Ser. A similar co-operative closure is also displayed with multipeptide solid solutions, for which the combination of different sequences of amino acids (AAs) controls the framework's response to the presence of guests in a nonlinear way; they compared the functional control to the effect of single-point mutations in proteins in which exchange of single AAs radically alters structure and function.

CALCULATION RESULTS AND DISCUSSION

The equilibrium difference between the Gibbs free energies of interaction of an SWNT with its surroundings in solid phase and cluster volume or on surface, *cf.* Figure 8, shows that on going from C_{60} (droplet) to SWNT (*bundlet*) the minimum is less marked (68% of C_{60}), causing a lesser number of units in SWNT ($n_{min} \approx 2$) than in C_{60} clusters (≈8) and a longer abscissa in C_{60} ($n_{abs} \approx 28$) than in SWNT (≈9). Thinner NT-BUD bundles (bundlet) result less stable while wider ones appear more stable than SWNT packages. The minimum of NT-BUD appears 55% of C_{60}. The minimum of GR (*columnlet*, 67%) is similar to SWNT but with fewer units (≈1) and shorter abscissa ($n_{abs} \approx 3$). Shorter GR-BUD stackings (columnlet) result less stable while longer ones appear more stable than GR columns. The minimum of GR-BUD (49% of C_{60}) is alike NT-BUD.

The solubility of SWNT/NT-BUD and GR/GR-BUD *vs.* temperature (*cf.* Figure 9) shows a solubility decay because of cluster formation. At $T \approx 260$K, C_{60}-FCC presents a phase transition to simple cubic (SC). The solubility drops with temperature result less marked for SWNT/NT-BUD and even GR/GR-BUD, in agreement with lesser numbers of units in clusters (Figure 8). At $T = 260$K from C_{60} (droplet) to SWNT, NT-BUD (bundlet), GR and GR-BUD (columnlet) the solubility (Figure 9) decreases to 3%, 2%, 0.3% and 0.1% of C_{60}, respectively.

The cluster distribution function by size in CS_2, calculated for saturation concentration at solvent temperature $T = 298.15$K (*cf.* Figure 10), shows that on going from C_{60} (droplet) to SWNT (bundlet) the maximum aggregate size decays from $n_{max} \approx 8$ to ≈2 and spreading is narrowed, in agreement with lesser number of units in clusters (Figure 8). The dispersal of NT-BUDs (bundlet) is somewhat enlarged to wider bundles with respect to SWNT. The dissemination of GRs (columnlet) is strongly narrowed in concordance with the fewest units ($n_{max} \approx 1$). The scattering of GR-BUDs (columnlet) is rather increased to longer stacks with regard to GR.

The concentration dependence for the heat of solution in toluene, benzene and CS_2 calculated at solvent temperature $T = 298.15$K (*cf.* Figure 11) shows that for C_{60} (droplet), on going from $C < 0.1\%$

Figure 8. C_{60}–SWNT–SWNC interaction energy with its surroundings in cluster volume or surface

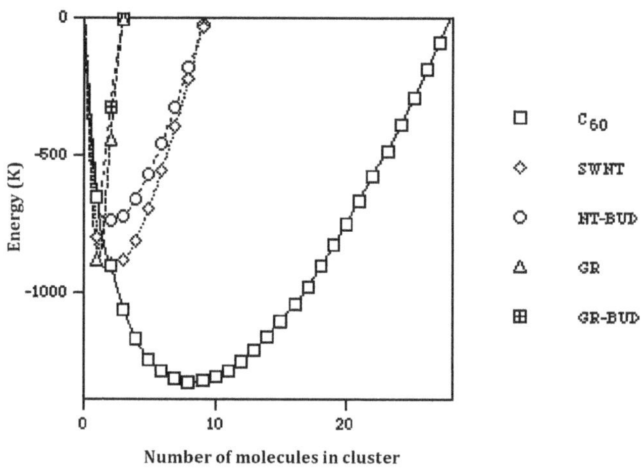

Figure 9. Temperature dependence of solubility of C_{60} (droplet)–SWNT/SWNC (bundlet)

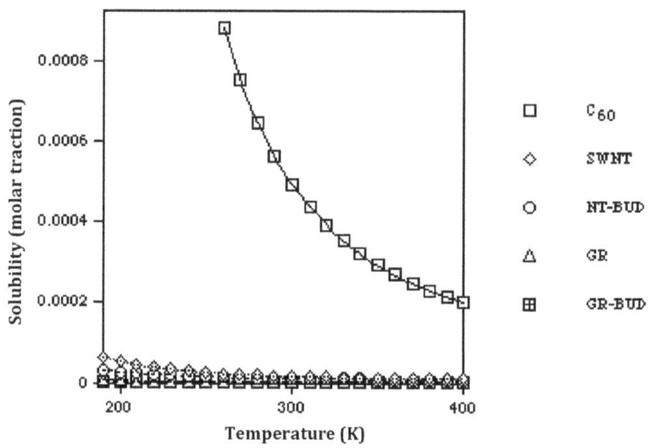

Figure 10. Cluster distribution saturated in CS_2 at 298.15K of C_{60}–SWNT/SWNC

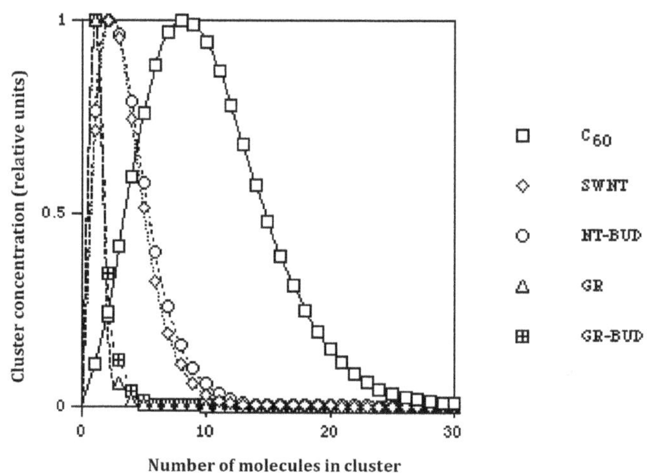

Figure 11. Heat of solution vs. concentration of C_{60}–SWNT/SWNC in toluene/benzene/CS_2 at 298.15K

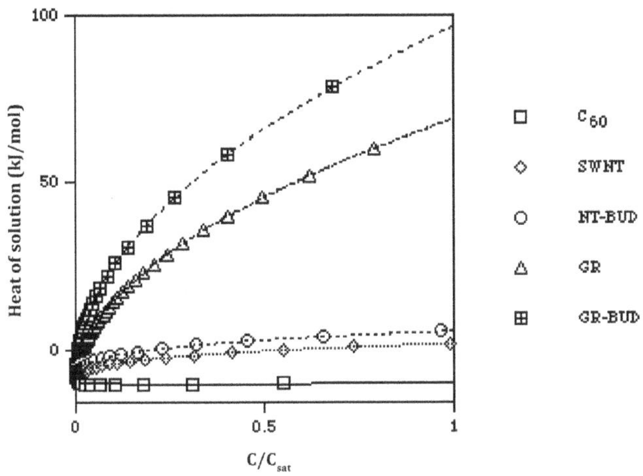

Figure 12. Heat of solution vs. temperature of C_{60}–SWNT/SWNC in toluene/benzene/CS_2 for saturation

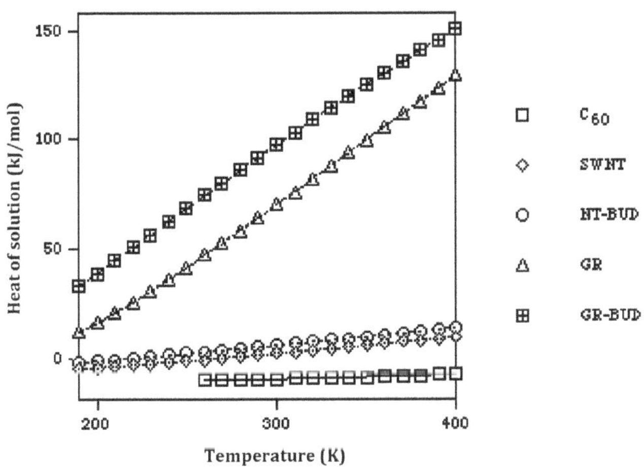

Figure 13. Diffusion coefficient vs. concentration of C_{60}–SWNT/SWNC in toluene at 298.15K

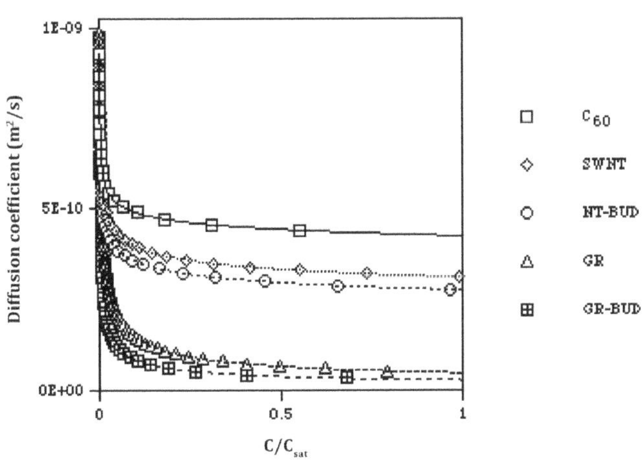

of saturated ($<n> \approx 1$) to $C = 15\%$ ($<n> \approx 7$) the heat of solution decays by 73%. In turn, for SWNT (bundlet) the heat of solution rises by 54% in the same range in agreement with a lesser number of units in clusters (Figures 8 and 10). Moreover, in NT-BUD (bundlet), GR and GR-BUD (columnlet) the heat of solution increases by 98%, 392% and 680%, respectively. The discrepancy between experimental data for the heat of solution of fullerenes, CNT/NT-BUDs and GR/GR-BUDs is ascribed to the sharp concentration dependence for the heat of solution.

In the temperature dependence for the heat of solution in toluene, *etc.* calculated for saturation concentration (*cf.* Figure 12) C_{60} results are plotted for $T > 260K$ after FCC/SC transition. For C_{60} (droplet) on going from $T = 260K$ to $T = 400K$ the heat of solution rises 2.7kJ·mol^{-1}. The heat of solution of SWNT (bundlet) increases 10.4kJ·mol^{-1} in the same range and becomes endergonic. For NT-BUD (bundlet) the heat of solution augments 11.3kJ·mol^{-1} similarly endergonic. Moreover, GR and GR-BUD (columnlet) heats of solution enlarge 82 and 76kJ·mol^{-1}, respectively, likewise endergonic.

The diffusion coefficient *vs.* concentration in toluene at $T = 298.15K$ (*cf.* Figure 13) shows that cluster formation, close to saturation, decreases the diffusion coefficients by 56%, 69%, 73%, 95% and 97% for C_{60} (droplet), SWNT, NT-BUD (bundlet), GR and GR-BUD (columnlet), respectively, compared with $(C_{60})_1$. Diffusion coefficients of SWNT, NT-BUD, GR and GR-BUD decay by 29%, 37%, 88% and 93% contrasted with $(C_{60})_n$ in agreement with lesser number of units in clusters.

CONCLUSION

From the discussion of the present results the following conclusions can be drawn.

1. Questions remain regarding behaviour of electronic structural quantities [highest occupied molecular orbital (HOMO), lowest unoccupied molecular orbital (LUMO), electronic density, electronic correlation and exchange, aromaticity, *etc.*] and developing most intimate relation between nanostructure bonding and symmetry in growth and diffusion. Experimental exploration is necessary for establishing a cluster nature of nanostructure solubility. Infrared absorption spectra of a nanostructure solution at various temperatures were analyzed; dependence will indicate presence of nanoclusters. According to Raoult's law, saturation vapour pressure of a solvent above solution differs from a pure solvent by a value proportional to solute-particle concentration. Solvent vapour flow determined by pressure difference enables dependence of solute-particle concentration: a non-linear dependence will indicate solution nanoclusters. Packing efficiencies and interaction-energy parameters of nanocones are intermediate between fullerene and single-wall C-nanotube clusters.
2. The nanoworld structural diversity is a consequence of its quantum nature. Several criteria reduced the analysis to a manageable number of magnitudes, *viz.* closeness, curvature, dimension and efficiency. Our non-computationally intensive approach, *i.e.*, object clustering plus property prediction, assessed reliability. Type, dimensions and producer selection of fullerenes, nanotubes and graphenes must be chosen to ensure the transfer of methods developed between laboratories.
3. Our interaction energy parameters for nanotube buds are derived from C_{60}. For nanotube bud a C_{60}/tube intermediate behaviour was expected. However, nanotube-bud properties result closer to tubes. A nanotube-like behaviour is calculated and bud properties are computed closer to tubes. Thinner nanotube-bud bundles appear less stable but wider ones are more stable than tube packages. The solubility decays with temperature result smaller for nanotube and bud than C_{60} in agreement with

lesser numbers of units in clusters. The diffusion coefficient drops with temperature result greater for nanotube bud than tube than C_{60}, which corresponds to lesser number of units in clusters.

4. Association energy parameters of nanographene are obtained from C_{60}. A nanotube-bud behaviour or further was expected. For $C_{graphene}$ nanobud an $C_{60}/C_{graphene}$ in-between behaviour was anticipated. Notwithstanding, nanobud features appear closer to $C_{graphene}$. Shorter nanobud stacks result less stable but longer ones appear more stable than $C_{graphene}$ columns. The solubility decays with temperature result smaller for $C_{graphene}$/bud than nanotube/bud than C_{60}, in agreement with lesser numbers of units in clusters. The discrepancy between experimental data for the heat of solution of fullerenes, nanotubes, graphenes and their buds is ascribed to the sharp concentration dependence for the heat of solution. The diffusion coefficient drops with temperature result greater for $C_{graphene}$/bud than nanotube/bud than C_{60}, corresponding to lesser number of units in aggregates. Clusters $(C_{60})_{13}$, nanotube/tube-bud$_7$ and $C_{graphene}/C_{graphene}$-bud$_3$ are representative of droplet, *bundlet* and *columnlet* models.
5. Some systems are dominated by the isolated pentagon rule, some others are not. Further work will explore similar nanostructures nature: generalization to systems more complex; *e.g.*, a way of bypassing weak homonuclear bonding exists in closed B_xN_x (involving replacement of 5-membered by 4-ring B_2N_2 with heteroatom alternation), BN/AlN tubes/heterojunctions, silicene, germanene and carbine. The C-nanostructures are more controllable while heterostructures present richer behaviour, especially for transtition-metal compounds, showing lubricant and electronic uses. It is challenging to understand the mechanisms in C-based materials, which contain *s*/*p* electrons, in contrast to hetero and heterometallostructures based on 3*d*/4*f* electrons.

ACKNOWLEDGMENT

The authors want to dedicate this manuscript to Dr. Luis Serrano-Andrés, who was greatly interested in this research and would have loved to see its conclusion. Francisco Torrens belongs to the Institut Uni versitari de Ciència Molecular, Universitat de València. Gloria Castellano belongs to the Departamento de Ciencias Experimentasles y Matemáticas, Facultad de Veterinaria y Ciencias Experimentales, Universidad Católica de Valencia "San Vicente Mártir". One of us, F. T., thanks support from the Spanish Ministerio de Economía y Competitividad (Project No. BFU2013-41648-P) and EU ERDF.

REFERENCES

Abbasalipourkabir, R., Salehzadeh, A., & Abdullah, R. (2011). Delivering tamoxifen within solid lipid nanoparticles. *Pharmaceutical Technology Europe*, *23*(4), 22–32.

Abe, S., Handa, H., Takahashi, R., Imaizumi, K., Fukidome, H., & Suemitsu, M. (2010). Surface chemistry involved in epitaxy of graphene on 3C-SiC(111)/Si(111). *Nanoscale Research Letters*, *5*, 1888–1891. PMID:21170403

Abellán, G., Carrasco, J. A., & Coronado, E. (2013). Room temperature magnetism in layered double hydroxides due to magnetic nanoparticles. *Inorganic Chemistry*, *52*(14), 7828–7830. doi:10.1021/ic400883k PMID:23795549

Abellán, G., Carrasco, J. A., Coronado, E., Romero, J., & Varela, M. (2014). Alkoxide-intercalated CoFe-layered double hydroxides as precursors of colloidal nanosheet suspensions: Structural, magnetic and electrochemical properties. *Journal of Materials Chemistry C*, *2*(19), 3723–3731. doi:10.1039/c3tc32578d

Abellán, G., Coronado, E., Martí-Gastaldo, C., Pinilla-Cienfuegos, E., & Ribera, A. (2010). Hexagonal nanosheets from the exfoliation of Ni^{2+}-Fe^{3+} LDHs: A route towards layered multifunctional materials. *Journal of Materials Chemistry*, *20*(35), 7451–7455. doi:10.1039/c0jm01447h

Abellán, G., Coronado, E., Martí-Gastaldo, C., Ribera, A., & Sánchez-Royo, J. F. (2012). Layered double hydroxide (LDH)–organic hybrids as precursors for low-temperature chemical synthesis of carbon nanoforms. *Chemical Science*, *3*(5), 1481–1485. doi:10.1039/c2sc01064j

Abrefah, J., Olander, D. R., Balooch, M., & Siekhaus, W. J. (1992). Vapor pressure of Buckminsterfullerene. *Applied Physics Letters*, *60*(11), 1313–1314. doi:10.1063/1.107327

Abrikosov, A. A. (1957). On the magnetic properties of superconductors of the second group. [English Translation]. *Soviet Physics, JETP*, *5*, 1174–1182.

Adini, A. R., Redlich, M., & Tenne, R. (2011). Medical applications of inorganic fullerene-like nanoparticles. *Journal of Materials Chemistry*, *21*(39), 15121–15131. doi:10.1039/c1jm11799h

Afzal, S., Daoud, W. A., & Langford, S. J. (2012). Self-cleaning cotton by porphyrin-sensitized visible-light photocatalysis. *Journal of Materials Chemistry*, *22*(9), 4083–4088. doi:10.1039/c2jm15146d

Afzal, S., Daoud, W. A., & Langford, S. J. (2013). Photostable self-cleaning cotton by a copper(II) porphyrin/TiO_2 visible-light photocatalytic system. *ACS Applied Materials & Interfaces*, *5*(11), 4753–4759. doi:10.1021/am400002k PMID:23465549

Aguilera-Sigalat, J., Casas-Solvas, J. M., Morant-Miñana, M. C., Vargas-Berenguel, A., Galian, R. E., & Pérez-Prieto, J. (2012b). Quantum dot/*cyclo*dextrin supramolecular systems based on efficient molecular recognition and their use for sensing. *Chemical Communications*, *48*(20), 2573–2575. doi:10.1039/C1CC15312A PMID:22080219

Aguilera-Sigalat, J., Rocton, S., Galian, R. E., & Pérez-Prieto, J. (2011). Fluorescence enhancement of amine-capped CdSe/ZnS quantum dots by thiol addition. *Canadian Journal of Chemistry*, *89*(3), 359–363. doi:10.1139/V10-160

Aguilera-Sigalat, J., Rocton, S., Sánchez-Royo, J. F., Galian, R. E., & Pérez-Prieto, J. (2012a). Highly fluorescent and photostable organic- and water-soluble CdSe/ZnS core–shell quantum dots capped with thiols. *RSC Advances*, *2*(4), 1632–1638. doi:10.1039/C1RA01005K

Aguilera-Sigalat, J., Sanchez-SanMartín, J., Agudelo-Morales, C. E., Zaballos, E., Galian, R. E., & Pérez-Prieto, J. (2012c). Further insight into the photostability of the pyrene fluorophore in halogenated solvents. *ChemPhysChem*, *13*(3), 835–844. doi:10.1002/cphc.201100843 PMID:22271708

Agustí, G., Cobo, S., Gaspar, A. B., Molnár, G., Moussa, N. O., Szilágyi, P. Á., & Bousseksou, A. et al. (2008). Thermal and light-induced spin crossover phenomena in new 3D Hofman-like microporous metalorganic frameworks produced as bulk materials and nanapatterned thin films. *Chemistry of Materials*, *20*(21), 6721–6732. doi:10.1021/cm8019878

Aimé, J.-P. (2010). BioInspired nanomaterials. *E-Nano Newsletter*, (19), 30-34.

Algar, W. R., & Krull, U. J. (2007). Towards multi-colour strategies for the detection of oligonucleotide hybridization using quantum dots as energy donors in fluorescence resonance energy transfer (FRET). *Analytica Chimica Acta*, *581*(2), 193–201. doi:10.1016/j.aca.2006.08.026 PMID:17386444

Ali, P. A., Reza, M. M., & Hossein, S. M. (2010). Removal of dissolved organic carbon by multi-walled carbon nanotubes, powdered activated carbon and granular activated carbon. *Research Journal of Chemistry and Environment*, *14*(4), 59–66.

Anderson, P. W., Halperin, B. I., & Varma, C. M. (1972). Anomalous low-temperature thermal properties of glasses and spin glasses. *Philosophical Magazine*, *25*(1), 1–9. doi:10.1080/14786437208229210

Anisimov, A. S., Nasibulin, A. G., Jiang, H., Launois, P., Cambedouzou, J., Shandakov, S. D., & Kauppinen, E. I. (2010). Mechanistic investigations of single-walled carbon nanotube synthesis by ferrocene vapor decomposition in carbon monoxide. *Carbon*, *48*(2), 380–388. doi:10.1016/j.carbon.2009.09.040

Anpo, M., & Kamat, P. V. (2010). *Environmentally benign photocatalysts: Applications of titanium oxide-based materials*. Berlin: Springer. doi:10.1007/978-0-387-48444-0

Arakelova, E., Khachatryan, A., Avjyan, K., Farmazyan, Z., Mirzoyan, A., Savchenko, L., & Arsenyan, F. et al. (2010). Zinc oxide nanocomposites with antitumour activity. *Natural Science*, *2*(12), 1341–1348. doi:10.4236/ns.2010.212163

Arora, P. S., & Kirshenbaum, K. (2004). Nano-tailoring: Stitching alterations on viral coats. *Chemistry & Biology*, *11*(4), 418–420. doi:10.1016/j.chembiol.2004.04.003 PMID:15123234

Artacho, E., Machado, M., Sánchez-Portal, D., Ordejón, P., & Soler, J. M. (2003). Electrons in dry DNA from density functional calculations. *Molecular Physics*, *101*(11), 1587–1594. doi:10.1080/0026897031000068587

Artacho, E., Sánchez-Portal, D., Ordejón, P., García, A., & Soler, J. M. (1999). Linear-scaling *ab-initio* calculations for large and complex systems. *Physica Status Solidi*, *215*, 809–817. doi:10.1002/(SICI)1521-3951(199909)215:1<809::AID-PSSB809>3.0.CO;2-0

Asenjo, A., Jaafar, M., Navas, D., & Vázquez, M. (2006). Quantitative magnetic force microscopy analysis of the magnetization process in nanowrire arrays. *Journal of Applied Physics*, *100*(2), 023909–1–6. doi:10.1063/1.2221519

Ashcroft, N. W. (1993). Elusive diffusive liquids. *Nature*, *365*(6445), 387–388. doi:10.1038/365387a0

Atwater, H. A., & Polman, A. (2010). Plasmonics for improved photovoltaic devices. *Nature Materials*, *9*(3), 205–213. doi:10.1038/nmat2629 PMID:20168344

Baba, A., Xia, C., Knoll, W., & Advincula, R. C. (2010). Electrochemical surface plasmon resonance and field-enhanced light scattering: Monomer copolymerization with a polysiloxane-conjugated polythiophene network precursor. *Macromolecular Chemistry and Physics*, *211*(24), 2624–2635. doi:10.1002/macp.201000471

Baba, M. S., Narasimhan, T. S. L., Balasubramanian, R., Sivaraman, N., & Mathews, C. K. (1994). Studies on the thermodynamics of the fullerene C_{60}–C_{70} binary system. *Journal of Physical Chemistry*, *98*(4), 1333–1340. doi:10.1021/j100055a047

Babiç, M., Horák, D., Trchová, M., Jendelová, P., Glogarová, K., Lesný, P., & Syková, E. et al. (2008). Poly(L-lysine)-modified iron oxide nanoparticles for stem cell labeling. *Bioconjugate Chemistry*, *19*(3), 740–750. doi:10.1021/bc700410z PMID:18288791

Bachelot, R., H'Dhili, F., Barchiesi, D., Lérondel, G., Fikri, R., Royer, P., ... Lahlil, K. (2003). Apertureless near-field optical microscopy: A study of the local tip field enhancement using photosensitive azobenzene-containing films. *J. Appl. Phys.*, *94*, 2060.

Bakry, R., Vallant, R. M., Najam-ul-Haq, M., Rainer, M., Szabo, Z., Huck, C. W., & Bonn, G. K. (2007). Medicinal applications of fullerenes. *International Journal of Nanomedicine*, *2*, 639–649. PMID:18203430

Balogh, L. P. (2009). The future of nanomedicine and the future of Nanomedicine: NBM. *Nanomedicine (London)*, *5*, 1.

Balu, A. M., Pineda, A., Yoshida, K., Campelo, J. M., Gai, P. L., Luque, R., & Romero, A. A. (2010). Fe/Al synergy in Fe_2O_3 nanoparticles supported on porous aluminosilicate materials: Excelling activities in oxidation reactions. *Chemical Communications (Cambridge)*, *46*(41), 7825–7827. doi:10.1039/c0cc02015j

Balzani, V., Credi, A., & Venturi, M. (2008). Molecular Devices and Machines, Weinheim: Wiley-VCH. doi:10.1002/9783527621682

Bar-Nahum, G., Epshtein, V., Ruckenstein, A. E., Rafikov, R., Mustaev, A., & Nudler, E. (2005). A ratchet mechanism of transcription elongation and its control. *Cell*, *120*(2), 183–193. doi:10.1016/j.cell.2004.11.045 PMID:15680325

Bard, A. J., Zhou, H., & Kwon, S. J. (2010). Electrochemistry of single nanoparticles *via* electrocatalytic amplification. *Israel Journal of Chemistry*, *50*(3), 267–276. doi:10.1002/ijch.201000014

Barnard, A. S. (2009). How can *ab initio* simulations address risks in nanotech? *Nature Nanotechnology*, *4*(6), 332–335. doi:10.1038/nnano.2009.126 PMID:19498383

Barnes, W. L., Dereux, A., & Ebbesen, T. (2003). Surface plasmon subwavelength optics. *Nature*, *424*(6950), 824–830. doi:10.1038/nature01937 PMID:12917696

Bartual-Murgui, C., Akou, A., Salmon, L., Molnár, G., Thibault, C., Real, J. A., & Bousseksou, A. (2011c). Guest effect on nanopatterned spin-crossover thin films. *Small*, *7*(23), 3385–3391. doi:10.1002/smll.201101089 PMID:21997948

Bartual-Murgui, C., Akou, A., Shepherd, H. J., Molnár, G., Real, J. A., Salmon, L., & Bousseksou, A. (2013). Tunable spin-crossover behavior of the Hofmann-like network {$Fe(bpac)[Pt(CN)_4]$} through host–guest chemistry. *Chemistry (Weinheim an der Bergstrasse, Germany)*, *19*(44), 15036–15043. doi:10.1002/chem.201300227 PMID:24105972

Bartual-Murgui, C., Ortega-Villar, N. A., Shepherd, H. J., Muñoz, M. C., Salmon, L., Molnár, G., & Real, J. A. et al. (2011a). Enhanced porosity in a new 3D Hofmann-like network exhibiting humidity sensitive cooperative spin transitions at room temperature. *Journal of Materials Chemistry*, *21*(20), 7217–7222. doi:10.1039/c0jm04387g

Bartual-Murgui, C., Salmon, L., Akou, A., Ortega-Villar, N. A., Shepherd, H. J., Muñoz, M. C., & Bousseksou, A. et al. (2012). Synergetic effect of host–guest chemistry and spin crossover in 3D Hofmann-like metal–organic frameworks [Fe(bpac)M(CN)$_4$] (M = Pt, Pd, Ni). *Chemistry (Weinheim an der Bergstrasse, Germany)*, *18*(2), 507–516. doi:10.1002/chem.201102357 PMID:22147670

Bartual-Murgui, C., Salmon, L., Akou, A., Thibault, C., Molnár, G., Mahfoud, T., & Bousseksou, A. et al. (2011b). High quality nano-patterned thin films of the coordination compound {Fe(pyrazine)[Pt(CN)$_4$]} deposited layer-by-layer. *New Journal of Chemistry*, *35*(10), 2089–2094. doi:10.1039/c1nj20212j

Berhanu, W. M., Pillai, G. G., Oliferenko, A. A., & Katritzky, A. R. (2012). Quantitative structure–activity/property relationships: The ubiquitous links between cause and effect. *ChemPlusChem*, *77*(7), 507–517. doi:10.1002/cplu.201200038

Bertolazzi, S., Brivio, J., & Kis, A. (2011). Stretching and breaking of ultrathin MoS_2. *ACS Nano*, *5*(12), 9703–9709. doi:10.1021/nn203879f PMID:22087740

Bertolazzi, S., Brivio, J., Radenovic, A., Kis, A., Wilson, H., Prisbrey, L., … Proksch, R. (2013) Exploring flatland: AFM of mechanical and electrical properties of graphene, MoS_2 and other low-dimensional materials. *Microscopy and Analysis, 2013*(5), 21-24.

Bezmel'nitsyn, V. N. (1994). Article. *Химическая физика*, *13*(12), 156–156.

Bezmel'nitsyn, V. N. (1996b). Article. *Technical Physics*, *41*, 986–986.

Bezmel'nitsyn, V. N., Eletskii, A. V., & Okun', M. V. (1998). Fullerenes in solutions. *Physics-Uspekhi*, *41*(11), 1091–1114. doi:10.1070/PU1998v041n11ABEH000502

Bezmel'nitsyn, V. N., Eletskii, A. V., Okun, M. V., & Stepanov, E. V. (1996a). Diffusion of aggregated fullerenes in solution. *Physica Scripta*, *53*(3), 364–367. doi:10.1088/0031-8949/53/3/017

Bezmel'nitsyn, V. N., Eletskii, A. V., Okun, M. V., & Stepanov, E. V. (1996c). Thermal diffusion of fullerenes in solutions. *Physica Scripta*, *53*(3), 368–370. doi:10.1088/0031-8949/53/3/018

Bezmel'nitsyn, V. N., Eletskii, A. V., & Stepanov, E. V. (1994). Cluster origin of fullerene solubility. *Journal of Physical Chemistry*, *98*(27), 6665–6667. doi:10.1021/j100078a001

Bezmel'nitsyn, V. N., Eletskii, A. V., & Stepanov, E. V. (1995). Article. *Журнал физической химии*, *69*, 735–735.

Bhar, K., Khan, S., Sánchez Costa, J., Ribas, J., Roubeau, O., Mitra, P., & Ghosh, B. K. (2012). Crystallographic evidence for reversible symmetry breaking in a spin-crossover d^7 cobalt(II) coordination polymer. *Angewandte Chemie International Edition*, *51*(9), 2142–2145. doi:10.1002/anie.201107116 PMID:22271674

Binning, G., Quate, C. F., & Gerber, C. (1986). Atomic force microscope. *Physical Review Letters*, *56*(9), 930–933. doi:10.1103/PhysRevLett.56.930 PMID:10033323

Bittner, A. M. (2005). Biomolecular rods and tubes in nanotechnology. *Naturwissenschaften*, *92*(2), 51–64. doi:10.1007/s00114-004-0579-8 PMID:15558225

Blomley, M., Cooke, J., Unger, E., Monaghan, M., & Cosgrove, D. (2001). Microbubble contrast agents: A new era in ultrasound. *British Medical Journal*, *322*(7296), 1222–1225. doi:10.1136/bmj.322.7296.1222 PMID:11358777

Boillot, M. L., Pillet, S., Tissot, A., Rivière, E., Claiser, N., & Lecomte, C. (2009). Ligand-driven light-induced spin change activity and bidirectional photomagnetism of styrylpyridine iron(II) complexes in polymeric media. *Inorganic Chemistry*, *48*(11), 4729–4736. doi:10.1021/ic802319c PMID:19374370

Boillot, M. L., Zarembowitch, J., & Sour, A. (2004). Ligand-driven light-induced spin change (LD-LISC): A promising photomagnetic effect. In Spin crossover in transition metal compounds II (pp. 261-276). Berlin: Springer.

Boldog, I., Gaspar, A. B., Martínez, V., Pardo-Ibañez, P., Ksenofontov, V., Bhattacharjee, A., & Real, J. A. et al. (2008). Spin-crossover nanocrystals with magnetic, optical, and structral bistability near room temperature. *Angewandte Chemie International Edition*, *47*, 6433–6437. PMID:18623300

Bong, D. T., Clark, T. D., Granja, J. R., & Ghadiri, M. R. (2001). Self-assembling organic nanotubes. *Angewandte Chemie International Edition*, *40*(6), 988–1011. doi:10.1002/1521-3773(20010316)40:6<988::AID-ANIE9880>3.0.CO;2-N PMID:11268062

Bonhommeau, S., Guillon, T., Daku, L. M. L., Demont, P., Sanchez Costa, J., Létard, J. F., & Bousseksou, A. et al. (2006). Photoswitching of the dielectric constant of the spin-crossover complex [Fe(L)$(CN)_2$]·H_2O. *Angewandte Chemie International Edition*, *45*(10), 1625–1629. doi:10.1002/anie.200503252 PMID:16470763

Bonnet, S., Molnár, G., Sanchez Costa, J., Siegler, M. A., Spek, A. L., Bousseksou, A., & Reedijk, J. et al. (2009). Influence of sample preparation, temperature, light, and pressure on the two-step spin crossover mononuclear compound [Fe(bapbpy)(NCS)2]. *Chemistry of Materials*, *21*(6), 1123–1136. doi:10.1021/cm803414q

Bonnet, S., Siegler, M. A., Sánchez Costa, J., Molnár, G., Bousseksou, A., Spek, A. L., & Reedijk, J. et al. (2008). A two-step spin crossover mononuclear iron(II) complex with a [HS–LS–LS] intermediate phase. *Chemical Communications*, *2008*(43), 5619–5621. doi:10.1039/b811746b PMID:18997971

Bosch-Navarro, C., Busolo, F., Coronado, E., Duan, Y., Martí-Gastaldo, C., & Prima-Garcia, H. (2013a). Influence of the covalent grafting of organic radicals to graphene on its magnetoresistance. *Journal of Materials Chemistry C*, *1*(30), 4590–4598. doi:10.1039/c3tc30799a

Bosch-Navarro, C., Coronado, E., & Martí-Gastaldo, C. (2013b). Controllable coverage of chemically modified graphene sheets with gold nanoparticles by thermal treatment of graphite oxide with N,N-dimethylformamide. *Carbon*, *54*, 201–207. doi:10.1016/j.carbon.2012.11.027

Bosch-Navarro, C., Coronado, E., Martí-Gastaldo, C., Sánchez-Royo, J. F., & Gómez Gómez, M. (2012). Influence of the pH on the synthesis of reduced graphene oxide under hydrothermal conditions. *Nanoscale*, *4*(13), 3977–3982. doi:10.1039/c2nr30605k PMID:22653666

Bouhelier, A., Bachelot, R., Lérondel, G., Kostcheev, S., Royer, P., & Wiederrecht, G. P. (2005). Surface plasmon characteristics of tunable photoluminescence in single gold nanorods. *Physical Review Letters*, *95*(26), 267405–1–4. doi:10.1103/PhysRevLett.95.267405 PMID:16486405

Boyd, B. J. (2008). Past and future evolution in colloidal drug delivery systems. *Expert Opinion on Drug Delivery*, *5*(1), 69–85. doi:10.1517/17425247.5.1.69 PMID:18095929

Brandbyge, M., Mozos, J. L., Ordejón, P., Taylor, J., & Stokbro, K. (2002). Density-functional method for non-equilibrium electron transport. *Physical Review B: Condensed Matter and Materials Physics*, *65*(16), 165401–1–17. doi:10.1103/PhysRevB.65.165401

Brown, S. (1992). Engineered iron oxide-adhesion mutants of the *Escherichia coli* phage lambda receptor. *Proceedings of the National Academy of Sciences of the United States of America*, *89*(18), 8651–8655. doi:10.1073/pnas.89.18.8651 PMID:1528875

Bruker. (2013). Atomic force microscopy reveals DNA double helix. *Microscopy and Analysis, 2013*(3).

Bueno-Alejo, C. J., D'Alfonso, C., Pacioni, N. L., González-Béjar, M., Grenier, M., Lanzalunga, O., & Scaiano, J. C. et al. (2012). Ultraclean derivatized monodisperse gold nanoparticles through laser drop ablation customization of polymorph gold nanostructures. *Langmuir*, *28*(21), 8183–8189. doi:10.1021/la3010689 PMID:22591001

Bueno Alejo, C. J., Fasciani, C., Grenier, M., Netto-Ferreira, J. C., & Scaiano, J. C. (2011). Reduction of resazurin to resorufin catalyzed by gold nanoparticles: Dramatic reaction acceleration by laser or LED plasmon excitation. *Catalysis Science & Technology*, *1*(8), 1506–1511. doi:10.1039/c1cy00236h

Buléon, A., Colonna, P., Planchot, V., & Ball, S. (1998). Starch granules: Structure and biosynthesis. *International Journal of Biological Macromolecules*, *23*(2), 85–112. doi:10.1016/S0141-8130(98)00040-3 PMID:9730163

Burrell, R. E. (1997). Anti-microbial coating for medical devices. *Patent Application, 1203*.

Buzdin, A. I. (1993). Multiple-quanta vortices at columnar defects. *Physical Review B: Condensed Matter and Materials Physics*, *47*(17), 11416–11419. doi:10.1103/PhysRevB.47.11416 PMID:10005280

Cañizo, J. A. (2005). Asymptotic behavior of solutions to the generalized Becker-Döring equations for general initial data. *Proceedings of the Royal Society of London. Series A*, *461*, 3731–3745.

Cañizo, J. A., López, J. L., & Nieto, J. (2004). Global L^1 theory and regularity for the 3D nonlinear Wigner–Poisson–Fokker–Planck system. *Journal of Differential Equations*, *198*(2), 356–373. doi:10.1016/j.jde.2003.07.004

Carbonera, C., Sánchez Costa, J., Money, V. A., Elhaïk, J., Howard, J. A. K., Halcrow, M. A., & Létard, J. F. (2006). Photomagnetic properties of iron(II) spin crossover complexes of 2,6-dipyrazolylpyridine and 2,6-dipyrazolylpyrazine ligands. *Dalton Transactions (Cambridge, England)*, *2006*(25), 3058–3066. doi:10.1039/B601366J PMID:16786064

Carl Zeiss Microscopy. (2013). *Cover story. Microscopy and Analysis Directory.*

Carroll, D. L., Redlich, P., Ajayan, P. M., Charlier, J. C., Blase, X., de Vita, A., & Car, R. (1997). Electronic structure and localized states at carbon nanotube tips. *Physical Review Letters*, *78*(14), 2811–2814. doi:10.1103/PhysRevLett.78.2811

Cartoixà, X., & Rurali, R. (2010). Simulating the structural, electronic and transport properties of silicon nanowires. *E-Nano Newslett.*, (19), 5-18.

Cassie, A. B. D., & Baxter, S. (1944). Wettability of porous surfaces. *Transactions of the Faraday Society*, *40*, 546–551. doi:10.1039/tf9444000546

Castellano, G., González-Santander, J. L., Lara, A., & Torrens, F. (2013). Classification of flavonoid compounds by using entropy of information theory. *Phytochemistry*, *93*, 182–191. doi:10.1016/j.phytochem.2013.03.024 PMID:23642389

Castellano, G., Lara, A., & Torrens, F. (2014). Classification of stilbenoid compounds by entropy of artificial intelligence. *Phytochemistry*, *97*, 62–69. doi:10.1016/j.phytochem.2013.10.010 PMID:24239224

Castellano, G., Tena, J., & Torrens, F. (2012). Classification of polyphenolic compounds by chemical structural indicators and its relation to antioxidant properties of *Posidonia oceanica* (L.) Delile. *MATCH: Communications in Mathematical and in Computer Chemistry*, *67*, 231–250.

Castellano, G., & Torrens, F. (2009). Local anaesthetics classified using chemical structural indicators. *Nereis*, (2), 7-17.

Castellano, G., & Torrens, F. (2015). Information entropy-based classification of triterpenoids and steroids from *Ganoderma. Phytochemistry*, 116, 305-313. PMID:26024957

Çervenka, J., Katsnelson, M. I., & Flipse, C. F. J. (2009). Room-temperature ferromagnetism in graphite driven by two-dimensional networks of point defects. *Nature Physics*, *5*(11), 840–844. doi:10.1038/nphys1399

Chatelier, R. C., & Minton, A. P. (1996). Adsorption of globular proteins on locally planar surfaces: Models for the effect of excluded surface area and aggregation of adsorbed protein on adsorption equilibria. *Biophysical Journal*, *71*(5), 2367–2374. doi:10.1016/S0006-3495(96)79430-4 PMID:8913577

Chatterjee, R. (2008). The challenge of regulating nanomaterials. *Environmental Science & Technology*, *42*(2), 339–343. doi:10.1021/es0870909 PMID:18284127

Chen, H. S., Kortan, A. R., Haddon, R. C., & Fleming, D. A. (1992). Thermodynamics of C_{60} in pure O_2, N_2 and Ar. *Journal of Physical Chemistry*, *96*, 1016–1018. doi:10.1021/j100182a003

Cheng, A., Klein, M. L., & Caccamo, C. (1993). Prediction of the phase diagram of rigid C_{60} molecules. *Physical Review Letters*, *71*(8), 1200–1203. doi:10.1103/PhysRevLett.71.1200 PMID:10055475

Chong, E. Z., Matthews, D. R., Summers, H. D., Njoh, K. L., Errington, R. J., & Smith, P. J. (2007). Development of FRET-based assays in the far-red using CdTe quantum dots. *Journal of Biomedicine and Biotechnology, 2007*, 54169.

Chule, K., Chule, A. V., Chen, B.-J., & Ling, Y.-C. (2006). Preparation and characterization of ZnO nanoparticles coated paper and its antibacterial activity study. *Green Chemistry*, *8*(12), 1034–1041. doi:10.1039/b605623g

Cioffi, C., Campidelli, S., Brunetti, F. G., Meneghetti, M., & Prato, M. (2006). Functionalisation of carbon nanohorns. *Chemical Communications*, (20), 2129–2131. doi:10.1039/b601176d PMID:16703130

Cioffi, C., Campidelli, S., Sooambar, C., Marcaccio, M., Marcolongo, G., Meneghetti, M., & Prato, M. et al. (2007). Synthesis, characterization, and photoinduced electron transfer in functionalized single wall carbon nanohorns. *Journal of the American Chemical Society*, *129*(13), 3938–3945. doi:10.1021/ja068007p PMID:17343379

Clapp, A. R., Medintz, I. L., & Mattoussi, H. (2006). Förster resonance energy transfer investigations using quantum dot fluorophores. *ChemPhysChem*, *7*(1), 47–57. doi:10.1002/cphc.200500217 PMID:16370019

Corma, A., & Garcia, H. (2008). Supported gold nanoparticles as catalysts for organic reactions. *Chemical Society Reviews*, *37*(9), 2096–2126. doi:10.1039/b707314n PMID:18762848

Coronado, E., Galán-Mascarós, J. R., Gómez-García, C. J., & Laukhin, V. (2000). Coexistence of ferromagnetism and metallic conductivity in a molecule-based layered compound. *Nature*, *408*(6811), 447–449. doi:10.1038/35044035 PMID:11100721

Coronado, E., Galán-Mascarós, J. R., Martí-Gastaldo, C., & Ribera, A. (2006). Insertion of magnetic bimetallic oxalate complexes into layered double hydroxides. *Chemistry of Materials*, *18*(26), 6112–6114. doi:10.1021/cm062054z

Coronado, E., Galán-Mascarós, J. R., Monrabal-Capilla, M., García-Martínez, J., & Pardo-Ibáñez, P. (2007). Bistable spin-crossover nanoparticles showing magnetic thermal hysteresis near room temperature. *Advanced Materials*, *19*(10), 1359–1361. doi:10.1002/adma.200700559

Coronado, E., Giménez-Marqués, M., Mínguez Espallargas, G., Rey, F., & Vitórica-Yrezábal, I. J. (2013). Spin-crossover modification through selective CO_2 sorption. *Journal of the American Chemical Society*, *135*(43), 15986–15989. doi:10.1021/ja407135k PMID:24125096

Coronado, E., Martí-Gastaldo, C., Navarro-Moratalla, E., Burzurí, E., Camón, A., & Luis, F. (2011). Hybrid magnetic/superconducting materials obtained by insertion of a single-molecule magnet into TaS_2 layers. *Advanced Materials*, *23*(43), 5021–5026. doi:10.1002/adma.201102730 PMID:21956436

Coronado, E., Martí-Gastaldo, C., Navarro-Moratalla, E., Ribera, A., Blundell, S. J., & Baker, P. J. (2010). Coexistence of superconductivity and magnetism by chemical design. *Nature Chemistry*, *2*(12), 1031–1036. doi:10.1038/nchem.898 PMID:21107366

D'Oliveira, J.-C., Minero, C., Pelizzetti, E., & Pichat, P. (1993). Photodegradation of dichlorophenols and trichlorophenols in TiO_2 aqueous suspensions: Kinetic effects of the positions of the Cl atoms and identification of the intermediates. *Journal of Photochemistry and Photobiology A Chemistry*, *72*(3), 261–267. doi:10.1016/1010-6030(93)80022-2

De Pablo, P. J., Moreno-Herrero, F., Colchero, J., Gómez Herrero, J., Herrero, P., Baró, A. M., & Artacho, E. et al. (2000). Absence of dc-conductivity in λ-DNA. *Physical Review Letters*, *85*(23), 4992–4995. doi:10.1103/PhysRevLett.85.4992 PMID:11102169

Dean, L. (2012). Size matters: Measurement helps solve nanoparticle toxicity challenges. *Chemistry International*, *34*(4), 6–9.

Deeb, C., Huang, L., Plain, J., Bouhelier, A., Soppera, O., Bachelot, R., & Royer, P. (2010). Nanophotopolymerization triggered by the enhanced optical near-field of metallic nanoparticles. *Lebanese Science Journal*, *11*(2), 105–115.

Deerink, T. J. (2008). The application of fluorescent quantum dots to confocal, multiphoton, and electron microscopic imaging. *Toxicologic Pathology*, *36*(1), 112–116. doi:10.1177/0192623307310950 PMID:18337229

Dimitratos, N., Lopez-Sanchez, J. A., Lennon, D., Porta, F., Prati, L., & Villa, A. (2006). Effect of particle size on monometallic and bimetallic (Au,Pd)/C on the liquid phase oxidation of glycerol. *Catalysis Letters*, *108*(3-4), 147–153. doi:10.1007/s10562-006-0036-8

Dîrtu, M. M., Neuhausen, C., Naik, A. D., Rotaru, A., Spinu, L., & Garcia, Y. (2010). Insights into the origin of cooperative effects in the spin transitions of $[Fe(NH_2trz)_3](NO_3)_2$: The role of supramolecular interactions evidenced in the crystal structure of $[Cu(NH_2trz)_3](NO_3)_2{\cdot}H2O$. *Inorganic Chemistry*, *49*(12), 5723–5736. doi:10.1021/ic100667f PMID:20507075

Douglas, T., & Stark, V. T. (2000). Nanophase cobalt oxyhydroxide mineral synthesized within the protein cage of ferritin. *Inorganic Chemistry*, *39*(8), 1828–1830. doi:10.1021/ic991269q PMID:12526579

Douglas, T., & Young, M. (2006). Viruses: Making friends with old foes. *Science*, *312*(5775), 873–875. doi:10.1126/science.1123223 PMID:16690856

Dreher, K. L. (2004). Health and environmental impact of nanotechnology: Toxicological assessment of manufactured nanoparticles. *Toxicological Sciences*, *77*(1), 3–5. doi:10.1093/toxsci/kfh041 PMID:14756123

Dresselhaus, M. S., Jorio, A., Hofmann, M., Dresselhaus, G., & Saito, R. (2010). Perspectives on carbon nanotubes and graphene Raman spectroscopy. *Nano Letters*, *10*(3), 751–758. doi:10.1021/nl904286r PMID:20085345

Drexler, E. (1986). *Engines of Creation: The Coming Era of Nanotechnology*. New York: Anchor Bools.

Duncan, R. (2007). Designing polymer conjugates as lysosomotropic nanomedicines. *Biochemical Society Transactions*, *35*(1), 56–60. doi:10.1042/BST0350056 PMID:17233601

Duncan, R., & Spreafico, F. (1994). Polymer conjugates. Pharmacokinetic considerations for design and development. *Clinical Pharmacokinetics*, *27*(4), 290–306. doi:10.2165/00003088-199427040-00004 PMID:7834965

Eletskii, A. V., Okun', M. V., & Smirnov, B. M. (1997). Growth of fractal structures in fullerene solutions. *Physica Scripta*, *55*(3), 363–366. doi:10.1088/0031-8949/55/3/016

Eletskii, A. V., & Smirnov, B. M. (1991). The C₆₀ cluster as a new form of carbon. *Uspekhi Fizicheskikh Nauk*, *161*(7), 173–173. doi:10.3367/UFNr.0161.199107e.0173

Enache, D. I., Edwards, J. K., Landon, P., Solsona-Espriu, B., Carley, A. F., Herzing, A. A., & Hutchings, G. J. et al. (2006). Solvent-free oxidation of primary alcohols to aldehydes using Au-Pd/TiO_2 catalysts. *Science*, *311*(5759), 362–365. doi:10.1126/science.1120560 PMID:16424335

Erickson, L. E., Koodali, R. T., & Richards, R. M. (Eds.). (2010). ACS Symp. Ser.: Vol. 1045. *Nanoscale Materials in Chemistry: Environmental Applications*. Washington, DC: American Chemical Society. doi:10.1021/bk-2010-1045

Escrig, J., Altbir, D., Jaafar, M., Navas, D., Asenjo, A., & Vázquez, M. (2007). Remanence of Ni nanowire arrays: Influence of size and labyrinth magnetic structure. *Physical Review B: Condensed Matter and Materials Physics*, *75*(18), 184429–1–5. doi:10.1103/PhysRevB.75.184429

Euliss, L. E., Grancharov, S. G., O'Brien, S., Deming, T. J., Stucky, G. D., Murray, C. B., & Held, G. A. (2003). Cooperative assembly of magnetic nanoparticles and block copolypeptides in aqueous media. *Nano Letters*, *3*(11), 1489–1493. doi:10.1021/nl034472y

Fan, Z., & Govorov, A. O. (2010). Plasmonic circular dichroism of chiral metal nanoparticle assemblies. *Nano Letters*, *10*(7), 2580–2587. doi:10.1021/nl101231b PMID:20536209

Fang, L., Wang, Y., Zou, P. Y., Tang, L., Xu, Z., Chen, H., Dong, C., Shan, L., & Wen, H. H. (2005). Fabrication and superconductivity of Na_xTaS_2 crystals. *Physical Review B*, *72*, 14534.

Faraday, M. (1857). The Bakerian Lecture: Experimental relations of gold (and other metals) to light. *Philosophical Transactions of the Royal Society of London*, *147*(0), 145–181. doi:10.1098/rstl.1857.0011

Fasciani, C., Bueno Alejo, C. J., Grenier, M., Netto-Ferreira, J. C., & Scaiano, J. C. (2011). High-temperature organic reactions at room temperature using plasmon excitation: Decomposition of dicumyl peroxide. *Organic Letters*, *13*(2), 204–207. doi:10.1021/ol1026427 PMID:21142017

Feynman, R. P. (1960). *There's plenty of room at the bottom: An invitation to enter a new field of physics*. Conference.

Flenniken, M., Allen, M., & Douglas, T. (2004). Microbe manufacturers of semiconductors. *Chemistry & Biology*, *11*(11), 1478–1480. doi:10.1016/j.chembiol.2004.11.004 PMID:15555996

Flenniken, M. L., Liepold, L. O., Crowley, B. E., Willits, D. A., Young, M. J., & Douglas, T. (2005). Selective attachment and release of a chemotherapeutic agent from the interior of a protein cage architecture. *Chemical Communications*, *2005*(4), 447–449. doi:10.1039/b413435d PMID:15654365

Flenniken, M. L., Willits, D. A., Brumfield, S., Young, M. J., & Douglas, T. (2003). The small heat shock protein cage from *Methanococcus jannaschii* is a versatile nanoscale platform for genetic and chemical modification. *Nano Letters*, *3*(11), 1573–1576. doi:10.1021/nl034786l

Flynn, C. E., Lee, S.-W., Peelle, B. R., & Belcher, A. M. (2003). Viruses as vehicles for growth, organization and assembly of materials. *Acta Materialia*, *51*(19), 5867–5880. doi:10.1016/j.actamat.2003.08.031

Forestier, T., Mornet, S., Daro, N., Nishihara, T., Mouri, S.-i., Tanaka, K., & Létard, J.-F. et al. (2008). Nanoparticles of iron(II) spin-crossover. *Chemical Communications*, *2008*(36), 4327–4329. doi:10.1039/b806347h PMID:18802559

Fourches, D., Pu, D. Q. Y., Tassa, C., Weissleder, R., Shaw, S. Y., Mumper, R. J., & Tropsha, A. (2010). Quantitative nanostructure–activity relationship modeling. *ACS Nano*, *4*(10), 5703–5712. doi:10.1021/nn1013484 PMID:20857979

Frangopol, P. T., & Morariu, V. V. (Eds.). (1988). *Seminars in biophysics* (Vol. 5). Bucharest: CIP.

Fuertes, G., García-Sáez, A. J., Esteban-Martín, S., Giménez, D., Sánchez-Muñoz, O. L., Schwille, P., & Salgado, J. (2010). Pores formed by Baxα5 relax to a smaller size and keep at equilibrium. *Biophysical Journal*, *99*(9), 2917–2925. doi:10.1016/j.bpj.2010.08.068 PMID:21044589

Fuertes, G., Giménez, D., Esteban-Martín, S., Sánchez-Muñoz, O. L., & Salgado, J. (2011). A lipocentric view of peptide-induced pores. *European Biophysics Journal*, *40*(4), 399–415. doi:10.1007/s00249-011-0693-4 PMID:21442255

Fujishima, A., Rao, T. N., & Tryk, D. A. (2000). Titanium dioxide photocatalysis. *Journal of Photochemistry and Photobiology A Chemistry*, *1*(1), 1–21. doi:10.1016/S1389-5567(00)00002-2

Fürst, J. A., Hashemi, J., Markussen, T., Brandbyge, M., Jauho, A. P., & Nieminen, R. M. (2009). Electronic transport properties of fullerene functionalized carbon nanotubes: *Ab initio* and tight-binding calculations. *Physical Review B*, *80*, 35427.

Gadet, V., Mallah, T., Castro, I., Verdaguer, M., & Veillet, P. (1992). High-T_c molecular-based magnets: A ferromagnetic bimatallic chromium(III)–nickel(II) cyanide with T_c = 90K. *Journal of the American Chemical Society*, *114*(23), 9213–9214. doi:10.1021/ja00049a078

Gai, S., Yang, P., Li, C., Wang, W., Dai, Y., Niu, N., & Lin, J. (2010). Synthesis of magnetic, up-conversion luminescent, and mesoporous core–shell-structured nanocomposites as drug carriers. *Advanced Functional Materials*, *20*(7), 1166–1172. doi:10.1002/adfm.200902274

Gaita-Ariño, A., & Schechter, M. (2011). Identification of strong and weak interacting two-level systems in KBr:CN. *Physical Review Letters*, *107*(10), 105504–1–5. doi:10.1103/PhysRevLett.107.105504 PMID:21981511

Galán-Mascarós, J. R., Coronado, E., Forment-Aliaga, A., Monrabal-Capilla, M., Pinilla-Cienfuegos, E., & Ceolin, M. (2010). Tuning size and thermal histeresis in bistable spin crossover nanoparticles. *Inorganic Chemistry*, *49*(12), 5706–5714. doi:10.1021/ic100751a PMID:20503990

Galbiati, M., Barraud, C., Tatay, S., Bouzehouane, K., Deranlot, C., Jacquet, E., & Petroff, F. et al. (2012). Unveiling self-assembled monolayers' potential for molecular spintronics: Spin transport at high voltage. *Advanced Materials*, *24*(48), 6429–6432. doi:10.1002/adma.201203136 PMID:23055410

Gasser, U., Weeks, E. R., Schofield, A., Pusey, P. N., & Weitz, D. A. (2001). Real-space imaging of nucleation and growth in colloidal crystallization. *Science*, *292*(5515), 258–262. doi:10.1126/science.1058457 PMID:11303095

Geim, A. K. (2009). Graphene: Status and prospects. *Science*, *324*(5934), 1530–1534. doi:10.1126/science.1158877 PMID:19541989

Geim, A. K., & Novoselov, K. S. (2007). The rise of graphene. *Nature Materials*, *6*(3), 183–191. doi:10.1038/nmat1849 PMID:17330084

Ghorbani, M. (2012). Enumeration of heterofullerenes: A survey. *MATCH: Communications in Mathematical and in Computer Chemistry*, *68*, 381–414.

Gilmore, J. L., Yi, X., Quan, L., & Kabanov, A. V. (2008). Nobel nanomaterials for clinical neuroscience. *Journal of Neuroimmune Pharmacology*, *3*(2), 83–94. doi:10.1007/s11481-007-9099-6 PMID:18210200

Girifalco, L. A. (1991). Interaction potential for C_{60} molecules. *Journal of Physical Chemistry*, *95*(14), 5370–5371. doi:10.1021/j100167a002

Girifalco, L. A. (1992). Molecular properties of C_{60} in the gas and solid phases. *Journal of Physical Chemistry*, *96*(2), 858–861. doi:10.1021/j100181a061

González, J., Guinea, F., & Vozmediano, M. A. H. (1992). Continuum approximation to fullerene molecules. *Physical Review Letters*, *69*(1), 172–175. doi:10.1103/PhysRevLett.69.172 PMID:10046217

Gonzalez-Arellano, C., Balu, A. M., Luque, R., & Macquarrie, D. J. (2010a). Catalytically active self-assembled silica-based nanostructures containing supported nanoparticles. *Green Chemistry*, *12*(11), 1995–2002. doi:10.1039/c0gc00282h

Gonzalez-Arellano, C., Luque, R., & Macquarrie, D. J. (2009). Nanotubular self-assembly of *n*-dodecylamine–TEOS–water–acetonitrile mixtures. *Chemical Communications*, (30), 4581–4583. doi:10.1039/b903368h PMID:19617990

Gonzalez-Arellano, C., Yoshida, K., Luque, R., & Gai, P. L. (2010b). Highly active and selective supported iron oxide nanoparticles in microwave-assisted *N*-alkylations of amines with alcohols. *Green Chemistry*, *12*(7), 1281–1287. doi:10.1039/c003410j

Gonzalez-Valls, I., & Lira-Cantu, M. (2009). Vertically-aligned nanostructures of ZnO for excitonic solar cells: A review. *Energy and Environmental Science*, *2*(1), 19–34. doi:10.1039/B811536B

Gonzalez-Valls, I., & Lira-Cantu, M. (2010). Dye sensitized solar cells based on vertically-aligned ZnO nanorods: Effect of UV light on power conversion efficiency and lifetime. *Energy and Environmental Science*, *3*(6), 789–795. doi:10.1039/b922354a

Gratzel, M. (2001). Photoelectrochemical cells. *Nature*, *414*(6861), 338–344. doi:10.1038/35104607 PMID:11713540

Guccione, S., Li, K., & Bednarski, M. (2004). Vascular-targeted nanoparticles for molecular imaging and therapy. *Methods in Enzymology*, *386*, 219–236. doi:10.1016/S0076-6879(04)86010-5 PMID:15120254

Guerrero-Martínez, A., Pérez-Juste, J., Carbó-Argibay, E., Tardajos, G., & Liz-Marzán, L. M. (2009). Gemini-surfactant-directed self-assembly of monodisperse gold nanorods into standing superlattices. *Angewandte Chemie International Edition*, *48*(50), 9484–9488. doi:10.1002/anie.200904118 PMID:19802865

Guerrero-Martínez, A., Pérez-Juste, J., & Liz-Marzán, L. M. (2010). Recent progress on silica coating of nanoparticles and related nanomaterials. *Advanced Materials*, *22*(11), 1182–1195. doi:10.1002/adma.200901263 PMID:20437506

Guionneau, P., Le Gac, F., Kaiba, A., Sánchez Costa, J., Chasseau, D., & Létard, J. F. (2007). A reversible metal–ligand bond break associated to a spin-crossover. *Chemical Communications*, *2007*(36), 3723–3725. doi:10.1039/b707836f PMID:17851607

Guldi, D. M., & Martin, N. (Eds.). (2010). Carbon nanotubes and related structures. Weinheim: Wiley-VCH. doi:10.1002/9783527629930

Guo, X., Zhang, F., Evans, D. G., & Duan, X. (2010). Layered double hydroxide films: Synthesis, properties and applications. *Chemical Communications*, *46*(29), 5197–5210. doi:10.1039/c0cc00313a PMID:20549015

Gupta, A. K., & Curtis, A. S. G. (2004). Lactoferrin and ceruloplasmin derivatized superparamagnetic iron oxide nanoparticles for trageting cell surface receptors. *Biomaterials*, *25*(15), 3029–3040. doi:10.1016/j.biomaterials.2003.09.095 PMID:14967536

Gupta, A. K., & Gupta, M. (2005). Synthesis and surface engineering of iron oxide nanoparticles for biomedical applications. *Biomaterials*, *26*(18), 3995–4021. doi:10.1016/j.biomaterials.2004.10.012 PMID:15626447

Haag, M. (2011). Calculating and understanding particulate contamination risk. *Pharmaceutical Technology Europe*, *23*(3), 38–41.

Haes, A. W., Hall, W. P., Chang, L., Klein, W. L., & Duyne, R. P. (2004). A localized surface plasmon resonance biosensor: First steps toward an assay for Alzheimer's disease. *Nano Letters*, *4*(6), 1029–1034. doi:10.1021/nl049670j

Hagen, M. H. J., Meijer, E. J., Mooij, G. C. A. M., Frenkel, D., & Lekkerkerker, H. N. W. (1993). Does C_{60} have a liquid phase? *Nature*, *365*(6445), 425–426. doi:10.1038/365425a0

Haluska, C. K., Riske, K. A., Marchi-Artzner, V., Lehn, J.-M., Lipowsky, R., & Dimova, R. (2006). Time scales of membrane fusion revealed by direct imaging of vesicle fusion with high temporal resolution. *Proceedings of the National Academy of Sciences of the United States of America*, *103*(43), 15841–15846. doi:10.1073/pnas.0602766103 PMID:17043227

Han, J., & Jaffe, R. (1998). Energetics and geometries of carbon nanocone tips. *The Journal of Chemical Physics*, *108*(7), 2817–2823. doi:10.1063/1.475672

Harada, K., Kamimura, O., Kasai, H., Matsuda, T., Tonomura, A., & Moshchalkov, V. V. (1996). Direct observation of vortex dynamics in superconducting films with regular arrays of defects. *Science*, *274*(5290), 1167–1170. doi:10.1126/science.274.5290.1167 PMID:8895460

Harris, S. (2014). Compound benefits. *Scientific Computing World, 2014*(134), 31-33.

Hartmann, U. (1985). *Theory of noncontact force microscopy: Scanning tunneling microscopy III. Theory of STM and related scanning techniques* (R. Wiensendanger & J. H. Guntherodt, Eds.). Berlin: Springer.

Hawker, C., & Wooley, K. (2005). The convergence of synthetic organic and polymer chemistries. *Science*, *309*(5738), 1200–1205. doi:10.1126/science.1109778 PMID:16109874

Hawthorne, M. F., Zink, J. I., Skelton, J. M., Bayer, M. J., Liu, C., Livshits, E., & Neuhauser, D. et al. (2004). Electrical or photocontrol of the rotary motion of a metallacarborane. *Science*, *303*(5665), 1849–1851. doi:10.1126/science.1093846 PMID:15031500

Hebard, A. F. (1993). Buckminsterfullerene. *Annual Review of Materials Science*, *23*(1), 159–191. doi:10.1146/annurev.ms.23.080193.001111

Heimburg, T. (2010). Lipid ion channels. *Biophysical Chemistry*, *150*(1-3), 2–22. doi:10.1016/j.bpc.2010.02.018 PMID:20385440

Hernandez, Y., Nicolosi, V., Lotya, M., Blighe, F. M., Sun, Z., De, S., & Coleman, J. N. et al. (2008). High-yield production of graphene by liquid-phase exfoliation of graphite. *Nature Nanotechnology*, *3*(9), 563–568. doi:10.1038/nnano.2008.215 PMID:18772919

Herrmann, J.-M. (1999). Heterogeneous photocatalysis: Fundamentals and applications to the removal of various types of aqueous pollutants. *Catalysis Today*, *53*(1), 115–129. doi:10.1016/S0920-5861(99)00107-8

Herrmann, J.-M. (2005). Heterogeneous photocatalysis: State of the art and present applications. *Topics in Catalysis*, *34*(1-4), 49–65. doi:10.1007/s11244-005-3788-2

Herrmann, J.-M. (2006a). Photocatalysis. In *Kirk-Othmer encyclopedia of chemical technology*. New York, NY: Wiley. doi:10.1002/0471238961.1608152019051816.a01.pub2

Herrmann, J.-M. (2006b). From catalysis by metals to bifunctional photocatalysis. *Topics in Catalysis*, *39*(1-2), 3–10. doi:10.1007/s11244-006-0032-7

Herrmann, J.-M. (2010). Fundamentals and misconceptions in photocatalysis. *Journal of Photochemistry and Photobiology A Chemistry*, *216*(2-3), 85–93. doi:10.1016/j.jphotochem.2010.05.015

Herrmann, J.-M., Guillard, C., & Pichat, P. (1993). Heterogeneous photocatalysis: An emerging technology for water treatment. *Catalysis Today*, *17*(1-2), 7–20. doi:10.1016/0920-5861(93)80003-J

Hirsch, L., Gobin, A., Lowery, A., Tam, F., Drezek, R., Halas, N., & West, J. (2006). Metal nanoshells. *Annals of Biomedical Engineering*, *34*(1), 15–22. doi:10.1007/s10439-005-9001-8 PMID:16528617

Hoffmann, M. R., Martin, S. T., Choi, W., & Bahnemann, D. W. (1995). Environmental applications of semiconductor photocatalysis. *Chemical Reviews*, *95*(1), 69–96. doi:10.1021/cr00033a004

Horák, D., Rittich, B., Španová, A., & Beneš, M. J. (2005). Magnetic microparticulate carriers with immobilized selective ligands in DNA disgostics. *Polymer*, *46*(4), 1245–1255. doi:10.1016/j.polymer.2004.11.049

Horsell, D. W., Hale, P. J., & Savchenko, A. K. (2011). Mechanical manipulation measurement of graphene by atomic force microscopy. *Microscopy and Analysis*, *25*(1), 15–17.

Hu, X., & Zhou, Q. (2013). Health and ecosystem risks of graphene. *Chemical Reviews*, *113*(5), 3815–3835. doi:10.1021/cr300045n PMID:23327673

Huskinson, B., Marshak, M. P., Suh, C., Er, S., Gerhardt, M. R., Galvin, C. J., & Aziz, M. J. et al. (2014). A metal-free organic–inorganic aqueous flow battery. *Nature*, *505*(7482), 195–198. doi:10.1038/nature12909 PMID:24402280

Ibn El Ahrach, H., Bachelot, R., Vial, A., Lérondel, G., Plain, J., Royer, P., & Soppera, O. (2007). Spectral degeneracy breaking of the plasmon resonance of single metal nanoparticles by nanoscale near-field photopolymerization. *Physical Review Letters*, *98*(10), 107402–1–4. doi:10.1103/PhysRevLett.98.107402 PMID:17358565

Iordache, O. (2011). *Modeling multi-level systems*. Berlin: Springer. doi:10.1007/978-3-642-17946-4

Iordache, O. (2012). *Self-evolvable systems: machine learning in social media*. Berlin: Springer. doi:10.1007/978-3-642-28882-1

Iordache, O. (2014). *Polytope Projects*. Boca Raton, FL: CRC.

Iordache, O., Apostol, M., & Frangopol, P. T. (1988). Non-exponential relaxation of drug–membrane interaction. Effects of cross-linking aldehydes and procaine. *Revue Roumaine de Biochimie*, *25*, 275–281.

Jaafar, M., Gómez-Herrero, J., Gil, A., Ares, P., Vázquez, M., & Asenjo, A. (2009). Variable-field magnetic force microscopy. *Ultramicroscopy*, *109*(6), 693–699. doi:10.1016/j.ultramic.2009.01.007 PMID:19250752

Jaafar, M., Iglesias-Freire, O., Serrano-Ramón, L., Ibarra, M. R., de Teresa, J. M., & Asenjo, A. (2011). Distinguishing magnetic and electrostatic interactions by a Kelvin probe force microscopy–magnetic force microscopy combination. *Beilstein Journal of Nanotechnology*, *2*, 552–560. doi:10.3762/bjnano.2.59 PMID:22003461

Jaafar, M., Martínez-Martín, D., Cuenca, M., Melcher, J., Raman, A., & Gómez-Herrero, J. (2012). Drive-amplitude-modulation atomic force microscopy: From vacuum to liquids. *Beilstein Journal of Nanotechnology*, *3*, 336–344. doi:10.3762/bjnano.3.38 PMID:22563531

Jacquemin, D., Michaux, C., Perpète, E. A., Maurel, F., & Perrier, A. (2010a). Photochromic molecular wires: Insights from theory. *Chemical Physics Letters*, *488*(4-6), 193–197. doi:10.1016/j.cplett.2010.02.017

Jacquemin, D., Perpète, E. A., Maurel, F., & Perrier, A. (2010b). *Ab initio* investigation of the electronic properties of coupled dithienylethenes. *The Journal of Physical Chemistry Letters*, *1*(1), 434–438. doi:10.1021/jz900293g

Jacquemin, D., Perpète, E. A., Maurel, F., & Perrier, A. (2010c). Doubly closing or not? Theoretical analysis for coupled photochromes. *The Journal of Physical Chemistry C*, *114*(20), 9489–9497. doi:10.1021/jp102118w

Jacquemin, D., Perpète, E. A., Maurel, F., & Perrier, A. (2010d). Simulation of the properties of a photochromic triad. *The Journal of Physical Chemistry Letters*, *1*(14), 2104–2108. doi:10.1021/jz1006753

Jacquemin, D., Perpète, E. A., Maurel, F., & Perrier, A. (2010e). TD-DFT simulations of the electronic properties of star-shaped photochromes. *Physical Chemistry Chemical Physics*, *12*(28), 7994–8000. doi:10.1039/b927323a PMID:20517564

Jacquemin, D., Perpète, E. A., Maurel, F., & Perrier, A. (2010f). Hybrid dithienylethene-naphthopyran multi-addressable photochromes: An *ab initio* analysis. *Physical Chemistry Chemical Physics*, *12*(40), 13144–13152. doi:10.1039/c0cp00400f PMID:20838697

Jacquez, J. A. (1985). *Compartmental Analysis in Biology and Medicine*. Ann Arborm MI: The University of Michigan Press.

Jia, C., & Guo, X. (2013). Molecule–electrode interfaces in molecular electronic devices. *Chemical Society Reviews*, *42*(13), 5642–5660. doi:10.1039/c3cs35527f PMID:23571285

Jimoda, L. A., Oke, E. O., & Salam, K. K. (2013). Modelling of mass transfer rate during biocoagulation-flocculation of coal-rich wastewater. *Journal of Scientific Research and Reports*, *2*(1), 376–390. doi:10.9734/JSRR/2013/3492

Jones, V. (2011). Challenges of particle characterisation. *Pharmaceutical Technology Europe*, *23*(4), 40–43.

Jongh, P. E., & Adelhelm, P. (2010). Nanosizing and nanoconfinement: New strategies towards meeting hydrogen storage goals. *ChemSusChem*, *3*(12), 1332–1348. doi:10.1002/cssc.201000248 PMID:21080405

Jordan, A., Scholz, R., Maier-Hauff, K., Johannsen, M., Wust, P., Nadobny, J., & Felix, R. et al. (2001). Presentation of a new magnetic field therapy system for the treatment of human solid tumors with magnetic fluid hyperthermia. *Journal of Magnetism and Magnetic Materials*, *225*(1-2), 118–126. doi:10.1016/S0304-8853(00)01239-7

Kagan, C. R., Murray, C. B., Nirmal, M., & Bawendi, M. G. (1996). Electronic energy transfer in CdSe quantum dot solids. *Physical Review Letters*, *76*(9), 1517–1520. doi:10.1103/PhysRevLett.76.1517 PMID:10061743

Kahn, O., & Martinez, C. J. (1998). Spin transition polymers: From molecular materials toward memory devices. *Science*, *279*(5347), 44–48. doi:10.1126/science.279.5347.44

Kaiser, U., Schwarz, A., & Wiesendanger, R. (2007). Magnetic exchange force microscopy with atomic resolution. *Nature*, *446*(7135), 522–525. doi:10.1038/nature05617 PMID:17392782

Kataria, S., Goyal, S., Dash, S., Sandhya, R., Mathew, M. D., & Tyagi, A. K. (2012). Evaluation of nanomechanical properties of hard coatings on a soft surface. *Thin Solid Films*, *522*, 297–303. doi:10.1016/j.tsf.2012.09.001

Khan, U., O'Neill, A., Lotya, M., De, S., & Coleman, J. N. (2010). High-concentration solvent exfoliation of graphene. *Small*, *6*(7), 864–871. doi:10.1002/smll.200902066 PMID:20209652

Khosravi, H., & Moradi, A. (2007). Comment on: *Electromagnetic wave propagation in single-wall carbon nanotubes. Physics Letters. [Part A]*, *364*(6), 515–516. doi:10.1016/j.physleta.2007.02.059

Kim, H., Abdala, A. A., & Mocosko, C. W. (2010). Graphene/polymer nanocomposites. *Macromolecules*, *43*(16), 6515–6530. doi:10.1021/ma100572e

Kim, P., Odom, T. W., Huang, J.-L., & Lieber, C. M. (1999). Electronic density of states of atomically resolved single-walled carbon nanotubes: Van Hove singularities and end states. *Physical Review Letters*, *82*(6), 1225–1228. doi:10.1103/PhysRevLett.82.1225

Kishino, K., Sekiguchi, H., & Kikuchi, A. (2009). Improved Ti-mask selective-area growth (SAG) by rf-plasma-assisted molecular beam epitaxy demonstrating extremely uniform GaN nanocolumn arrays. *Journal of Crystal Growth*, *311*(7), 2063–2068. doi:10.1016/j.jcrysgro.2008.11.056

Klem, M. T., Willits, D., Solis, D. J., Belcher, A. M., Young, M., & Douglas, T. (2005). Bio-inspired synthesis of protein-encapsulated CoPt nanoparticles. *Advanced Functional Materials*, *15*(9), 1489–1494. doi:10.1002/adfm.200400453

Knorr, N., & Vinzelberg, S. (2012). Charge writing and detection by EFM and KPFM scanning probe techniques. *Microscopy and Analysis*, *26*(5), 7–12.

Kobayashi, Y., Ke, X., Hata, H., Schiffer, P., & Mallouk, T. E. (2008). Soft chemical conversion of layered double hydroxides to superparamagnetic spinel platelets. *Chemistry of Materials*, *20*(6), 2374–2381. doi:10.1021/cm703443q

Kolodney, E., Tsipinyuk, B., & Budrevich, A. (1994). The thermal stability and fragmentation of C_{60} molecule up to 2000 K on the milliseconds time scale. *The Journal of Chemical Physics*, *100*(11), 8542–8545. doi:10.1063/1.466755

Korobov, M. V., & Sidorov, L. N. (1994).. . *Zhurnal Khimicheskoi Termodinamiki*, *26*, 61–61.

Kowshik, M., Deshmukh, N., Vogel, W., Urban, J., Kulkarni, S. K., & Paknikar, K. M. (2002). Microbial synthesis of semiconductor CdS nanoparticles, their characterization, and their use in the fabrication of an ideal diode. *Biotechnology and Bioengineering*, *78*(5), 583–588. doi:10.1002/bit.10233 PMID:12115128

Kozek, K. A., Kozek, K. M., Wu, W.-C., Mishra, S. R., & Tracy, J. B. (2013). Large-scale synthesis of gold nanorods through continuous secondary growth. *Chemistry of Materials*, *25*(22), 4537–4544. doi:10.1021/cm402277y PMID:24415848

Kramer, R. M., Li, C., Carter, D. C., Stone, M. O., & Naik, R. R. (2004). Engineered protein cages for nanomaterial synthesis. *Journal of the American Chemical Society*, *126*(41), 13282–13286. doi:10.1021/ja046735b PMID:15479082

Kreibig, U., & Vollner, M. (1995). Optical Properties of Metal Clusters. Berlin: Springer. doi:10.1007/978-3-662-09109-8

Krishnan, A., Dujardin, E., Treacy, M. M. J., Hugdahl, J., Lynum, S., & Ebbesen, T. W. (1997). Photoisomerization in dendrimers by harvesting of low-energy photons. *Nature*, *388*(6641), 451–454. doi:10.1038/41284

Kroto, H. W. (1987). The stability of the fullerenes C_n, with $n = 24, 28, 32, 36, 50, 60$ and 70. *Nature*, *329*(6139), 529–531. doi:10.1038/329529a0

Ksibi, M., Zemzemi, A., & Boukchina, R. (2003). Photocatalytic degradability of substituted phenols over UV irradiated TiO_2. *Journal of Photochemistry and Photobiology A Chemistry*, *159*(1), 61–70. doi:10.1016/S1010-6030(03)00114-X

Kumar, D. N. T., & Wei, Q. (2013). Analysis of quantum dots for nano–bio applications as the technological platform of the future. *Research Journal of Biotechnology*, *8*(5), 78–82.

Kumari, K., Kumar, V., & Singh, K. (2014). Non-lithographic fabrication of Ni-Se heterojunction nanowires and their electrical characterization. *Advances in Research*, *2*, 332-337.

Kuykendall, T., Ulrich, P., Aloni, S., & Yang, P. (2007). Complete composition tunability of InGaN nanowrires using a combinatorial approach. *Nature Materials*, *6*(12), 951–956. doi:10.1038/nmat2037 PMID:17965718

Kuznetsov, V. N., & Serpone, N. (2006). Visible light absorption by various titanium dioxide specimens. *The Journal of Physical Chemistry B*, *110*(50), 25203–25209. doi:10.1021/jp064253b PMID:17165964

Kuznetsov, V. N., & Serpone, N. (2007). Photoinduced coloration and photobleaching of titanium dioxide in TiO_2/polymer compositions upon UV and visible-light excitation of color centers' absorption bands: Direct experimental evidence negating band gap narrowing in anion-/cation-doped TiO_2. *The Journal of Physical Chemistry C*, *111*(42), 15277–15288. doi:10.1021/jp073511h

Lakowicz, J. R. (1999). *Principles of Fluorescence Spectroscopy*. New York, NY: Kluwer Academic. doi:10.1007/978-1-4757-3061-6

Lau, K. K. S., Bico, J., Teo, K. B. K., Chhowalla, M., Amaratunga, G. A. J., Milne, W. I., & Gleason, K. K. et al. (2003). Superhydrophobic carbon nanotube forests. *Nano Letters*, *3*(12), 1701–1705. doi:10.1021/nl034704t

Law, M., Greene, L. E., Johnson, J. C., Saykally, R., & Yang, P. (2005). Nanowire dye-sensitized solar cells. *Nature Materials*, *4*(6), 455–459. doi:10.1038/nmat1387 PMID:15895100

Le Corre, D., Bras, J., & Dufresne, A. (2010). Starch nanoparticles: A review. *Biomacromolecules*, *11*(5), 1139–1153. doi:10.1021/bm901428y PMID:20405913

Lee, J., Mahendra, S., & Alvarez, P. J. J. (2010). Nanomaterials in the construction industry: A review of their applications and environmental health and safety considerations. *ACS Nano*, *4*(7), 3580–3590. doi:10.1021/nn100866w PMID:20695513

Legrini, O., Oliveros, E., & Braun, A. M. (1993). Photochemical processes for water treatment. *Chemical Reviews*, *93*(2), 671–698. doi:10.1021/cr00018a003

Leite, F. L., Bueno, C. C., Da Róz, A. L., Ziemath, E. C., & Oliveira, O. N. Jr. (2012). Theoretical models for surface forces and adhesion and their measurement using atomic force microscopy. *International Journal of Molecular Sciences*, *13*(12), 12773–12856. doi:10.3390/ijms131012773 PMID:23202925

Leszczynski, J., & Shukla, M. K. (Eds.). (2009). *Practical aspects of computational chemistry*. Berlin: Springer.

Létard, J. F., Guionneau, P., Nguyen, O., Sánchez Costa, J., Marcén, S., Chastanet, G., & Goux-Capes, L. et al. (2005). A guideline to the design of molecular-based materials with long-lived photomagnetic lifetimes. *Chemistry (Weinheim an der Bergstrasse, Germany)*, *11*(16), 4582–4589. doi:10.1002/chem.200500112 PMID:15861388

Levy, N., Burke, S. A., Meaker, K. L., Panlasigui, M., Zettl, A., Guinea, F., & Crommie, M. F. et al. (2010). Strain-induced pseudo-magnetic fields greater than 300 tesla in graphene nanobubbles. *Science*, *329*(5991), 544–547. doi:10.1126/science.1191700 PMID:20671183

Li, M., Wong, K. K. W., & Mann, S. (1999). Organization of inorganic nanoparticles using biotin–streptavidin connectors. *Chemistry of Materials*, *11*(1), 23–26. doi:10.1021/cm980610m

Li, X., Liu, L., Qin, Y., Wu, W., Guo, Z. X., Dai, L., & Zhu, D. (2003). C_{60} modified single-walled carbon nanotubes. *Chemical Physics Letters*, *377*(1-2), 32–36. doi:10.1016/S0009-2614(03)01088-1

Li, Z., Li, J., & Luo, X. (2013). Application of phosphonic acid self-assembled monolayer in organic field-effect transistors. *Applied Surface Science*, *282*, 487–491. doi:10.1016/j.apsusc.2013.05.158

Linke, H. (2002). Special issue on Ratchets and Brownian Motors: Basics, Experiments and Applications. *Applied Physics. A, Materials Science & Processing*, *75*(2), 167–354. doi:10.1007/s003390201401

Linke, H., Alemán, B. J., Melling, L. D., Taormina, M. J., Francis, M. J., Dow-Hygelund, C. C., & Stout, A. et al. (2006). Self-propelled Leidenfrost droplets. *Physical Review Letters*, *96*(15), 154502–1–4. doi:10.1103/PhysRevLett.96.154502 PMID:16712160

Liu, I., Niu, T.-S., Zhang, L., & Yang, J.-S. (2010). Review on nano-drugs. *Natural Science*, *2*(01), 41–48. doi:10.4236/ns.2010.21006

Liu, N., & Giessen, H. (2010). Coupling effect in optical metamaterials. *Angewandte Chemie International Edition*, *49*(51), 9838–9852. doi:10.1002/anie.200906211 PMID:21154486

Liu, X. G., Li, B., Geng, D. Y., Cui, W. B., Yang, F., Xie, Z. G., & Zhang, Z. D. et al. (2009). (Fe, Ni)/C nanocapsules for electromagnetic-wave-absorber in the whole Ku-band. *Carbon*, *47*(2), 470–474. doi:10.1016/j.carbon.2008.10.028

Liu, Z., Sun, X., Nakayama-Ratchford, N., & Dai, H. (2007). Supramolecular chemistry on water-soluble carbon nanotubes for drug loading and delivery. *ACS Nano*, *1*(1), 50–56. doi:10.1021/nn700040t PMID:19203129

Longfeng, L., Xiangdong, M., Maolin, Z., Guiying, L., & Taicheng, A. (2010). Low-temperature synthesis of TiO_2 photo catalyst in homogeneous hydrolysis system for dye degradation. *Research Journal of Chemistry and Environment*, *14*(4), 40–43.

Love, J. C., Estroff, L. A., Kriebel, J. K., Nuzzo, R. G., & Whitesides, G. M. (2005). Self-assembled monolayers of thiolates on metals as a form of nanotechnology. *Chemical Reviews*, *105*(4), 1103–1169. doi:10.1021/cr0300789 PMID:15826011

Machado, M., Ordejón, P., Sánchez-Portal, D., Artacho, E., & Soler, J. M., (submitted for publication). Density functional calculations of planar DNA base-pairs. *Journal of Physical Chemistry B*.

Mailänder, V., & Landfester, K. (2009). Interaction of nanoparticles with cells. *Biomacromolecules*, *10*(9), 2379–2400. doi:10.1021/bm900266r PMID:19637907

Makiura, R., Motoyama, S., Umemura, Y., Yamanaka, H., Sakata, O., & Kitigawa, H. (2010). Surface nano-architecture of a metal–organic framework. *Nature Materials*, *9*(7), 565–571. doi:10.1038/nmat2769 PMID:20512155

Manuel, P. (2010). Computational aspects of carbon and boron nanotubes. *Molecules (Basel, Switzerland)*, *15*(12), 8709–8722. doi:10.3390/molecules15128709 PMID:21119566

Marin, M. L., McGilvray, K. L., & Scaiano, J. C. (2008). Photochemical strategies for the synthesis of gold nanoparticles from Au(III) and Au(I) using photoinduced free radical generation. *Journal of the American Chemical Society*, *130*(49), 16572–16584. doi:10.1021/ja803490n PMID:19049456

Martí-Gastaldo, C., Antypov, D., Warren, J. E., Briggs, M. E., Chater, P. A., Wiper, P. V., & Rosseinsky, M. J. et al. (2014). Side-chain control of porosity closure in single- and multiple-peptide-based porous materials by cooperative folding. *Nature Chemistry*, *6*(4), 343–351. doi:10.1038/nchem.1871 PMID:24651203

Martin, C. R., & Kohli, P. (2003). The emerging field of nanotube biotechnology. *National Review*, *2*, 29–37. PMID:12509757

Martin, Y., & Wickramasinghe, H. K. (1987). Magnetic imaging by *force microscopy* with 1000Å resolution. *Applied Physics Letters*, *50*(20), 1455–1457. doi:10.1063/1.97800

Martínez, V., Boldog, I., Gaspar, A. B., Ksenofontov, V., Bhattacharjee, A., Gütlich, P., & Real, J. A. (2010). Spin crossover phenomenon in nanocrystals and nanoparticles of $[Fe(3\text{-}Fpy)_2M(CN)_4]$ (M^{II} = Ni, Pd, Pt) two-dimensional coordination polymers. *Chemistry of Materials*, *22*(14), 4271–4281. doi:10.1021/cm101022u

Martínez-Martín, D., Jaafar, M., Pérez, R., Gómez-Herrero, J., & Asenjo, A. (2010). Upper bound for the magnetic force gradient in graphite. *Physical Review Letters*, *105*(25), 257203–1–4. doi:10.1103/PhysRevLett.105.257203 PMID:21231621

Martins, S., Sarmento, B., Ferreira, D. C., & Souto, E. B. (2007). Lipid-based colloidal carriers for peptide and protein delivery – liposomes versus lipid nanoparticles. *International Journal of Nanomedicine*, *2*, 595–607. PMID:18203427

Mathews, C. K., Baba, M. S., Narasimhan, T. S. L., Balasubramanian, R., Sivaraman, N., Srinivasan, T. G., & Rao, P. R. V. (1992). Vaporization studies on buckminsterfullerene. *Journal of Physical Chemistry*, *96*(9), 3566–3568. doi:10.1021/j100188a002

McGilvray, K. L., Decan, M. R., Wang, D., & Scaiano, J. C. (2006). Facile photochemical synthesis of unprotected aqueous gold nanoparticles. *Journal of the American Chemical Society*, *128*(50), 15980–15981. doi:10.1021/ja066522h PMID:17165719

Medintz, I. L., Clapp, A. R., Mattoussi, H., Goldman, E. R., Fisher, B., & Mauro, J. M. (2003). Self-assembled nanoscale biosensors based on quantum dot FRET donors. *Nature Materials*, *2*(9), 630–638. doi:10.1038/nmat961 PMID:12942071

Meldrum, F. C., Heywood, B. R., & Mann, S. (1992). Magnetoferritin: *In vitro* synthesis of a novel magnetic protein. *Science*, *257*(5069), 522–523. doi:10.1126/science.1636086 PMID:1636086

Meng, T., Wang, C. Y., & Wang, S. Y. (2008). First-principles study of a hybrid carbon material: Imperfect fullerenes covalently bonded to defective single-walled carbon nanotubes. *Physical Review B*, *77*, 33415.

Menkiti, M. C., & Onukwuli, O. D. (2010). Coag-flocculation studies of *Moringa oleifera* coagulant (MOC) in brewery effluent: Nephelometric approach. *Journal of American Science*, *6*(12), 788–806.

Meyer, J. C., Geim, A. K., Katsnelson, M. I., Novoselov, K. S., Booth, T. J., & Roth, S. (2007). The structure of suspended graphene sheets. *Nature*, *446*(7131), 60–63. doi:10.1038/nature05545 PMID:17330039

Minton, A. P. (2000). Effects of excluded surface area and adsorbate clustering on surface adsorption of proteins I. Equilibrium models. *Biophysical Chemistry*, *86*(2-3), 239–247. doi:10.1016/S0301-4622(00)00151-4 PMID:11026688

Minton, A. P. (2001). Effects of excluded surface area and adsorbate clustering on surface adsorption of proteins. II. Kinetic models. *Biophysical Journal*, *80*(4), 1641–1648. doi:10.1016/S0006-3495(01)76136-X PMID:11259279

Mirkin, C. A., Letsinger, R. L., Mucic, R. C., & Storhoff, J. J. (1996). A DNA-based method for rationally assembling nanoparticles into macroscopic materials. *Nature*, *382*(6592), 607–609. doi:10.1038/382607a0 PMID:8757129

Miyamachi, T., Gruber, M., Davesne, V., Bowen, M., Boukari, S., Joly, L., Scheurer, F., Rogez, G., Yamada, T. K., Ohresser, P., Beaurepaire, E., & Wulfhekel, W. (2012). Robust spin crossover and memristance across a single molecule. *Nature Communications, 3*, 938–1-6.

Mkrtchyan, G. S., & Shmidt, V. V. (1972). Interaction between a cavity and a vortex in a superconductor of the second kind. *Soviet Physics, JETP*, *34*, 195–197.

Molina-Sánchez, A., Segura-Ruiz, J., Garro, N., García-Cristóbal, A., Cantarero, A., Iikawa, F., & Rizzi, A. et al. (2012). Inhomogeneous electron distribution in InN nanowires: Influence on the optical properties. *Physica Status Solidi*, *9*(c), 1001–1004.

Mondal, A., Li, Y., Herson, P., Seuleiman, M., Boillot, M. L., Rivière, E., & Lescouëzec, R. et al. (2012). Photomagnetic effect in a cyanide-bridged mixed-valence $\{Fe^{II}_2Fe^{III}_2\}$ molecular square. *Chemical Communications*, *48*(45), 5653–5655. doi:10.1039/c2cc17835d PMID:22550634

Mondal, A., Li, Y., Seuleiman, M., Julve, M., Toupet, L., Buron-Le Cointe, M., & Lescouëzec, R. (2013). On/off photoswitching in a cyanide-bridged $\{Fe_2Co_2\}$ magnetic molecular square. *Journal of the American Chemical Society*, *135*(5), 1653–1656. doi:10.1021/ja3087467 PMID:23321056

Moniruzzaman, M., & Winey, K. I. (2006). Polymer nanocomposites containing carbon nanotubes. *Macromolecules*, *39*(16), 5194–5205. doi:10.1021/ma060733p

Moradi, A. (2010). Theory of carbon nanotubes as optical nano waveguides. *Journal of Electromagnetic Analysis and Applications*, *2*(12), 672–676. doi:10.4236/jemaa.2010.212088

Morant-Miñana, M. C. (personal communication).

Muhlen, C., Elverfeldt, D., Bassler, N., Neudorfer, I., Steitz, B., Petri-Fink, A., & Peter, K. et al. (2007). Superparamagnetic iron oxide binding and uptake as imaged by magnetic resonance is mediated by the integrin receptor Mac-1 (CD11b/CD18): Implications on imaging of atherosclerotic plaques. *Atherosclerosis*, *193*(1), 102–111. doi:10.1016/j.atherosclerosis.2006.08.048 PMID:16997307

Mukhina, M. V., Maslov, V. G., Baranov, A. V., Artemyev, M. V., & Fedorov, A. V. (2013). Anisotropic absorption of CdSe/ZnS quantum rods embedded in polymer film. *Advances in Nano Research*, *1*, 153-158.

Müller-Meskamp, L., Karthäuser, S., Zandvliet, H. J. W., Homberger, M., Simon, U., & Waser, R. (2009). Field-emission resonances at tip/α,ω-mercaptoalkyl ferrocene/Au interfaces studied by STM. *Small, 5*(4), 496–502. doi:10.1002/smll.200800802 PMID:19197965

Müller-Meskamp, L., Karthäuser, S., Zandvliet, H. J. W., Homberger, M., Simon, U., & Waser, R. (2010). Field emission resonances at tip/mercaptoalkylferrocene/Au interfaces. *E-Nano Newsletter*, (19), 28-29.

Murphy, C. J., & Coffer, J. L. (2002). Quantum dots: A primer. *Applied Spectroscopy*, *56*(1), 16A–27A. doi:10.1366/0003702021954214

Murphy, C. J., Thompson, L. B., Alkilany, A. M., Sisco, P. N., Boulos, S. P., Sivapalan, S. T., & Huang, J. et al. (2010). The many faces of gold nanorods. *The Journal of Physical Chemistry Letters*, *1*(19), 2867–2875. doi:10.1021/jz100992x

Nacci, A., & Cioffi, N. (2011). Special issue: Nano-catalysts and nano-technologies for green organic synthesis. *Molecules (Basel, Switzerland)*, *16*(12), 1452–1453. doi:10.3390/molecules16021452 PMID:21307822

Nair, R. R., Ren, W., Jalil, R., Riaz, I., Kravets, V. G., Britnell, L., & Geim, A. K. et al. (2010). Fluorographene: A two-dimensional counterpart of Teflon. *Small*, *6*(24), 2877–2884. doi:10.1002/smll.201001555 PMID:21053339

Nasibulin, A. G., Anisimov, A. S., Pikhitsa, P. V., Jiang, H., Brown, D. P., Choi, M., & Kauppinen, E. I. (2007a). Investigations of NanoBud formation. *Chemical Physics Letters*, *446*(1-3), 109–114. doi:10.1016/j.cplett.2007.08.050

Nasibulin, A. G., Pikhitsa, P. V., Jiang, H., Brown, D. P., Krasheninnikov, A. V., Anisimov, A. S., & Kauppinen, E. I. et al. (2007b). A novel hybrid carbon material. *Nature Nanotechnology*, *2*(3), 156–161. doi:10.1038/nnano.2007.37 PMID:18654245

Nasibulin, A. G., Shandakov, S. D., Anisimov, A. S., Gonzalez, D., Jiang, H., Pudas, M., & Kauppinen, E. I. et al. (2008). Charging of aerosol products during ferrocene vapor decomposition in N_2 and CO atmospheres. *The Journal of Physical Chemistry C*, *112*(15), 5762–5769. doi:10.1021/jp7118026

Nel, A., Zhao, Y., & Mädler, L. (2013). Environmental health and safety considerations for nanotechnology. *Accounts of Chemical Research*, *46*(3), 605–606. doi:10.1021/ar400005v PMID:23964654

Neu, J. C., Cañizo, J. A., & Bonilla, L. L. (2002). Three eras of micellization. *Phys. Rev. E, 66*, 61406.

Neuberger, T., Schöpf, B., Hofmann, H., Hofmann, M., & Rechenberg, B. (2005). Superparammagnetic nanoparticles for biomedical applications. Possibilities and limitations of a new drug delivery system. *Journal of Magnetism and Magnetic Materials*, *293*(1), 483–496. doi:10.1016/j.jmmm.2005.01.064

Niemeyer, C. M. (2001). Nanoparticles, proteins, and nucleic acids: Biotechnology meets materials science. *Angewandte Chemie International Edition*, *40*(22), 4128–4158. doi:10.1002/1521-3773(20011119)40:22<4128::AID-ANIE4128>3.0.CO;2-S

Niemeyer, C. M., & Mirkin, C. A. (Eds.). (2004). Nanobiotechnlogy: Concepts, Applications and Perspectives. Weinheim: Wiley-VCH. doi:10.1002/3527602453

Noginov, M. A., Zhu, G., Belgrave, A. M., Bakker, R., Shalaev, V. M., Narimanov, E. E., & Wiesner, U. et al. (2009). Demonstration of a spaser-based nanolaser. *Nature*, *460*(7259), 1110–1112. doi:10.1038/nature08318 PMID:19684572

Notman, R., Noro, M., O'Malley, B., & Anwar, J. (2006). Molecular basis for dimethylsulfoxide (DMSO) action on lipid membranes. *Journal of the American Chemical Society*, *128*(43), 13982–13983. doi:10.1021/ja063363t PMID:17061853

Novoselov, K. S., Geim, A. K., Morozov, S. V., Jiang, D., Zhang, Y., Dubonos, S. V., & Firsov, A. A. et al. (2004). Electric field effect in atomically thin carbon films. *Science*, *306*(5696), 666–669. doi:10.1126/science.1102896 PMID:15499015

Novoselov, K. S., Jiang, D., Schedin, F., Booth, T. J., Khotkevich, V. V., Morozov, S. V., & Geim, A. K. (2005). Two-dimensional atomic crystals. *Proceedings of the National Academy of Sciences of the United States of America*, *102*(30), 10451–10453. doi:10.1073/pnas.0502848102 PMID:16027370

Oberdörster, G., Maynard, A., Donaldson, K., Castranova, V., Fitzpatrick, J., Ausman, K., … Yang, H. (2005). A report from the ILSI Research Foundation/Risk Science Institute Nanomaterial Toxicity Screening Working Group: Particle and fibre toxicology. *Particle and Fibre Toxicology, 2*, 8.

Okamoto, K., Iyi, N., & Sasaki, T. (2007). Factors affecting the crystal size of the MgAl-LDH (layered double hydroxide) prepared by using ammonia-releasing reagents. *Applied Clay Science*, *37*(1-2), 23–31. doi:10.1016/j.clay.2006.10.008

Oliva, J. M., Allan, N. L., Schleyer, P. R., Viñas, C., & Teixidor, F. (2005). Strikingly long C···C distances in 1,2-disubstituted *ortho*-carboranes and their dianions. *Journal of the American Chemical Society*, *127*(39), 13538–13547. doi:10.1021/ja052091b PMID:16190717

Oliva, J. M., Klein, D. J., Schleyer, P. R., & Serrano-Andrés, L. (2009a). Design of carborane molecular architectures with electronic structure computations: From endohedral and polyradical systems to multidimensional networks. *Pure and Applied Chemistry*, *81*(4), 719–729. doi:10.1351/PAC-CON-08-09-18

Oliva, J. M., Serrano-Andrés, L., Havlas, Z., & Michl, J. (2009b). On the electronic structure of a dianion, a radiacal anion, and a neutral biradical $(HB)_{11}C{-}C{\equiv}C{-}C(BH)_{11}$ carborane dimer. *Journal of Molecular Structure THEOCHEM*, *912*(1-3), 13–20. doi:10.1016/j.theochem.2009.01.033

Oppenländer, T. (2003). Photochemical purification of water and air. Weinheim: Wiley-VCH.

Ordejón, P., Artacho, E., & Soler, J. M. (1996). Mixed approach to incorporate self-consistency into order-N LCAO methods. *Materials Research Society Symposium Proceedings*, *408*, 85-90.

Ortega-Villar, N., Thompson, A. L., Muñoz, M. C., Ugalde-Saldívar, V. M., Goeta, A. E., Moreno-Esparza, R., & Real, J. A. (2005). Solid- and solution-state studies of the novel μ-dicyanamide-bridged dinuclear spin-crossover system $\{[(Fe(bztpen)]_2[\mu\text{-}N(CN)_2]\}(PF_6)_3{\cdot}nH_2O$. *Chemistry (Weinheim an der Bergstrasse, Germany)*, *11*(19), 5721–5734. doi:10.1002/chem.200500171 PMID:16028299

Ozin, G. A., & Arsenault, A. C. (2005). *Nanochemistry*. Cambridge, UK: The Royal Society of Chemistry.

Pagona, G., Fan, J., Maignè, A., Yudasaka, M., Iijima, S., & Tagmatarchis, N. (2007b). Aqueous carbon nanohorn–pyrene–porphyrin nanoensembles: Controlling charge-transfer interactions. *Diamond and Related Materials*, *16*(4-7), 1150–1153. doi:10.1016/j.diamond.2006.11.071

Pagona, G., Mountrichas, G., Rotas, G., Karousis, N., Pispas, S., & Tagmatarchis, N. (2009). Properties, applications and functionalisation of carbon nanohorns. *International Journal of Nanotechnology*, *6*(1/2), 176–195. doi:10.1504/IJNT.2009.021715

Pagona, G., Sandanayaka, A. S. D., Araki, Y., Fan, J., Tagmatarchis, N., Charalambidis, G., & Ito, O. et al. (2007a). Covalent functionalization of carbon nanohorns with porphyrins: Nanohybrid formation and photoinduced electron and energy transfer. *Advanced Functional Materials*, *17*(10), 1705–1711. doi:10.1002/adfm.200700039

Pagona, G., Sandanayaka, A. S. D., Araki, Y., Fan, J., Tagmatarchis, N., Yudasaka, M., & Ito, O. et al. (2006b). Electronic interplay on illuminated aqueous carbon nanohorn–porphyrin ensembles. *The Journal of Physical Chemistry B*, *110*(42), 20729–20732. doi:10.1021/jp064685m PMID:17048875

Pagona, G., Tagmatarchis, N., Fan, J., Yudasaka, M., & Iijima, S. (2006a). Cone-end functionalization of carbon nanohorns. *Chemistry of Materials*, *18*(17), 3918–3920. doi:10.1021/cm0604864

Park, Y. K., Bold, B., Lee, W. K., Jeon, M. H., An, K. H., Jeong, S. Y., & Shim, Y. K. (2011). D-(+)-Galactose-conjugated single-walled carbon nanotubes as new chemical probes for electrochemical biosensors for the cancer marker galectin-3. *International Journal of Molecular Sciences*, *12*(12), 2946–2957. doi:10.3390/ijms12052946 PMID:21686160

Peelle, B. R., Krauland, E. M., Wittrup, K. D., & Belcher, A. M. (2005). Probing the interface between biomolecules and inorganic materials using yeast surface display and genetic engineering. *Acta Biomaterialia*, *1*(2), 145–154. doi:10.1016/j.actbio.2004.11.004 PMID:16701791

Pendry, J. B., Holden, A. J., Stewart, W. J., & Youngs, I. (1996). Extremely low frequency plasmons in metallic mesostructures. *Physical Review Letters*, *76*(25), 4773–4776. doi:10.1103/PhysRevLett.76.4773 PMID:10061377

Perdew, J. P., Burke, K., & Ernzerhof, M. (1996). Generalized gradient approximation made simple. *Physical Review Letters*, *77*(18), 3865–3868. doi:10.1103/PhysRevLett.77.3865 PMID:10062328

Perrier, A., Maurel, F., Ciofini, I., & Jacquemin, D. (2011). A theoretical spectroscopy investigation of bifunctional platinum-bridged diarylethenes. *Chemical Physics Letters*, *502*(1-3), 77–81. doi:10.1016/j.cplett.2010.12.044

Petta, J. R., Slater, S. K., & Ralph, D. C. (2004). Spin-dependent transport in molecular tunnel junctions. *Physical Review Letters*, *93*(13), 136601–1–4. doi:10.1103/PhysRevLett.93.136601 PMID:15524747

Phillips, W. A. (1972). Tunneling states in amorphous solids. *Journal of Low Temperature Physics*, *7*(3-4), 351–360. doi:10.1007/BF00660072

Pignataro, B. (Ed.). (2010a). Tomorrow's Chemistry Today. Weinheim: Wiley-VCH.

Pignataro, B. (Ed.). (2010b). Ideas in Chemistry and Molecular Sciences: Advances in Nanotechnology, Materials and Devices. Weinheim: Wiley-VCH.

Pino, E., & Encinas, M. V. (2012). Photocatalytic degradation of chlorophenols on TiO_2-325mesh and TiO_2-P25. An extended kinetic study of photodegradation under competitive conditions. *Journal of Photochemistry and Photobiology A Chemistry*, *242*, 20–27. doi:10.1016/j.jphotochem.2012.05.019

Pong, B. K., Trout, B. L., & Lee, J. Y. (2008). Modified ligand-exchange for efficient solubilization of CdSe/ZnS quantum dots in water: A procedure guided by computational studies. *Langmuir*, *24*(10), 5270–5276. doi:10.1021/la703431j PMID:18412382

Prins, F., Monrabal-Capilla, M., Osorio, E. A., Coronado, E., & Zant, H. S. J. (2011). Room-temperature electrical addressing of a bistable spin-crossover molecular system. *Advanced Materials*, *23*(13), 1545–1549. doi:10.1002/adma.201003821 PMID:21449059

Pushkarova, Y., & Kholin, Y. (2014). A procedure for meaningful unsupervised clustering and its application for solvent classification. *Central European Journal of Chemistry*, *12*(5), 594–603. doi:10.2478/s11532-014-0514-6

Puzyn, T., Leszczynska, D., & Leszczynski, J. (2009). Toward the development of *nano-QSARs*: Advances and challenges. *Small*, *5*(22), 2494–2509. doi:10.1002/smll.200900179 PMID:19787675

Puzyn, T., Rasulev, B., Gajewicz, A., Hu, X., Dasari, T. P., Michalkova, A., & Leszczynski, J. et al. (2011). Using nano-QSAR to predict the cytotoxicity of metal oxide nanoparticles. *Nature Nanotechnology*, *6*(3), 175–178. doi:10.1038/nnano.2011.10 PMID:21317892

Quiñones, R., Raman, A., & Gawalt, E. S. (2007). An approach to differentiating between multi- and monolayers using MALDI–TOF MS. *Surface and Interface Analysis*, *39*(7), 593–600. doi:10.1002/sia.2561

Rallo, R., France, B., Liu, R., Nair, S., George, S., Damoiseaux, R., & Cohen, Y. et al. (2011). Self-organizing map analysis of toxicity-related cell signaling pathways for metal and metal oxide nanoparticles. *Environmental Science & Technology*, *45*(4), 1695–1702. doi:10.1021/es103606x PMID:21250674

Reines, D., Conaway, R. C., & Conaway, J. W. (1999). Mechanism and regulation of transcriptional elongation by RNA polymerase II. *Current Opinion in Cell Biology*, *11*(3), 342–346. doi:10.1016/S0955-0674(99)80047-7 PMID:10395562

Richter, H., Dereux, A., Grilles, J. M., Guillaume, C., Thiry, P. A., Lucas, A. A., & de Hoffmann, E. et al. (1994). Sublimation of pure C60 fullerene and of C60 adsorbed on MgO or graphite powders. *Berichte der Bunsengesellschaft fur Physikalische Chemie*, *98*(10), 1329–1329. doi:10.1002/bbpc.19940981019

Rieger, R., & Müllen, K. (2010). Forever young: Polycyclic aromatic hydrocarbons as model cases for structural and optical studies. *Journal of Physical Organic Chemistry*, *23*, 315–325.

Rivado-Casas, L., Sampedro, D., Campos, P. J., Fusi, S., Zanirato, V., & Olivucci, M. (2009). Fluorenylidene–pyrroline biomimetic light-driven molecular switches. *The Journal of Organic Chemistry*, *74*(13), 4666–4674. doi:10.1021/jo802792j PMID:19485347

Rohrschneider, L. (1966). Eine methode zur chrakterisierung von gaschromatographischen trennflüssigkeiten. *Journal of Chromatography. A*, *22*, 6–22. doi:10.1016/S0021-9673(01)97064-5

Roubeau, O., Alcazar Gomez, J. M., Balskus, E., Kolnaar, J. J. A., Haasnoot, J. G., & Reedijk, J. (2001). Spin-transition behaviour in chains of FeII bridged by 4-substituted 1,2,4-triazoles carrying alkyl tails. *New Journal of Chemistry*, *25*(1), 144–150. doi:10.1039/b007094g

RSNA. (2005). Nanoparticles show promise in cancer detection and treatment. *RSNA News*, (10), 3-5.

Rubesova, E., Berger, F., Wendland, M., Hong, K., Stevens, K., Gooding, C., & Lang, P. (2002). Gd-labeled liposomes for monitoring liposome-encapsulated chemotherapy: Quantification of regional uptake in tumor and effect on drug delivery. *Academic Radiology*, *9*(2), S525–S527. doi:10.1016/S1076-6332(03)80283-0 PMID:12188328

Rugar, D., Mamin, H. J., Guethner, P., Lambert, S. E., Stern, J. E., McFadyen, I., & Yogi, T. (1990). Magnetic force microscopy: General principles and application to longitudinal recording media. *Journal of Applied Physics*, *68*(3), 1169–1183. doi:10.1063/1.346713

Saleh, S., Taha, M. O., Haddadin, R. N., Marzooqa, D., & Hodali, H. (2011). Preparation of silver- and zinc-doped mullite-based ceramics showing anti-bacterial biofilm properties. *Molecules (Basel, Switzerland)*, *16*(12), 2862–2870. doi:10.3390/molecules16042862

Sampedro, D., Migani, A., Pepi, A., Busi, E., Basosi, R., Latterini, L., & Olivucci, M. et al. (2004). Design and photochemical characterization of a biomimetic light-driven Z/E switcher. *Journal of the American Chemical Society*, *126*(30), 9349–9359. doi:10.1021/ja038859e PMID:15281826

Sánchez Costa, J., Lappalainen, K., de Ruiter, G., Quesada, M., Tang, J., Mutikainen, I., & Reedijk, J. et al. (2007). Remarkable steric effects and influence of monodentate axial ligands L on the spin-crossover properties of *trans*-[Fe^{II}(N_4 ligand)L] coplexes. *Inorganic Chemistry*, *46*(10), 4079–4089. doi:10.1021/ic0624017 PMID:17441713

Sánchez-Portal, D., Ordejón, P., Artacho, E., & Soler, J. M. (1997). Density-functional method for very large systems with LCAO basis sets. *International Journal of Quantum Chemistry*, *65*, 453–461. doi:10.1002/(SICI)1097-461X(1997)65:5<453::AID-QUA9>3.0.CO;2-V

Sanvito, S. (2010). Molecular spintronics: The rise of spinterface science. *Nature Physics*, *6*(8), 562–564. doi:10.1038/nphys1714

Sato, O., Iyoda, T., Fujishima, A., & Hashimoto, K. (1996). Photoinduced magnetization of a cobalt–iron cyanide. *Science*, *272*(5262), 704–705. doi:10.1126/science.272.5262.704 PMID:8662564

Sato, O., Tao, J., & Zhang, Y. Z. (2007). Control of magnetic properties through external stimuli. *Angewandte Chemie International Edition*, *46*(13), 2152–2187. doi:10.1002/anie.200602205 PMID:17348074

Scaiano, J. C., Stamplecoskie, K. G., & Hallett-Tapley, G. L. (2012). Photochemical Norrish type I reaction as a tool for metal nanoparticle synthesis: Importance of proton coupled electron transfer. *Chemical Communications*, *48*(40), 4798–4808. doi:10.1039/c2cc30615h PMID:22498952

Schellman, J. A. (1975). Macromolecular binding. *Biopolymers*, *14*(5), 999–1018. doi:10.1002/bip.1975.360140509

Schnakenberg, L. (1977). *Thermodynamic network analysis of biological systems*. Berlin: Springer. doi:10.1007/978-3-642-96394-0

Schuller, J. A., Barnard, E. S., Cai, W., Jun, Y. C., White, J. S., & Brongersma, M. L. (2010). Plamonics for extreme light concentration and manipulation. *Nature Materials*, *9*(3), 193–204. doi:10.1038/nmat2630 PMID:20168343

Schwartz, D. K. (2001). Mechanisms and kinetics of self-assembled monolayer formation. *Annual Review of Physical Chemistry*, *52*(1), 107–137. doi:10.1146/annurev.physchem.52.1.107 PMID:11326061

Schwarz, J. A., Contescu, C. I., & Putyera, K. (Eds.). (2004). *Dekker Encyclopedia of Nanoscience and Nanotechnology*. New York, NY: Marcel Dekker.

Schwarzenbach, G. (1952). Der chelateffekt. *Helvetica Chimica Acta*, *35*(7), 2344–2359. doi:10.1002/hlca.19520350721

Science Special Issue. (2002). Supramolecular chemistry and self-assembly. *Science, 295*, 2395-2421.

Seeman, N. C. (2005). Structural DNA nanotechnology: An overview. *Methods in Molecular Biology (Clifton, N.J.)*, *303*, 143–166. PMID:15923682

Segura-Ruiz, J., Garro, N., Cantarero, A., Denker, C., Malindretos, J., & Rizzi, A. (2009). Optical studies of MBE-grown InN nanocolumns: Evidence of surface electron accumulation. *Physical Review B: Condensed Matter and Materials Physics*, *79*(11), 115305–1–9. doi:10.1103/PhysRevB.79.115305

Segura-Ruiz, J., Martinez-Criado, G., Sans, J. A., Tucoulou, R., Cloetens, P., Snigireva, I., & Cantarero, A. et al. (2011). Direct observation of elemental segregation in InGaN nanowires by X-ray nanoprobe. [RRL]. *Physica Status Solidi*, *5*(3), 95–97. doi:10.1002/pssr.201004527

Segura-Ruiz, J., Molina-Sánchez, A., Garro, N., García-Cristóbal, A., Cantarero, A., Iikawa, F., & Rizzi, A. et al. (2010). Inhomogeneous free-electron distribution in InN nanowires: Photoluminescence excitation experiments. *Physical Review B: Condensed Matter and Materials Physics*, *82*(12), 125319–1–9. doi:10.1103/PhysRevB.82.125319

Seker, U. O. S., & Demir, H. V. (2011). Material binding peptides for nanotechnology. *Molecules (Basel, Switzerland)*, *16*(12), 1426–1451. doi:10.3390/molecules16021426 PMID:21307821

Sepúlveda, B., Angelomé, P. C., Lechuga, L. M., & Liz-Marzán, L. M. (2009). LSPR-based nanobiosensors. *Nano Today*, *4*(3), 244–251. doi:10.1016/j.nantod.2009.04.001

Serpone, N. (2006). Is the band gap of pristine TiO_2 narrowed by anion- and cation-doping of titanium dioxide in second-generation photocatalysts? *The Journal of Physical Chemistry B*, *110*(48), 24287–24293. doi:10.1021/jp065659r PMID:17134177

Serrano-Andrés, L., Klein, D. J., Schleyer, P. R., & Oliva, J. M. (2008). What electronic structures and geometries of carborane mono- and *ortho-*, *meta-*, and *para-*diradicals are preferred? *Journal of Chemical Theory and Computation*, *4*(8), 1338–1347. doi:10.1021/ct800150h PMID:26631709

Serrano-Andrés, L., & Oliva, J. M. (2006). Photochemical window mechanism for controlled atom release in carborane endohedral boxes: Theoretical evidence. *Chemical Physics Letters*, *432*(1-3), 235–239. doi:10.1016/j.cplett.2006.10.077

Seydel, J. K., & Wiese, M. (2002). Drug-Membrane Interactions: Analysis, Drug Distribution, Modeling. Weinheim: Wiley-VCH. doi:10.1002/3527600639

Shanthi, S., & Priya, K. S. (2012). Photo degradation of dyes from their aqueous solutions of their binary mixture, using TiO_2 as the oxidant with different sources of energy. *Journal of Chemistry and Chemical Engineering*, *6*, 951–955.

Sharifi, M., Marschall, R., Wilhelm, M., Wallacher, D., & Wark, M. (2011). Detection of homogeneous distribution of functional groups in mesoporous silica by small angle neutron scattering and in situ adsoprtion of nitrogen or water. *Langmuir*, *27*(9), 5516–5522. doi:10.1021/la2000188 PMID:21480601

Sheetal, O., & Pragati, T. (2010). Kinetics of photocatalytic degradation of methylene blue in a TiO_2 slurry reactor. *Research Journal of Chemistry and Environment*, *14*(4), 9–13.

Shevchenko, V. Y., Madison, A. E., & Shudegov, V. E. (2003). The structural diversity of the nanoworld. *Glass Physics and Chemistry*, *29*(6), 577–582. doi:10.1023/B:GPAC.0000007934.93203.f3

Shilatifard, A., Conaway, J. W., & Conaway, R. C. (1997). Mechanism and regulation of transcriptional elongation and termination by RNA polymerase II. *Current Opinion in Genetics & Development*, *7*(2), 199–204. doi:10.1016/S0959-437X(97)80129-3 PMID:9115429

Shukla, D., Ahearn, W. G., & Farid, S. (2005). Chain amplification in photoreactions of *N*-alkoxypyridinium salts with alcohols: Mechanism and kinetics. *The Journal of Organic Chemistry*, *70*(17), 6809–6819. doi:10.1021/jo050726j PMID:16095300

Shukla, D., Ahearn, W. G., & Farid, S. (2006). Enhancement of chain amplification in photoreactions of *N*-methoxypyridinium salts with alcohols. *Photochemistry and Photobiology*, *82*(1), 146–151. doi:10.1562/2005-06-28-RA-594 PMID:16178662

Singh, P., Gonzalez, M. J., & Manchester, M. (2006). Viruses and their uses in nanotechnology. *Drug Development Research*, *67*(1), 23–41. doi:10.1002/ddr.20064

Slinker, J. D., Muren, N. B., Renfrew, S. E., & Barton, J. K. (2011). DNA charge transport over 34 nm. *Nature Chemistry*, *3*(3), 228–233. doi:10.1038/nchem.982 PMID:21336329

Smirnov, B. M. (1992). Systems of atoms with a short-range interaction. *Uspekhi Fizicheskikh Nauk*, *162*(12), 97–97. doi:10.3367/UFNr.0162.199212b.0097

Smith, B. R., Heverhagen, J., Knopp, M., Schmalbrock, P., Shapiro, J., Shiomi, M., & Lee, S. C. et al. (2007). Localization to atherosclerotic plaque and biodistribution of biochemically derivatized superparamagnetic iron oxide nanoparticles (SPIONs) contrast particles for magnetic resonance imaging (MRI). *Biomedical Microdevices*, *9*(5), 719–727. doi:10.1007/s10544-007-9081-3 PMID:17562181

Snyder, L. R. (1974). Classification of the solvent properties of common liquids. *Journal of Chromatography. A*, *92*(2), 223–230. doi:10.1016/S0021-9673(00)85732-5

Snyder, L. R. (1997). Changing reversed-phase high performance liquid chromatography selectivity: Which variables should be tried first? *Journal of Chromatography. B, Biomedical Sciences and Applications*, *689*(1), 105–115. doi:10.1016/S0378-4347(96)00351-9 PMID:9061486

Som, C., Berges, M., Chaudhry, Q., Dusinska, M., Fernandes, T. F., Olsen, S. I., & Nowack, B. (2010). The importance of life cycle concepts for the development of safe nanoproducts. *Toxicology*, *269*(2-3), 160–169. doi:10.1016/j.tox.2009.12.012 PMID:20025922

Son, S. J., Bai, X., & Lee, S. B. (2007). Inorganic hollow nanoparticles and nanotubes in nanomedicine. Part 1. Drug/gene delivery applications. *Drug Discovery Today*, *12*(15-16), 650–656. doi:10.1016/j.drudis.2007.06.002 PMID:17706547

Stamplecoskie, K. G., Fasciani, C., & Scaiano, J. C. (2012). Dual-stage lithography from a light-driven, plasmon-assisted process: A hierarchical approach to subwavelength features. *Langmuir*, *28*(30), 10957–10961. doi:10.1021/la301728r PMID:22803690

Stamplecoskie, K. G., Pacioni, N. L., Larson, D., & Scaiano, J. C. (2011). Plasmon-mediated photopolymerization maps plasmon fields for silver nanoparticles. *Journal of the American Chemical Society*, *133*(24), 9160–9163. doi:10.1021/ja201139z PMID:21615121

Stamplecoskie, K. G., & Scaiano, J. C. (2010). Light emitting diode irradiation can control the morphology and optical properties of silver nanoparticles. *Journal of the American Chemical Society*, *132*(6), 1825–1827. doi:10.1021/ja910010b PMID:20102152

Stinaff, E., Scheibner, M., Bracker, A., Ponomarev, I., Korenev, V., Ware, M., & Gammon, D. et al. (2006). Optical signatures of coupled quantum dots. *Science*, *311*(5761), 636–639. doi:10.1126/science.1121189 PMID:16410487

Stoecklein, W., Parkin, S. S. P., & Scott, J. C. (1988). Ferromagnetic resonance studies of exchange-biased Permalloy thin films. *Pysical Review B*, *38*(10), 6847–6854. doi:10.1103/PhysRevB.38.6847 PMID:9945364

Strichartz, G. R. (Ed.). (1987). Handbook of experimental pharmacology *Local Anesthetics No. 81*. Heidelberg, Germany: Springer. doi:10.1007/978-3-642-71110-7

Sun, X., Liu, Z., Welsher, K., Robinson, J. T., Goodwin, A., Zaric, S., & Dai, H. (2008). Nano-graphene oxide for cellular imaging and drug delivery. *Nano Research*, *1*(3), 203–212. doi:10.1007/s12274-008-8021-8 PMID:20216934

Sweeney, R. Y., Mao, C., Gao, X., Burt, J. L., Belcher, A. M., Georgiou, G., & Iverson, B. L. (2004). Bacterial biosynthesis of cadmium sulfide nanocrystals. *Chemistry & Biology*, *11*(11), 1553–1559. doi:10.1016/j.chembiol.2004.08.022 PMID:15556006

Tagmatarchis, N., Maigné, A., Yudasaka, M., & Iijima, S. (2006). Functionalization of carbon nanohorns with azomethine ylides: Towards solubility enhancement and electron-transfer processes. *Small*, *2*(4), 490–494. doi:10.1002/smll.200500393 PMID:17193072

Takezawa, N., & Fukushima, K. (1994). Optimal size of a cylindrical insulating inclusion acting as a pinning center for magnetic flux in superconductors. *Physica. C, Superconductivity*, *228*(1-2), 149–159. doi:10.1016/0921-4534(94)90186-4

Takezawa, N., & Fukushima, K. (1997). Optimal size of an insulating inclusion acting as a pinning center for magnetic flux in superconductors: Calculation of pinning force. *Physica. C, Superconductivity, 290*(1-2), 31–37. doi:10.1016/S0921-4534(97)01574-8

Tamura, R., & Tsukada, M. (1995). Electronic states of the cap structure in the carbon nanotube. *Physical Review B: Condensed Matter and Materials Physics, 52*(8), 6015–6026. doi:10.1103/PhysRevB.52.6015 PMID:9981793

Tan, M. C., Chow, G. M., & Ren, L. (Ed.) (2010). Nanostructured Materials for Biomedical Applications. Trivandrum, India: Research Signpost.

Tang, J., Sánchez Costa, J., Smulders, S., Molnár, G., Bousseksou, A., Teat, S. J., & Reedijk, J. et al. (2009). Two-step spin-transition iron(III) compound with a wide [high spin-low spin] plateau. *Inorganic Chemistry, 48*(5), 2128–2135. doi:10.1021/ic801973x PMID:19235971

Taniguchi, N. (1974). *Proceedings of the International Conference on Production Engineering, Part II*. Tokyo: Japan Society of Precision Engineering.

Tenne, R., & Redlich, M. (2010). Recent progress in the research of inorganic fullerene-like nanoparticles and inorganic nanotubes. *Chemical Society Reviews, 39*(5), 1423–1434. doi:10.1039/B901466G PMID:20419198

Teresawa, M., Takezawa, N., Fukushima, K., Mitamura, T., Fan, X., Tsubakino, H., & Tatara, G. et al. (1998). Flux pinning and flux creep in $La_{2-x}Sr_xCuO_4$ with splayed columnar defects. *Physica. C, Superconductivity, 296*(1-2), 57–64. doi:10.1016/S0921-4534(97)01822-4

Thorek, D. L. J., Chen, A. K., Czupryna, J., & Tsourkas, A. (2006). Superparamagnetic iron oxide nanoparticle probes for molecular imaging. *Annals of Biomedical Engineering, 34*(1), 23–28. doi:10.1007/s10439-005-9002-7 PMID:16496086

Tiberj, A., Camara, N., Godignon, P., & Camassel, J. (2011). Micro-Raman and micro-transmission imaging of epitaxial graphene grown on the Si and C faces of 6H-SiC. *Nanoscale Research Letters, 6*, 478.

Tikhonenko, F. V., Horsell, D. W., Gorbachev, R. V., & Savchenko, A. K. (2008). Weak localization in graphene flakes. *Physical Review Letters, 100*, 56802.

Tokmakoff, A., Haynes, D. R., & George, S. M. (1991). Desorption kinetics of C_{60} multilayers from Al_2O_3 (0001). *Chemical Physics Letters, 186*(4-5), 450–455. doi:10.1016/0009-2614(91)90207-P

Torrens, F. (2003). Valence topological charge-transfer indices for dipole moments. *Molecules (Basel, Switzerland), 8*(1), 169–185. doi:10.3390/80100169 PMID:18007514

Torrens, F. (2004a). Periodic table of carbon nanotubes based on the chiral vector. *Internet Electronic Journal of Molecular Design, 3*, 514–527.

Torrens, F. (2004b). Valence topological charge-transfer indices for dipole moments: Percutaneous enhancers. *Molecules (Basel, Switzerland), 9*(12), 1222–1235. doi:10.3390/91201222 PMID:18007514

Torrens, F. (2004c). Fractal dimension of transdermal-delivery drug models. *Lebanese Science Journal, 5*(1), 61–70.

Torrens, F. (2005a). Periodic properties of carbon nanotubes based on the chiral vector. *Internet Electronic Journal of Molecular Design*, *4*, 59–81.

Torrens, F. (2005b). Calculations on cyclopyranoses as co-solvents of single-wall carbon nanotubes. *Molecular Simulation*, *31*(2-3), 107–114. doi:10.1080/08927020412331308494

Torrens, F. (2005c). Calculations on solvents and co-solvents of single-wall carbon nanotubes: Cyclopyranoses. *Journal of Molecular Structure THEOCHEM*, *757*(1-3), 183–191. doi:10.1016/j.theochem.2005.03.023

Torrens, F. (2005d). Calculations on solvents and co-solvents of single-wall carbon nanotubes: Cyclopyranose. *Nanotechnology*, *16*(5), S181–S189. doi:10.1088/0957-4484/16/5/009

Torrens, F. (2005e). Some calculations on single-wall carbon nanotubes. *Problems of Nonlinear Analysis in Engineering Systems*, *11*(2), 1–16.

Torrens, F. (2006a). Calculations of organic-solvent dispersions of single-wall carbon nanotubes. *International Journal of Quantum Chemistry*, *106*(3), 712–718. doi:10.1002/qua.20835

Torrens, F. (2006b). Corrigendum: Effect of type, size and deformation on polarizability of carbon nanotubes from atomic increments. *Nanotechnology*, *17*(5), 1541–1541. doi:10.1088/0957-4484/17/5/C01

Torrens, F., & Castellano, G. (2005). Cluster origin of the solubility of single-wall carbon nanotubes. *Computing Letters*, *1*(4), 331–336. doi:10.1163/157404005776611303

Torrens, F., & Castellano, G. (2006). Periodic classification of local anaesthetics (procaine analogues). *International Journal of Molecular Sciences*, *7*(1), 12–34. doi:10.3390/i8010012

Torrens, F., & Castellano, G. (2007a). Cluster nature of the solvation features of single-wall carbon nanotubes. *Current Research in Nanotechnology*, *1*, 1–29.

Torrens, F., & Castellano, G. (2007b). Effect of packing on the cluster nature of C nanotubes: An information entropy analysis. *Microelectronics Journal*, *38*(12), 1109–1122. doi:10.1016/j.mejo.2006.04.004

Torrens, F., & Castellano, G. (2007c). Cluster origin of the transfer phenomena of single-wall carbon nanotubes. *Journal of Computational and Theoretical Nanoscience*, *4*, 588–603.

Torrens, F., & Castellano, G. (2007d). Asymptotic analysis of coagulation-fragmentation equations of carbon nanotube clusters. *Nanoscale Research Letters*, *2*(7), 337–349. doi:10.1007/s11671-007-9070-8

Torrens, F., & Castellano, G. (2008a). Properties of fullerite and other symmetric carbon forms: Similarity laws. *Symmetry*, *19*, 341–370.

Torrens, F., & Castellano, G. (2008b). Fractal dimension of transdermal-delivery drug models: 4-Alkylanilines. *Journal of Liquid Chromatography & Related Technologies*, *31*(15), 2337–2347. doi:10.1080/10826070802281877

Torrens, F., & Castellano, G. (2009a). Asymptotic analysis of coagulation–fragmentation equations. In F. Columbus (Ed.), *Carbon nanotubes: New research* (pp. 7–16). New York, NY: Nova.

Torrens, F., & Castellano, G. (2009b). Topological charge-transfer indices: From small molecules to proteins. *Current Proteomics*, *6*(4), 204–213. doi:10.2174/157016409789973770

Torrens, F., & Castellano, G. (2010a). Fullerite crystal thermodynamic characteristics and the law of corresponding states. *Journal of Nanoscience and Nanotechnology*, *10*(2), 1208–1222. doi:10.1166/jnn.2010.1858 PMID:20352780

Torrens, F., & Castellano, G. (2010b). Cluster nature of the solvent features of single-wall carbon nanohorns. *International Journal of Quantum Chemistry*, *110*(3), 563–570. doi:10.1002/qua.22054

Torrens, F., & Castellano, G. (2011a). (Co-)solvent selection for single-wall carbon nanotubes: *Best* solvents, acids, superacids and guest–host inclusion complexes. *Nanoscale*, *3*(6), 2494–2510. doi:10.1039/c0nr00922a PMID:21331393

Torrens, F., & Castellano, G. (2011b). Using chemical structural indicators for periodic classification of local anaesthetics. *International Journal of Chemoinformatics and Chemical Engineering*, *1*(2), 15–35. doi:10.4018/ijcce.2011070102

Torrens, F., & Castellano, G. (2012). *Bundlet* model for single-wall carbon nanotubes, nanocones and nanohorns. *International Journal of Chemoinformatics and Chemical Engineering*, *2*(1), 48–98. doi:10.4018/IJCCE.2012010105

Torrens, F., & Castellano, G. (2013a). *Bundlet* model of single-wall carbon, BC_2N and BN nanotubes, cones and horns in organic solvents. *Journal of Nanomaterials & Molecular Nanotechnology, 2*, 1000107.

Torrens, F., & Castellano, G. (2013b). Solvent features of cluster single-wall C, BC_2N and BN nanotubes, cones and horns. *Microelectronic Engineering*, *108*, 127–133. doi:10.1016/j.mee.2013.02.046

Torrens, F., & Castellano, G. (2013c). C-nanostructures cluster models in organic solvents: Fullerenes, tubes, buds and graphenes. *Journal of Chemistry and Chemical Engineering*, *7*, 1026–1035.

Torrens, F., & Castellano, G. (2013d). Elementary polarizability of Sc/*fullerene*/*graphene* aggregates and di/graphene–cation interactions. *Journal of Nanomaterials & Molecular Nanotechnology, S1*, 001.

Torrens, F., & Castellano, G. (2014a). Cluster *bundlet* model of single-wall C, BC_2N and BN nanotubes, cones and horns. In A. G. Mercader, E. A. Castro, & A. K. Haghi (Eds.), *Nanoscience and computational chemistry: Research progress* (pp. 271–307). Waretown, NJ: Apple Academic.

Torrens, F., & Castellano, G. (2014b). Nanostructures cluster models in solution: Extension to C, BC_2N and BN fullerenes, tubes and cones. In M. Khosrow-Pour (Ed.), *Contemporary advancements in information technology development in dynamic environments* (pp. 221–253). Hershey, PA: IGI Global. doi:10.4018/978-1-4666-6252-0.ch012

Torrens, F., & Castellano, G. (2014c). Cluster model expanded to C-nanostructures: Fullerenes, tubes, graphenes and their buds. *Austin Journal of Nanomedicine & Nanotechnology, 2*(2), 7–1-7.

Torrens, F., & Castellano, G. (in press). Solvents and co-solvents of single-wall carbon nanotubes: New best solvents, superacids, nitric acid and guest–host inclusion complexes. In: E. A. Castro (Ed.), New Developments and Applications of QSAR-QSPR Theory. Trivandrum, India: Research Signpost.

Torrens, F., & Castellano, G. (2014d). Cluster solvation models of carbon nanostructures: Extension to fullerenes, tubes and buds. *Journal of Molecular Modeling*, 20, 2263-1-9. PMID:24869779

Türkarslan, Ö., Böyükbayram, A. E., & Toppare, L. (2010). Amperometric alcohol biosensors based on conducting polymers: Polypyrrole, poly(3,4-ethylenedioxythiophene) and poly(3,4-ethylenedioxypyrrole). *Synthetic Metals*, *160*(7-8), 808–813. doi:10.1016/j.synthmet.2010.01.027

Uptain, S. M., Kane, C. M., & Chamberlin, M. J. (1997). Basic mechanisms of transcript elongation and its regulation. *Annual Review of Biochemistry*, *66*(1), 117–172. doi:10.1146/annurev.biochem.66.1.117 PMID:9242904

Vajpeyi, A. P., Ajagunna, A. O., Tsagaraki, K., Androulidaki, M., & Georgakilas, A. (2009). InGaN nanopillars grown on silicon substrate using plasma assisted molecular beam epitaxy. *Nanotechnology*, *20*(32), 325605–1–5. doi:10.1088/0957-4484/20/32/325605 PMID:19620761

Volatron, F., Catala, L., Rivière, E., Gloter, A., Stéphan, O., & Mallah, T. (2008). Spin-crossover coordination nanoparticles. *Inorganic Chemistry*, *47*(15), 6584–6586. doi:10.1021/ic800803w PMID:18590329

Voliani, V., Ricci, F., Signore, G., Nifosi, R., Luin, S., & Beltram, F. (2011). Multiphoton molecular photorelease in click-chemistry-functionalized gold nanoparticles. *Small*, *7*(23), 3271–3275. doi:10.1002/smll.201101753 PMID:22012898

Vorob'ev, V. S., & Eletskii, A. V. (1995).. . *Теплофизика высоких температур*, *33*, 862–862.

Vorob'ev, V. S., & Eletskii, A. V. (1996). Similarity laws and thermodynamic quantities for fullerite C_{60}. *Chemical Physics Letters*, *254*(3-4), 263–267. doi:10.1016/0009-2614(96)00329-6

Wadhavane, P. D., Galian, R. E., Izquierdo, M. A., Aguilera-Sigalat, J., Galindo, F., Schmidt, L., & Luis, S. V. et al. (2012). Photoluminiscence enhancement of CdSe quantum dots: A case of organogel–nanoparticle symbiosis. *Journal of the American Chemical Society*, *134*(50), 20554–20563. doi:10.1021/ja310508r PMID:23214451

Wang, F., Wang, J., & Liu, X. (2010). Direct evidence of a surface quenching effect on size-dependent luminescence of upconversion nanoparticles. *Angewandte Chemie International Edition*, *49*(41), 7456–7460. doi:10.1002/anie.201003959 PMID:20809558

Wang, M., & Li, C. M. (2011). Magnetic properties of all-carbon graphene-fullerene nanobuds. *Physical Chemistry Chemical Physics*, *13*(13), 5945–5951. doi:10.1039/c0cp02433c PMID:21336407

Wang, Q. H., Kalantar-Zadeh, K., Kis, A., Coleman, J. N., & Strano, M. S. (2012). Electronics and optoelectronics of two-dimensional transition metal dichalcogenides. *Nature Nanotechnology*, *7*(11), 699–712. doi:10.1038/nnano.2012.193 PMID:23132225

Wang, W., & Richter, C. A. (2006). Spin-polarized inelastic electron tunneling spectroscopy of a molecular magnetic tunnel junction. *Applied Physics Letters*, *89*, 153105–1–3.

Weber, G. (1975). Energetics of ligand binding to proteins. *Advances in Protein Chemistry*, *29*, 1–83. doi:10.1016/S0065-3233(08)60410-6 PMID:1136898

Wei, L., & Wang, Y. N. (2004). Electromagnetic wave propagation in single-wall carbon nanotubes. *Physics Letters. [Part A]*, *333*(3-4), 303–309. doi:10.1016/j.physleta.2004.10.048

Wen, Y., Xu, J., He, H., Lu, B., Li, Y., & Dong, B. (2009). Electrochemical polymerization of 3,4-ethylenedioxythiphene in aqueous micellar solution containing biocompatible amino acid-based surfactant. *Journal of Electroanalytical Chemistry*, *634*(1), 49–58. doi:10.1016/j.jelechem.2009.07.012

Werner, D., Hashimoto, S., & Uwada, T. (2010). Remarkable photothermal effect of interband excitation on nanosecond laser-induced reshaping and size reduction of pseudospherical gold nanoparticles in aqueous solution. *Langmuir*, *26*(12), 9956–9963. doi:10.1021/la100015t PMID:20210316

Werner, F., Limbach, F., Carsten, N., Denker, C., Malindretos, J., & Rizzi, A. (2009). Electrical conductivity of InN nanowires and the influence of the native indium oxide formed at their surface. *Nano Letters*, *9*(4), 1567–1571. doi:10.1021/nl8036799 PMID:19290610

White, R. J., Luque, R., Budarin, V. L., Clark, J. H., & Macquarrie, D. J. (2009). Supported metal nanoparticles on porous materials. Methods and applications. *Chemical Society Reviews*, *38*(2), 481–494. doi:10.1039/B802654H PMID:19169462

Whitesides, G. M. (2005). Nanoscience, nanotechnology and chemistry. *Small*, *1*(2), 172–179. doi:10.1002/smll.200400130 PMID:17193427

Wickline, S., Neubauer, A., Winter, P., Caruthers, S., & Lanza, G. (2006). Applications of nanotechnology to artherosclerosis, thrombosis, and vascular biology. *Arteriosclerosis, Thrombosis, and Vascular Biology*, *26*(3), 435–441. doi:10.1161/01.ATV.0000201069.47550.8b PMID:16373609

Willets, K. A., & van Duyne, R. P. (2007). Localized surface plasmon resonance spectroscopy and sensing. *Annual Review of Physical Chemistry*, *58*(1), 267–297. doi:10.1146/annurev.physchem.58.032806.104607 PMID:17067281

Willner, I., Willner, B., & Tel-Vered, R. (2011). Electroanalytical applications of metallic nanoparticles and supramolecular nanostructures. *Electroanalysis*, *23*(1), 13–28. doi:10.1002/elan.201000506

Wu, J., El Hamaoui, B., Li, J., Zhi, L., Kolb, U., & Müllen, K. (2005). Solid-state synthesis of "bamboo-like" and straight carbon nanotubes by thermolysis of hexa-*peri*-hexabenzocoronene–cobalt complexes. *Small*, *1*(2), 210–212. doi:10.1002/smll.200400049 PMID:17193432

Wu, X., & Zeng, X. C. (2008). First-principles study of a carbon nanobud. *ACS Nano*, *2*(7), 1459–1465. doi:10.1021/nn800256d PMID:19206315

Wu, X., & Zeng, X. C. (2009). Periodic graphene nanobuds. *Nano Letters*, *9*(1), 250–256. doi:10.1021/nl802832m PMID:19072128

Wuister, S. F., de Mello Donegá, C., & Meijerink, A. (2004). Influence of thiol capping on the exciton liminiscence and decay kinetics of CdTe and CdSe quantum dots. *The Journal of Physical Chemistry B*, *108*(45), 17393–17397. doi:10.1021/jp047078c

Xing, M., Shen, H., Zhao, W., Liu, Y., Du, Y., Yu, Z., & Chen, X. (2011). dsDNA-coated quantum dots. *BioTechniques*, *50*(4), 259–261. PMID:21548911

Xu, L., Pirollo, K., Tang, W., Rait, A., & Chang, E. (1999). Transferrin-liposome-mediated systemic p53 gene therapy in combination with radiation results in regression of human head and neck cancer xenografts. *Human Gene Therapy*, *10*(18), 2941–2952. doi:10.1089/10430349950016357 PMID:10609655

Yu, J., & Wehrly, T. E. (2004). An approach to the residence time distribution for stochastic multi-compartment models. *Mathematical Biosciences*, *191*(2), 185–205. doi:10.1016/j.mbs.2004.06.005 PMID:15363653

Yu, K., & Chen, J. (2014). Graphene-based transparent conductive electrodes. *Material Matters*, *9*(1), 6–13.

Yudasaka, M., Iijima, S., & Crespi, V. H. (2008). Single-wall carbon nanohorns and nanocones. *Topics in Applied Physics*, *111*, 605–629. doi:10.1007/978-3-540-72865-8_19

Zayat, A. V., & Smolyaninov, I. I. (2003). Near-field photonics: Surface plasmon polaritons and localized surface plasmons. *Journal of Optics A*, *5*(4), 16–50. doi:10.1088/1464-4258/5/4/353

Zhang, B. L., Wang, C. Z., Ho, M., & Chan, C. T. (1993). Melting of carbon cages. *Zeitschrift für Physik D*, *26*(S1), 285–285. doi:10.1007/BF01425692

Zhang, K., Zhang, L. L., Zhao, X. S., & Wu, J. (2010). Graphene/polyaniline nanofiber composites as supercapacitor electrodes. *Chemistry of Materials*, *22*(4), 1392–1401. doi:10.1021/cm902876u

Zhang, X., Coleman, A. C., Katsonis, N., Browne, W. R., van Wees, B. J., & Feringa, B. L. (2010). Dispersion of graphene in ethanol using a simple solvent exchange method. *Chemical Communications*, *46*(40), 7539–7541. doi:10.1039/c0cc02688c PMID:20848021

Zhao, M.-Q., Zhang, Q., Huang, J.-Q., & Wei, F. (2012). Hierarchical nanocomposites derived from nanocarbons and layered double hydroxides – Properties, Synthesis, and applications. *Advanced Functional Materials*, *22*(4), 675–694. doi:10.1002/adfm.201102222

Zhao, X., Xu, S., Wang, L., Duan, X., & Zhang, F. (2010). Exchange-biased $NiFe_2O_4$/NiO nanocomposites derived from NiFe-layered double hydroxides as a single precursor. *Nano Research*, *3*(3), 200–210. doi:10.1007/s12274-010-1023-3

Zhou, W., & Fu, W. (2013). Mesoporous TiO_2: Preparation, doping, and as a composite for photocatalysis. *ChemCatChem*, *5*(4), 885–894. doi:10.1002/cctc.201200519

Zhu, J., Kase, D., Shiba, K., Kasuya, D., Yudasaka, M., & Iijima, S. (2003). Binary nanomaterials based on nanocarbons: A case for probing carbon nanohorns' biorecognition properties. *Nano Letters*, *3*(8), 1033–1036. doi:10.1021/nl034266q

Zhu, S., & Xu, G. (2010). Single-walled carbon nanohorns and their applications. *Nanoscale*, *2*(12), 2538–2549. doi:10.1039/c0nr00387e PMID:20957266

Zhu, X., & Su, H. (2009). Magnetism in hybrid carbon nanostructures: Nanobuds. *Physical Review B: Condensed Matter and Materials Physics*, *79*(16), 165401–1–5. doi:10.1103/PhysRevB.79.165401

Zhu, Y., Li, W., Li, Q., Li, Y., Li, Y., Zhang, X., & Huang, Q. (2009). Effects of serum proteins on intracellular uptake and cytotoxicity of carbon nanoparticles. *Carbon*, *47*(5), 1351–1358. doi:10.1016/j.carbon.2009.01.026

Zhu, Y., Zhang, X., Zhu, J., Zhao, Q., Li, Y., Li, W., & Huang, Q. et al. (2012). Cytotoxicity of phenol red in toxicity assays for carbon nanoparticles. *International Journal of Molecular Sciences*, *13*(12), 12336–12348. doi:10.3390/ijms131012336 PMID:23202901

KEY TERMS AND DEFINITIONS

Bundlet Model for Cluster: A representation of a group of cylindrical particles joined by their sides.

Columnlet Model for Cluster: A representation of a stack of flat particles.

Droplet Model for Cluster: A representation of a group of particles as a small round mass of liquid.

Fullerene: Cluster of C_{60} with a truncated icosahedral structure (black-and-white-television soccer ball) formed by 20 hexagonal and 12 pentagonal faces, which resembles a geodesic dome from Buckmisnterfuller.

Nanocone: A conical sub-100nm-sized particle.

Nanographene: A sub-100nm graphitic sheet.

Nanographene Bud: A sub-100nm fullerene-functionalized graphene.

Nanohorn: A low-apex conical sub-100nm-sized particle.

Nanotube: A hollow round sub-100nm-thick pipe.

Nanotube Bud (Nanobud): A sub-100nm-thick fullerene-functionalized pipe.

Chapter 9
Mirroring Nature:
Symmetrical and Crystalline Structures Derived from Natural Forms

Anna Ursyn
University of Northern Colorado, USA

ABSTRACT

This chapter provides discussion of the visual ways of learning basic physical and chemical concepts related to symmetry and the crystalline structures. All kinds of symmetrical structures are present in substances and their various molecular compositions that are researched in fields related to pharmacology. Great part of technologies, methodologies, tools, and applications require knowledge visualization skills to understand and present concepts and processes. Exploration of science-based concepts and nature-related processes supports attaining visualization literacy, which is needed for explaining physical and chemical notions, clinical procedures, and publicizing clinical and mobile medical informatics. This chapter discusses the ways of preparing to this task artists, graphic designers, computer graphics students, as well as people in charge of hospitals, medical centers, and pharmaceutical industries who hire designers. The chapter offers exercises in visualization of scientific concepts by providing two projects about basic science-related themes: (1) Symmetry and pattern in animal world: geometry and art, and (2) Crystals and crystal caves. Each project invites the reader to create visual presentation of the theme.

INTRODUCTION

We cannot overstate the role of knowledge visualization (Bertschi, Bresciani, Crawford, Goebel, Kienreich, Lindner, Sabol, & Moere, 2011) in building trust and cooperation between patients and physicians. Knowledge visualization uses visual representation to transfer knowledge between individuals. It happens often that people who are in charge of hospitals, medical centers, and pharmaceutical industries hire artists to produce visual and time based presentations of particular disease developments and related symptoms and appropriate medical procedures in order to inform patients, prepare them for treatment, and gain their informed cooperation in fighting illness. Researchers and teachers seek ways to visualize often hard-to-capture abstract concepts for better understanding and communication. The designing of

DOI: 10.4018/978-1-5225-0248-7.ch009

assistive technologies, adaptive devices, and rehabilitative appliances, aimed at supporting good condition and improving capabilities of ailing or disabled people, requires of the designers preparation in several domains. Rehabilitation may comprise both physical and cognitive training involving, for example the use of virtual reality (Khetrapal, 2011). Virtual humans offer potential for interactive training; they may also serve for developing diagnostic skills of developing clinicians (Kenny, Parsons, Gratch, Leusli, & Rizzo, 2007). Apart from direct physical actions, the use of "mobile medical informatics, combined with new techniques for discovering patterns in complex clinical situations" (Gasmelseid, 2011) serves as a means for improving clinical practice. Many times physicians and therapists perform some tasks on behalf of patients with the use of software agents or programs, using multiple information and communication platforms (Gasmelseid, 2008). A great part of technologies, methodologies, tools and applications require visualization skills from the part of designers and users. For this reason, visualization literacy gains in importance, both in the area of explaining clinical procedures and publicizing mobile medical informatics.

Artists, graphic designers, and computer graphics students may play an important role in making the ground for informed collaboration with specialists in the medicine and pharmacy domains. This chapter discusses the ways of helping them to prepare to this aim, taking into detailed consideration notions related to symmetry and crystalline structures. Medical illustration is an applied art discipline comprised of professional medical illustrators. Association of Medical Illustrators (2014) is an international organization promoting the power of visual media to advance science and healthcare. The majority of medical illustrators in the profession have a master's degree from an accredited two-year graduate program

Current means of delivering knowledge, for example, technologies involving 3D printing (Mercuri & Meredith, 2014), augmented reality (Bredl, Groß, Hünniger, & Fleischer, 2012), and open source printers may support research in pharmaceutical institutions. 3D printing technologies have already arrived to the school environment (Irwin, Opplinger, Pearce, & Anzalone, 2015); Schelly, Anzalone, Wijnen & Pearce, 2015). In 3D printing, based on the rapid prototyping process, additive processes are used, with successive layers of thermoplastic material laid down according to a computer program (Grujović, Radović, Kanjevac, Borota, Grujović, & Divac, 2011; Ravikumar, Khan, Mohanty, Sageer, & Aigali, 2015).

Basic knowledge about therapeutic procedures and understanding the mechanisms of drug action may support such cooperation and prevent making wrong decisions. For example, children not fully aware about the nature of their sickness and the plan for a treatment might hide pills and then throw them away instead of taking a full series of an antibiotic. Adults often enter a vicious circle when they are suffering because of the drugs' side effects imposed on other organs and so they take painkillers or psychotropic drugs. As a rule, medications have their descriptions attached, printed on their containers and as leaflets. They have the drugs' chemical names included, along with chemical formulas expressed both as the single lines of chemical element symbols and as the structural formulas – graphical representation of spatial relationship between atoms. While this information is understandable to professionals, in many cases it does not help the patients.

Artists and designers are often hired to develop posters, develop or record short movies, and build physical models, which would be later presented to patients. This work may present much difficulty, as several distinct abilities and skills may be needed to carry out the task as required. For this reason art and design students may need some training in scientific visualization and scientific illustration to become prepared for fulfilling the demands of the society and cooperate with medical and pharmaceutical professionals. Scientific visualization deals with physically-based data that are defined, selected, transformed, and represented according to space coordinates, such as geographic data or computer tomography data of a body for medical use (e.g., Acosta, 2012). Abstract or model-based scientific visualizations present real objects in a digital way directly from the data. Art-science cooperative learning projects make knowledge comprehensible to a wide audience. For example, in the TV medical drama "House, M.D." dynamic visualizations presented the interior of human organism. Scientific illustration are made by selecting, transforming, and thus representing the data, often using simulation of experiments, thus advancing analysis, study, and understanding the data. In the hospital environment, "it will give the operator a useful tool for communication to the best for the patient" (Viola, Birkeland, Solteszova, Helljesen, Hauser, Kotopoulis, Nylund, Ulvang, Øye, Hausken, & Gilja, 2013). Technical illustrations are especially useful in communication with a nontechnical audience. Illustrative technical illustrations are often applied in visualization; they employ high level of abstraction. Viola & Gröller (2007) examined expressive visualization techniques that uncover important information through dynamic changes (cut-away views), deformations (ghosted views), or spatial modifications of the parts of the data (exploded views).

Digital art and computer graphics students may thus secure for themselves an attractive line of work after completion of a course of study, finding their niche in the supply and demand based competitive job market. Thus, the artists might combine artistic talent with knowledge, do what they like to do while being useful and needed in a society. One may say such training for knowledge visualization should support understanding of a physical and chemical side of pharmaceutical ingredients, along with familiarity with basic anatomy (such as did Leonardo da Vinci when drawing his Vitruvian Man in 1487) and essential physiological processes. Some skill in cognitive thinking about abstract concepts would be helpful in knowledge visualization to understand the concepts and processes from several points of view:

- As a child who has little or no experience and responds more freely when knowledge is presented with the use of avatars or animations. As described by Li, Sofra, & Power (2008, p. 2), "they use their 3D avatar appearance to project their chosen personalities and characteristics to others within a virtual world and, at the same time, can maintain their chosen degree of anonymity."
- As an adult patient who needs information about what happens with their body in general terms and how medical procedures and medicines may have beneficial effects at the macro, micro, and nano level, by improving the functioning of the whole organs, acting at single cells, or restraining bacteria and viruses from damaging the organism
- As a physician who finds help in explaining the facts and what's going on with the patient and seeking their cooperation in fighting a disease
- As an artist who applies knowledge visualization techniques to present metaphorically what they are expected to show, and does it in a beautiful, attractive way. Classic works of Edward Tufte (1983, 1990) may support artists in the task, telling how "graphical display of information enhances density, complexity, dimensionality, and even sometimes beauty of communication."

This chapter provides discussion of the visual ways of learning some of the basic physical and chemical concepts and processes. Knowledge of physical and chemical principles and properties of drugs and chemical compounds used as drugs is essential for medical and pharmacy students (Ma & Hadzija, 2012). The chapter offers exercises in visualization of scientific concepts by providing two projects about basic science-related themes:

1. Symmetry and pattern in animal world: geometry and art, and
2. Crystals and crystal caves.

Figure 1. Anna Ursyn, Double Symmetry
(©2010, A. Ursyn. Used with permission).

Each project invites the reader to create visual presentation of the theme. Figure 1 presents a study of symmetry carried out as an abstract image – a printout of a work developed as the 3D computer program: VAX mainframe, FORTRAN 77, Interactive Graphic Library (IGL) and PPC.

In order to examine paintings containing representations of plants, animals, symmetry, and pattern, and then make these themes in art more familiar, we can find them in Google Images, and also on web art galleries and collections; examples of links are provided in the Additional Reading section.

SYMMETRY AND PATTERN IN ANIMAL WORLD: GEOMETRY AND ART

Preliminary Comments

The objective of this project is to look over living forms in nature and in art, and examine how the nature-related concepts of symmetry (bilateral, radial, or helical), asymmetry, and patterns pertain to the general design of animal bodies, the animals' look, and behavior. A project "Symmetry and pattern in animal world" integrates several art concepts such as symmetry, pattern, tessellation (natural or human-made filling a surface with figures with no overlaps or gaps), and general composition of the artwork, with issues related to geometry (symmetry, tessellation), biology (animal shapes and forms, their symmetry and patterns, adaptation), art and art history (patterns in artistic or decorative design, animals in art, and general composition of the artwork), and computer graphic skills. We can examine symmetry (or lack of symmetry) in animal surface patterns, for example in fish scales or bird feather. Working on symmetry for artistic projects promotes application of our spatial visualization skills. Then, after some study about the use of symmetry and pattern in the art works, some readers may want to combine this knowledge with art and computer graphic skills and create an artistic representation of the animal world using symmetry and patterns. One may also want to apply tessellation in order to design a background for the artwork. Figure 2 shows my student's impression about animal symmetry and pattern.

Figure 2. Megan Brockriede, Animal Symmetry
(© 2011, M. Brockriede. Used with permission).

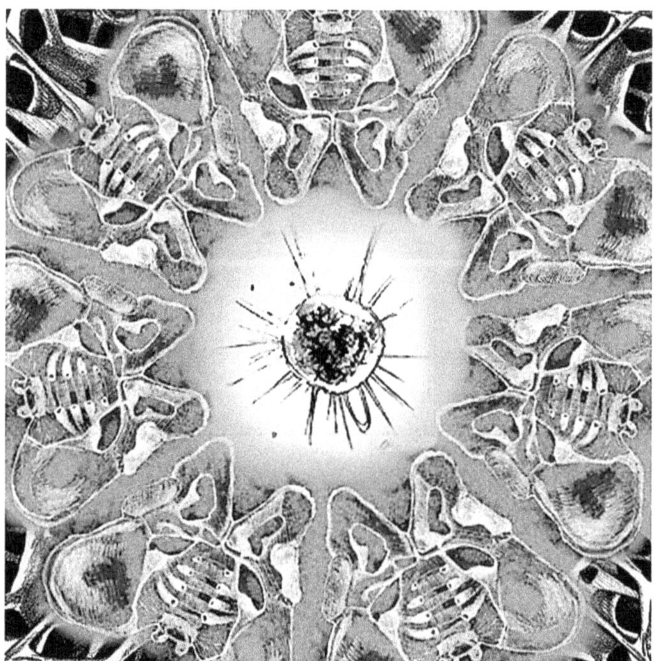

A Concept of Symmetry Related to Geometry, Physical Chemistry, and Pharmacy

There are many kinds of symmetry existing in geometry, in natural world, and human works. For example, water, when in liquid state, has bilateral symmetry, with the symmetric stretch of the two O-H bonds and some molecular vibrations (Rankin, Mitzel, & Morrison, 2013; Kettle, 2007); when frozen, water usually becomes symmetrical in various ways, developing hexagonal crystals. Ice, snowflakes, feather ice on the twigs, hail, sleet, icicles, glaciers, and polar caps, all develop their own order and various arrangements of symmetry axes; we will discuss them in a project about crystals and snowflakes. Scientists and artists see a purpose in symmetry investigations, for example, mathematicians, anthropologists, artists, designers, architects who conduct computer analysis of the facades, friezes, and some architectural details, as well as researchers in many fields of natural sciences, medicine, pharmacology, biology, geology, or chemistry. Many artists have created masterpieces this way. The theme of symmetry can certainly be considered inspirational to create biologically inspired art, because symmetrical forms and shapes possess an aesthetic beauty and an order reflected by their geometry. We can appreciate these forms finding the importance of adaptations that animals develop as an answer to the conditions of life, examining mathematical order in natural forms, and re-creating it in our own artwork.

Geometry studies properties of space: shapes, sizes, positions of figures, and distance concepts. As a starting point from which we can develop further statements we accept some axioms, universally established propositions, truths that are accepted without proof. For example, a symmetry axiom tells that for all points A and B, AB = BA. Axioms seem to be self-evident for human reasoning; however, many of the inventive and significant discoveries, such as the Heisenberg's uncertainty principle in quantum mechanics, resulted from the questioning of axioms and mixing a subject and an object of investigation. Julian Voss-Andreae wrote, in accordance with Einstein, Podolsky, & Rosen (1935), "Quantum theory remains philosophically problematic because 'objective realism' turns out to be incompatible with quantum theory There is no accurate space-time representation of, say, an electron: It is neither a particle nor a wave or any other "thing" (Voss-Andreae, 2011, p.14).

Bioimaging explores biological structures and functions to create information visualizations in two, three, or four dimensions, drawing information from sources such as light waves, nuclear magnetic resonance, x-rays, or ultrasounds. Researchers can now find symmetrical structures at the nano level, when the super resolved fluorescence microscopy brought optical microscopy into the nano dimension, winning the 2014 Nobel Prize (Chang, 2014). The total internal reflection fluorescence system serves for imaging just below the surface of a specimen; it allows imaging adhesion, hormone binding, neurotransmitter secretion, and membrane dynamics. Several new bioimaging systems do not require slicing and thus damaging tissues, and thus can be used to provide pharmaceutical, neurobiological, and other information about deep, intact structures. Ultrafast, dual-wavelength lasers with ultrafast (femtosecond) pulses provide simultaneous activation of multiple neurons (Arrigoni & Gallaher, 2015).

Many molecules such as carbon dioxide, benzene, or carbontetrachloride are perfectly symmetrical. Symmetry is present when similar parts of an object are arranged on the opposite sides of a point, line (axis), or plane. Elements of symmetry in a crystalline molecule include the center, axes, and planes of symmetry. The center of symmetry is a point where a line drawn through it is equally distant from the opposite parallel faces of the crystal. An axis of symmetry is a line passing through the center of symmetry about which we can rotate a crystal and see its same appearance several times before it rotates 360°. A plane of symmetry divides a crystal into two identical mirror halves. As an example of a biological structure, a chicken eggshell has an asymmetric shape close to an oval. In geometry, an oval is a two dimensional closed curve that may have one or two axes of symmetry. The three dimensional surface is called ovoid; it can be made by rotating an oval curve around one of its axes of symmetry. With two axes of symmetry, the figure looks similar from both ends, while with an only one axis of symmetry it resembles a chicken eggshell, having one end somehow pointed.

There are several types of geometrical symmetry, for example, bilateral (reflection or mirror), rotational (when an object looks the same after rotation), cylindrical, spherical, and helical symmetry (like in a drill bit), not to mention some kinds not so obviously seen in everyday objects, such as translational (where a particular translation - moving in a specified direction does not change the object), glide reflection (in a line or plane combined with a translation), or rotoreflection symmetry, which presents rotation about an axis, combined with reflection in a plane perpendicular to that axis.

A figure that has bilateral symmetry has two mirror-like halves, which correspond exactly if folded along its line of symmetry. The halves are congruent, they are the same size, shape, and coincide exactly when superimposed. In nature, optical isomers are symmetrical around a plane. Molecules of some sugar isomers are deflecting the rays of light in right or left direction. Symmetrical objects show several elements of symmetry, for example, a crystal may show rotation axes, a center of symmetry, or mirror planes - the imaginary planes that separate an object into halves.

An object has a rotational (radial) symmetry when it can be rotated around an imaginary line called the rotation axis and retain the same appearance as before rotating, repeating itself several times during a complete rotation. For example, with a six-fold rotation axis the crystal repeats itself each 60°. A center of symmetry is equally distant from any point on the surface of a symmetrical object.

Objects with helical symmetry combine symmetry such as that of a circle with the translation of this object along a long axis. For example, springs, screws, drill bits, or slinky toys have helical symmetry. An infinite helical symmetry appears when a cross section of a helix doesn't change after every small rotation. The N-fold helical symmetry exists when a cross section repeats itself regularly. Double helix is the case when it repeats after every full rotation, when it returns to its initial position. A molecule of DNA, a deoxyribonucleic acid, has a non-repeating helical symmetry.

Symmetry in Living Organisms

Humans and animals evolve and adapt to a continuously changing environment. In the course of indefinitely long periods of time the wonderful ways of animal adaptation resulted in a variety of extraordinary shapes and forms. The general design of animal bodies, their symmetry, and patterns has also been changing in time. Symmetry and its changes are used in biology to introduce the concept of classification.

Many of the earliest organisms lacked symmetry. Then, animals sitting mostly in one place at the bottom of the ocean, as well as some floating animals developed radial symmetry. Having the rotational symmetry, such creatures did not change their appearance after rotating a certain number of degrees around the center of their body. Radial arrangement of body parts strengthens an animal's skeleton. Also, animals with rotational symmetry can sense the danger from all directions. Most of these animals, often in relation to the times past, have a three-rayed symmetry, but other, for example, jellyfish, retain a pentagonal symmetry of their skeleton, major organ systems, and rays radiating from the mouth in five tube feet. There are also seven-armed starfish, while sea lilies develop 10 to 200 arms resembling fern fronds. Ernst von Haeckel (1834-1919, Haeckel, Breidbach, Hartman, & Eibl-Eibesfeldt, 1998), German naturalist, philosopher, physician, and artist analyzed patterns when he described thousands of

Figure 3. Marine animals: Pheodaria, Circogoniaicosahedra
(Ernst von Haeckel, 1904. Image in the public domain, available at Wikimedia Commons).

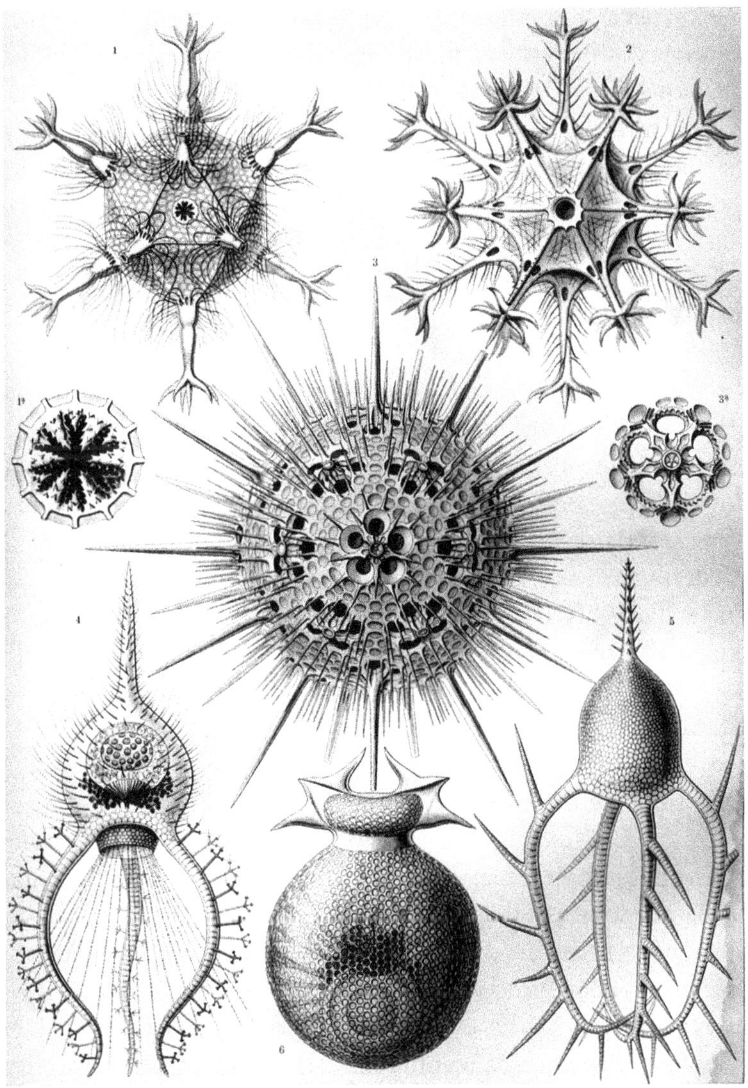

new species and life development processes through evolution, stressing their aesthetic value. His work from 1904 "Kunstformen der Natur" (Art Forms in Nature) contains images of marine animals, often displaying symmetry of various kinds (Figure 3).

Cylindrical symmetry around the vertical axis (such as in balloon-shaped creatures swimming in the oceans, and also in pollens) facilitates forward motion. In any body with the spherical symmetry, for example in spherical pollen, everything is the same on all sides. Fish and some other marine animals including snakes support their locomotion using tails or vertical caudal fins called a tailfins, which re-direct the flow of water and reduce the retarding force of hydrodynamic drag of water. One may say the action of a rudder can be somehow compared to a fish tail, when a rudder in a ship or an aircraft serves to control yaw. Also, a tail of a swimming dog serves for this purpose.

Bilateral symmetry occurs in all living groups and is especially marked in the larval stages. During evolution, muscle contraction replaced the beating of cilia; streamlined bodies with bilateral symmetry (similar to line symmetry) became dominant. This type of symmetry occurs when the animal has the front and the rear parts developed differently and it displays the forward motion. In bilaterally symmetrical animals, for example in butterflies, the halves of their bodies, when seen along the horizontal or the vertical axis (x or y axis) form each other's mirror images. By moving their tailfins from side to side, fish get a thrust by setting water in motion, which pushes them forward in water. Some species such as whales, dolphins, and porpoises – small toothed whales get thrust using their pectoral fins and moving horizontal tail fins up and down.

Patterns in Animals

We may see many patterns in our surroundings on everyday basis; they are of natural such as these sculptured in sand by wind or of an accidental origin. Patterns we can see in plants and animals have consistent forms, traits, or features characteristic of an individual or a group. Animal patterns, such as the designs on scales of fish, or on feathers of birds, enable animals to recognize individuals of their own species, scare predators, or to camouflage. Simple mathematical equations describe and explain the development of curves, stripes and spots in plants and animals (D'Arcy Thompson, 1917/2011). Humans can describe and classify these species on the basis of these characteristics (Neville, 1977). Fractal geometry explains origination of patterns in living organisms.

Asymmetry in Living Organisms

Patterns on animal exteriors are not always symmetrical. For example, some species of salamanders, frogs, vipers, fly larvae, and fish have asymmetrical patterns. Asymmetry of a pattern may be useful in terms of the camouflage possibilities: they aid to conceal their bodies from enemies by making them appear to be part of the natural surroundings. An asymmetrical pattern can help an animal with bilateral symmetry to change its appearance, to hide and become invisible against the background. In later stages of evolutionary development, body forms became not so simple. Most animal bodies, including people, cannot be divided exactly into two halves, even when they look symmetrical from external appearance. Other forms, for example marine animals holothurians (Holothurian, 2011) also called sea cucumbers (echinoderms), show bilateral symmetry outside and radial symmetry of hemal sinuses inside.

We can find many examples of asymmetry in animals. It can be expressed as a zigzag pattern down the back of a viper, or as a position of a liver or a heart. Not only internal organs, but the skull and the brain are also asymmetrical in humans, as well as in some animals and birds. Organisms develop not only a structural but also a behavioral asymmetry. Two halves of the human brain display different abilities, ways of learning, and thinking. Human and animal handedness is an asymmetry in skill development. For example, Fiddler crabs have one claw much bigger than another. Earthworms, when put in a maze, display a behavioral asymmetry when they can recognize the difference between right and left. Some insects use their jaws like left-handed or right-handed scissors, depending on the species they belong to.

Some technologies are applied in zoology to explain existing symmetry or asymmetry. For example nuclear transplantation or application of a mutagen drug may induce biased symmetry. The notion of asymmetry pertains also to several other disciplines. For example, in organic chemistry chirality refers most often to molecules: chiral molecules display lack of an internal plane of symmetry, usually because of the presence of an asymmetric carbon atom. Chirality means, in simple terms, the existence of left/right opposition (IUPAC, 2014; Leffingwell, 2003). Chirality is also present in biological structures; it often determines bioactivity (for example, odor perception) and is an important factor in drug efficacy. As was stated at the 2011 Congress of the European Society for Evolutionary Biology, "asymmetric shapes in animals and plants can come in two mirror-image forms. However, such chiral dimorphism is found in some structures but not in others. This makes chirality one of the very few developmental traits that can be studied consistently across all multicellular organisms, offering a goldmine of research questions in evo-devo (evolutionary developmental biology), evolutionary ecology, and macro-evolution" (Schilthuizen & Gravendeel, 2011).

Asymmetry as Seen by Humans

The art theorist and perceptual psychologist Rudolf Arnheim (1988) stated that gravity makes the space asymmetrical, not in a geometrical but in dynamical sense, because an upward movement requires energy, whereas downward movement can be done by removing any support that keeps an object from falling. We perceive this asymmetry by two senses, with kinesthesia (awareness of the tension in the muscles and joints of the body) and vision. Michael Leyton (2006) discussed his asymmetry principle. He proposed in his "The Foundations of Aesthetics" that the two principles, maximization of transfer and maximization of recoverability are the basic principles both of geometry and aesthetics. They are fundamental to aesthetic judgment in the arts (painting, music, and poetry), the sciences (general relativity and quantum mechanics), and computer programming (object-oriented programming). Recoverability of the backward history is possible due to the asymmetry principle: to ensure recoverability of the past, any asymmetry in the present must go back to symmetry in the past. According to the author, in mathematics and physics "asymmetry" really means distinguishability, and "symmetry" means indistinguishability. Thus, the asymmetry principle really says that, to ensure recoverability, any distinguishability in the present must go back to indistinguishability in the past.

Humans, often in a similar way as animals, apply asymmetrical patterns as a camouflage, a way of blending with the environment; they use hunting camouflage clothing in colors (for example, bright orange) that are perceived as dull by the game animals, and battledresses or camouflage netting for military purposes. Artists and designers contribute with their projects. For example, for warplane and ship camouflage, in order to confuse the periscope view of the submarine gunners during World War I, Norman Wilkinson, Everett L. Warner, and other artists painted the high-contrast, asymmetric shapes on ship surfaces, thus confusing the periscope view of the German submarines (Berens, 1999). Other, non-military applications include art. Artists, for example, Liu Bolin (2014) or Bev Doolittle (2014) create camouflage art, while architects and designers construct unusual camouflage restaurants (2014) of different kinds. Figure 4 presents an artwork about uneven symmetry entitled "Focal Reflection."

A glass wall creates city symmetry. This one-sided, asymmetric reflection makes a fragmental composition when we pass by, which ends with the end of the glass wall. We feel forces of such enchanting details that determine the whole city's identity.

Figure 4. Anna Ursyn, Focal Reflection
(© 2009, A. Ursyn. Used with permission).

The Aesthetic Values of Symmetry and Asymmetry

In a book entitled "Aesthetic Measure" George David Birkhoff (1884-1944) proposed a mathematical theory of aesthetics: in an equation M=O/C, Aesthetic Measure (M) is a function of Order (O) divided by Complexity (C.) The Gestalt psychology theory of mind postulated that brain has self-organizing tendencies and recognizes the whole of a figure rather than its individual parts (Birkhoff, 2003). Aesthetics has been investigated as part of cognitive science, since Semir Zeki (1993, 1999), professor of neuroaesthetics at London University College associated perception of the great works of art with working principles of the brain. Artists, acting like instinctive neuroscientists, capture in their art works the essence of things in a similar way as the brain acts when it captures the essential information about the world from a stream of sensory input. Neuroesthetics explores the visual brain using anatomical, electro-physiological, psychological methods and imaging techniques.

Symmetry plays a remarkable role in most human visual endeavors. Natural objects displaying symmetry evoke wonder and surprise because their intricacy. For example in architecture, such architectural details as stain-glass windows, mosaics, and friezes, visual arts, pottery and ceramics, quilts, textiles, and carpets many times make a varied use of symmetry as an important principle in their design. Maybe for that reason symmetry is so often seen not only beautiful but also conducive to visual communication.

One may discuss the aesthetics of visualization as related to the visual competence in the art, design, and technological solutions in visualization. In a growing number of publications the aesthetics concept refers to design effectiveness, efficiency, and easiness to understand (a low cognitive cost) of visual presentation, not exclusively the beauty of an image. Researchers who deal with advancements in visualization associate aesthetics with readability, and readability with understanding. Intensive research on the optimal layout aesthetics has been conducted in the field of graph drawing and the aesthetics of graph drawing *algorithms* (for example, Purchase, 2010; Lau & Vande Moere, 2007). Spatial relationships between nodes and edges and the overall layout, including graph's symmetry, area, flow, and aspect ratio, determine the aesthetics of a graph (Bennett, Ryall, Spalteholz, & Gooch, 2007). Methods of measuring the aesthetics of a layout and the results obtained suggested reducing the number of edge crossings and making the best use of symmetry, e.g., by maximizing symmetry of subgraphs.

A mathematician, an anthropologist, and an architect conducted together a computer analysis of friezes covering the facades of hundreds of buildings in the village of Pirgi, a small town on the Greek island of Chios in the eastern Aegean Sea (James, James, & Kalisperis 2004). Gray and white friezes contain squares, triangles and rhomboids, different on each house. Mathematic analysis of friezes' symmetry revealed five basic actions of moving a frieze (translation, vertical and horizontal mirror, half-turn and glide reflection). The aesthetic value of asymmetry has been appreciated in Chinese art and art criticism. In Chinese calligraphy, the "dynamic asymmetry' proportion of a Chinese character, with a subtle discrimination, is prized by Chinese artists and art theorists (Huang & Balsys, 2009).

Patterns in Art

Patterns – artistic or decorative designs made of lines, which are omnipresent in art and design domains, make a basis of ornaments; they are specific for different cultures. Pattern is based on the repetition of units coming from the natural or artificial origin. Patterns that are used in decorative arts and architecture have distinctive styles, which are specific for each culture or ethnic group. A great part of ornaments display motifs as elements of patterns. Owen Jones made a huge collection of ornaments typical for different countries. In 1856 he wrote a monographic book entitled "The Grammar of Ornament" (Jones, 2001). Works of art created according to various art movements or artistic groups display patterns characteristic of this group, for example, decorative patterns typical of the Art Nouveau or Secession styles. In cultural anthropology, a study of decorative patterns, ornaments, and their characteristics support the detailed analysis of an individual culture. Some common leading ideas on which patterns of ornaments are based are in accordance with the arrangement typical of the distribution of form in nature.

Animals in Art

We may easily find wonderful pictures of animals when we look at the cave walls (regardless of a debate as to whether the cave paintings were created for artistic intent), ancient or antique paintings, as well as the paintings from the Renaissance or modern times. Real and fantastic animals served in the medieval paintings (from most of the Gothic period of the 12th through the 15th century) as a part of decorative vocabulary and vehicles for religious allegory and moral instruction (Animals in Medieval Art, 2011). One can also find fantastic animals and beasts in classical paintings of ancient Egypt, Greece, and Rome, and in the dreamlike or nightmarish pictures from the times of the Surrealism movement (originated in France in the 1920s). Such bizarre, fictional beasts were derived both from mythology and pure fantasy.

In order to examine paintings containing representations of animals, and make the animal theme in art more familiar, we can find them in Google Images, and on web art galleries and collections. We may surely find artwork in books. Many of the titles of art works collected below come from The Art Book (2012), The 20th Century Art Book (1996), and The American Art Book (1999).

While looking at the artwork, it may be interesting to fix one's eyes on the painting and give serious thought to a real cause or a motive the artist might have in mind to create it. It may not necessarily be an accurate and detailed representation of a scene with the animals but a message to convey or an artistic statement that provides a reason for bringing this work into being. You are requested to figure out the content of the artwork, and then, in the same way, to make an effort to include some meaning into your animal representations. Below, some approaches to the animal theme are shown in works from various periods of time.

- Jacopo Bassano, c1590, "The Animals Entering the Ark," "Sacrificio di Noe".
- Jan Breugel, c1620, "The Garden of Eden," "Paradise".
- Frans Snyder, 1620s, "A Game Stall," "Still Life of Dead Game".
- Aelbert Cuyp, c1650, "Cattle with Horseman and Peasants," "Cows in Water".
- Carel Fabritius, 1654, "The Goldfinch".
- Edward Hicks, 1833/4, "The Peaceable Kingdom".
- John James Audubon, 1835-8, "Roseate Spoonbill".
- George Catlin, c.1857, "Ambush for Flamingoes".
- Sir Stanley Spencer, 1935, "Saint Francis and the Birds".
- Henri Rousseau, 1906, "The Monkeys".
- Giacomo Balla, 1913, "Flight of Swallows".
- Raoul Dufy, c1926, "The Paddock".
- Marcel Broodthaers, 1964-5, "Casserole and Closed Mussels".
- Arthur Boyd, 1987, "The Australian Skapegoat".

The Visual Appeal and Aesthetic Concerns about Mathematics-Derived Art Forms

We can realize how much the mathematical way of thinking is a visual process and how often the beauty of forms derived from mathematical formulas becomes an inspiration for art and computer graphics majors for learning, thinking, and creating. Visual mathematics may be examined as the language of space design and is often used for the composition of an artwork. Historically, this idea provided a common language for mathematicians, visual artists, architects, musicians, crystallographers, cartographers, and many other professionals. In his work *On Growth and Form*, Scottish embryologist D'Arcy Wentworth Thompson (1860-1948) looked for explanation of evolutionary changes in mathematical and physical laws of 'economy and transformation' (D'Arcy Thompson, 1917/2011). Intuition, along with knowledge and calculations, has been stressed as a leading force in problem solving. A Dutch scientist and crystallographer Arthur Loeb (1993), in his book entitled, "Concepts & Images – Visual Mathematics," claimed that intuition is a form of non-verbalized knowledge. Some scientists and artists use their knowledge and intuition in such a way that their abstract reasoning takes form of images, rather than words or formulas. A perceptual psychologist and film theoretician Rudolph Arnheim (1969) stressed the role of the visual thinking, which allows gifted minds with intuitive wisdom to avoid troubles with formalistic thought operations due to their brilliant cross-circuits.

Several topics of research in visual mathematics provide inspiration for artists. For example, geometry provides unlimited possibilities for creating patterns and ornaments that are present in the variety of design styles characteristic of all of cultures of the world. Artists from different periods, such as the Italian Renaissance painter Piero della Francesca (1410/20-1492) or the Italian Surrealist artist Giorgio de Chirico (1888-1978) explored symmetry and perspective, some of them making exciting tricks, such as the Northern Renaissance artist and printmaker Hans Holbein the Younger (1497-1543) who applied *anamorphosis* in "The Ambassadors" (1533), or the Dutch graphic artist Maurits Cornelis Escher (1898-1972) who created impossible sceneries. Some shapes and forms that contain mathematical regularities

Figure 5. a. Amy Wickstrom, "Symmetry"; b. Gabe Eltaeb, "Symmetry"
(a: © 2010, A. Wickstrom. b: © 2010, G. Eltaeb. Used with permission.)

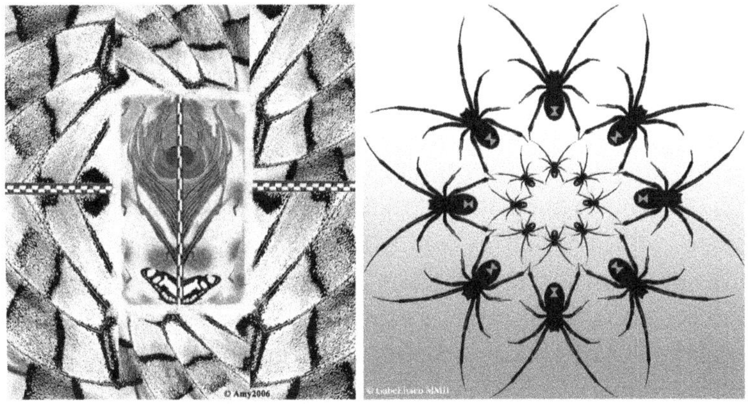

and can be described by equations seem to be especially inspirational and aesthetically appealing. For example, complex polygons, tessellations of strings and lattices, the geometry displayed by soap bubble films, tiles, Oriental mosaics, Roman mosaics, or mazes motivated mathematicians and artists to create art- and design-related works.

Study of geometric two-dimensional shapes and three-dimensional forms resulted in works of art done in various media, both traditional and digital, including painting, sculpture, printmaking, and architecture. Computer programs and computer graphics produced with the use of software packages give a boost to such studies. They make a separate category of digital art and, at the same time, serve as a tool for creating works in other media. Teams working on visualization or simulation projects usually include a mathematician, and the direct results of their work look many times like artistic productions.

Figure 5 (a and b) shows the student ways of exploring the theme of symmetry and asymmetry in insects.

One may ask, does the beauty and the aesthetic value of a presented object make its representation in an artwork? The mathematician and sculptor Helaman Ferguson (1994) claims that mathematics is an invisible art form. As he states, "computer graphics make mathematics visible." What are the distinguishing characteristics of natural forms, their mathematical descriptions, and mathematics-derived artwork? The shell of a mollusk contains a logarithmic spiral. It is a natural phenomenon, not the result of some mathematical program in the nautilus. A drawing of a logarithmic spiral is a biology-inspired, mathematics-derived form. Spiral forms have been explored, discovered and modified by people, but they are hardly an artwork. In the same way, when we visualize the laws of nature and develop their representations, can we consider those representations an artwork? Which criteria concerning the artistic quality of mathematically developed art may be accepted by art criticism?

A need for embellishing things seems to be universal among people who are not necessarily educated in mathematics. Intricate designs of old embroideries and patterns for laces seem to be derived from mathematical principles in an intuitive way. In the same way, we knowingly employ mathematics in developing complex polygons and fractal images to create an artwork. Artistic expressions based on the beauty of natural and calculated patterns, exquisiteness of mathematics, or a tessellation principle had been applied both by old Islamic artists and by the twentieth-century artists, for example M. C. Escher. Other concerns about determining artistic quality of mathematically developed art may also relate to its precision. While designing an artwork, a mathematician is careful to avoid ambiguity. Everybody can see the exact and precisely defined formula representation. On the contrary, visual communication in fine arts is often based on metaphors. Everyone may receive different, unique, and individual messages from the same artwork. It may well be that a representation of mathematical formulas cannot inevitably become an art form, but it is their intentional transformation that may become an artwork.

Laces of complex ornamentation, geometric representations of polygons, and interlaced patterns, all evoke in us the sense of beauty. Some paintings and photographs explore the beauty of magnified or reduced natural objects. However, Rudolph Arnheim (1990) wrote that photographs might affect our observation by singling out accidentals as readily as essentials, and thus making everything equally important. In a similar way, the exquisite perfection of soap bubble geometries, as well as fractal shapes rose to the form of art because of their natural beauty. A question arises, whether the art critics' evaluation of the mathematics-derived artwork depends on the aesthetic quality of patterns, the beauty of underlying mathematical formulas defining those ornamental arrangements, or the human intervention.

Graphic elegance, which can be found in the simplicity of the image demands, "every bit of ink on a graphic requires a reason" (Tufte, 1983/2001). The secret of perfect design may be in its simplicity or its elaborate intricateness.

Bodies of water give us reflections, and thus create symmetrical images. However, reflected colors differ from the original ones, so the landscape becomes less real to the eye. We often examine and then show the difference between the real and the reflected objects when we admire a landscape. Then, we may think about a gossip, a paraphrase, a pastiche, a metaphor, a story, a fable, and a performance in terms of the changes in a message. We go to the online, virtual worlds, visit a theater, an opera, or a ballet performance to see the two events mentally and physically pictured, one going in a real and another one in a reflected, virtual space. Even the accompanying audience can be real or virtual, with lots of possible ways to affect our perception.

Concepts and Processes Related to Fractals

Fractals represent a form of scale symmetry that appears when the objects magnified or reduced in size have the same properties. Mathematicians name some objects symmetrical with respect to a given mathematical operation applied to this object, when this operation preserves some property of the object. Such operations form a symmetry group of the object. Fractal geometry describes objects in a different way than the traditional Euclidian geometry. It takes for granted that objects are self-similar so, when magnified, their parts are similar to the whole, and the likeness continues when the parts are magnified more and more. In 1975, Benoit Mandelbrot, a Polish-born French mathematician (who moved to the United States in 1958), developed the mathematical theory of fractals. The concept of self-similarity is forming the groundwork for fractal geometry that describes objects that are self-similar. Fractal objects have the same structure, whatever their scale. They can be subdivided into parts, each of which is a smaller copy of the whole. The smaller parts are like the larger ones. When fractal objects are magnified, their parts are similar to the whole, the likeness continuing with the parts of the parts and so on, forever. That means, the part, whatever its size, has the same topology as the whole. Self-similarity can be seen in many plants, including the fern leaves. Each leaf branching off the fern is a smaller version of the entire plant. The Cantor Set, the simplest fractal, may be constructed using the following algorithm: we may build a simple fractal by dividing a line in 3 parts and removing the middle part. When we repeat

Figure 6. Anna Ursyn, "Symmetrical Colors"
(© 2013, A. Ursyn. Used with permission.).

this series of steps, first on the 2 remaining parts, then on the 4 parts produced by that operation, and so on, the fractal will have an immeasurably great number of infinitely small parts.

Some types of fractals that are in use today include:

- **Iterated Function System:** (Random and deterministic), where the process is repeated and results in a self-similarity independent of a scale, with all the frames being simply iterations of the original frame. *The Sierpinski Gasket*, one of the most basic types of fractals built from equilateral (or other) triangles can be formed as an iterated function system.
- **Mandelbrot Sets:** The nonlinear fractals where the relationship between their parts is subject to change. They become the source of stunning color computer graphics images. They are also important in the dynamical system theory.
- **Julia Sets:** With objects generated through an iterative process called after a French mathematician, Gaston Julia. By counting the number of times each point goes through the function before it gets a certain distance away, we can determine what color to make that point. Therefore all points that are equally far away will be the same color.
- **The L-System:** A parallel rewriting system used to generate self-similar fractals. "Rewriting is a technique for defining complex objects by successively replacing parts of a simple initial object using a set of rewriting rules or productions" (Prusinkiewicz, Lindenmeyer, Hanan, Fracchia, Fowler, de Boer, & Mercer, L., 1991, p. 1). The Hungarian biologist and botanist Aristid Lindenmayer introduced a string-rewriting mechanism called L-systems in 1968 for the study of biological models for plant growth.

We can find fractals everywhere in the natural world – in the eroded landscapes of mountain surfaces, crystalline caves in the crust, in the patterns of frost on a cold window, as well as on the surface of a virus. Fractal nature of many materials can be put into service of technology, so scientists investigate the fractal character of objects and events in biology, geology, meteorology, and a great number of other domains. Using a fractal theory, physicists can describe the chaotic course of events occurring in dynamical systems, Fractals are useful in explaining the atomic structure of materials such as glasses, gels, and other amorphous materials, as well as the changes in physical features such as heat- and electrical conductivity. Visual models based on concepts of self-similarity and of kinetic growth describe particles which, when added to a structure, neither come off nor rearrange themselves. In biology, polymer chains building proteins have the fractal shape, as well as patterns of atoms on the protein surface. Fractals are used to model soil erosion and to analyze seismic patterns. Coastlines and surfaces of mountains can be modeled with fractal geometry; for example, Okubo & Aki (1987), examined fractal dimensions of seismic activity related to the brittle crust of the California's San Andreas Fault. Small earthquakes are fractal in time as well as space. Fractals are able to describe complex figures, such as seashores, clouds, or trees. They are used in a variety of applications ranging from special film and TV effects to economy and physics. Fractal models serve in chemistry, physics, and materials science to describe a passage from one state in which matter can exist (solid, liquid, gas, or plasma) to another, which depends on temperature and pressure. Researchers control the growth of polymers and colloids in liquid and gaseous phases.

Fractal geometry is a branch of bio-inspired computing, and involves cellular automata, L-systems, iterated function systems, particle systems, Brownian motion, and fractal proteins for landscape design, growth processes, and modeling. Inspiration for designing cellular automata comes from complex natural systems, such as colonies of insects or the nervous system, and pertains decentralized systems, such as sensory networks and networks showing the intrinsic connectivity and conductance (Packard, 2001). With evolutionary approach, scientists can solve a variety of problems, for example, the Rubik's cube (El-Sourani, Hauke, & Borschbach, 2010).

Artists have often applied fractal design in their work. Fractals can be used to create visual art or visual music, where the results of calculations can be presented graphically or acoustically. They are produced by algorithms, which are used to calculate a graphic or acoustic fractal. Final results may vary depending on odd or even, left or right values and situations chosen. Fractals, especially fractal music, result from iteration (repeated application of an equation) in the recursive operations, where the items are repeated in a self-similar way during the process. By using fractal-generating applets we may bring into being images of plants, snowflakes, landscapes, and many other forms, including fractal art, music, and fractal music generators. Using a recursive splitting technique, artists produce fractal images of great statistical complexity. Landscapes with trees and other branching structures made with the use of fractal geometry have been used as backgrounds in many motion pictures, still life works, and animations.

As they often elicit an aesthetic response in us, fractals are often given the status of a work of art. Fractals are exceedingly common in nature, and fractal art can imitate nature. On the interface of science and art, computer-graphics artists and specialists have produced fractal images of great statistical complexity.

Fractal art, which came about as a consequence of the developments in fractal geometry, has been presented in tens of handsomely printed books and displayed at art exhibitions. Computer-graphics artists create fractal images of great statistical complexity. For example, fractal landscapes serve as backgrounds in motion pictures, while trees and other branching structures can be seen in animations. Decorative arts are often produced with fractals.

The appearance of fractal designs may resemble various styles in art, depending on the method of calculation. With all their decorative merits, fractals have often evoked discussions about their aesthetic values in terms of fine art. In one of his books Claude Lévi-Strauss, French philosopher and anthropologist associated with the development of structuralism, characterized the essence of fractals in art. As he noticed, in 1854 the French Romantic artist Eugène Delacroix (1798 - 1863) observed in nature the distinctive property that objects have the same invariant structure whatever their scale. Lévi-Strauss recalled the opinion of a German philosopher Immanuel Kant (1724-1804) about an aesthetic judgment. In an aesthetic theory developed by Kant, judgments about beauty rest on feeling but they should be validated in harmony with mental structure, so they are not merely statements of taste or opinion. According to Kant, there are judgments of taste that are subjective and judgments of reason that are universally valid. Aesthetic judgment falls somewhere between these two kinds. Lévi-Strauss (1997) stated that, in this intermediary space, fractals are given the status of a work of art, because they are appealing and at the same time, objectively governed by reason. Innumerable individual art works reside on the Internet, for example when we go to Google Images, such as Helaman Ferguson's "Umbilic" or Clifford Pickover's "Lensflare VTT."

When we look (for example, at the Google Images) at the photographs taken by the American photographer and environmentalist Anselm Adams (1902-1984), such as his "Moon and Mount McKinley, Denali National Park" or the "Oak Tree, Sunset City," we can see a striking resemblance between the fractal and photographic rendering of natural scenery. Artists saw such resemblance and drew analogies with their own fractal art. For example, an algorithmic artist Kerry Mitchell (2011) wrote, "I varied the thickness of the curve in accordance with the grayscale level of the target photograph, Ansel Adams' *Oak Tree, Sunset City, California*. I contrasted the straight, precise lines of the curve with the natural branching of the tree and tried to pay homage to Adams' image of a fractal with one of my own." A photography artist Julia Jones (2005) wrote, "Great photographs are like living things, subject to the laws of life. They are made up of cells or parts, each as significant as the whole. There are photos within photos: fractals. Fractals work and exist in nature and in Adams' photographs; these fractals play out to recreate what Adams saw in the scene, making it alive again to the viewer. Adams saw and captured pictures within pictures; masterpieces within the one masterpiece. Adams once argued, no note is beautiful on its own: it is the repetition, variation, and the note's place in the whole that allows one to hear its beauty. "Frozen Fractal" (Figure 7) refers to the unity of the natural, mathematical, and artistic manifestations of fractals.

A PROJECT ABOUT SYMMETRY

A Composition with Symmetry

Now you may want to make an abstract composition made of concrete geometric representations using simple geometric forms. Draw on the computer screen some forms showing several types of symmetry: the bilateral, the radial, cylindrical, or spherical symmetry, copy them and arrange on the screen. It may be useful in further work on your composition. Explain a relation between the type of symmetry and the animal moving capacity, drawing two- and three-dimensional models and applying visualization made with the use of color and pattern coding. You may also want to construct models out of paper to explore principles of symmetry.

Figure 7. Anna Ursyn, "Frozen Fractal"
(© 2013, A. Ursyn. Used with permission.)

To create an body with a bilateral symmetry, draw a left or right side of a shape you plan to design, then perform a "Save as ..." operation, and create it's another side using the "Brush mirror" option. It can be a geometric form, for example, an acute triangle (having angles less than 90°), and the right-angle triangle or, if you prefer so, it may also be a butterfly or a flower. Organize several such shapes; for example, create a steel bridge applying copying, flipping, and mirroring triangles.

To show an object with radial symmetry, for example an open umbrella or a flower, draw a segment of your object: an isosceles triangle (having two sides and two angles the same) or a petal of a flower, then rotate it three, four, five or six times around an imaginary line called the rotation axis, so your segment will repeat itself each 120°, 90°, 72°, or 60°. Now you may find a center of symmetry equally distant from any point on the exterior surface of your object. You can also see that your shape will retain the same appearance after rotation.

Make a sketch of an object with a cylindrical symmetry, for example, a glass, a top hat or a tube.

To show a spherical symmetry, design a spherical object, for example, a space ship that fulfills the space explorers' needs and demands. How will you divide the sphere interior? Draw a plan of the interior. Name your file and perform a "Save as ..." operation.

Tessellation

Applying tessellation may now create another abstract art piece. Starting with a geometric form, begin working on a tessellated composition – a Virtual Garden. You may want to apply tessellation in order to design a background for your artwork. Look at art works created with the use of tessellation, for example at the works of Maurits Cornelis Escher, and then create your own tessellation. To create tessellation, choose a rectangle or a geometric shape you have already sketched. Take away a fragment from this rectangle by cutting out a shape of your choice from its left side. Slide this fragment so that it becomes

Figure 8. Pete Fadner, "Tessellation"
(© 2010, P. Fadner. Used with permission.)

closely attached with its straight-line edge to the bottom line of your rectangle. It would be sensible to save your work often. Now, repeat this procedure, now on the other side of the rectangle: select and subtract a fragment on the right side, and attach its straight-line side to the upper border of the rectangle. Copy and paste the new shape you have just created, making several copies. Select one color for a half of these shapes, choose another color for the rest of them, and color them all using the paint bucket tool. Arrange all fragments as a two-colored whole. In some graphic programs (for example, Adobe Photoshop) the stages of a work are being organized in layers, so you may color shapes from each layer separately, then flatten the image and arrange the shapes into a tessellation pattern. Figure 8 presents a student's tessellation.

Virtual Garden: Art Production

Continue creating an imaginative drawing of the virtual garden built out of previously created images of animals and plants. Using your tessellation as a background, create a virtual garden. Create a landscape with animals, emphasizing symmetry and patterns in nature. Looking at small, plastic toy animals make several images of animals on the screen. You have many options in doing this. You may create hand drawings in pencil and then scan them, or draw with a mouse or graphic tablet straight on the screen. You may scan pictures or your photos of animals and paste the images into your work. Also, you may import animal pictures from the web (being careful to obey copyright rules!) or from clip-art, but the artwork would be more genuine, original, and authentic with your hand-drawn animals. Remember that no two animals look alike, even when they belong to the same population, so avoid copying and pasting the images of animals several times.

Add some plants: trees, flowers, and grass, carefully composing them together with your tessellation background. Create a sense of perspective by placing bigger objects in the foreground, partially covering farther objects by the nearer ones, and applying lighter shades in front than in the back of the space of your virtual garden. Try to imagine the impact of the environment you depict on each of the animals, resulting from changes and events you invented. How is your virtual garden experienced and felt by all of the creatures living there?

In your work, use patterns to convey the sense of variety. Assign a specific pattern to each of the specimens. Select one pattern that indicates that this individual belongs to its class. Arrange patterns (that means, animals with patterns) in order to bring harmony in your composition. Make your landscape purposeful, and organize it into a whole, not resembling a piece of fabric that had been cut at random from a bale to make a tablecloth. Remember one general principle, which is usually expressed sententiously as "form follows function." Each animal shape you draw should indirectly convey the characteristic of movement, which is specific to the animal you choose to portray, while each symmetrical or asymmetrical pattern may tell about animal's locomotion or its possible ways of disguising its true appearance.

When you are ready with the general design of your picture, examine it in a detailed and complete manner: does the artwork convey the message you intended to express by art, or is it just an accurate depiction of the object of your attention you have been working on? After saving your image under a new name, select the best and most meaningful part of your work and crop it, so you can endeavor now toward completing a new, blown-up segment. Work on it, having in mind both the meaning you want to convey and the aesthetic matters of artistic beauty and taste. Your new work may be less descriptive than your initial composition, but you may find it more engaging and intent. After you feel you have completed your work, compare it critically with the first rendering in terms of your initial goals and your actual answer to the issues you were working on.

CRYSTALS AND CRYSTAL CAVES

Preliminary Comments

The objective of this project is to use our knowledge derived from the inquiries to create beautiful projects about minerals: gems and other more mundane earth baubles. Project about crystals relates to several areas of interest: physics, geometry, chemistry, visual arts, and computer graphic skills. We will examine a concept of crystallization, combine it with art and graphic skills, and then create a composition of crystalline forms by making a geometric composition with lines and patterns. Finally, we will draw a crystal and then, through copying and pasting to selection, create a three-dimensional representation of the interior of a cave using symmetry, repetition, and pattern. Then we will discuss rock art and its preservation problems, and examine modern and contemporary art. Thus, this project makes connection with mathematics (by examining crystals' angles and symmetry or making fractal design), chemistry (by studying how crystals take on a precise crystalline form; we will give special attention to water in a solid state: ice and snowflakes), geology (learning about crystal caves), design principles (for example, when making a geometric composition with lines and patterns, and applying tessellation) and also art and art history (by contemplating rock art, cave art, preservation problems, and at the same time looking at modern and contemporary art). This interactive project involves also an application of spatial visualization skills of the reader.

Physical Concepts and Processes Related to the Structure of Crystals

Before working on the project it may be useful to examine the properties of crystals, such as their chemical composition and structure. We will first examine a crystal in terms of its physical properties, such as size, shape, texture, flexibility, and color. We can model crystalline structures using geometric concepts and describing crystals by means of the horizontal and vertical coordinates. In two and three dimensions, we can draw the geometric relationships in crystals and allocate their points and edges along x, y, and z-axes. Transformations of drawings, such as reflections, translations, and rotations may help extend our knowledge about geometric relationships. The concept of fractal design is helpful in understanding a structure of natural phenomena such as crystals. It allows a pictorial representation of objects that involve patterns of self-similar shapes. It could be helpful to examine the web-based tutorials about how to create fractals with the use of a computer and learn about various kinds of software for creating fractals.

Some substances don't have any order in the array of atoms; we call them amorphous. Many other materials have a crystalline structure, where atoms are set in a three-dimensional pattern. Crystals display such repeating pattern of atoms, *ions*, or molecules, with constant distances between these parts. A set of atoms makes a unit cell of such a pattern and is repeated in a regular way. Some crystals, such as table salt or copper are built of few kinds of atoms and have a simple structure. Some proteins, such as those building teeth, have complex crystalline structure and many kinds of molecules. Physical tests made with the use of light, electricity, pressure, ultrasound, and other factors show the internal symmetry of crystals. Most of the minerals that form the Earth's crust have a crystalline structure. Thus, most rocks, metals, and ice are made of crystals. The most widely known materials have crystalline structure, for example, sand and clay that are used for pottery and ceramic tiles, gypsum (used for finishing wall surfaces), as well as many edible substances, such as sugar and salt. Human-made forms, sculptures, and architectural designs often reveal crystalline structure.

More diamonds are grown rather than produced by mining. People produce crystals for industrial applications and for the purpose of scientific research. Geologists grow crystals in their laboratories and learn about processes involved in building up the mountains. Rubies, sapphires and many synthetic gems are grown for jewelry and industry. Industrially grown diamonds, the crystalline forms of carbon (C), are the hardest of solids. Crystals are very useful in electronics, semiconductor industry, optics, the production of television screens and tapes for videocassettes.

The computer industry uses crystals showing superconductivity. Silicon (Si) crystals are grown in layers to produce integrated circuits. Grown quartz (SiO_2) crystals show piezoelectricity: when their shape is changed by rapid stress, they produce the electric current that is used in piezoelectric circuits. A seeded growth method serves for the fabrication of high-permeance membranes. Grown quartz crystals change shape when a voltage is applied, so they are used for radio oscillators. In watches, quartz crystals modulate the time-keeping circuits. Many kinds of interactions can produce changes in a system, although the total quantities of matter and energy remain unchanged. When we learn about the structure of a crystal, we recognize that solids, unlike liquids and gases, tend to retain memory of the events that changed them, because they preserve their shapes, while liquids and gases keep the shape of the container as long as they are inside.

Crystalline material has been separated from the tobacco-mosaic virus protein in 1933 (Stanley, 1937). We cannot ascribe the crystalline structure to inanimate forms only. A closer look into organic forms on the nano level revealed that some viruses have an ability to organize themselves into liquid crystals.

Thus, for several reasons, "current knowledge about nanostructures makes difficult defining the distinction between organic and inorganic, living and inanimate, natural and artificial, or human and machine" (Cheetham, 2010).

Crystal Caves

On the web and in books we may find pictures of caves built out of crystalline minerals such as calcite, a form of calcium carbonate $CaCO_3$. Examples of such caves come from different parts of the world. Caves are the roofed cavities in the rocks produced by underground water or by the waves of the sea. The biggest caves in North America are: the Carlsbad Caverns in New Mexico (2011), more than 117 limestone caverns located on 46,766.45 acres, that is, 189.26 km^2, and the Mammoth Cave in Kentucky, which is 25 feet high and 4 miles long. The Big Room in the Carlsbad Caverns National Park is the third largest chamber in North America and the seventh largest in the world. The largest in the world is the Sarawak Chamber in Malaysia. Chambers formed when sulfuric acid dissolved the surrounding limestone. Calcite from limestone dissolved and later deposited by water forms beautiful, white, pink or yellow structures called stalactites (which grow downward from the wet roof, like big icicles) and stalagmites (growing upward, like trunks of trees or stacks of saucers). Sometimes they meet, making columns. In many places paintings dating from prehistoric times cover the cave walls and ceilings.

The Stone Age

The Stone Age is a prehistoric period of human culture before the use of metals, and before the use of writing. It is divided into the Old, or Paleolithic age, which lasted from about 3.5 million to 10,000 years ago, the Middle or Mesolithic age, which means "middle stone," and the New, or Neolithic age. The Paleolithic was the stage of hunting and gathering, and the only domesticated animals were dogs. Realistic and symbolic cave paintings created by early humans from before 20,000 years were found in many places: for example, in Altamira (Spain), Lascaux (France), Africa, India, and Australia. Mesolithic started about 8300 BC. Farming and herding practices characterized Neolithic Period, and also the use of small, chipped stone tools. Isolated Stone Age cultures survived on many continents until the 19th and 20th centuries. Estimation of the age of the findings involves many up-to-date technologies, such as radiometric age dating used to measure the amount of radioactive decay of isotopes present in the rocks and in organic tissues. Isotopes are pairs of atoms that have the same number of protons (atomic numbers) but a different number of neutrons. In radioactive decay, one atom is converted into another one with emission of radiation. The amount of changed isotopes in a rock gives a measure of its age.

Art in the Caves

Prehistoric paintings and engravings produced on rock surfaces of caves by their early non-literate inhabitants have been discovered mostly in the 19th and 20th centuries. Thousands of sites in America, Europe, Asia, Africa, and Oceania contain ancient art, with thousands, and even millions of figures carved, engraved, or painted on the walls of rock shelters and caves. Drawings of bison, horses, deer, and cattle are in the caves of Altamira in northern Spain, Lascaux in France, and in New Mexico in the USA. The authenticity of the Paleolithic, Mesolithic, or Neolithic paintings is accepted many years after the discovery of a cave, after examination with the use of many methods; for example, with X-ray analyses. The earliest cave art was created some 32,000 years ago by the nomadic tribes. North American Indians and the tribes who traveled on the plains and in the woods of Europe were hunting animals, eating their meat, employing the skins for clothing and building shelters, and using bones and antlers as tools, weapons, and domestic utensils. Maybe they created their beautiful artwork to communicate with other people or the gods.

Pictographs – pictorial symbols for phrases and objects were used in prehistoric times, for example, for creating drawings and paintings on rock walls. Ancient people created beautiful images on the cave walls and ceilings, providing an artistic expression and an account of everyday human life. However, the basic subject is the animal. Some images were engraved with stone tools or bones on the rock surfaces. Decorated bones, dear-antlers and stone tools were found nearby in the caves. Ancient artists made their drawings with charcoal, and created paintings often using water solution of iron oxides and the golden-yellow earth pigment called ochre, so the rocks absorbed the paints. When the treasures of ancient art were discovered after tens of thousands years, they were well preserved in the stable humidity and temperature of the caves. When caves were opened to the public, many problems of conservation arose due to the physical, chemical, and microbiological reason. Rocks began to fall, fade and peel of the paint due to the cave lighting, and sometimes acts of vandalism happened as well. For this reason, cave art specialists built in many locations the replicas of the caves and reproductions of the cave paintings, while the original caves have been closed to the public or they are open for a very limited number of people interested in scientific research and education. In his film "Roma," Fellini shows a moment of discovery of Roman frescoes, which fade in front of the viewers' eyes the moment the flashlight affects them.

Modern and Contemporary Art Connections

Visual communication in fine arts is often based on a metaphor, where one thing developed in the mind represents another thing, so each viewer may receive the unique and individual message from the same artwork. For example, abstract sculptures created by the American artist Joel Shapiro (born 1941, accessible at the Google Images) in the Minimalist style from regular blocks, arranged in larger-than-life, human-like forms may evoke a feeling of movement.

On the web or using books on art we may examine reproductions of works of artists inspired by the beauty of the natural order. For example, natural forms served the American artist Aaron Siskind (1903-1992), as a theme for almost abstract photography, such as "Martha's Vineyard." We can see how often geometric shapes inspired artists as a source for creating art. Fascination with patterns and regular arrangements inspired a great many artists in the 20th century. For example, Frantisek Kupka (1871-1957) composed his imaginative "Cathedral" of geometrical shapes filled with colors. Lyonel Feininger (871-1956) composed his "Sailing Boats" from overlapping triangles of color. In the Constructivism style, Ljubov Popova (1889-1924) set up the "Space-Force Construction" and Naum Gabo (1890-1977), created a delicate constructivist sculpture "Linear Construction in Space No.2" from a complex 3-dimensional arrangement of nylon strings. William Latham is a creator of the Organic Art that can be seen on his website at http://doc.gold.ac.uk/~latham/. Latham and Todd created in the early 1990s the FormGrow system, which translated DNA data to 3D computer art forms. They continue their work (Latham, Shaw, Todd, Leymarie, Jefferys, & Kelley, 2008) by developing biological software that performs analysis of the biochemical properties of the proteins encoded by genes in DNA, and then controlling the parameters of a fixed FormGrow structure.

Crystallization: Snowflakes Have a Crystalline Design

Crystals take diverse kinds of forms, many times evoking our wonder because of their beauty. Crystallography is a study of crystals: their form, growth, structure, chemistry, bonding, and physical properties. The nature of binding among atoms, compounds, and molecules determines the way they form crystals. When crystals grow, they develop various arrangements of symmetry axes. For example, snowflakes grow with great variety of designs, with their shape depending on the internal pattern. In most cases, the outer surface of a crystalline matter (for example, of an ice cube) does not reveal the arrangement of crystals. Only certain angles of rotational symmetry are possible for crystals: 60 degrees, 90 degrees, 120 degrees, 180 degrees, and 360 degrees. Due to the possible combinations, there are 32 crystal classes, which are grouped into the 7 crystallographic systems.

Sometimes water in the atmosphere changes directly from gas to solid, omitting the liquid phase; we can see the result as the snow. Snowflakes, looking like the six-armed stars or feathery jewels, are actually ice crystals, sometimes as large as six inches (15 centimeters) from one side to the other. Snowflakes develop when the supercooled cloud droplets, which can remain liquid even at 0^0 F, start to freeze around an ice lattice arranged around a few molecules of different kinds acting as a nucleus (Glossary of Meteorology, 2013). When examining a snowflake, we may think about the elements and principles of design, such as shape, or symmetry, and see the uniqueness of their structures. We can see rotational (mostly six-fold) symmetry. This kind of symmetry occurs in nature in flowers, fruits, starfish, jellyfish, and other sea animals. Leonardo da Vinci conducted studies on rotational symmetry. It occurs universally in decorative arts, prehistoric and contemporary.

Most of us accept that regular patterns are perceived as beauty in nature: it may be a natural meander of a river; it may also be a fractal design of a mollusk shell, a snowflake, a pinecone, a sunflower, or a foliage arrangement on a tree, many of them displaying properties of cellular automata. Common, primary, simple elements of our environment, such as raindrops, snowflakes, and simplified animal shapes might serve as a carrier of a notion. The use of natural metaphors might yield art works that are massively structured in cognitive terms. We may find an extensive use of metaphors in literature, visual arts, and music. In a visual mode of communication, images that have iconic properties or serve as generally accepted symbols. They are considered crucial to visualization of the hard to explain objects and notions.

Simple subjects, like snowflakes, are often being pictured in art works, serving as metaphorical statements about the theme itself and the artist's thoughts on the theme. It might be interesting to understand

Figure 9. Emily Andrus (left), Leonard Rodriguez (right), "Snowflakes"
(© 2010, E. Andrus, © 2011, L. Rodriguez. Used with permission.)

rules that control movements of a single snowflake, then compare and contrast them with dynamics of a composition created for the ballet dancers, and examine to what degree nature might inspire visual and performing artists. Years ago, in the times when photography was quite new, and black-and-white only, Wilson A. Bentley (1865-1931) created more than 5000 photographs of snow crystals, not finding any two alike. Bentley had done his pioneering work on photomicrography by connecting a folding (bellow) camera to a microscope. Many versions of his book entitled "Snow Crystals" (Bentley, 1995/1931) are still available in bookstores and on the web. Images of crystals, never repeating their awesome shapes, can be seen at the Google Images after typing 'Bentley snowflake.' Figure 9 shows variants of the snowflake shapes created by students Emily Andrus and Leonard Rodriguez.

Natural phenomena such as the snowflakes, involve patterns. Fractals, which are usually generated by computers, represent such patterns. Fractal-generating applets may be useful in bringing into being images of snowflakes or snowy landscapes. You may then examine your final product made up by grouping of the single structures. The pattern of snow falling on the city street transforms the way we see an order of windows and the whole town. By lowering the brightness and contrast, snowflakes intensify each other to the point of dizziness of the viewer. To picture this effect, you may want to experiment with the shades of white and to show depth in your composition. In other respect, you may want to develop an interior scene with a pattern made with snowflakes, with light passing through a window.

Now it's a good time to create images of snowflakes, on paper or on the computer. Apply transformations and effects in Adobe Illustrator, or simply, cutouts of snowflakes: choose an image as a background and then paste multiplied and resized images of snowflakes as a new layer onto the background image.

Creating a "Crystal Cave" Project

On the web or in illustrated books examine pictures of various kinds of crystals. Some of them are famous jewels; some, for example quartz crystals, may be found easily both in the countryside and in stores. Create a drawing in your sketchbook of a crystal using traditional tools, such as pencil or chalk. Then, make a simple three-dimensional model out of wooden sticks or matches in order to explain the crystalline structure of matter. Explain properties and composition of crystals through two- and three-dimensional models and visualizations made with the use of color and pattern coding.

Geometric Composition Derived from an Image of a Crystal

While working on the project, apply your spatial visualization skills – spatial reasoning skills to draw upon your understanding of the three-dimensional world. Build a three- dimensional representation of a cave out of two-dimensional pictures. Figure 10 shows stages of work in progress, first by drawing a single crystal, and then copying an image and repeating it over the surface of the drawing.

Looking at a sample of a crystal, draw an angle of your choice. While working on the project, apply geometric concepts and examine shapes of crystals and their symmetry, thus recognizing geometry in

Figure 10. a. Tory Wagoner, Mike Wyckoff; b. Loren Music, Peter Vrazsity
(a: © 1992, T. Wagoner and M. Wyckoff; b: © 1992, L. Music and P. Vrazsity. Used with permission.)

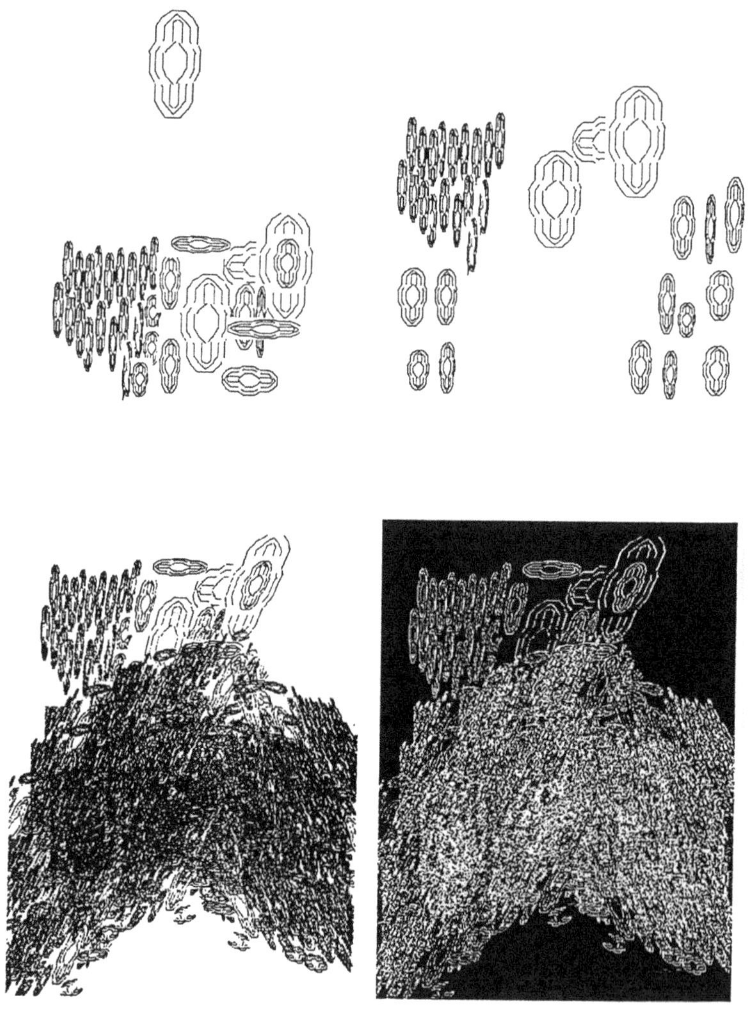

nature. Draw a single crystal, then copy and repeat the image over the surface of your drawing. Look, the angle in a crystal does not change! An angle between any pair of crystal faces remains constant; it is the same for all minerals. The crystal may change the size but not an angle, because in a crystal, the angle between any pair of crystal faces is constant, the same for all pieces of a mineral, regardless of its overall shape or size. Measure the angles on the drawings and check the crystals to see whether they keep the same angle. This structure cannot be seen in amorphous minerals: that is, in minerals displaying no pattern to the arrangement of atoms.

You may also generate this project in another way, with any graphic software or by writing a program. Create an image of a crystal with the use of any graphic software, or, if you have some experience in programming, you may write a program for developing a structure of a crystal. When you write a program, remember that a small change in a program makes a big difference in a crystal design. This is the way the connoisseurs appreciate the works of art. Geometric composition derived from an image of a crystal may also serve as a starting point for creating an animation about growing crystals.

Abstract Composition Showing Symmetry and Pattern

While working on the project, recognize and use your art as a form of communication, so by applying visual art materials, tools, techniques, and processes, you will make connection between visual arts and other disciplines. Explore the concepts of symmetry (as a repetition of units) and pattern while drawing direct representations of crystals and crystalline materials. Use and modify patterns, change proportions, and repeat selected details, changing their scale and slant, distorting or rotating them, and applying perspective. Manipulate the image by using color selections, adding lines, applying textures, and enhancing the depth of the picture.

Combine these skills with concepts related to crystals to create artwork that visually conveys your understanding of the crystallography-related concepts, and communicate your experiences gained in this assignment. Using samples of crystals and drawings of crystal lattices as the stimulus for creating art, work on your composition of crystals applying symmetry and pattern. Transform the image of the crystal the same way as nature does.

Make an imaginative drawing of a crystal cave built out of previously created crystal rocks. Contrary to the Paleolithic tradition, create an interior of a cave in the spirit of contemporary art. Explore the concept of symmetry. Copy a small part of your abstract composition and use symmetry and pattern to create a composition that would embellish the cave and reflect its crystalline structure. Everybody can create their own solution to this art production and create a different image of the cave interior, because everybody began with selecting different angle and building different crystals. You may experiment with a few different angles and compare the results. Show the effect of the movement of the tectonic plates. How does it look when rocks are heated, bent, twisted, squeezed, stretched, and the crusted fragments collide? Try to imagine what kind of light might be found in a cave, and how it may affect your image.

Now you may want to write a story in your journal, on your listserv, or elsewhere on the web about your imaginative trip to a cave, and then you may incorporate this story into your picture using the story itself as a pattern for the background. Now, you may want to create an animated story about growing crystals in a crystal cave. You can also apply a design with the flipbook approach, where, in order to put things in motion, each next image has added lines. Again, you may use a concept of fractal design. You may also want to participate in a web discussion that you may initiate, about different uses of pattern (as a repetition of units) in the design created in this project, as well as about the use of symmetry.

A Fractal Design Composition

While working on this project, it would be helpful to examine some information about the developments in electronic media for art and learn about utilizing fractals as an artistic tool in art. To create a cave interior, build a composition out of the small elements. For this project, you may use a concept of fractal design. We may build a simple fractal by dividing a line in 3 parts and removing the middle part. When we repeat this series of steps, first on the 2 remaining parts, then on the 4 parts produced by that operation, and so on, the object has a great number of parts, each of which is very small. On the web, there are many sites about fractals, helpful fractal tutorials, and beautiful examples of fractal art.

Artwork resulting from writing a program is often as pleasing and impressive as nature itself, despite any possible remarks and disagreements about the essence of art and about conditions that identify what it is. You may also prefer to design a fractal image, with geometric shapes that are self-similar and have fractional dimensions.

CONCLUSION

The chapter "Mirroring Nature: Symmetrical and Crystalline Structures Derived from Natural Forms" discusses possibilities of making visualizations about connection between science, computing, and art. This chapter provides two projects about science-related themes:

1. Symmetry and pattern in animal world – geometry and art, and
2. Crystals and crystal caves.

Projects are aimed at supporting the readers' skills in visualization and scientific illustration for the purpose of medicine and pharmacology applications. The readers are invited to engage in creating visual presentation of each theme. The projects are aimed at enhancing the readers' skills in using visual communication media to gather interesting facts and ideas and actively respond to them in a pictorial way.

REFERENCES

Acosta, D. (2012). *WebGL Demo – The X Toolkit for Scientific Visualization*. Retrieved March 10, 2014, from http://www.webgl.com/2012/03/webgl-demo-the-x-toolkit-for-scientific-visualization/

Animals in Medieval Art. (2011). *Heilbrun Timeline of Art History*. Department of Medieval Art and The Cloisters, The Metropolitan Museum of Art. Retrieved March 14, 2014, from http://www.metmuseum.org/toah/hd/best/hd_best.htm

Arnheim, R. (1969). Visual Thinking. Berkeley, CA: University of California Press. (Also: London: Faber and Faber, 1969).

Arnheim, R. (1988). *The Power of the Center, A Study of Composition in the Visual Arts*. Berkeley, CA: University of California Press.

Arnheim, R. (1990). Language and the Early Cinema. *Leonardo, Digital Image Digital Cinema.*

Arrigoni, M., & Gallaher, N. (2015, February-March). Diverse Applications Drive Lasers for Multiphoton Microscopy. *BioPhotonics*, 32-36.

Association of Medical Illustrators. (2014). Retrieved March 16, 2014, from www.ami.org/medical-illustration/education.html

Bennett, C., Ryall, J., Spalteholz, L., & Gooch, A. (2007). *The Aesthetics of Graph Visualization. In Computational Aesthetics in Graphics, Visualization, and Imaging* (pp. 57–64). Eurographics Association.

Bentley, W. A. (1995). *Snowflakes in Photographs*. Mineola, NY: Dover Publications. (Original work published 1931)

Berens, R. R. (1999). The Role of Artists in Ship Camouflage During World War I. *Leonardo, 32*(1), 53-59.

Bertschi, S., Bresciani, S., Crawford, T., Goebel, R., Kienreich, W., Lindner, M., & Moere, A. V. et al. (2011). What is Knowledge Visualization? Perspectives on an Emerging Discipline. *Proceedings of the Information Visualisation 15th International Conference*. doi:10.1109/IV.2011.58

Birkhoff, G. D. (2003). *Aesthetic Measure*. Kessinger Publishing. (Original work published 1933)

Bolin, L. (2014). *Camouflage art*. Retrieved March 15, 2014, from http://www.toxel.com/inspiration/2009/10/04/camouflage-art-by-liu-bolin/

Bredl, K., Groß, A., Hünniger, J., & Fleischer, J. (2012). The Avatar as a Knowledge Worker? How Immersive 3D Virtual Environments may Foster Knowledge Acquisition. *Electronic Journal of Knowledge Management*, *10*(1), 15–25.

Camouflage Restaurants. (2014). Retrieved March 15, 2014, from http://www.toxel.com/inspiration/2009/06/20/10-unusual-and-creative-restaurants/

Carlsbad Caverns National Park, New Mexico. (n.d.). National Park Service, U.S. Department of Interior. Retrieved March 15, 2014, from http://www.nps.gov/cave/index.htm

Chang. (2014). Nobel Laureates Pushed Limits of Microscopes. *The New York Times*. Retrieved February 8, 2015, from http://www.nytimes.com/2014/10/09/science/nobel-prize-chemistry.html?_r=0

Cheetham, M. A. (2010). The Crystal interface in contemporary art: Metaphors of the organic and inorganic. *Leonardo*, *43*(3), 251–255. doi:10.1162/leon.2010.43.3.250

D'Arcy, W., & Thompson. (1917/2011). On Growth and Form. CreateSpace Independent Publishing Platform.

Doolittle, B. (2014). *Camouflage Art*. Retrieved March 15, 2014, from http://www.bnr-art.com/doolitt/

Einstein, A., Podolsky, B., & Rosen, N. (1935). Can Quantum-Mechanical Description of Physical Reality Be Considered Complete? *Physical Review, 47*, 777-780.

El-Sourani, N., Hauke, S., & Borschbach, M. (2010). An Evolutionary Approach for Solving the Rubik's Cube Incorporating Exact Methods. In C. Di Chio et al. (Eds.), *Applications of Evolutionary Computing, LNCS 6024 EvoApplications 2010 Proceedings* (pp. 80–89). Berlin: Springer.

Ferguson, C., & Ferguson, H. (1994). *Mathematics in Stone and Bronze*. Erie, PA: Meridian Creative Group.

Gasmelseid, T. M. (2008). Engineering Multi-agent Systems. In M. Quigley (Ed.), *Encyclopedia of Information Ethics and Security* (pp. 187–193). IGI Global.

Gasmelseid, T. M. (2011). Improving Medical Practice through Mobile Medical Informatics In Clinical Technologies: Concepts, Methodologies, Tools and Applications. IGI Global.

Glossary of Meteorology. (2013). Retrieved March 15, 2014, from http://glossary.ametsoc.org/?p=1&query=cloud+seeding&submit=Search

Grujović, N., Radović, M., Kanjevac, V., Borota, J., Grujović, G., & Divac, D. (2011). 3D printing technology in education environment. *34th International Conference on Production Engineering* (pp. 29-30).

Haeckel, E., Breidbach, O., Hartman, R., & Eibl-Eibesfeldt, I. (1998). *Art Forms in Nature: The Prints of Ernst Haeckel (Monographs)*. München: Prestel Publishing. (Original work published 1904)

Holothurian. (2011). In *Encyclopædia Britannica*. Retrieved March 14, 2014, from http://www.britannica.com/EBchecked/topic/269645/holothurian

Huang, Q., & Balsys, R. J. (2009). Applying Fractal and Chaos Theory to Animation in the Chinese Literati Tradition. *Sixth International Conference on Computer Graphics, Imaging and Visualization* (pp.112-122). Los Alamitos, CA: IEEE Computer Society. doi:10.1109/CGIV.2009.56

Irwin, J. L., Opplinger, D. E., Pearce, J. M., & Anzalone, G. (2015). Evaluation of RepRap 3D Printer Workshops in K-12 STEM (Program/Curriculum Evaluation). *122nd ASEE Annual Conference & Exposition*. doi:10.18260/p.24033

IUPAC. (2014). *Compendium of Chemical Terminology* (2nd ed.). Retrieved March 15, 2014, from http://www.iupac.org/home/publications/e-resources/nomenclature-and-terminology.html

James, A. V., James, D. A., & Kalisperis, L. N. (2004). A unique art form: The friezes of Pirgi. *Leonardo*, *37*(3), 234–242. doi:10.1162/0024094041139409

Jones, J. (2005). *Fractals & Frequencies*. Retrieved March 15, 2014, from http://fractalsandfrequencies.blogspot.com/

Jones, O. (2001, 1856). The Grammar of Ornament: Illustrated by Examples from Various Styles of Ornament. London: A Dorling Kindersley Book, The Ivy Press, Limited.

Kenny, P., Parsons, T. D., Gratch, J., Leusli, A., & Rizzo, A. (2007). Virtual Patients for Clinical Therapist Skills Training. In *Proceedings of the 7th International Working Conference, IVA 2007*. doi:10.1007/978-3-540-74997-4_19

Kettle, S. F. A. (2007). Symmetry and Structure: Readable Group Theory for Chemists (3rd ed.). Wiley-Blackwell.

Khetrapal, N. (2011). Cognition Meets Assistive Technology. In Clinical Technologies: Concepts, Methodologies, Tools and Applications, (pp. 779-791). IGI Global. doi:10.4018/978-1-60960-561-2.ch310

Latham, W., Shaw, M., Todd, S., Leymarie, F. F., Jefferys, B., & Kelley, L. (2008). Using DNA to Generate 3D Organic Art Forms. In M. Giacobini et al. (Eds.), *Applications of Evolutionary Computing, LNCS 4974 2008 EvoWorkshops Proceedings* (pp. 433–442). Berlin: Springer. doi:10.1007/978-3-540-78761-7_46

Lau, A., & Vande Moere, A. (2007). Towards a Model of Information Aesthetics in Information Visualization. In *Proceedings of 11th International Conference on Information Visualisation* (pp. 87-92). Los Alamitos, CA: IEEE Computer Society. doi:10.1109/IV.2007.114

Leffingwell, J. C. (2003). Chirality and Bioactivity I: Pharmacology. *Leffingwell Reports, 3*(1). Retrieved March 15, 2014, from http://www.leffingwell.com/download/chirality-phamacology.pdf

Lévi-Strauss, C. (1997). Look, Listen, Learn. Basic Books.

Leyton, M. (2006). The Foundations of Aesthetics. In Aesthetic Computing (pp. 289-314). Leonardo Books.

Li, K.-Y. R., Sofra, J., & Power, M. (2008). 3D Avatars and Collaborative Virtual Environments. In M. Quigley (Ed.), *Encyclopedia of Information Ethics and Security* (pp. 1–6). IGI Global.

Loeb, A. L. (1993). *Concepts & Images: Visual Mathematics. Design Science Collection.* Boston: Birkhäuser. doi:10.1007/978-1-4612-0343-8

Ma, J. K. H., & Hadzija, B. (2012). Basic Physical Pharmacy. Jones & Bartlett Learning.

Mercuri, R., & Meredith, K. (2014). An educational venture into 3D Printing. *Integrated STEM Education Conference (ISEC).* IEEE.

Mitchell, K. (2011). *A Celebration of Beauty.* Digital Art Guild; Art through Technology. Retrieved March 15, 2014, from http://www.digitalartguild.com/content/view/56/26/

Neville, A. C. (1977). Symmetry and Asymmetry Problems in Animals. In R. Duncan & M. Weston-Smith (Eds.), *The Encyclopedia of Ignorance: Everything you ever wanted to know about the unknown* (pp. 331–338). Oxford, UK: Paragon Press.

Okubo, P. G., & Aki, K. (1987). Fractal Geometry in the San Andreas Fault System. *Journal of Geophysical Research, 92*(B1), 345–355. doi:10.1029/JB092iB01p00345

Packard, A. (2001). A 'neural' net that can be seen with the naked eye. In Neuronal coding of perceptual systems (pp. 397-402). World Scientific.

Prusinkiewicz, P., Lindenmeyer, A., Hanan, J. S., Fracchia, F. D., Fowler, D. R., de Boer, M. J. M., & Mercer, L. (1991). *The Algorithmic Beauty of Plants (The Virtual Laboratory)* (1st ed.). Springer.

Purchase, H. (2010). Graph Drawing Aesthetics in User-sketched Graph Layouts. In *AUIC'10 Proceedings of the Eleventh Australasian Conference on User Interface.*

Rankin, D. W. H., Mitzel, N., & Morrison, C. (2013). *Structural Methods in Molecular Inorganic Chemistry (Inorganic Chemistry: A Textbook Series)* (1st ed.). Wiley.

Ravikumar, R., & Mohsin Khan, I. (2015). Design & Development of a 3D Printer. *Proceedings of 12th IRF International Conference.*

Schelly, C., Anzalone, G., Wijnen, B., & Pearce, J. M. (2015). Open-Source 3-D Printing Technologies for Education: Bringing Additive Manufacturing to the Classroom. *Journal of Visual Languages and Computing*, *28*, 226–237. doi:10.1016/j.jvlc.2015.01.004

Schilthuizen, M., & Gravendeel, B. (2011). Evolution of Chirality Symposium. *The 13th Congress of the European Society for Evolutionary Biology*. Retrieved August 26, 2011, from http://www.eseb2011.de/

Stanley, W. M. (1937). Crystalline Tobacco-mosaic Virus Protein. *American Journal of Botany*, *24*(2), 59–68. doi:10.2307/2436720

The 20th Century Artbook. (1996). San Francisco, CA: Chronicle Books/Phaidon Press.

The American Art Book. (1999). London: Phaidon Press.

The Artbook. (2012). Phaidon Press Inc.

Tufte, E. R. (1983/2001). *The Visual Display of Quantitative Information* (2nd ed.). Cheshire, CT: Graphics Press.

Tufte, E. R. (1990). *Envisioning Information* (2nd ed.). Cheshire, CT: Graphics Press.

Viola, I., Birkeland, Å., Solteszova, V., Helljesen, L., Hauser, H., Kotopoulis, S., Nylund, K., Ulvang, D. M., Øye,O. K., Hausken, T., & Gilja, O. H. (2013). *High-Quality 3D Visualization of In-Situ Ultrasonography*. DOI: 10.2312/conf/EG2013/med/001-004

Viola, I., & Gröller, E. (2007). On the Role of Topology in Focus+Context Visualization. In H. Hauser, H. Hagen, & H. Theisel (Eds.), *Topology-based Methods in Visualization: Mathematics and Visualization* (pp. 183–199). Springer. doi:10.1007/978-3-540-70823-0_12

Voss-Andreae. (2011). Quantum Sculpture: Art Inspired by the Deeper Nature of Reality. *Leonardo, Journal of the International Society for the Arts, Sciences and Technology*, *44*(1), 14–20.

Zeki, S. (1993). Vision of the Brain. Wiley-Blackwell.

Zeki, S. (1999). Art and the Brain. *Journal of Conscious Studies: Controversies in Science and the Humanities*, *6*(6/7), 76–96.

ADDITIONAL READING

Arnheim, R. (1974). *Art and Visual Perception*. Berkeley: University of California Press.

Briggs, J. (1992). *Fractals, The Patterns of Chaos. A New Aesthetic of Art, Science, and Nature*. Touchstone Books.

Falconer, K. (2003). Fractal Geometry: Mathematical Foundations and Applications. John Wiley & Sons, Ltd. xxv. ISBN 0470848626. doi:10.1002/0470013850

Flake, G. W. (2000). *The Computational Beauty of Nature: Computer Explorations of Fractals, Chaos, Complex Systems, and Adaptation*. Bradford Books.

Gleick, J. (2000). *Faster: The Acceleration of Just About Everything*. Vintage Books.

Harrison, A. (1996). *Fractals in Chemistry (Oxford Chemistry Primers, 22)*. Oxford Univ. Press.

Henisch, H. K. (1988). *Crystals in Gels and Liesegang Rings*. Cambridge University Press. doi:10.1017/CBO9780511525223

Henisch, H. K. (1996). *Crystals in Gels*. Dover Pubs.

Holden, A. (2012). *Shapes, Space, and Symmetry. Dover Books on Mathematics*. Dover Publications.

Holden, A., & Morrison, P. S. (Photographer), (1982). Crystals and Crystal Growing. The MIT Press; 1st MIT editon. ISBN 0262580500.

Novak, M. M. (Ed.). (2004). *Thinking in Patterns: Fractals and Related Phenomena in Nature*. World Scientific Pub Co Inc.

Sands, D. E. (1994). *Introduction to Crystallography (Dover Classics of Science and Mathematics)*. Dover Pubs.

Schroeder, M. (1992). *Fractals, Chaos, Power Laws. Minutes from an Infinite Paradise*. W H Freeman & Co.

Stevens, R. (2005). Creating Fractals (Graphics Series). Cengage Learning; 1 edition. ISBN 1584504234.

The 20th Century Artbook. (1996). Phaidon Press.

The American Art Book. (1999). London: Phaidon Press.

The Art Book. (1994). London: Phaidon Press. ISBN 0714829846 (available in a big format – hard or soft cover, or a pocketsize format).

The Art Book. (2012, first published 1994). Phaidon Press Inc. ISBN 9780714864679.

Tufte, E. R. (1983/2001). *The Visual Display of Quantitative Information* (2nd ed.). Cheshire, Connecticut: Graphics Press.

Tufte, E. R. (1990). *Envisioning Information* (2nd ed.). Graphics Press.

Tufte, E. R. (1997). *Visual Explanations: Images and Quantities, Evidence and Narrative*. Graphics Press.

Tufte, E. R. (2006). *Beautiful Evidence*. Graphics Press.

Tukey, J. W. (1977). Exploratory Data Analysis (1 ed.). Pearson. ISBN 0201076160.

Wynn, C. M., & Wiggins, A. M. (1996). *The Five Biggest Ideas in Science*. Wiley.

KEY TERMS AND DEFINITIONS

Algorithm: A mathematical recipe, a sequence of instructions telling how to proceed computation to implement it as a program. Algorithms are actually mathematical equations used to create repetition. An algorithm is a procedure for solving a complicated problem by carrying out a fixed sequence of simpler, unambiguous steps. A recursive process means that an algorithm is applied multiple times to perform operations on its previous products. Such procedures are used in computer programs and in programmed learning.

Anamorphosis: A way of distorting perspective in such a way that an image seems to be meaningless until viewed from a particular angle. Anamorphic drawings have appeared in the Renaissance art. Leonardo da Vinci experimented with such optical illusions. Hans Holbein the Younger placed an anamorphic picture of a skull in his painting "The Ambassadors" (1533). It is visible only when viewed from close up and to one side of the painting.

Cellular Automaton: A discrete model consisting of a grid of cells, each in a finite number of states (such as 'on' or 'off'), in sets called neighborhoods. The fixed rules, such as mathematical functions, determine the creation of new generations of sets with the new states of cells. Plant leaves, seashells, and even neural networks are examples of naturally occurring cellular automata.

Crystal Binding: There are many types of crystal binding: covalent (sharing electrons in pairs), ionic (with a charge transfer from cations to anions), metallic (with cationic atoms weakly bound to the ion cores), and molecular (with van der Waals interactions between molecules). In most crystals, the binding is a mixture of ionic, covalent and metallic binding, with a three-dimensional array of unit cells.

Crystallography: The study of the crystal's form, growth, physical properties resulting from its structure, the nature of the bonding among its atoms, and its chemical composition. Molecular biologists and organic chemists are often crystallographers. They make use of crystallographic data to examine the structure of organic molecules and ways to concentrate and crystallize the molecules in plants and animals. For example, Rosalind Franklin and others examined DNA crystals using X-ray diffraction. In 1952, James D. Watson and Francis Crick proposed the double helix structure of the DNA molecule determined with the use of crystallographic data.

Evo-Devo: An informal term that means evolutionary developmental biology. This is a study of evolution and generation of form and pattern, through research on comparative gene function/expression, embryogenesis, genomics, phylogenetics, and paleontology, among other venues. Materials on this topic can be found in the EvoDevo Journal, (http://www.evodevojournal.com/).

Heisenberg's Uncertainty Principle: A part of quantum mechanics. This principle states that we cannot know precisely about certain pairs of physical properties, such as a particle's position and momentum (the product of the mass of a particle and its velocity) at the same time, because the measuring process involves interaction, which disturbs the particle. For example, a photon of light used in a measurement is bouncing off the particle. Thus, one cannot, even theoretically, predict the moment-to-moment behavior of a system consisting of the subject and the object of examination (somebody who makes an observation and an object observed). There is a theoretical limit for simultaneous measuring at an atomic scale because the more precisely is figured one amount, the more uncertain is the other one.

Ions – Cations and Anions: An atom or a molecule that has an electric charge is called an ion. An electric charge results from the presence of single, double, triple, or even higher negative electrons unequal to the number of positive protons in the nucleus of an atom. The removal or addition of one or more electrons changes a neutral atom into an ion. Cation is an ion or group of ions that have a positive charge. Anion is a negatively charged ion. Polyelectrolytes are large molecules with many charged groups. During electrolysis (produced in an electrolyte solution by applying an electric current) cations move toward the cathode (negative electrode) and anions migrate to an anode.

Isomers: Chemical compounds that have the same molecular formula and mass but different structural formulas. Isomerism may be structural (when atoms are bonded together in different orders) or spatial (when atoms are placed in different positions in space). They may some different physical or chemical properties caused by different arrangement of atoms in their molecules. For example, ethyl alcohol (CH_3CH_2OH) and methyl ether (CH_3OCH_3) contain the same atoms bonded in different ways (Retrieved November 18, 2010 from http://www.britannica.com/EBchecked/topic/378561/methyl-ether).

Julia Sets: The function for Julia sets is f (z)=zA2-i. Each point in the Julia set must be passed through a function many times to determine whether the point will eventually go to infinity. Julia set uses one equation that does not change, while in the Mandelbrot set equation changes with the point that is being plotted.

Mandelbrot Set: The function for the Mandelbrot set is z (n+ 1)+ z (2/n)+c (c is a constant). Fractal shapes are distorted from one length to another and retain some degree of self-similarity. In these nonlinear fractals, the more the set is magnified the more its unpredictability increases.

Pattern: An artistic or decorative design made of recurring lines or any repeated elements. We can see patterns everywhere in nature, mathematics, art, architecture, and design. A pattern makes a basis of ornaments, which are specific for different cultures. Owen Jones (1856) made a huge collection of ornaments typical for different countries. He wrote an amazing monographic book entitled "The Grammar of Ornament."

Permeance: Denotes the degree to which a material allows a flow of magnetic energy. In electromagnetic circuits permeance is usually larger for cross-sections.

Symmetry: The correspondence in size, form, and arrangement of parts on opposite sides of a plane, line, or point. A crystal shows symmetry when it has s a center of symmetry, rotation axes, or mirror planes (imaginary planes that divide it into halves). There are several types of symmetry: for example, line or mirror symmetry, radial, cylindrical, or spherical symmetry. A figure that has line symmetry has two identical halves when folded along its line of symmetry, and these halves are congruent, meaning they are the same size and shape. An object has a radial symmetry when it can be rotated around the rotation axis. For example, with a fourfold rotation axis the crystal repeats itself each 90°. Angles of rotational symmetry possible for crystals are: 60 degrees, 90 degrees, 120 degrees, 180 degrees, and 360 degrees. The halves of the bilaterally symmetrical animals, for example, butterflies, when seen along the axis, form each other's mirror images. Most animals and people cannot be divided into two identical halves, even when they look symmetrical from external appearance. Two halves of the human brain display different abilities and ways of learning and thinking.

Compilation of References

Abbasalipourkabir, R., Salehzadeh, A., & Abdullah, R. (2011). Delivering tamoxifen within solid lipid nanoparticles. *Pharmaceutical Technology Europe*, *23*(4), 22–32.

Abellán, G., Carrasco, J. A., & Coronado, E. (2013). Room temperature magnetism in layered double hydroxides due to magnetic nanoparticles. *Inorganic Chemistry*, *52*(14), 7828–7830. doi:10.1021/ic400883k PMID:23795549

Abellán, G., Carrasco, J. A., Coronado, E., Romero, J., & Varela, M. (2014). Alkoxide-intercalated CoFe-layered double hydroxides as precursors of colloidal nanosheet suspensions: Structural, magnetic and electrochemical properties. *Journal of Materials Chemistry C*, *2*(19), 3723–3731. doi:10.1039/c3tc32578d

Abellán, G., Coronado, E., Martí-Gastaldo, C., Pinilla-Cienfuegos, E., & Ribera, A. (2010). Hexagonal nanosheets from the exfoliation of Ni^{2+}-Fe^{3+} LDHs: A route towards layered multifunctional materials. *Journal of Materials Chemistry*, *20*(35), 7451–7455. doi:10.1039/c0jm01447h

Abellán, G., Coronado, E., Martí-Gastaldo, C., Ribera, A., & Sánchez-Royo, J. F. (2012). Layered double hydroxide (LDH)–organic hybrids as precursors for low-temperature chemical synthesis of carbon nanoforms. *Chemical Science*, *3*(5), 1481–1485. doi:10.1039/c2sc01064j

Abe, S., Handa, H., Takahashi, R., Imaizumi, K., Fukidome, H., & Suemitsu, M. (2010). Surface chemistry involved in epitaxy of graphene on 3C-SiC(111)/Si(111). *Nanoscale Research Letters*, *5*, 1888–1891. PMID:21170403

Abrefah, J., Olander, D. R., Balooch, M., & Siekhaus, W. J. (1992). Vapor pressure of Buckminsterfullerene. *Applied Physics Letters*, *60*(11), 1313–1314. doi:10.1063/1.107327

Abrikosov, A. A. (1957). On the magnetic properties of superconductors of the second group.[English Translation]. *Soviet Physics, JETP*, *5*, 1174–1182.

Acosta, D. (2012). *WebGL Demo – The X Toolkit for Scientific Visualization*. Retrieved March 10, 2014, from http://www.webgl.com/2012/03/webgl-demo-the-x-toolkit-for-scientific-visualization/

Adini, A. R., Redlich, M., & Tenne, R. (2011). Medical applications of inorganic fullerene-like nanoparticles. *Journal of Materials Chemistry*, *21*(39), 15121–15131. doi:10.1039/c1jm11799h

Afzal, S., Daoud, W. A., & Langford, S. J. (2012). Self-cleaning cotton by porphyrin-sensitized visible-light photocatalysis. *Journal of Materials Chemistry*, *22*(9), 4083–4088. doi:10.1039/c2jm15146d

Afzal, S., Daoud, W. A., & Langford, S. J. (2013). Photostable self-cleaning cotton by a copper(II) porphyrin/TiO_2 visible-light photocatalytic system. *ACS Applied Materials & Interfaces*, *5*(11), 4753–4759. doi:10.1021/am400002k PMID:23465549

Aguilera-Sigalat, J., Casas-Solvas, J. M., Morant-Miñana, M. C., Vargas-Berenguel, A., Galian, R. E., & Pérez-Prieto, J. (2012b). Quantum dot/*cyclo*dextrin supramolecular systems based on efficient molecular recognition and their use for sensing. *Chemical Communications*, *48*(20), 2573–2575. doi:10.1039/C1CC15312A PMID:22080219

Aguilera-Sigalat, J., Rocton, S., Galian, R. E., & Pérez-Prieto, J. (2011). Fluorescence enhancement of amine-capped CdSe/ZnS quantum dots by thiol addition. *Canadian Journal of Chemistry*, *89*(3), 359–363. doi:10.1139/V10-160

Aguilera-Sigalat, J., Rocton, S., Sánchez-Royo, J. F., Galian, R. E., & Pérez-Prieto, J. (2012a). Highly fluorescent and photostable organic- and water-soluble CdSe/ZnS core–shell quantum dots capped with thiols. *RSC Advances*, *2*(4), 1632–1638. doi:10.1039/C1RA01005K

Aguilera-Sigalat, J., Sanchez-SanMartín, J., Agudelo-Morales, C. E., Zaballos, E., Galian, R. E., & Pérez-Prieto, J. (2012c). Further insight into the photostability of the pyrene fluorophore in halogenated solvents. *ChemPhysChem*, *13*(3), 835–844. doi:10.1002/cphc.201100843 PMID:22271708

Agustí, G., Cobo, S., Gaspar, A. B., Molnár, G., Moussa, N. O., Szilágyi, P. Á., & Bousseksou, A. et al. (2008). Thermal and light-induced spin crossover phenomena in new 3D Hofman-like microporous metalorganic frameworks produced as bulk materials and nanapatterned thin films. *Chemistry of Materials*, *20*(21), 6721–6732. doi:10.1021/cm8019878

Aimé, J.-P. (2010). BioInspired nanomaterials. *E-Nano Newsletter*, (19), 30-34.

Akerlof, G. (1970). The market for lemons: Quality uncertainty and the market mechanism. *The Quarterly Journal of Economics*, *84*(3), 488–500. doi:10.2307/1879431

Algar, W. R., & Krull, U. J. (2007). Towards multi-colour strategies for the detection of oligonucleotide hybridization using quantum dots as energy donors in fluorescence resonance energy transfer (FRET). *Analytica Chimica Acta*, *581*(2), 193–201. doi:10.1016/j.aca.2006.08.026 PMID:17386444

Ali, P. A., Reza, M. M., & Hossein, S. M. (2010). Removal of dissolved organic carbon by multi-walled carbon nanotubes, powdered activated carbon and granular activated carbon. *Research Journal of Chemistry and Environment*, *14*(4), 59–66.

Anderson, P. W., Halperin, B. I., & Varma, C. M. (1972). Anomalous low-temperature thermal properties of glasses and spin glasses. *Philosophical Magazine*, *25*(1), 1–9. doi:10.1080/14786437208229210

Animals in Medieval Art. (2011). *Heilbrun Timeline of Art History*. Department of Medieval Art and The Cloisters, The Metropolitan Museum of Art. Retrieved March 14, 2014, from http://www.metmuseum.org/toah/hd/best/hd_best.htm

Anisimov, A. S., Nasibulin, A. G., Jiang, H., Launois, P., Cambedouzou, J., Shandakov, S. D., & Kauppinen, E. I. (2010). Mechanistic investigations of single-walled carbon nanotube synthesis by ferrocene vapor decomposition in carbon monoxide. *Carbon*, *48*(2), 380–388. doi:10.1016/j.carbon.2009.09.040

Anpo, M., & Kamat, P. V. (2010). *Environmentally benign photocatalysts: Applications of titanium oxide-based materials*. Berlin: Springer. doi:10.1007/978-0-387-48444-0

Anumba, C. J., & Newnham, L. N. (2000). Computer-based collaborative building design: Conceptual model. *Int J Construct Inform Technol*, *8*(1), 1–14.

Anumba, C., Ren, A., Thorpe, O., Ugwu, O., & Newnham, L. (2003). Negotiation within a multi-agent system for the collaborative design of light industrial buildings. *Advances in Engineering Software*, *34*(7), 389–401. doi:10.1016/S0965-9978(03)00038-3

Arakelova, E., Khachatryan, A., Avjyan, K., Farmazyan, Z., Mirzoyan, A., Savchenko, L., & Arsenyan, F. et al. (2010). Zinc oxide nanocomposites with antitumour activity. *Natural Science*, *2*(12), 1341–1348. doi:10.4236/ns.2010.212163

Archambault, P. M., Légaré, F., Lavoie, A., Gagnon, M.-P., Lapointe, J., St-Jacques, S., & Pham-Dinh, M. et al. (2010). Healthcare professionals' intentions to use wiki-based reminders to promote best practices in trauma care: A survey protocol. *Implementation Science; IS*, *5*(1), 45. doi:10.1186/1748-5908-5-45 PMID:20540775

Arnheim, R. (1969). Visual Thinking. Berkeley, CA: University of California Press. (Also: London: Faber and Faber, 1969).

Arnheim, R. (1990). Language and the Early Cinema. *Leonardo, Digital Image Digital Cinema.*

Arnheim, R. (1988). *The Power of the Center, A Study of Composition in the Visual Arts.* Berkeley, CA: University of California Press.

Arora, P. S., & Kirshenbaum, K. (2004). Nano-tailoring: Stitching alterations on viral coats. *Chemistry & Biology*, *11*(4), 418–420. doi:10.1016/j.chembiol.2004.04.003 PMID:15123234

Arrigoni, M., & Gallaher, N. (2015, February-March). Diverse Applications Drive Lasers for Multiphoton Microscopy. *BioPhotonics*, 32-36.

Artacho, E., Machado, M., Sánchez-Portal, D., Ordejón, P., & Soler, J. M. (2003). Electrons in dry DNA from density functional calculations. *Molecular Physics*, *101*(11), 1587–1594. doi:10.1080/0026897031000068587

Artacho, E., Sánchez-Portal, D., Ordejón, P., García, A., & Soler, J. M. (1999). Linear-scaling *ab-initio* calculations for large and complex systems. *Physica Status Solidi*, *215*, 809–817. doi:10.1002/(SICI)1521-3951(199909)215:1<809::AID-PSSB809>3.0.CO;2-0

Arundel, A., Sawaya, D., & Valeanu, I. (2009). *Human health biotechnologies to 2015.* OECD.

Asenjo, A., Jaafar, M., Navas, D., & Vázquez, M. (2006). Quantitative magnetic force microscopy analysis of the magnetization process in nanowire arrays. *Journal of Applied Physics*, *100*(2), 023909–1–6. doi:10.1063/1.2221519

Ashcroft, N. W. (1993). Elusive diffusive liquids. *Nature*, *365*(6445), 387–388. doi:10.1038/365387a0

Association of Medical Illustrators. (2014). Retrieved March 16, 2014, from www.ami.org/medical-illustration/education.html

Atwater, H. A., & Polman, A. (2010). Plasmonics for improved photovoltaic devices. *Nature Materials*, *9*(3), 205–213. doi:10.1038/nmat2629 PMID:20168344

Awuah, G. B., & Amal, M. (2011). Impact of globalization: The ability of less developed countries' (LDCs') firms to cope with opportunities and challenges. *European Business Review*, *23*(1), 120–132. doi:10.1108/09555341111098026

Baba, A., Xia, C., Knoll, W., & Advincula, R. C. (2010). Electrochemical surface plasmon resonance and field-enhanced light scattering: Monomer copolymerization with a polysiloxane-conjugated polythiophene network precursor. *Macromolecular Chemistry and Physics*, *211*(24), 2624–2635. doi:10.1002/macp.201000471

Baba, M. S., Narasimhan, T. S. L., Balasubramanian, R., Sivaraman, N., & Mathews, C. K. (1994). Studies on the thermodynamics of the fullerene C_{60}–C_{70} binary system. *Journal of Physical Chemistry*, *98*(4), 1333–1340. doi:10.1021/j100055a047

Babiç, M., Horák, D., Trchová, M., Jendelová, P., Glogarová, K., Lesný, P., & Syková, E. et al. (2008). Poly(L-lysine)-modified iron oxide nanoparticles for stem cell labeling. *Bioconjugate Chemistry*, *19*(3), 740–750. doi:10.1021/bc700410z PMID:18288791

Bachelot, R., H'Dhili, F., Barchiesi, D., Lérondel, G., Fikri, R., Royer, P., … Lahlil, K. (2003). Apertureless near-field optical microscopy: A study of the local tip field enhancement using photosensitive azobenzene-containing films. *J. Appl. Phys.*, *94*, 2060.

Bakry, R., Vallant, R. M., Najam-ul-Haq, M., Rainer, M., Szabo, Z., Huck, C. W., & Bonn, G. K. (2007). Medicinal applications of fullerenes. *International Journal of Nanomedicine*, *2*, 639–649. PMID:18203430

Balfour, F., Barrett, A., Brady, D., Capell, K., Magnusson, P., Matlack, C., . . . Wheatley, J. (2005). "Fakes!". *Business Week*. Retrieved September 5, 2014, from http://www.businessweek.com/magazine/content/05_06/b3919001_mz001.htm

Balogh, L. P. (2009). The future of nanomedicine and the future of Nanomedicine: NBM. *Nanomedicine (London)*, *5*, 1.

Balu, A. M., Pineda, A., Yoshida, K., Campelo, J. M., Gai, P. L., Luque, R., & Romero, A. A. (2010). Fe/Al synergy in Fe_2O_3 nanoparticles supported on porous aluminosilicate materials: Excelling activities in oxidation reactions. *Chemical Communications (Cambridge)*, *46*(41), 7825–7827. doi:10.1039/c0cc02015j

Balzani, V., Credi, A., & Venturi, M. (2008). Molecular Devices and Machines, Weinheim: Wiley-VCH. doi:10.1002/9783527621682

Bard, A. J., Zhou, H., & Kwon, S. J. (2010). Electrochemistry of single nanoparticles *via* electrocatalytic amplification. *Israel Journal of Chemistry*, *50*(3), 267–276. doi:10.1002/ijch.201000014

Bar-Nahum, G., Epshtein, V., Ruckenstein, A. E., Rafikov, R., Mustaev, A., & Nudler, E. (2005). A ratchet mechanism of transcription elongation and its control. *Cell*, *120*(2), 183–193. doi:10.1016/j.cell.2004.11.045 PMID:15680325

Barnard, A. S. (2009). How can *ab initio* simulations address risks in nanotech? *Nature Nanotechnology*, *4*(6), 332–335. doi:10.1038/nnano.2009.126 PMID:19498383

Barnes, W. L., Dereux, A., & Ebbesen, T. (2003). Surface plasmon subwavelength optics. *Nature*, *424*(6950), 824–830. doi:10.1038/nature01937 PMID:12917696

Barnett, J. M. (2005). Shopping for Gucci on Canal Street: Reflections on Status Consumption, Intellectual Property, and the Incentive Thesis. *Virginia Law Review*, *91*(6), 1381-1428.

Barsky, E. (2006). Introducing Web 2.0: RSS trends for health librarians. *Journal of Canadian Health Library Association*, *27*(1), 7–8. doi:10.5596/c06-001

Bartual-Murgui, C., Akou, A., Salmon, L., Molnár, G., Thibault, C., Real, J. A., & Bousseksou, A. (2011c). Guest effect on nanopatterned spin-crossover thin films. *Small*, *7*(23), 3385–3391. doi:10.1002/smll.201101089 PMID:21997948

Bartual-Murgui, C., Akou, A., Shepherd, H. J., Molnár, G., Real, J. A., Salmon, L., & Bousseksou, A. (2013). Tunable spin-crossover behavior of the Hofmann-like network {Fe(bpac)[Pt$(CN)_4$]} through host–guest chemistry. *Chemistry (Weinheim an der Bergstrasse, Germany)*, *19*(44), 15036–15043. doi:10.1002/chem.201300227 PMID:24105972

Bartual-Murgui, C., Ortega-Villar, N. A., Shepherd, H. J., Muñoz, M. C., Salmon, L., Molnár, G., & Real, J. A. et al. (2011a). Enhanced porosity in a new 3D Hofmann-like network exhibiting humidity sensitive cooperative spin transitions at room temperature. *Journal of Materials Chemistry*, *21*(20), 7217–7222. doi:10.1039/c0jm04387g

Bartual-Murgui, C., Salmon, L., Akou, A., Ortega-Villar, N. A., Shepherd, H. J., Muñoz, M. C., & Bousseksou, A. et al. (2012). Synergetic effect of host–guest chemistry and spin crossover in 3D Hofmann-like metal–organic frameworks [Fe(bpac)M$(CN)_4$] (M = Pt, Pd, Ni). *Chemistry (Weinheim an der Bergstrasse, Germany)*, *18*(2), 507–516. doi:10.1002/chem.201102357 PMID:22147670

Bartual-Murgui, C., Salmon, L., Akou, A., Thibault, C., Molnár, G., Mahfoud, T., & Bousseksou, A. et al. (2011b). High quality nano-patterned thin films of the coordination compound {Fe(pyrazine)[Pt$(CN)_4$]} deposited layer-by-layer. *New Journal of Chemistry*, *35*(10), 2089–2094. doi:10.1039/c1nj20212j

Baxter, P., & Jack, S. (2008). Qualitative case study methodology: Study design and implementation for novice researchers. *Qualitative Report*, *13*(4), 544–559.

Beatty, P. (2001). Standardization in Survey Research: How Did it Get Here and How Far Should it Go? In *Bulletin of the International Statistical Institute. 53rd Session Proceedings*. Seoul, South Korea: International Statistics Institute.

Beldarrain, Y. (2006). Distance Education Trends: Integrating New Technologies to Foster Student Interaction and Collaboration. *Distance Education*, *27*(2), 139–153. doi:10.1080/01587910600789498

Bennett, C., Ryall, J., Spalteholz, L., & Gooch, A. (2007). *The Aesthetics of Graph Visualization. In Computational Aesthetics in Graphics, Visualization, and Imaging* (pp. 57–64). Eurographics Association.

Bentley, W. A. (1995). *Snowflakes in Photographs*. Mineola, NY: Dover Publications. (Original work published 1931)

Berens, R. R. (1999). The Role of Artists in Ship Camouflage During World War I. *Leonardo, 32*(1), 53-59.

Berhanu, W. M., Pillai, G. G., Oliferenko, A. A., & Katritzky, A. R. (2012). Quantitative structure–activity/property relationships: The ubiquitous links between cause and effect. *ChemPlusChem*, *77*(7), 507–517. doi:10.1002/cplu.201200038

Bertalanffy, L. (2011). Teoría General de los Sistemas. Ed. Fondo de Cultura y Económica.

Bertolazzi, S., Brivio, J., Radenovic, A., Kis, A., Wilson, H., Prisbrey, L., … Proksch, R. (2013) Exploring flatland: AFM of mechanical and electrical properties of graphene, MoS_2 and other low-dimensional materials. *Microscopy and Analysis, 2013*(5), 21-24.

Bertolazzi, S., Brivio, J., & Kis, A. (2011). Stretching and breaking of ultrathin MoS_2. *ACS Nano*, *5*(12), 9703–9709. doi:10.1021/nn203879f PMID:22087740

Bertschi, S., Bresciani, S., Crawford, T., Goebel, R., Kienreich, W., Lindner, M., & Moere, A. V. et al. (2011). What is Knowledge Visualization? Perspectives on an Emerging Discipline.*Proceedings of the Information Visualisation 15th International Conference*. doi:10.1109/IV.2011.58

Bethesda Statement on Open Access Publishing. (2003, June 20). Retrieved March 11, 2014, from http://legacy.earlham.edu/~peters/fos/bethesda.htm

Bezmel'nitsyn, V. N. (1994). Article. *Химическая физика*, *13*(12), 156–156.

Bezmel'nitsyn, V. N. (1996b). Article. *Technical Physics*, *41*, 986–986.

Bezmel'nitsyn, V. N., Eletskii, A. V., & Okun', M. V. (1998). Fullerenes in solutions. *Physics-Uspekhi*, *41*(11), 1091–1114. doi:10.1070/PU1998v041n11ABEH000502

Bezmel'nitsyn, V. N., Eletskii, A. V., Okun, M. V., & Stepanov, E. V. (1996a). Diffusion of aggregated fullerenes in solution. *Physica Scripta*, *53*(3), 364–367. doi:10.1088/0031-8949/53/3/017

Bezmel'nitsyn, V. N., Eletskii, A. V., Okun, M. V., & Stepanov, E. V. (1996c). Thermal diffusion of fullerenes in solutions. *Physica Scripta*, *53*(3), 368–370. doi:10.1088/0031-8949/53/3/018

Bezmel'nitsyn, V. N., Eletskii, A. V., & Stepanov, E. V. (1994). Cluster origin of fullerene solubility. *Journal of Physical Chemistry*, *98*(27), 6665–6667. doi:10.1021/j100078a001

Bezmel'nitsyn, V. N., Eletskii, A. V., & Stepanov, E. V. (1995). Article. *Журнал физической химии*, *69*, 735–735.

Bhar, K., Khan, S., Sánchez Costa, J., Ribas, J., Roubeau, O., Mitra, P., & Ghosh, B. K. (2012). Crystallographic evidence for reversible symmetry breaking in a spin-crossover d^7 cobalt(II) coordination polymer. *Angewandte Chemie International Edition*, *51*(9), 2142–2145. doi:10.1002/anie.201107116 PMID:22271674

Bickman, L., & Rog, D. J. (Eds.). (2009). Applied Social Research Methods. Thousand Oaks, CA: SAGE.

Bieri, C. (2012). *Global Pharma & Biotech M&A Report*. IMAP.

Billings, D. M. (2009). Wikis and blogs: Consider the possibilities for continuing nursing education. *Journal of Continuing Education in Nursing*, *40*(12), 534–535. doi:10.3928/00220124-20091119-10 PMID:20000260

Binning, G., Quate, C. F., & Gerber, C. (1986). Atomic force microscope. *Physical Review Letters*, *56*(9), 930–933. doi:10.1103/PhysRevLett.56.930 PMID:10033323

Bioassociate. (2012). *Pharmerging markets: China - the next major innovative pharma market*. Bioassociate Consulting & Management Ltd.

Bird, J. (2008). *Monoclonal Antibodies Report: 2008 Update*. Academic Press.

Birkhoff, G. D. (2003). *Aesthetic Measure*. Kessinger Publishing. (Original work published 1933)

Bittner, A. M. (2005). Biomolecular rods and tubes in nanotechnology. *Naturwissenschaften*, *92*(2), 51–64. doi:10.1007/s00114-004-0579-8 PMID:15558225

Björk, B.-C. (2012). The hybrid model for open access publication of scholarly articles: A failed experiment? *Journal of the American Society for Information Science*, *63*, 1496–1504. doi:10.1002/asi.22709

Blomley, M., Cooke, J., Unger, E., Monaghan, M., & Cosgrove, D. (2001). Microbubble contrast agents: A new era in ultrasound. *British Medical Journal*, *322*(7296), 1222–1225. doi:10.1136/bmj.322.7296.1222 PMID:11358777

BOAI10. (2002, February 14). *Budapest Open Access Initiative | Read the Budapest Open Access Initiative*. Retrieved March 11, 2014, from http://www.budapestopenaccessinitiative.org/read

Bogdanich, W. (2008). Panama Releases Report on '06 Poisoning. *The New York Times*. Retrieved on September 6, 2014, from http://www.nytimes.com/2008/02/14/world/americas/14panama.html?ref=health

Bogdanich, W., & Hooker, J. (2007, May 6). A Poison's Path. *The New York Times*.

Boillot, M. L., Zarembowitch, J., & Sour, A. (2004). Ligand-driven light-induced spin change (LD-LISC): A promising photomagnetic effect. In Spin crossover in transition metal compounds II (pp. 261-276). Berlin: Springer.

Boillot, M. L., Pillet, S., Tissot, A., Rivière, E., Claiser, N., & Lecomte, C. (2009). Ligand-driven light-induced spin change activity and bidirectional photomagnetism of styrylpyridine iron(II) complexes in polymeric media. *Inorganic Chemistry*, *48*(11), 4729–4736. doi:10.1021/ic802319c PMID:19374370

Boldog, I., Gaspar, A. B., Martínez, V., Pardo-Ibañez, P., Ksenofontov, V., Bhattacharjee, A., & Real, J. A. et al. (2008). Spin-crossover nanocrystals with magnetic, optical, and structral bistability near room temperature. *Angewandte Chemie International Edition*, *47*, 6433–6437. PMID:18623300

Bolin, L. (2014). *Camouflage art*. Retrieved March 15, 2014, from http://www.toxel.com/inspiration/2009/10/04/camouflage-art-by-liu-bolin/

Bonarini, A., & Trianni, V. (2001). Learning fuzzy classifier systems for multi-agent coordination. *Information Sciences*, *136*(1–4), 215–239. doi:10.1016/S0020-0255(01)00149-9

Bong, D. T., Clark, T. D., Granja, J. R., & Ghadiri, M. R. (2001). Self-assembling organic nanotubes. *Angewandte Chemie International Edition*, *40*(6), 988–1011. doi:10.1002/1521-3773(20010316)40:6<988::AID-ANIE9880>3.0.CO;2-N PMID:11268062

Bonhommeau, S., Guillon, T., Daku, L. M. L., Demont, P., Sanchez Costa, J., Létard, J. F., & Bousseksou, A. et al. (2006). Photoswitching of the dielectric constant of the spin-crossover complex $[Fe(L)(CN)_2]\cdot H_2O$. *Angewandte Chemie International Edition*, *45*(10), 1625–1629. doi:10.1002/anie.200503252 PMID:16470763

Bonnet, S., Molnár, G., Sanchez Costa, J., Siegler, M. A., Spek, A. L., Bousseksou, A., & Reedijk, J. et al. (2009). Influence of sample preparation, temperature, light, and pressure on the two-step spin crossover mononuclear compound[Fe(bapbpy)(NCS)2]. *Chemistry of Materials*, *21*(6), 1123–1136. doi:10.1021/cm803414q

Bonnet, S., Siegler, M. A., Sánchez Costa, J., Molnár, G., Bousseksou, A., Spek, A. L., & Reedijk, J. et al. (2008). A two-step spin crossover mononuclear iron(II) complex with a [HS–LS–LS] intermediate phase. *Chemical Communications*, *2008*(43), 5619–5621. doi:10.1039/b811746b PMID:18997971

Booz & Company. (2012). *The potential for pharmaceutical quality services.* Available from: http://www.knmp.nl/downloads/nieuws/nieuws-2012/booz-co-the-potential-for-pharmaceutical-quality-services.pdf/view

Bosch-Navarro, C., Busolo, F., Coronado, E., Duan, Y., Martí-Gastaldo, C., & Prima-Garcia, H. (2013a). Influence of the covalent grafting of organic radicals to graphene on its magnetoresistance. *Journal of Materials Chemistry C*, *1*(30), 4590–4598. doi:10.1039/c3tc30799a

Bosch-Navarro, C., Coronado, E., & Martí-Gastaldo, C. (2013b). Controllable coverage of chemically modified graphene sheets with gold nanoparticles by thermal treatment of graphite oxide with N,N-dimethylformamide. *Carbon*, *54*, 201–207. doi:10.1016/j.carbon.2012.11.027

Bosch-Navarro, C., Coronado, E., Martí-Gastaldo, C., Sánchez-Royo, J. F., & Gómez Gómez, M. (2012). Influence of the pH on the synthesis of reduced graphene oxide under hydrothermal conditions. *Nanoscale*, *4*(13), 3977–3982. doi:10.1039/c2nr30605k PMID:22653666

Bouhelier, A., Bachelot, R., Lérondel, G., Kostcheev, S., Royer, P., & Wiederrecht, G. P. (2005). Surface plasmon characteristics of tunable photoluminescence in single gold nanorods. *Physical Review Letters*, *95*(26), 267405–1–4. doi:10.1103/PhysRevLett.95.267405 PMID:16486405

Bouldin, A. S., Holmes, E. R., & Fortenberry, M. L. (2006). "Blogging" About Course Concepts: Using Technology for Reflective Journaling in a Communications Class. *American Journal of Pharmaceutical Education*, *70*(4), 84. doi:10.5688/aj700484 PMID:17136203

Boulos, M. N. K., Hetherington, L., & Wheeler, S. (2007). Second Life: An overview of the potential of 3-D virtual worlds in medical and health education. *Health Information and Libraries Journal*, *24*(4), 233–245. doi:10.1111/j.1471-1842.2007.00733.x PMID:18005298

Boulos, M. N. K., Maramba, I., & Wheeler, S. (2006). Wikis, blogs and podcasts: A new generation of Web-based tools for virtual collaborative clinical practice and education. *BMC Medical Education*, *6*(1), 41–41. doi:10.1186/1472-6920-6-41 PMID:16911779

Boutellier, R., & Ullman, F. (2007). China's unique position in discovery and preclinical research. *Drug Discovery Today*, *12*(1-2), 4–7. doi:10.1016/j.drudis.2006.11.009 PMID:17198968

Boutin, S., & Therriault, P. (2011). The learner as a participant in the construction of content through the use of a wiki in a competency-based program of occupational therapy. In *INTED 2011 Proceedings*. Valencia: IATED. Retrieved from http://library.iated.org/view/BOUTIN2011THE

Boyd, B. J. (2008). Past and future evolution in colloidal drug delivery systems. *Expert Opinion on Drug Delivery*, *5*(1), 69–85. doi:10.1517/17425247.5.1.69 PMID:18095929

Brandbyge, M., Mozos, J. L., Ordejón, P., Taylor, J., & Stokbro, K. (2002). Density-functional method for non-equilibrium electron transport. *Physical Review B: Condensed Matter and Materials Physics*, *65*(16), 165401–1–17. doi:10.1103/PhysRevB.65.165401

Bratsas, C., Kapsas, G., Konstantinidis, S., Koutsouridis, G., & Bamidis, P. D. (2009). A semantic wiki within moodle for Greek medical education. In *22nd IEEE International Symposium on Computer-Based Medical Systems, 2009. CBMS 2009* (pp. 1–6). IEEE. doi:10.1109/CBMS.2009.5255417

Braun, J., & Sieper, J. (2004). Biological therapies in the spondyloarthritidesthe current state. *Oxford Journals Medicine & Health Rheumatology, 43*(9). Retrieved august 31, 2014, from http://rheumatology.oxfordjournals.org/content/43/9/1072.full.pdf+html

Bredl, K., Groß, A., Hünniger, J., & Fleischer, J. (2012). The Avatar as a Knowledge Worker? How Immersive 3D Virtual Environments may Foster Knowledge Acquisition. *Electronic Journal of Knowledge Management*, *10*(1), 15–25.

Bristol, T. J. (2010). Twitter: Consider the possibilities for continuing nursing education. *Journal of Continuing Education in Nursing*, *41*(5), 199–200. doi:10.3928/00220124-20100423-09 PMID:20481418

Brown, S. (1992). Engineered iron oxide-adhesion mutants of the *Escherichia coli* phage lambda receptor. *Proceedings of the National Academy of Sciences of the United States of America*, *89*(18), 8651–8655. doi:10.1073/pnas.89.18.8651 PMID:1528875

Brown, S. A. (1997). *Revolution at the Checkout Counter: The Explosion of the Bar Code*. Cambridge, MA: Harvard University Press.

Bruker. (2013). Atomic force microscopy reveals DNA double helix. *Microscopy and Analysis, 2013*(3).

Bueno Alejo, C. J., Fasciani, C., Grenier, M., Netto-Ferreira, J. C., & Scaiano, J. C. (2011). Reduction of resazurin to resorufin catalyzed by gold nanoparticles: Dramatic reaction acceleration by laser or LED plasmon excitation. *Catalysis Science & Technology*, *1*(8), 1506–1511. doi:10.1039/c1cy00236h

Bueno-Alejo, C. J., D'Alfonso, C., Pacioni, N. L., González-Béjar, M., Grenier, M., Lanzalunga, O., & Scaiano, J. C. et al. (2012). Ultraclean derivatized monodisperse gold nanoparticles through laser drop ablation customization of polymorph gold nanostructures. *Langmuir*, *28*(21), 8183–8189. doi:10.1021/la3010689 PMID:22591001

Bulćon, A., Colonna, P., Planchot, V., & Ball, S. (1998). Starch granules: Structure and biosynthesis. *International Journal of Biological Macromolecules*, *23*(2), 85–112. doi:10.1016/S0141-8130(98)00040-3 PMID:9730163

Buller, D. B., & Burgoon, J. K. (1996). Interpersonal Deception Theory. *Communication Theory*, *6*(3), 203–242. doi:10.1111/j.1468-2885.1996.tb00127.x

Burrell, R. E. (1997). Anti-microbial coating for medical devices. *Patent Application, 1203*.

Bussières, J.-F., Métras, M.-É., & Leclerc, G. (2012). Use of Moodle, ExamSoft, and Twitter® in a First-Year Pharmacy Course. *American Journal of Pharmaceutical Education*, *76*(5), 94. doi:10.5688/ajpe76594 PMID:22761535

Buzdin, A. I. (1993). Multiple-quanta vortices at columnar defects. *Physical Review B: Condensed Matter and Materials Physics*, *47*(17), 11416–11419. doi:10.1103/PhysRevB.47.11416 PMID:10005280

Cain, J., & Fox, B. I. (2009). *Web 2.0 and Pharmacy Education*. Academic Press.

Caire, G., Chainho, P., Evans, R., Garijo, F., Gomez Sanz, J., Kearney, P., & Stark, J. (2002). Agent Oriented Anlysis Using MESSAGE/UML. In Agent-Oriented Software Engineering II, (pp. 119-135). Springer.

Camiel, L. D., Goldman-Levine, J. D., Kostka-Rokosz, M. D., & McCloskey, W. W. (2014a). *Twitter® as a Medium for Pharmacy Students' Personal Learning Network Development*. Currents in Pharmacy Teaching and Learning. doi:10.1016/j.cptl.2014.04.008

Camiel, L. D., Goldman-Levine, J. D., Kostka-Rokosz, M. D., & McCloskey, W. W. (2014b). Twitter® as an In-Class Backchannel Tool in a Large Required Pharmacy Course. *American Journal of Pharmaceutical Education*, *78*(3), 67. doi:10.5688/ajpe78367 PMID:24761028

Camouflage Restaurants. (2014). Retrieved March 15, 2014, from http://www.toxel.com/inspiration/2009/06/20/10-unusual-and-creative-restaurants/

Cañete, J. (2008). *Manual S.E.R. de las Enfermedades Reumáticas. 3a edición. Sociedad Española de Reumatología*. Editorial Médica Panamericana.

Cañizo, J. A. (2005). Asymptotic behavior of solutions to the generalized Becker-Döring equations for general initial data. *Proceedings of the Royal Society of London. Series A*, *461*, 3731–3745.

Cañizo, J. A., López, J. L., & Nieto, J. (2004). Global L^1 theory and regularity for the 3D nonlinear Wigner–Poisson–Fokker–Planck system. *Journal of Differential Equations*, *198*(2), 356–373. doi:10.1016/j.jde.2003.07.004

Carbonel, J. (2006). *Semiología de las enfermedades reumáticas*. Madrid: Editorial Médica Panamericana.

Carbonera, C., Sánchez Costa, J., Money, V. A., Elhaïk, J., Howard, J. A. K., Halcrow, M. A., & Létard, J. F. (2006). Photomagnetic properties of iron(II) spin crossover complexes of 2,6-dipyrazolylpyridine and 2,6-dipyrazolylpyrazine ligands. *Dalton Transactions (Cambridge, England)*, *2006*(25), 3058–3066. doi:10.1039/B601366J PMID:16786064

Carl Zeiss Microscopy. (2013). *Cover story. Microscopy and Analysis Directory*.

Carlsbad Caverns National Park, New Mexico. (n.d.). National Park Service, U.S. Department of Interior. Retrieved March 15, 2014, from http://www.nps.gov/cave/index.htm

Carroll, D. L., Redlich, P., Ajayan, P. M., Charlier, J. C., Blase, X., de Vita, A., & Car, R. (1997). Electronic structure and localized states at carbon nanotube tips. *Physical Review Letters*, *78*(14), 2811–2814. doi:10.1103/PhysRevLett.78.2811

Cartoixà, X., & Rurali, R. (2010). Simulating the structural, electronic and transport properties of silicon nanowires. *E-Nano Newslett.*, (19), 5-18.

Carver, B. (2003). *Creating an institutional repository: A role for libraries*. Retrieved from http://marylaine.com/exlibris/xlib 181.html

Cassie, A. B. D., & Baxter, S. (1944). Wettability of porous surfaces. *Transactions of the Faraday Society*, *40*, 546–551. doi:10.1039/tf9444000546

Castellano, G., & Torrens, F. (2009). Local anaesthetics classified using chemical structural indicators. *Nereis*, (2), 7-17.

Castellano, G., González-Santander, J. L., Lara, A., & Torrens, F. (2013). Classification of flavonoid compounds by using entropy of information theory. *Phytochemistry*, *93*, 182–191. doi:10.1016/j.phytochem.2013.03.024 PMID:23642389

Castellano, G., Lara, A., & Torrens, F. (2014). Classification of stilbenoid compounds by entropy of artificial intelligence. *Phytochemistry*, *97*, 62–69. doi:10.1016/j.phytochem.2013.10.010 PMID:24239224

Castellano, G., Tena, J., & Torrens, F. (2012). Classification of polyphenolic compounds by chemical structural indicators and its relation to antioxidant properties of *Posidonia oceanica* (L.) Delile. *MATCH: Communications in Mathematical and in Computer Chemistry*, *67*, 231–250.

Castellano, G., & Torrens, F. (2015). Information entropy-based classification of triterpenoids and steroids from *Ganoderma. Phytochemistry*, 116, 305-313. PMID:26024957

Center for Medicines in the Public Interest. (2011). *Intellectual Property Rights Issues and Imported Counterfeit Good.* Center for Medicine in the Public Interest. Retrieved on September 5, 2014, from http://cmpi.org/in-the-news/testimony/

Çervenka, J., Katsnelson, M. I., & Flipse, C. F. J. (2009). Room-temperature ferromagnetism in graphite driven by two-dimensional networks of point defects. *Nature Physics*, *5*(11), 840–844. doi:10.1038/nphys1399

Chakraborty, G., Allred, A., Sukhdial, A. S., & Bristol, T. (1997). Use of negative cues to reduce demand for counterfeit products. In M. Brucks & D. J. MacInnis (Eds.), *Advances in Consumer Research* (Vol. 24). Provo, UT: Association for Consumer Research.

Chakravarty, R. (2011). Self-archiving in open access institutional repositories: Whose court is the ball in? *International Journal of Information Research*, *1*(1).

Chakravarty, R., & Mahajan, P. (2011). Open access journal initiatives in India. *International Journal of Information Dissemination and Technology*, *1*(1).

Chang. (2014). Nobel Laureates Pushed Limits of Microscopes. *The New York Times*. Retrieved February 8, 2015, from http://www.nytimes.com/2014/10/09/science/nobel-prize-chemistry.html?_r=0

Chatelier, R. C., & Minton, A. P. (1996). Adsorption of globular proteins on locally planar surfaces: Models for the effect of excluded surface area and aggregation of adsorbed protein on adsorption equilibria. *Biophysical Journal*, *71*(5), 2367–2374. doi:10.1016/S0006-3495(96)79430-4 PMID:8913577

Chatterjee, R. (2008). The challenge of regulating nanomaterials. *Environmental Science & Technology*, *42*(2), 339–343. doi:10.1021/es0870909 PMID:18284127

Chaudhry, P. E., & Walsh, M. G. (1996). An assessment of the impact of counterfeiting in international markets: The piracy paradox persists. *The Columbia Journal of World Business*, *3*(1), 34–49. doi:10.1016/S0022-5428(96)90039-3

Cheetham, M. A. (2010). The Crystal interface in contemporary art: Metaphors of the organic and inorganic. *Leonardo*, *43*(3), 251–255. doi:10.1162/leon.2010.43.3.250

Cheng, A., Klein, M. L., & Caccamo, C. (1993). Prediction of the phase diagram of rigid C_{60} molecules. *Physical Review Letters*, *71*(8), 1200–1203. doi:10.1103/PhysRevLett.71.1200 PMID:10055475

Cheng, C. J., & Shiu, E. C. C. (2008). Re-innovation: The construct, measurement, and validation. *Technovation*, *28*(10), 658–666. doi:10.1016/j.technovation.2007.08.002

Chen, H. S., Kortan, A. R., Haddon, R. C., & Fleming, D. A. (1992). Thermodynamics of C_{60} in pure O_2, N_2 and Ar. *Journal of Physical Chemistry*, *96*, 1016–1018. doi:10.1021/j100182a003

Chiesa, V., & Chiaroni, D. (2005). *Industrial Clusters in Biotechnology: Driving Forces, Development Processes and Management Practices*. London: Imperial College Press.

Chin, C. D. (2008). Biotechnology for global health: Solutions for the developing world. *Consilience*, *1*, 1–12.

Chong, E. Z., Matthews, D. R., Summers, H. D., Njoh, K. L., Errington, R. J., & Smith, P. J. (2007). Development of FRET-based assays in the far-red using CdTe quantum dots. *Journal of Biomedicine and Biotechnology, 2007*, 54169.

Chretien, K., Goldman, E., & Faselis, C. (2008). The reflective writing class blog: Using technology to promote reflection and professional development. *Journal of General Internal Medicine*, *23*(12), 2066–2070. doi:10.1007/s11606-008-0796-5 PMID:18830767

Chu, L. F., Young, C., Zamora, A., Kurup, V., & Macario, A. (2010). Anesthesia 2.0: Internet-based information resources and Web 2.0 applications in anesthesia education. *Current Opinion in Anaesthesiology*, *23*(2), 218–227. doi:10.1097/ACO.0b013e328337339c PMID:20090518

Chule, K., Chule, A. V., Chen, B.-J., & Ling, Y.-C. (2006). Preparation and characterization of ZnO nanoparticles coated paper and its antibacterial activity study. *Green Chemistry*, *8*(12), 1034–1041. doi:10.1039/b605623g

Cioffi, C., Campidelli, S., Brunetti, F. G., Meneghetti, M., & Prato, M. (2006). Functionalisation of carbon nanohorns. *Chemical Communications*, (20), 2129–2131. doi:10.1039/b601176d PMID:16703130

Cioffi, C., Campidelli, S., Sooambar, C., Marcaccio, M., Marcolongo, G., Meneghetti, M., & Prato, M. et al. (2007). Synthesis, characterization, and photoinduced electron transfer in functionalized single wall carbon nanohorns. *Journal of the American Chemical Society*, *129*(13), 3938–3945. doi:10.1021/ja068007p PMID:17343379

Clapp, A. R., Medintz, I. L., & Mattoussi, H. (2006). Förster resonance energy transfer investigations using quantum dot fluorophores. *ChemPhysChem*, *7*(1), 47–57. doi:10.1002/cphc.200500217 PMID:16370019

Clauw Daniel, J. (2005). *Fibromyalgia and Other Central Pain Syndromes* (1st ed.). Lippincott Williams and Wilkins.

Coase, R. (1937). The Nature of the Firm. *Economica*, *4*, 386–405.

Coase, R. (1960). The Problem of Social Cost. *The Journal of Law & Economics*, *3*(1), 1–44. doi:10.1086/466560

Colvin, R. (1999). Innovative technologies help thwart counterfeiting. *Modern Plastics*, *76*(7), 57–58.

Conn, V. S., Valentine, J. C., Cooper, H. M., & Rantz, M. J. (2003). Grey literature in meta-analyses. *Nursing Research*, *52*(4), 256–261. doi:10.1097/00006199-200307000-00008 PMID:12867783

Cooper, H. M. (1998). Synthesizing research: A guide for literature reviews. *Sage (Atlanta, Ga.)*.

Cordell, V. V., Wongtada, N., & Kieschnick, R. L. Jr. (1996). Counterfeit purchase intentions: Role of lawfulness attitudes and product traits as determinants. *Journal of Business Research*, *35*(1), 41–53. doi:10.1016/0148-2963(95)00009-7

Corma, A., & Garcia, H. (2008). Supported gold nanoparticles as catalysts for organic reactions. *Chemical Society Reviews*, *37*(9), 2096–2126. doi:10.1039/b707314n PMID:18762848

Coronado, E., Galán-Mascarós, J. R., Gómez-García, C. J., & Laukhin, V. (2000). Coexistence of ferromagnetism and metallic conductivity in a molecule-based layered compound. *Nature*, *408*(6811), 447–449. doi:10.1038/35044035 PMID:11100721

Coronado, E., Galán-Mascarós, J. R., Martí-Gastaldo, C., & Ribera, A. (2006). Insertion of magnetic bimetallic oxalate complexes into layered double hydroxides. *Chemistry of Materials*, *18*(26), 6112–6114. doi:10.1021/cm062054z

Coronado, E., Galán-Mascarós, J. R., Monrabal-Capilla, M., García-Martínez, J., & Pardo-Ibáñez, P. (2007). Bistable spin-crossover nanoparticles showing magnetic thermal hysteresis near room temperature. *Advanced Materials*, *19*(10), 1359–1361. doi:10.1002/adma.200700559

Coronado, E., Giménez-Marqués, M., Mínguez Espallargas, G., Rey, F., & Vitórica-Yrezábal, I. J. (2013). Spin-crossover modification through selective CO_2 sorption. *Journal of the American Chemical Society*, *135*(43), 15986–15989. doi:10.1021/ja407135k PMID:24125096

Coronado, E., Martí-Gastaldo, C., Navarro-Moratalla, E., Burzurí, E., Camón, A., & Luis, F. (2011). Hybrid magnetic/superconducting materials obtained by insertion of a single-molecule magnet into TaS_2 layers. *Advanced Materials*, *23*(43), 5021–5026. doi:10.1002/adma.201102730 PMID:21956436

Coronado, E., Martí-Gastaldo, C., Navarro-Moratalla, E., Ribera, A., Blundell, S. J., & Baker, P. J. (2010). Coexistence of superconductivity and magnetism by chemical design. *Nature Chemistry*, *2*(12), 1031–1036. doi:10.1038/nchem.898 PMID:21107366

Counterfeiting Facts and Stats. (2009). *Protection from Brand Infection*. CMO Council. Retrieved on July 21, 2011, from http://www.cmocouncil.org/programs/current/protection/protection_couterfeit_stats.asp

Coutinho, C. (2008). Web 2.0: uma revisão integrativa de estudos e investigações. In Actas do Encontro sobre Web 2.0. Braga: Universidade do Minho, CIEd.

Coutinho, C. (2011). *Metodologia de Investigação em Ciências Sociais e Humanas: Teoria e Prática*. Almedina.

D'Arcy, W., & Thompson. (1917/2011). On Growth and Form. CreateSpace Independent Publishing Platform.

D'Oliveira, J.-C., Minero, C., Pelizzetti, E., & Pichat, P. (1993). Photodegradation of dichlorophenols and trichlorophenols in TiO_2 aqueous suspensions: Kinetic effects of the positions of the Cl atoms and identification of the intermediates. *Journal of Photochemistry and Photobiology A Chemistry*, *72*(3), 261–267. doi:10.1016/1010-6030(93)80022-2

Daar, A. S., Thorsteinsdóttir, H., Martin, D. K., Smith, A. C., Nast, S., & Singer, P. A. (2006). *Top ten biotechnologies for improving health in developing countries*. UNESCO.

Davidsson, P., & Wernstedt, F. (2004). *A framework for evaluation of multi-agent system approaches to logistics network management. In Multi-Agent Systems: An Application Science*. The Netherlands: Kluwer.

De León, M. (2012). El aprendizaje de las enfermedades reumáticas desde una perspectiva tecnológica. *Revista cubana de Reumatología, 15*(19).

De Pablo, P. J., Moreno-Herrero, F., Colchero, J., Gómez Herrero, J., Herrero, P., Baró, A. M., & Artacho, E. et al. (2000). Absence of dc-conductivity in λ-DNA. *Physical Review Letters*, *85*(23), 4992–4995. doi:10.1103/PhysRevLett.85.4992 PMID:11102169

Dean, L. (2012). Size matters: Measurement helps solve nanoparticle toxicity challenges. *Chemistry International*, *34*(4), 6–9.

Deeb, C., Huang, L., Plain, J., Bouhelier, A., Soppera, O., Bachelot, R., & Royer, P. (2010). Nanophotopolymerization triggered by the enhanced optical near-field of metallic nanoparticles. *Lebanese Science Journal*, *11*(2), 105–115.

Deerink, T. J. (2008). The application of fluorescent quantum dots to confocal, multiphoton, and electron microscopic imaging. *Toxicologic Pathology*, *36*(1), 112–116. doi:10.1177/0192623307310950 PMID:18337229

DeLoach, S. A. (1999). Systems engineering: A methodology and language for designing agent systems. In *Proceedings of Agent Oriented Information Systems* (pp. 45–57). Academic Press.

DeLoach, S. A., Wood, M. F., & Sparkman, C. H. (2001). Multiagent system engineering. *International Journal of Software Engineering and Knowledge Engineering*, *11*(3), 231–258. doi:10.1142/S0218194001000542

DeMatos, C. A., Ituassu, C. T., & Rossi, C. A. V. (2007). Consumer attitudes toward counterfeits: A review and extension. *Journal of Consumer Marketing*, *24*(1), 36–47. doi:10.1108/07363760710720975

Dickey, M. D. (2005). Three-dimensional virtual worlds and distance learning: Two case studies of Active Worlds as a medium for distance education. *British Journal of Educational Technology*, *36*(3), 439–451. doi:10.1111/j.1467-8535.2005.00477.x

Dimitratos, N., Lopez-Sanchez, J. A., Lennon, D., Porta, F., Prati, L., & Villa, A. (2006). Effect of particle size on monometallic and bimetallic (Au,Pd)/C on the liquid phase oxidation of glycerol. *Catalysis Letters*, *108*(3-4), 147–153. doi:10.1007/s10562-006-0036-8

Dîrtu, M. M., Neuhausen, C., Naik, A. D., Rotaru, A., Spinu, L., & Garcia, Y. (2010). Insights into the origin of cooperative effects in the spin transitions of $[Fe(NH_2trz)_3](NO_3)_2$: The role of supramolecular interactions evidenced in the crystal structure of $[Cu(NH_2trz)_3](NO_3)_2$·H2O. *Inorganic Chemistry*, *49*(12), 5723–5736. doi:10.1021/ic100667f PMID:20507075

Doney, P., & Cannon, J. P. (1997). An examination of the nature of trust in buyer-seller relationships. *Journal of Marketing*, *61*(2), 35–51. doi:10.2307/1251829

Doolittle, B. (2014). *Camouflage Art*. Retrieved March 15, 2014, from http://www.bnr-art.com/doolitt/

Douglas, T., & Stark, V. T. (2000). Nanophase cobalt oxyhydroxide mineral synthesized within the protein cage of ferritin. *Inorganic Chemistry*, *39*(8), 1828–1830. doi:10.1021/ic991269q PMID:12526579

Douglas, T., & Young, M. (2006). Viruses: Making friends with old foes. *Science*, *312*(5775), 873–875. doi:10.1126/science.1123223 PMID:16690856

Dreher, K. L. (2004). Health and environmental impact of nanotechnology: Toxicological assessment of manufactured nanoparticles. *Toxicological Sciences*, *77*(1), 3–5. doi:10.1093/toxsci/kfh041 PMID:14756123

Dresselhaus, M. S., Jorio, A., Hofmann, M., Dresselhaus, G., & Saito, R. (2010). Perspectives on carbon nanotubes and graphene Raman spectroscopy. *Nano Letters*, *10*(3), 751–758. doi:10.1021/nl904286r PMID:20085345

Drexler, E. (1986). *Engines of Creation: The Coming Era of Nanotechnology*. New York: Anchor Bools.

Duncan, R. (2007). Designing polymer conjugates as lysosomotropic nanomedicines. *Biochemical Society Transactions*, *35*(1), 56–60. doi:10.1042/BST0350056 PMID:17233601

Duncan, R., & Spreafico, F. (1994). Polymer conjugates. Pharmacokinetic considerations for design and development. *Clinical Pharmacokinetics*, *27*(4), 290–306. doi:10.2165/00003088-199427040-00004 PMID:7834965

Ebner, M., Lienhardt, C., Rohs, M., & Meyer, I. (2010). Microblogs in Higher Education – A chance to facilitate informal and process-oriented learning? *Computers & Education*, *55*(1), 92–100. doi:10.1016/j.compedu.2009.12.006

EDUCAUSE. (Ed.). (2009). *7 Things You Should Know About Microblogging*. Retrieved from http://net.educause.edu/ir/library/pdf/ELI7051.pdf

Einstein, A., Podolsky, B., & Rosen, N. (1935). Can Quantum-Mechanical Description of Physical Reality Be Considered Complete? *Physical Review, 47*, 777-780.

Ekman, P. (1992). *Telling Lies: Clues to Deceit in the Marketplace, Politics, and Marriage*. New York: Norton.

Eletskii, A. V., Okun', M. V., & Smirnov, B. M. (1997). Growth of fractal structures in fullerene solutions. *Physica Scripta*, *55*(3), 363–366. doi:10.1088/0031-8949/55/3/016

Eletskii, A. V., & Smirnov, B. M. (1991). The C₆₀ cluster as a new form of carbon. *Uspekhi Fizicheskikh Nauk*, *161*(7), 173–173. doi:10.3367/UFNr.0161.199107e.0173

El-Sourani, N., Hauke, S., & Borschbach, M. (2010). An Evolutionary Approach for Solving the Rubik's Cube Incorporating Exact Methods. In C. Di Chio et al. (Eds.), *Applications of Evolutionary Computing, LNCS 6024 EvoApplications 2010 Proceedings* (pp. 80–89). Berlin: Springer.

Enache, D. I., Edwards, J. K., Landon, P., Solsona-Espriu, B., Carley, A. F., Herzing, A. A., & Hutchings, G. J. et al. (2006). Solvent-free oxidation of primary alcohols to aldehydes using Au-Pd/TiO_2 catalysts. *Science*, *311*(5759), 362–365. doi:10.1126/science.1120560 PMID:16424335

EPC Information Services (EPCIS) Version 1.0.1 Specification. (2010). Retrieved on September 5, 2014, from http://www.epcglobalinc.org/standards/epcis

EPCglobal. (2010). Retrieved on September 7, 2014, from http://www.epcglobalinc.org/about/

Epocrates. (2014). *Epocrates overview*. Available from: http://www.epocrates.com/products

Erickson, L. E., Koodali, R. T., & Richards, R. M. (Eds.). (2010). ACS Symp. Ser.: Vol. 1045. *Nanoscale Materials in Chemistry: Environmental Applications*. Washington, DC: American Chemical Society. doi:10.1021/bk-2010-1045

Escrig, J., Altbir, D., Jaafar, M., Navas, D., Asenjo, A., & Vázquez, M. (2007). Remanence of Ni nanowire arrays: Influence of size and labyrinth magnetic structure. *Physical Review B: Condensed Matter and Materials Physics*, *75*(18), 184429–1–5. doi:10.1103/PhysRevB.75.184429

Euliss, L. E., Grancharov, S. G., O'Brien, S., Deming, T. J., Stucky, G. D., Murray, C. B., & Held, G. A. (2003). Cooperative assembly of magnetic nanoparticles and block copolypeptides in aqueous media. *Nano Letters*, *3*(11), 1489–1493. doi:10.1021/nl034472y

Fan, E. X. (2003). Technological spillovers from foreign direct investment. *Asian Development Review*, *20*(1), 34–56.

Fang, L., Wang, Y., Zou, P. Y., Tang, L., Xu, Z., Chen, H., Dong, C., Shan, L., & Wen, H. H. (2005). Fabrication and superconductivity of Na_xTaS_2 crystals. *Physical Review B, 72*, 14534.

Fan, Z., & Govorov, A. O. (2010). Plasmonic circular dichroism of chiral metal nanoparticle assemblies. *Nano Letters*, *10*(7), 2580–2587. doi:10.1021/nl101231b PMID:20536209

Faraday, M. (1857). The Bakerian Lecture: Experimental relations of gold (and other metals) to light. *Philosophical Transactions of the Royal Society of London*, *147*(0), 145–181. doi:10.1098/rstl.1857.0011

Fasciani, C., Bueno Alejo, C. J., Grenier, M., Netto-Ferreira, J. C., & Scaiano, J. C. (2011). High-temperature organic reactions at room temperature using plasmon excitation: Decomposition of dicumyl peroxide. *Organic Letters*, *13*(2), 204–207. doi:10.1021/ol1026427 PMID:21142017

Ferber, J. (1999). *Multi-Agent Systems: An Introduction to Distributed Artificial Intelligence*. Reading, MA: Addison-Wesley.

Ferguson, C., & Ferguson, H. (1994). *Mathematics in Stone and Bronze*. Erie, PA: Meridian Creative Group.

Feynman, R. P. (1960). *There's plenty of room at the bottom: An invitation to enter a new field of physics*. Conference.

Fischer, M. A., Haley, H., Saarinen, C. L., & Chretien, K. C. (2011). Comparison of blogged and written reflections in two medicine clerkships. *Medical Education*, *45*(2), 166–175. doi:10.1111/j.1365-2923.2010.03814.x PMID:21208262

Flenniken, M. L., Liepold, L. O., Crowley, B. E., Willits, D. A., Young, M. J., & Douglas, T. (2005). Selective attachment and release of a chemotherapeutic agent from the interior of a protein cage architecture. *Chemical Communications*, *2005*(4), 447–449. doi:10.1039/b413435d PMID:15654365

Flenniken, M. L., Willits, D. A., Brumfield, S., Young, M. J., & Douglas, T. (2003). The small heat shock protein cage from *Methanococcus jannaschii* is a versatile nanoscale platform for genetic and chemical modification. *Nano Letters*, *3*(11), 1573–1576. doi:10.1021/nl034786l

Flenniken, M., Allen, M., & Douglas, T. (2004). Microbe manufacturers of semiconductors. *Chemistry & Biology*, *11*(11), 1478–1480. doi:10.1016/j.chembiol.2004.11.004 PMID:15555996

Flynn, C. E., Lee, S.-W., Peelle, B. R., & Belcher, A. M. (2003). Viruses as vehicles for growth, organization and assembly of materials. *Acta Materialia*, *51*(19), 5867–5880. doi:10.1016/j.actamat.2003.08.031

Forbes, M. O., & Hickey, M. T. (2008). Podcasting: Implementation and evaluation in an undergraduate nursing program. *Nurse Educator*, *33*(5), 224–227. doi:10.1097/01.NNE.0000334775.98018.e8 PMID:18769328

Forestier, T., Mornet, S., Daro, N., Nishihara, T., Mouri, S.-i., Tanaka, K., & Létard, J.-F. et al. (2008). Nanoparticles of iron(II) spin-crossover. *Chemical Communications*, *2008*(36), 4327–4329. doi:10.1039/b806347h PMID:18802559

Fourches, D., Pu, D. Q. Y., Tassa, C., Weissleder, R., Shaw, S. Y., Mumper, R. J., & Tropsha, A. (2010). Quantitative nanostructure–activity relationship modeling. *ACS Nano*, *4*(10), 5703–5712. doi:10.1021/nn1013484 PMID:20857979

Fox, B. I., & Varadarajan, R. (2011). Use of Twitter® to Encourage Interaction in a Multi-campus Pharmacy Management Course. *American Journal of Pharmaceutical Education*, *75*(5), 88. doi:10.5688/ajpe75588 PMID:21829262

Fralinger, L., & Bull, J. (2013). Measuring the international usage of US institutional repositories. *OCLC Systems & Services*, *29*(3), 134–150. doi:10.1108/OCLC-10-2012-0039

Frangopol, P. T., & Morariu, V. V. (Eds.). (1988). *Seminars in biophysics* (Vol. 5). Bucharest: CIP.

Frew, S. E., Sammut, S. M., Shore, A. F., Ramjist, J. K., Al-Bader, S., Rezaie, R., & Singer, P. A. et al. (2008). Chinese health biotech and the billion-patient market. *Nature Biotechnology*, *26*(1), 37–53. doi:10.1038/nbt0108-37 PMID:18183014

Frey, D., Stockheim, T., Woelk, P., & Zimmermann, R. (2003). Integrated Multi-agent-based Supply Chain Management. In *Proceedings of 1st International Workshop on Agent-based Computing for Enterprise Collaboration.*

Fuertes, G., García-Sáez, A. J., Esteban-Martín, S., Giménez, D., Sánchez-Muñoz, O. L., Schwille, P., & Salgado, J. (2010). Pores formed by Baxα5 relax to a smaller size and keep at equilibrium. *Biophysical Journal*, *99*(9), 2917–2925. doi:10.1016/j.bpj.2010.08.068 PMID:21044589

Fuertes, G., Giménez, D., Esteban-Martín, S., Sánchez-Muñoz, O. L., & Salgado, J. (2011). A lipocentric view of peptide-induced pores. *European Biophysics Journal*, *40*(4), 399–415. doi:10.1007/s00249-011-0693-4 PMID:21442255

Fujishima, A., Rao, T. N., & Tryk, D. A. (2000). Titanium dioxide photocatalysis. *Journal of Photochemistry and Photobiology A Chemistry*, *1*(1), 1–21. doi:10.1016/S1389-5567(00)00002-2

Fürst, J. A., Hashemi, J., Markussen, T., Brandbyge, M., Jauho, A. P., & Nieminen, R. M. (2009). Electronic transport properties of fullerene functionalized carbon nanotubes: *Ab initio* and tight-binding calculations. *Physical Review B*, *80*, 35427.

Fu, Y., & Piplani, R. (2000). Multi-agent enabled modelling and simulation towards collaborative inventory management in supply chains. In *Proceedings of the 2000 Winter Simulation Conference.*

Gadet, V., Mallah, T., Castro, I., Verdaguer, M., & Veillet, P. (1992). High-T_c molecular-based magnets: A ferromagnetic bimatallic chromium(III)–nickel(II) cyanide with T_c = 90K. *Journal of the American Chemical Society*, *114*(23), 9213–9214. doi:10.1021/ja00049a078

Gai, S., Yang, P., Li, C., Wang, W., Dai, Y., Niu, N., & Lin, J. (2010). Synthesis of magnetic, up-conversion luminescent, and mesoporous core–shell-structured nanocomposites as drug carriers. *Advanced Functional Materials*, *20*(7), 1166–1172. doi:10.1002/adfm.200902274

Gaita-Ariño, A., & Schechter, M. (2011). Identification of strong and weak interacting two-level systems in KBr:CN. *Physical Review Letters*, *107*(10), 105504–1–5. doi:10.1103/PhysRevLett.107.105504 PMID:21981511

Galán-Mascarós, J. R., Coronado, E., Forment-Aliaga, A., Monrabal-Capilla, M., Pinilla-Cienfuegos, E., & Ceolin, M. (2010). Tuning size and thermal histeresis in bistable spin crossover nanoparticles. *Inorganic Chemistry*, *49*(12), 5706–5714. doi:10.1021/ic100751a PMID:20503990

Galbiati, M., Barraud, C., Tatay, S., Bouzehouane, K., Deranlot, C., Jacquet, E., & Petroff, F. et al. (2012). Unveiling self-assembled monolayers' potential for molecular spintronics: Spin transport at high voltage. *Advanced Materials*, *24*(48), 6429–6432. doi:10.1002/adma.201203136 PMID:23055410

Gasmelseid, T. (2006). Multi Agent Negotiation Framework in Resource Bounded Environments. In *Proceedings of the Information and Communication Technologies, 2006. ICTTA '06*. (pp. 465 – 470). Retrieved from http://ieeexplore.ieee.org/xpl/freeabs_all.jsp?tp=&arnumber=1684414&isnumber=35470

Gasmelseid, T. (2009a). On the design of multi-agent, context aware and mobile systems. In Handbook on Modern System Analysis and Design Applications and Technologies. Idea Group Publishing.

Gasmelseid, T. (2014). Managing Stakeholder Concerns in Large-Scale Multi-Agent Information Systems. *International Journal of Agent Technologies and Systems, 5*(4), 68-86.

Gasmelseid, T. M. (2011). Improving Medical Practice through Mobile Medical Informatics In Clinical Technologies: Concepts, Methodologies, Tools and Applications. IGI Global.

Gasmelseid, T. (2009). Improving clinical practice through mobile medical informatics. In U. Bhuvan (Ed.), *Handbook of Research on Mobile Business: Technical, Methodological and Social perspective, Project HRMB2* (pp. 604–614). IGI Publishing. doi:10.4018/978-1-60566-156-8.ch056

Gasmelseid, T. M. (2008). Engineering Multi-agent Systems. In M. Quigley (Ed.), *Encyclopedia of Information Ethics and Security* (pp. 187–193). IGI Global.

Gasser, U., Weeks, E. R., Schofield, A., Pusey, P. N., & Weitz, D. A. (2001). Real-space imaging of nucleation and growth in colloidal crystallization. *Science*, *292*(5515), 258–262. doi:10.1126/science.1058457 PMID:11303095

Gassmann, O., Reepmeyer, G., & Zedtwitz, M. v. (2008). *Leading Pharmaceutical Innovation: Trends and Drivers for Growth in the Pharmaceutical Industry*. Berlin: Springer. doi:10.1007/978 3 540 77636 9

Gaurav, J., Christian, R., & Robert, H. (2009). *A Multi-Agent Simulation (MAS) of the Pharmaceutical Supply Chain (PSC)*. POMS 20th Annual Conference, Orlando, FL.

Geim, A. K. (2009). Graphene: Status and prospects. *Science*, *324*(5934), 1530–1534. doi:10.1126/science.1158877 PMID:19541989

Geim, A. K., & Novoselov, K. S. (2007). The rise of graphene. *Nature Materials*, *6*(3), 183–191. doi:10.1038/nmat1849 PMID:17330084

George, D. R., & Dellasega, C. (2011). Use of social media in graduate-level medical humanities education: Two pilot studies from Penn State College of Medicine. *Medical Teacher*, *33*(8), e429–e434. doi:10.3109/0142159X.2011.586749 PMID:21774639

Geyer, H., Beylefeld, A., & Alwyn, H. (2008). To Podcast or not to Podcast? Students' Feedback on a Different Learning Experience in Histology. In R. Williams & D. Remenyi (Eds.), *The Proceedings of the 7th European Conference on e-Learning*. Agia Napa, Cyprus: Academic Conferences Limited.

Ghauri, P. N., & Rao, P. M. (2009). Intellectual property, pharmaceutical MNEs and the developing world. *Journal of World Business*, *44*(2), 206–215. doi:10.1016/j.jwb.2008.05.008

Ghorbani, M. (2012). Enumeration of heterofullerenes: A survey. *MATCH: Communications in Mathematical and in Computer Chemistry*, *68*, 381–414.

GIA. (2012a). *Human Vaccines: A Global Strategic Business Report*. Global Industry Analysts, Inc.

GIA. (2012b). *Nanobiotechnology: A Global Strategic Business Report*. Global Industry Analysts, Inc.

Gilmore, J. L., Yi, X., Quan, L., & Kabanov, A. V. (2008). Nobel nanomaterials for clinical neuroscience. *Journal of Neuroimmune Pharmacology*, *3*(2), 83–94. doi:10.1007/s11481-007-9099-6 PMID:18210200

Giniat, E., Fung, P., Weir, A., & Meyring, N. (2011). *China's Pharmaceutical Industry - Poised for the Giant Leap*. KPMG Advisory (China) Limited.

Girifalco, L. A. (1991). Interaction potential for C_{60} molecules. *Journal of Physical Chemistry*, *95*(14), 5370–5371. doi:10.1021/j100167a002

Girifalco, L. A. (1992). Molecular properties of C_{60} in the gas and solid phases. *Journal of Physical Chemistry*, *96*(2), 858–861. doi:10.1021/j100181a061

Giuliani Partners LLC. (2004). *Examination and Assessment of Prescription Drug Importation from Foreign Sources to the United States. Interim Findings*. Author.

Giunchiglia, F., Perini, A., & Mylopoulus, J. (2002). The Tropos software development methodology: Processes, models and diagrams. In *Proceedings of the First International Joint Conference on Autonomous Agents and Multi agent Systems* (pp. 63-74). Bologna, Italy: ACM Press. doi:10.1145/544741.544748

Giunchiglia, F., Perini, A., & Sannicol, F. (2001). Knowledge level software engineering. In *Intelligent Agents VIII (LNCS)* (Vol. 2333, pp. 6–20). Seattle, WA: Springer-Verlag. doi:10.1007/3-540-45448-9_2

Glossary of Meteorology . (2013). Retrieved March 15, 2014, from http://glossary.ametsoc.org/?p=1&query=cloud+seeding&submit=Search

Godo, L., Puyol-Gruart, J., Sabater, J., Torra, V., Barrufet, P., & Fàbregas, F. (2003). A multi agent approach for monitoring the prescription of restricted use antibiotics. *Artificial Intelligence in Medicine*, *27*(3), 259–282. doi:10.1016/S0933-3657(03)00006-X PMID:12667739

Goldman, R. H., Cohen, A. P., & Sheahan, F. (2008). Using Seminar Blogs to Enhance Student Participation and Learning in Public Health School Classes. *American Journal of Public Health*, *98*(9), 1658–1663. doi:10.2105/AJPH.2008.133694 PMID:18633075

Gomes, M. J., & Coutinho, C. (2008). Meta-análise da investigação realizada no âmbito do mestrado em Tecnologia Educativa da UM. In *As TIC na Educação em Portugal. Concepções e Práticas* (pp. 60–70). Porto: Porto Editora.

Gonzalez-Arellano, C., Balu, A. M., Luque, R., & Macquarrie, D. J. (2010a). Catalytically active self-assembled silica-based nanostructures containing supported nanoparticles. *Green Chemistry*, *12*(11), 1995–2002. doi:10.1039/c0gc00282h

Gonzalez-Arellano, C., Luque, R., & Macquarrie, D. J. (2009). Nanotubular self-assembly of *n*-dodecylamine–TEOS–water–acetonitrile mixtures. *Chemical Communications*, (30), 4581–4583. doi:10.1039/b903368h PMID:19617990

Gonzalez-Arellano, C., Yoshida, K., Luque, R., & Gai, P. L. (2010b). Highly active and selective supported iron oxide nanoparticles in microwave-assisted *N*-alkylations of amines with alcohols. *Green Chemistry*, *12*(7), 1281–1287. doi:10.1039/c003410j

González, J., Guinea, F., & Vozmediano, M. A. H. (1992). Continuum approximation to fullerene molecules. *Physical Review Letters*, *69*(1), 172–175. doi:10.1103/PhysRevLett.69.172 PMID:10046217

Gonzalez-Valls, I., & Lira-Cantu, M. (2009). Vertically-aligned nanostructures of ZnO for excitonic solar cells: A review. *Energy and Environmental Science*, *2*(1), 19–34. doi:10.1039/B811536B

Gonzalez-Valls, I., & Lira-Cantu, M. (2010). Dye sensitized solar cells based on vertically-aligned ZnO nanorods: Effect of UV light on power conversion efficiency and lifetime. *Energy and Environmental Science*, *3*(6), 789–795. doi:10.1039/b922354a

Google Play. (2014a). *Epocrates*. Available from: https://play.google.com/store/apps/details?id=com.epocrates&hl=en

Google Play. (2014b). *Micromedex Drug Interactions*. Available from: https://play.google.com/store/apps/details?id=com.thomson.druginteractions&hl=en

Govindu, R., & Chinnam, R. (2007). MASCF: A generic process-cantered methodological framework for analysis and design of multi-agent supply chain systems. *Computers & Industrial Engineering*, *53*(4), 584–609. doi:10.1016/j.cie.2007.06.003

Grassley, J. S., & Bartoletti, R. (2009). Wikis and blogs: Tools for online interaction. *Nurse Educator*, *34*(5), 209–213. doi:10.1097/NNE.0b013e3181b2b59b PMID:19726963

Gratzel, M. (2001). Photoelectrochemical cells. *Nature*, *414*(6861), 338–344. doi:10.1038/35104607 PMID:11713540

Grazioli, S., & Jarvenpaa, S. (2003). Consumer and business deception on the Internet: Content analysis of documentary evidence. *International Journal of Electronic Commerce*, *7*(4), 93–118.

Grosseck, G., & Holotesku, C. (2008). Can we use Twitter® for educational activities? In *4th Scientific Conference eLSE "elearning and Software for Education"*. Retrieved from http://www.scribd.com/doc/2286799/Can-we-use-Twitter-for-educational-activities

Grossman, G. M., & Shapiro, C. (1988a). Counterfeit product trade. *The American Economic Review*, *78*(1), 59–75.

Grossman, G. M., & Shapiro, C. (1988b). Foreign counterfeiting of status goods. *The Quarterly Journal of Economics*, *103*(1), 79–100. doi:10.2307/1882643

Grujović, N., Radović, M., Kanjevac, V., Borota, J., Grujović, G., & Divac, D. (2011). 3D printing technology in education environment.*34th International Conference on Production Engineering* (pp. 29-30).

Guccione, S., Li, K., & Bednarski, M. (2004). Vascular-targeted nanoparticles for molecular imaging and therapy. *Methods in Enzymology*, *386*, 219–236. doi:10.1016/S0076-6879(04)86010-5 PMID:15120254

Guerrero-Martínez, A., Pérez-Juste, J., Carbó-Argibay, E., Tardajos, G., & Liz-Marzán, L. M. (2009). Gemini-surfactant-directed self-assembly of monodisperse gold nanorods into standing superlattices. *Angewandte Chemie International Edition*, *48*(50), 9484 9488. doi:10.1002/anie.200904118 PMID:19802865

Guerrero-Martínez, A., Pérez-Juste, J., & Liz-Marzán, L. M. (2010). Recent progress on silica coating of nanoparticles and related nanomaterials. *Advanced Materials*, *22*(11), 1182–1195. doi:10.1002/adma.200901263 PMID:20437506

Guionneau, P., Le Gac, F., Kaiba, A., Sánchez Costa, J., Chasseau, D., & Létard, J. F. (2007). A reversible metal–ligand bond break associated to a spin-crossover. *Chemical Communications*, *2007*(36), 3723–3725. doi:10.1039/b707836f PMID:17851607

Guldi, D. M., & Martin, N. (Eds.). (2010). Carbon nanotubes and related structures. Weinheim: Wiley-VCH. doi:10.1002/9783527629930

Guo, X., Zhang, F., Evans, D. G., & Duan, X. (2010). Layered double hydroxide films: Synthesis, properties and applications. *Chemical Communications*, *46*(29), 5197–5210. doi:10.1039/c0cc00313a PMID:20549015

Gupta, A. K., & Curtis, A. S. G. (2004). Lactoferrin and ceruloplasmin derivatized superparamagnetic iron oxide nanoparticles for trageting cell surface receptors. *Biomaterials*, *25*(15), 3029–3040. doi:10.1016/j.biomaterials.2003.09.095 PMID:14967536

Gupta, A. K., & Gupta, M. (2005). Synthesis and surface engineering of iron oxide nanoparticles for biomedical applications. *Biomaterials*, *26*(18), 3995–4021. doi:10.1016/j.biomaterials.2004.10.012 PMID:15626447

Haag, M. (2011). Calculating and understanding particulate contamination risk. *Pharmaceutical Technology Europe*, *23*(3), 38–41.

Haeckel, E., Breidbach, O., Hartman, R., & Eibl-Eibesfeldt, I. (1998). *Art Forms in Nature: The Prints of Ernst Haeckel (Monographs)*. München: Prestel Publishing. (Original work published 1904)

Haes, A. W., Hall, W. P., Chang, L., Klein, W. L., & Duyne, R. P. (2004). A localized surface plasmon resonance biosensor: First steps toward an assay for Alzheimer's disease. *Nano Letters*, *4*(6), 1029–1034. doi:10.1021/nl049670j

Hagen, M. H. J., Meijer, E. J., Mooij, G. C. A. M., Frenkel, D., & Lekkerkerker, H. N. W. (1993). Does C_{60} have a liquid phase? *Nature*, *365*(6445), 425–426. doi:10.1038/365425a0

Haluska, C. K., Riske, K. A., Marchi-Artzner, V., Lehn, J.-M., Lipowsky, R., & Dimova, R. (2006). Time scales of membrane fusion revealed by direct imaging of vesicle fusion with high temporal resolution. *Proceedings of the National Academy of Sciences of the United States of America*, *103*(43), 15841–15846. doi:10.1073/pnas.0602766103 PMID:17043227

Han, J., & Jaffe, R. (1998). Energetics and geometries of carbon nanocone tips. *The Journal of Chemical Physics*, *108*(7), 2817–2823. doi:10.1063/1.475672

Han, P. (2009). China's growing biomedical industry. *Biologicals*, *37*(3), 169–172. doi:10.1016/j.biologicals.2009.02.010 PMID:19427231

Hansen, M. M., Murray, P. J., & Erdley, W. S. (2009). The potential of 3-D virtual worlds in professional nursing education. *Studies in Health Technology and Informatics*, *146*, 582–586. PMID:19592909

Harada, K., Kamimura, O., Kasai, H., Matsuda, T., Tonomura, A., & Moshchalkov, V. V. (1996). Direct observation of vortex dynamics in superconducting films with regular arrays of defects. *Science*, *274*(5290), 1167–1170. doi:10.1126/science.274.5290.1167 PMID:8895460

Haridasan, S., & Rani, K. (2012). Institutional Repository Initiatives in Indian Universities: An Evaluative Study. *Journal of Library and Information Science*, *37*(2), 71–87.

Harris, G. (2011, January 23). Federal Research Center Will Help Develop Medicines. *New York Times,* p. 1.

Harris, S. (2014). Compound benefits. *Scientific Computing World, 2014*(134), 31-33.

Hartmann, U. (1985). *Theory of noncontact force microscopy: Scanning tunneling microscopy III. Theory of STM and related scanning techniques* (R. Wiensendanger & J. H. Guntherodt, Eds.). Berlin: Springer.

Ha, S., & Lennon, S. J. (2006). Purchase intent for fashion counterfeit products: Ethical ideologies, ethical judgments, and perceived risks. *Clothing & Textiles Research Journal*, *24*, 297–315.

Hawker, C., & Wooley, K. (2005). The convergence of synthetic organic and polymer chemistries. *Science*, *309*(5738), 1200–1205. doi:10.1126/science.1109778 PMID:16109874

Hawthorne, M. F., Zink, J. I., Skelton, J. M., Bayer, M. J., Liu, C., Livshits, E., & Neuhauser, D. et al. (2004). Electrical or photocontrol of the rotary motion of a metallacarborane. *Science*, *303*(5665), 1849–1851. doi:10.1126/science.1093846 PMID:15031500

Hebard, A. F. (1993). Buckminsterfullerene. *Annual Review of Materials Science*, *23*(1), 159–191. doi:10.1146/annurev.ms.23.080193.001111

Heimburg, T. (2010). Lipid ion channels. *Biophysical Chemistry*, *150*(1-3), 2–22. doi:10.1016/j.bpc.2010.02.018 PMID:20385440

Hernandez, Y., Nicolosi, V., Lotya, M., Blighe, F. M., Sun, Z., De, S., & Coleman, J. N. et al. (2008). High-yield production of graphene by liquid-phase exfoliation of graphite. *Nature Nanotechnology*, *3*(9), 563–568. doi:10.1038/nnano.2008.215 PMID:18772919

Herrmann, J.-M. (1999). Heterogeneous photocatalysis: Fundamentals and applications to the removal of various types of aqueous pollutants. *Catalysis Today*, *53*(1), 115–129. doi:10.1016/S0920-5861(99)00107-8

Herrmann, J.-M. (2005). Heterogeneous photocatalysis: State of the art and present applications. *Topics in Catalysis*, *34*(1-4), 49–65. doi:10.1007/s11244-005-3788-2

Herrmann, J.-M. (2006a). Photocatalysis. In *Kirk-Othmer encyclopedia of chemical technology*. New York, NY: Wiley. doi:10.1002/0471238961.1608152019051816.a01.pub2

Herrmann, J.-M. (2006b). From catalysis by metals to bifunctional photocatalysis. *Topics in Catalysis*, *39*(1-2), 3–10. doi:10.1007/s11244-006-0032-7

Herrmann, J.-M. (2010). Fundamentals and misconceptions in photocatalysis. *Journal of Photochemistry and Photobiology A Chemistry*, *216*(2-3), 85–93. doi:10.1016/j.jphotochem.2010.05.015

Herrmann, J.-M., Guillard, C., & Pichat, P. (1993). Heterogeneous photocatalysis: An emerging technology for water treatment. *Catalysis Today*, *17*(1-2), 7–20. doi:10.1016/0920-5861(93)80003-J

Hilton, B., Choi, C. J., & Chen, S. (2004). The ethics of counterfeiting in the fashion industry: Quality, credence and profit issues. *Journal of Business Ethics*, *55*(4), 343–352. doi:10.1007/s10551-004-0989-8

Hirsch, L., Gobin, A., Lowery, A., Tam, F., Drezek, R., Halas, N., & West, J. (2006). Metal nanoshells. *Annals of Biomedical Engineering*, *34*(1), 15–22. doi:10.1007/s10439-005-9001-8 PMID:16528617

History of RFID. (2010). *RFID Journal*. Retrieved on September 6, 2014, from http://www.rfidjournal.com/article/articleview/1338/1/129/

Hoffmann, M. R., Martin, S. T., Choi, W., & Bahnemann, D. W. (1995). Environmental applications of semiconductor photocatalysis. *Chemical Reviews*, *95*(1), 69–96. doi:10.1021/cr00033a004

Holothurian. (2011). In *Encyclopædia Britannica*. Retrieved March 14, 2014, from http://www.britannica.com/EBchecked/topic/269645/holothurian

Honey, M. L., Diener, S., Connor, K., Veltman, M., & Bodily, D. (2009). Teaching in Virtual Space: An interactive session demonstrating Second Life® simulation for haemorrhage management. In *Same places, different spaces*. Retrieved from. http://www.ascilite.org.au/conferences/auckland09/procs/honey-interactivesession.pdf

Horák, D., Rittich, B., Španová, A., & Beneš, M. J. (2005). Magnetic microparticulate carriers with immobilized selective ligands in DNA disgostics. *Polymer*, *46*(4), 1245–1255. doi:10.1016/j.polymer.2004.11.049

Horn, W. (1999). *Artificial intelligence in medicine.* Joint European Conference on Artificial Intelligence in Medicine and Medical Decision Making, AIMDM'99, Aalborg, Denmark.

Horsell, D. W., Hale, P. J., & Savchenko, A. K. (2011). Mechanical manipulation measurement of graphene by atomic force microscopy. *Microscopy and Analysis*, *25*(1), 15–17.

Huang, Q., & Balsys, R. J. (2009). Applying Fractal and Chaos Theory to Animation in the Chinese Literati Tradition. *Sixth International Conference on Computer Graphics, Imaging and Visualization* (pp.112-122). Los Alamitos, CA: IEEE Computer Society. doi:10.1109/CGIV.2009.56

Hu, J., & Weliman, M. (2001). Learning about other agents in a dynamic multi-agent system. *Cognitive Systems Research*, *2*(1), 67–79. doi:10.1016/S1389-0417(01)00016-X

Huskinson, B., Marshak, M. P., Suh, C., Er, S., Gerhardt, M. R., Galvin, C. J., & Aziz, M. J. et al. (2014). A metal-free organic–inorganic aqueous flow battery. *Nature*, *505*(7482), 195–198. doi:10.1038/nature12909 PMID:24402280

Hu, X., Ma, Q., & Zhang, S. (2006). Biopharmaceuticals in China. *Biotechnology Journal*, *1*(11), 1215–1224. doi:10.1002/biot.200600083 PMID:17089435

Hu, X., & Zhou, Q. (2013). Health and ecosystem risks of graphene. *Chemical Reviews*, *113*(5), 3815–3835. doi:10.1021/cr300045n PMID:23327673

Hvistendahl, M. (2012). China's Push in Tissue Engineering. *Science*, *338*(6109), 900–902. doi:10.1126/science.338.6109.900 PMID:23161989

Hyman, R. (1989). The psychology of deception. *Annual Review of Psychology*, *40*(1), 133–154. doi:10.1146/annurev.ps.40.020189.001025

IBM RFID Information Center. (2010). Retrieved on September 8, 2014, from http://en.wikipedia.org/wiki/IBM_RFID_Information_Center

Ibn El Ahrach, H., Bachelot, R., Vial, A., Lérondel, G., Plain, J., Royer, P., & Soppera, O. (2007). Spectral degeneracy breaking of the plasmon resonance of single metal nanoparticles by nanoscale near-field photopolymerization. *Physical Review Letters*, *98*(10), 107402–1–4. doi:10.1103/PhysRevLett.98.107402 PMID:17358565

Iglesias, C., Garrijo, M., & Gonzalez, J. (1998). A survey of agent-oriented methodologies. In *Intelligent Agents V – Proceedings of the 1998 Workshop on Agent Theories, Architectures and Languages*. Academic Press. doi:10.1007/3-540-49057-4_21

IMPACT. (2011). *The Handbook: Facts-Activities-Documents 2006-2010.* Rome: Agenzia Italiana del Farmaco.

IMS. (2011). *Shaping the biosimilars opportunity: A global perspective on the evolving biosimilars landscape*. IMS Health.

Inefuku, H. W. (2013). Whatever Happened to Art and Design?: Using Archival Practice to Manage the Impact of Academic Restructuring on Institutional Repositories. *Journal of Library Administration*, *53*(4), 209–222. doi:10.1080/01930826.2013.865383

Information on Heparin. (2010). US Food and Drug Administration. Retrieved on September 6, 2014, from http://www.fda.gov/Drugs/DrugSafety/PostmarketDrugSafetyInformationforPatientsandProviders/UCM112597

Inspectie voor de Gezondheidszorg. (2009). *Het resultaat telt.* Available from: http://www.farmaactueel.nl/beleidsstukken/IGZ2009.pdf

International Anti-Counterfeiting Coalition. (2008). *Get real: The truth about counterfeiting*. Retrieved on September 5, 2014 from www.iacc.org/aboutcounterfeiting/the-truth-about-counterfeiting.php)

Internet Repository. (2010). In *Wikipedia*. Retrieved December 17, 2010, from http://en.wikipedia.org/w/index.php?title=Institutional_repository&oldid=402861088

Iordache, O. (2011). *Modeling multi-level systems*. Berlin: Springer. doi:10.1007/978-3-642-17946-4

Iordache, O. (2012). *Self-evolvable systems: machine learning in social media*. Berlin: Springer. doi:10.1007/978-3-642-28882-1

Iordache, O. (2014). *Polytope Projects*. Boca Raton, FL: CRC.

Iordache, O., Apostol, M., & Frangopol, P. T. (1988). Non-exponential relaxation of drug–membrane interaction. Effects of cross-linking aldehydes and procaine. *Revue Roumaine de Biochimie*, *25*, 275–281.

Irwin, J. L., Opplinger, D. E., Pearce, J. M., & Anzalone, G. (2015). Evaluation of RepRap 3D Printer Workshops in K-12 STEM (Program/Curriculum Evaluation).*122nd ASEE Annual Conference & Exposition*. doi:10.18260/p.24033

iTunes. (2014a). *Epocrates*. Available from: https://itunes.apple.com/en/app/epocrates/id281935788?mt=8

iTunes. (2014b). *Micromedex Drug Interactions*. Available from: https://itunes.apple.com/us/app/micromedex-drug-interactions/id391763035?mt=8

IUPAC. (2014). *Compendium of Chemical Terminology* (2nd ed.). Retrieved March 15, 2014, from http://www.iupac.org/home/publications/e-resources/nomenclature-and-terminology.html

Jaafar, M., Gómez-Herrero, J., Gil, A., Ares, P., Vázquez, M., & Asenjo, A. (2009). Variable-field magnetic force microscopy. *Ultramicroscopy*, *109*(6), 693–699. doi:10.1016/j.ultramic.2009.01.007 PMID:19250752

Jaafar, M., Iglesias-Freire, O., Serrano-Ramón, L., Ibarra, M. R., de Teresa, J. M., & Asenjo, A. (2011). Distinguishing magnetic and electrostatic interactions by a Kelvin probe force microscopy–magnetic force microscopy combination. *Beilstein Journal of Nanotechnology*, *2*, 552–560. doi:10.3762/bjnano.2.59 PMID:22003461

Jaafar, M., Martínez-Martín, D., Cuenca, M., Melcher, J., Raman, A., & Gómez-Herrero, J. (2012). Drive-amplitude-modulation atomic force microscopy: From vacuum to liquids. *Beilstein Journal of Nanotechnology*, *3*, 336–344. doi:10.3762/bjnano.3.38 PMID:22563531

Jacquemin, D., Michaux, C., Perpète, E. A., Maurel, F., & Perrier, A. (2010a). Photochromic molecular wires: Insights from theory. *Chemical Physics Letters*, *488*(4-6), 193–197. doi:10.1016/j.cplett.2010.02.017

Jacquemin, D., Perpète, E. A., Maurel, F., & Perrier, A. (2010b). *Ab initio* investigation of the electronic properties of coupled dithienylethenes. *The Journal of Physical Chemistry Letters*, *1*(1), 434–438. doi:10.1021/jz900293g

Jacquemin, D., Perpète, E. A., Maurel, F., & Perrier, A. (2010c). Doubly closing or not? Theoretical analysis for coupled photochromes. *The Journal of Physical Chemistry C*, *114*(20), 9489–9497. doi:10.1021/jp102118w

Jacquemin, D., Perpète, E. A., Maurel, F., & Perrier, A. (2010d). Simulation of the properties of a photochromic triad. *The Journal of Physical Chemistry Letters*, *1*(14), 2104–2108. doi:10.1021/jz1006753

Jacquemin, D., Perpète, E. A., Maurel, F., & Perrier, A. (2010e). TD-DFT simulations of the electronic properties of star-shaped photochromes. *Physical Chemistry Chemical Physics*, *12*(28), 7994–8000. doi:10.1039/b927323a PMID:20517564

Jacquemin, D., Perpète, E. A., Maurel, F., & Perrier, A. (2010f). Hybrid dithienylethene-naphthopyran multi-addressable photochromes: An *ab initio* analysis. *Physical Chemistry Chemical Physics*, *12*(40), 13144–13152. doi:10.1039/c0cp00400f PMID:20838697

Jacquez, J. A. (1985). *Compartmental Analysis in Biology and Medicine*. Ann Arborm MI: The University of Michigan Press.

Jadad, A. R., Moher, D., & Klassen, T. P. (1998). Guides for reading and interpreting systematic reviews: II. How did the authors find the studies and assess their quality? *Archives of Pediatrics & Adolescent Medicine*, *152*(8), 812–817. doi:10.1001/archpedi.152.8.812 PMID:9701144

Jain, P. (2011). New trends and future applications/directions of institutional repositories in academic institutions. *Library Review*, *60*(2), 125–141. doi:10.1108/00242531111113078

James, A. V., James, D. A., & Kalisperis, L. N. (2004). A unique art form: The friezes of Pirgi. *Leonardo*, *37*(3), 234–242. doi:10.1162/0024094041139409

Janodia, M. D., Sreedhar, D., Ligade, V. S., Pise, A., & Udupa, N. (2008). Facets of Technology Transfer: A Perspective of Pharmaceutical Industry. *Journal of Intellectual Property Rights*, *13*, 28–34.

Jeffery, K. G. (2006). Open Access: An Introduction. *Ercim News, 64*.

Jesus, Â., Gomes, M. J., & Cruz, A. (2012). *A B-learning strategy for Therapeutics at the Bachelor Level*. Presented at the FIP World Centennial Congress of Pharmacy and Pharmaceutical Sciences, Amsterdam: International Pharmaceutical Federation.

Jesus, Â., Gomes, M. J., & Cruz, A. (2013). *Case Based Learning Digital - Proposta para Estruturação da Formação*. Presented at the XII Congresso Internacional Galego-Português de Psicopedagogia, Braga.

Jesus, Â., Cruz, A., & Gomes, M. J. (2011). Case Based, Learner Centered Approach to Pharmacotherapy. In *Proceedings from EDULEARN 2011* (pp. 6074–6080). Barcelona: IATED.

Jia, C., & Guo, X. (2013). Molecule–electrode interfaces in molecular electronic devices. *Chemical Society Reviews*, *42*(13), 5642–5660. doi:10.1039/c3cs35527f PMID:23571285

Jiang, Y., Wang, Y., & Yan, X. (2001). Chinese pharmaceutical companies: An emerging industry. *Drug Discovery Today*, *6*(12), 610–612. doi:10.1016/S1359-6446(01)01842-6 PMID:11408192

Jimoda, L. A., Oke, E. O., & Salam, K. K. (2013). Modelling of mass transfer rate during biocoagulation-flocculation of coal-rich wastewater. *Journal of Scientific Research and Reports*, *2*(1), 376–390. doi:10.9734/JSRR/2013/3492

Jin, C. (2005). *Towards Indigenous Innovation: Pathways for Chinese Firms.Workshop of Technology Innovation and Economic Development*.

Johnson, K. U. S. (2010, April 30). Seizes Big Batches of Fake Goods. *Wall Street Journal*.

Johnson, P. E., Grazioli, S., Jamal, K., & Berryman, R. G. (2001). Jamal, K. & Berryman, R.G. Detecting deception: Adversarial problem solving in a low base-rate world. *Cognitive Science*, *25*(3), 355–392. doi:10.1207/s15516709cog2503_2

Joint, N. (2006). Institutional repositories, self archiving and the role of the library. *Library Review*, *55*(2), 81–84. doi:10.1108/00242530610649576

Jones, J. (2005). *Fractals & Frequencies*. Retrieved March 15, 2014, from http://fractalsandfrequencies.blogspot.com/

Jones, O. (2001, 1856). The Grammar of Ornament: Illustrated by Examples from Various Styles of Ornament. London: A Dorling Kindersley Book, The Ivy Press, Limited.

Jones, V. (2011). Challenges of particle characterisation. *Pharmaceutical Technology Europe*, *23*(4), 40–43.

Jongh, P. E., & Adelhelm, P. (2010). Nanosizing and nanoconfinement: New strategies towards meeting hydrogen storage goals. *ChemSusChem*, *3*(12), 1332–1348. doi:10.1002/cssc.201000248 PMID:21080405

Jordan, A., Scholz, R., Maier-Hauff, K., Johannsen, M., Wust, P., Nadobny, J., & Felix, R. et al. (2001). Presentation of a new magnetic field therapy system for the treatment of human solid tumors with magnetic fluid hyperthermia. *Journal of Magnetism and Magnetic Materials*, *225*(1-2), 118–126. doi:10.1016/S0304-8853(00)01239-7

Kagan, C. R., Murray, C. B., Nirmal, M., & Bawendi, M. G. (1996). Electronic energy transfer in CdSe quantum dot solids. *Physical Review Letters*, *76*(9), 1517–1520. doi:10.1103/PhysRevLett.76.1517 PMID:10061743

Kahn, O., & Martinez, C. J. (1998). Spin transition polymers: From molecular materials toward memory devices. *Science*, *279*(5347), 44–48. doi:10.1126/science.279.5347.44

Kaiser, U., Schwarz, A., & Wiesendanger, R. (2007). Magnetic exchange force microscopy with atomic resolution. *Nature*, *446*(7135), 522–525. doi:10.1038/nature05617 PMID:17392782

Kaldoudi, E., Bamidis, P., Papaioakeim, M., & Vargemezis, V. (2008). *Problem-Based Learning via Web 2.0 Technologies*. IEEE; doi:10.1109/CBMS.2008.136

Kamel Boulos, M. N., & Wheeler, S. (2007). The emerging Web 2.0 social software: An enabling suite of sociable technologies in health and health care education. *Health Information and Libraries Journal*, *24*(1), 2–23. doi:10.1111/j.1471-1842.2007.00701.x PMID:17331140

Kataria, S., Goyal, S., Dash, S., Sandhya, R., Mathew, M. D., & Tyagi, A. K. (2012). Evaluation of nano-mechanical properties of hard coatings on a soft surface. *Thin Solid Films*, *522*, 297–303. doi:10.1016/j.tsf.2012.09.001

Kats, W. A. (1977). *Rheumatic Diseases: diagnosis and management*. Lippincott.

Kendall, E. (1999). Role modeling for agent system analysis, design, and implementation. In *Proceedings of Third International Symposium on Mobile Agents* (MA'99). Palm Springs, FL: Academic Press. doi:10.1109/ASAMA.1999.805405

Kenny, P., Parsons, T. D., Gratch, J., Leusli, A., & Rizzo, A. (2007). Virtual Patients for Clinical Therapist Skills Training. In *Proceedings of the 7th International Working Conference, IVA 2007*. doi:10.1007/978-3-540-74997-4_19

Kettle, S. F. A. (2007). Symmetry and Structure: Readable Group Theory for Chemists (3rd ed.). Wiley-Blackwell.

Khan, U., O'Neill, A., Lotya, M., De, S., & Coleman, J. N. (2010). High-concentration solvent exfoliation of graphene. *Small*, *6*(7), 864–871. doi:10.1002/smll.200902066 PMID:20209652

Khetrapal, N. (2011). Cognition Meets Assistive Technology. In Clinical Technologies: Concepts, Methodologies, Tools and Applications, (pp. 779-791). IGI Global. doi:10.4018/978-1-60960-561-2.ch310

Khosravi, H., & Moradi, A. (2007). Comment on: *Electromagnetic wave propagation in single-wall carbon nanotubes. Physics Letters. [Part A]*, *364*(6), 515–516. doi:10.1016/j.physleta.2007.02.059

Kim, H., Abdala, A. A., & Mocosko, C. W. (2010). Graphene/polymer nanocomposites. *Macromolecules*, *43*(16), 6515–6530. doi:10.1021/ma100572e

Kim, L. (1997). *Imitation to Innovation: The Dynamics of Korea's Technological Learning*. Cambridge, MA: Harvard Business School Press.

Kim, P., Odom, T. W., Huang, J.-L., & Lieber, C. M. (1999). Electronic density of states of atomically resolved single-walled carbon nanotubes: Van Hove singularities and end states. *Physical Review Letters*, *82*(6), 1225–1228. doi:10.1103/PhysRevLett.82.1225

King, B., & Zhang, X. Securing the pharmaceutical supply chain using RFID. In *International Conference on Multimedia and ubiquitous Engineering*. doi:10.1109/MUE.2007.186

Kirmani, A., & Rao, A. R. (2000). No pain, no gain: A critical review of the literature on signaling unobservable product quality. *Journal of Marketing*, *64*(2), 66–79. doi:10.1509/jmkg.64.2.66.18000

Kirriemuir, J. (2008). Second Life® in higher education, medicine and health. *Health Information on the Internet*, (64), 6–8.

Kishino, K., Sekiguchi, H., & Kikuchi, A. (2009). Improved Ti-mask selective-area growth (SAG) by rf-plasma-assisted molecular beam epitaxy demonstrating extremely uniform GaN nanocolumn arrays. *Journal of Crystal Growth*, *311*(7), 2063–2068. doi:10.1016/j.jcrysgro.2008.11.056

Klein, S., Higgins, A., Kipp, A., & Mangan, A. (2010). Drug Living Lab – Information Infrastructures to Increase the Security of Pharmaceutical Supply Chains. Unpublished.

Klem, M. T., Willits, D., Solis, D. J., Belcher, A. M., Young, M., & Douglas, T. (2005). Bio-inspired synthesis of protein-encapsulated CoPt nanoparticles. *Advanced Functional Materials*, *15*(9), 1489–1494. doi:10.1002/adfm.200400453

KNMP. (2013). *Medicatiebeoordeling*. Available from http://www.knmp.nl/downloads/organisatie-regelgeving/organisatie-regelgeving-normen-en-richtlijnen/richtlijnmedicatiebeoordeling.pdf

Knorr, N., & Vinzelberg, S. (2012). Charge writing and detection by EFM and KPFM scanning probe techniques. *Microscopy and Analysis*, *26*(5), 7–12.

Kobayashi, Y., Ke, X., Hata, H., Schiffer, P., & Mallouk, T. E. (2008). Soft chemical conversion of layered double hydroxides to superparamagnetic spinel platelets. *Chemistry of Materials*, *20*(6), 2374–2381. doi:10.1021/cm703443q

Kolkata, A. (2012). *Biologics in Rheumatology 2012. Medicine Update 2012 Vol. 22*. Retrieved august 30, 2014, from http://www.apiindia.org/pdf/medicine_update_2012/rheumatology_05.pdf

Kolodney, E., Tsipinyuk, B., & Budrevich, A. (1994). The thermal stability and fragmentation of C_{60} molecule up to 2000 K on the milliseconds time scale. *The Journal of Chemical Physics*, *100*(11), 8542–8545. doi:10.1063/1.466755

Koppensteiner, G., Merdan, M., Lepuschitz, W., Moser, T., & Reinprecht, C. (2011). *Multi Agent Systems combined with Semantic Technologies for Automated Negotiation in Virtual Enterprises. In Multi-Agent Systems - Modelling, Control, Programming, Simulations and Applications* (pp. 221–242). InTech.

Korobov, M. V., & Sidorov, L. N. (1994).. . *Zhurnal Khimicheskoi Termodinamiki*, *26*, 61–61.

Kowshik, M., Deshmukh, N., Vogel, W., Urban, J., Kulkarni, S. K., & Paknikar, K. M. (2002). Microbial synthesis of semiconductor CdS nanoparticles, their characterization, and their use in the fabrication of an ideal diode. *Biotechnology and Bioengineering*, *78*(5), 583–588. doi:10.1002/bit.10233 PMID:12115128

Kozek, K. A., Kozek, K. M., Wu, W.-C., Mishra, S. R., & Tracy, J. B. (2013). Large-scale synthesis of gold nanorods through continuous secondary growth. *Chemistry of Materials*, *25*(22), 4537–4544. doi:10.1021/cm402277y PMID:24415848

Kramer, R. M., Li, C., Carter, D. C., Stone, M. O., & Naik, R. R. (2004). Engineered protein cages for nanomaterial synthesis. *Journal of the American Chemical Society*, *126*(41), 13282–13286. doi:10.1021/ja046735b PMID:15479082

Kreibig, U., & Vollner, M. (1995). Optical Properties of Metal Clusters. Berlin: Springer. doi:10.1007/978-3-662-09109-8

Krishnamurthy, M. M., & Kemparaju, T. D. (2011). Institutional repositories in Indian universities and research institutes: A study. *Program: Electronic Library & Information Systems*, *45*(2), 185–198. doi:10.1108/00330331111129723

Krishnan, A., Dujardin, E., Treacy, M. M. J., Hugdahl, J., Lynum, S., & Ebbesen, T. W. (1997). Photoisomerization in dendrimers by harvesting of low-energy photons. *Nature*, *388*(6641), 451–454. doi:10.1038/41284

Kroto, H. W. (1987). The stability of the fullerenes C_n, with n = 24, 28, 32, 36, 50, 60 and 70. *Nature*, *329*(6139), 529–531. doi:10.1038/329529a0

Krska, J., Cromarty, J. A., Arris, F., Jamieson, D., Hansford, D., Duffus, P. R., & Seymour, D. G. et al. (2001). Pharmacist-led medication review in patients over 65: A randomized, controlled trial in primary care. *Age and Ageing*, *30*(3), 205–211. doi:10.1093/ageing/30.3.205 PMID:11443021

Ksibi, M., Zemzemi, A., & Boukchina, R. (2003). Photocatalytic degradability of substituted phenols over UV irradiated TiO_2. *Journal of Photochemistry and Photobiology A Chemistry*, *159*(1), 61–70. doi:10.1016/S1010-6030(03)00114-X

Kumar, N. (2002). *Intellectual property rights, technology and economic development: Experiences of Asian countries.* RIS Discussion Paper, 25.

Kumar, D. N. T., & Wei, Q. (2013). Analysis of quantum dots for nano–bio applications as the technological platform of the future. *Research Journal of Biotechnology*, *8*(5), 78–82.

Kumari, K., Kumar, V., & Singh, K. (2014). Non-lithographic fabrication of Ni-Se heterojunction nanowires and their electrical characterization. *Advances in Research*, *2*, 332-337.

Kumar, N., & Aggarwal, A. (2005). Liberalization, outward orientation and in-house R&D activity of multinational and local firms: A quantitative exploration for Indian manufacturing. *Research Policy*, *34*(4), 441–460. doi:10.1016/j.respol.2005.01.010

Kumar, R. (2005). *Research Methodology*. Thousand Oaks, CA: SAGE.

Kuykendall, T., Ulrich, P., Aloni, S., & Yang, P. (2007). Complete composition tunability of InGaN nanowrires using a combinatorial approach. *Nature Materials*, *6*(12), 951–956. doi:10.1038/nmat2037 PMID:17965718

Kuznetsov, V. N., & Serpone, N. (2006). Visible light absorption by various titanium dioxide specimens. *The Journal of Physical Chemistry B*, *110*(50), 25203–25209. doi:10.1021/jp064253b PMID:17165964

Kuznetsov, V. N., & Serpone, N. (2007). Photoinduced coloration and photobleaching of titanium dioxide in TiO_2/polymer compositions upon UV and visible-light excitation of color centers' absorption bands: Direct experimental evidence negating band gap narrowing in anion-/cation-doped TiO_2. *The Journal of Physical Chemistry C*, *111*(42), 15277–15288. doi:10.1021/jp073511h

Lakowicz, J. R. (1999). *Principles of Fluorescence Spectroscopy*. New York, NY: Kluwer Academic. doi:10.1007/978-1-4757-3061-6

Latham, W., Shaw, M., Todd, S., Leymarie, F. F., Jefferys, B., & Kelley, L. (2008). Using DNA to Generate 3D Organic Art Forms. In M. Giacobini et al. (Eds.), *Applications of Evolutionary Computing, LNCS 4974 2008 EvoWorkshops Proceedings* (pp. 433–442). Berlin: Springer. doi:10.1007/978-3-540-78761-7_46

Lau, A., & Vande Moere, A. (2007). Towards a Model of Information Aesthetics in Information Visualization. In *Proceedings of 11th International Conference on Information Visualisation* (pp. 87-92). Los Alamitos, CA: IEEE Computer Society. doi:10.1109/IV.2007.114

Lau, K. K. S., Bico, J., Teo, K. B. K., Chhowalla, M., Amaratunga, G. A. J., Milne, W. I., & Gleason, K. K. et al. (2003). Superhydrophobic carbon nanotube forests. *Nano Letters*, *3*(12), 1701–1705. doi:10.1021/nl034704t

Law, M., Greene, L. E., Johnson, J. C., Saykally, R., & Yang, P. (2005). Nanowire dye-sensitized solar cells. *Nature Materials*, *4*(6), 455–459. doi:10.1038/nmat1387 PMID:15895100

Lawrence, R. (2005). New opportunities in Asia: A focus on India and China. *Drug Discovery Today*, *10*(2), 89–91. doi:10.1016/S1359-6446(04)03315-X PMID:15718156

Lazear, E. (1995). Bait and switch. *Journal of Political Economy*, *103*(4), 813–830. doi:10.1086/262004

Lazonick, W. (2004). Indigenous innovation and economic development: Lessons from China's leap into the information age. *Industry and Innovation*, *11*(4), 273–297. doi:10.1080/1366271042000289360

Le Corre, D., Bras, J., & Dufresne, A. (2010). Starch nanoparticles: A review. *Biomacromolecules*, *11*(5), 1139–1153. doi:10.1021/bm901428y PMID:20405913

Lee, C., & Wang, M. (2006). *An Ontology Based Intelligent Agent for Respiratory Waveform Classification.* Advances in applied artificial intelligence: 19th international conference on industrial and engineering applications of artificial intelligence and expert systems, IEA/AIE, Annecy, France. doi:10.1007/11779568_131

Lee, J., Mahendra, S., & Alvarez, P. J. J. (2010). Nanomaterials in the construction industry: A review of their applications and environmental health and safety considerations. *ACS Nano*, *4*(7), 3580–3590. doi:10.1021/nn100866w PMID:20695513

Lee, P. (2008). Opening the door to the Chinese pharmaceutical market. *Chemistry Today*, *26*, 76–81.

Leffingwell, J. C. (2003). Chirality and Bioactivity I: Pharmacology. *Leffingwell Reports*, *3*(1). Retrieved March 15, 2014, from http://www.leffingwell.com/download/chirality-phamacology.pdf

Legrini, O., Oliveros, E., & Braun, A. M. (1993). Photochemical processes for water treatment. *Chemical Reviews*, *93*(2), 671–698. doi:10.1021/cr00018a003

Leite, F. L., Bueno, C. C., Da Róz, A. L., Ziemath, E. C., & Oliveira, O. N. Jr. (2012). Theoretical models for surface forces and adhesion and their measurement using atomic force microscopy. *International Journal of Molecular Sciences*, *13*(12), 12773–12856. doi:10.3390/ijms131012773 PMID:23202925

Leszczynski, J., & Shukla, M. K. (Eds.). (2009). *Practical aspects of computational chemistry*. Berlin: Springer.

Létard, J. F., Guionneau, P., Nguyen, O., Sánchez Costa, J., Marcén, S., Chastanet, G., & Goux-Capes, L. et al. (2005). A guideline to the design of molecular-based materials with long-lived photomagnetic lifetimes. *Chemistry (Weinheim an der Bergstrasse, Germany)*, *11*(16), 4582–4589. doi:10.1002/chem.200500112 PMID:15861388

Lévi-Strauss, C. (1997). Look, Listen, Learn. Basic Books.

Levy, N., Burke, S. A., Meaker, K. L., Panlasigui, M., Zettl, A., Guinea, F., & Crommie, M. F. et al. (2010). Strain-induced pseudo-magnetic fields greater than 300 tesla in graphene nanobubbles. *Science*, *329*(5991), 544–547. doi:10.1126/science.1191700 PMID:20671183

Leyton, M. (2006). The Foundations of Aesthetics. In Aesthetic Computing (pp. 289-314). Leonardo Books.

Liberati, A., Altman, D. G., Tetzlaff, J., Mulrow, C., Gøtzsche, P. C., Ioannidis, J. P., & Moher, D. et al. (2009). The PRISMA statement for reporting systematic reviews and meta-analyses of studies that evaluate health care interventions: Explanation and elaboration. *Annals of Internal Medicine*, *151*(4), W-65. doi:10.7326/0003-4819-151-4-200908180-00136 PMID:19622512

Lightner, N. J., & Eastman, C. (2002). User preference for product information in remote purchase environments. *Journal of Electronic Commerce Research*, *3*(3), 174–186.

Li, K.-Y. R., Sofra, J., & Power, M. (2008). 3D Avatars and Collaborative Virtual Environments. In M. Quigley (Ed.), *Encyclopedia of Information Ethics and Security* (pp. 1–6). IGI Global.

Li, M., Wong, K. K. W., & Mann, S. (1999). Organization of inorganic nanoparticles using biotin–streptavidin connectors. *Chemistry of Materials*, *11*(1), 23–26. doi:10.1021/cm980610m

Lind, J. (1999). *A review of multiagent systems development methods: Technical report*. Martlesham Heath, UK: British Telecom, Adastral Park Labs.

Linke, H. (2002). Special issue on Ratchets and Brownian Motors: Basics, Experiments and Applications. *Applied Physics. A, Materials Science & Processing*, *75*(2), 167–354. doi:10.1007/s003390201401

Linke, H., Alemán, B. J., Melling, L. D., Taormina, M. J., Francis, M. J., Dow-Hygelund, C. C., & Stout, A. et al. (2006). Self-propelled Leidenfrost droplets. *Physical Review Letters*, *96*(15), 154502–1–4. doi:10.1103/PhysRevLett.96.154502 PMID:16712160

Lisa, M., Hogg, L., & Jennings, N. (2001). Socially intelligent reasoning for autonomous agents. *IEEE Transactions on Systems, Man, and Cybernetics. Part A, Systems and Humans*, *31*(5), 381–393. doi:10.1109/3468.952713

Liu, I., Niu, T.-S., Zhang, L., & Yang, J.-S. (2010). Review on nano-drugs. *Natural Science*, *2*(01), 41–48. doi:10.4236/ns.2010.21006

Liu, N., & Giessen, H. (2010). Coupling effect in optical metamaterials. *Angewandte Chemie International Edition*, *49*(51), 9838–9852. doi:10.1002/anie.200906211 PMID:21154486

Liu, X. G., Li, B., Geng, D. Y., Cui, W. B., Yang, F., Xie, Z. G., & Zhang, Z. D. et al. (2009). (Fe, Ni)/C nanocapsules for electromagnetic-wave-absorber in the whole Ku-band. *Carbon*, *47*(2), 470–474. doi:10.1016/j.carbon.2008.10.028

Liu, X., & Buck, T. (2007). Innovation performance and channels for international technology spillovers: Evidence from Chinese high-tech industries. *Research Policy*, *36*(3), 355–366. doi:10.1016/j.respol.2006.12.003

Liu, Z., Sun, X., Nakayama-Ratchford, N., & Dai, H. (2007). Supramolecular chemistry on water-soluble carbon nanotubes for drug loading and delivery. *ACS Nano*, *1*(1), 50–56. doi:10.1021/nn700040t PMID:19203129

Li, X., Liu, L., Qin, Y., Wu, W., Guo, Z. X., Dai, L., & Zhu, D. (2003). C_{60} modified single-walled carbon nanotubes. *Chemical Physics Letters*, *377*(1-2), 32–36. doi:10.1016/S0009-2614(03)01088-1

Li, Y., Meijer, E., Duysters, G., & Rochemont, M. (2011). Patent transactions with China in a new era: A European perspective. *Journal of Knowledge-based Innovation in China*, *3*(2), 136–156. doi:10.1108/17561411111138973

Li, Z., Li, J., & Luo, X. (2013). Application of phosphonic acid self-assembled monolayer in organic field-effect transistors. *Applied Surface Science*, *282*, 487–491. doi:10.1016/j.apsusc.2013.05.158

Llambí, L., Esteves, E., Martinez, E., Forster, T., García, S., Miranda, N., & Margolis, A. et al. (2011). Teaching tobacco cessation skills to Uruguayan physicians using information and communication technologies. *The Journal of Continuing Education in the Health Professions*, *31*(1), 43–48. doi:10.1002/chp.20100 PMID:21425359

Loeb, A. L. (1993). *Concepts & Images: Visual Mathematics. Design Science Collection.* Boston: Birkhäuser. doi:10.1007/978-1-4612-0343-8

Longfeng, L., Xiangdong, M., Maolin, Z., Guiying, L., & Taicheng, A. (2010). Low-temperature synthesis of TiO_2 photo catalyst in homogeneous hydrolysis system for dye degradation. *Research Journal of Chemistry and Environment*, *14*(4), 40–43.

Love, J. C., Estroff, L. A., Kriebel, J. K., Nuzzo, R. G., & Whitesides, G. M. (2005). Self-assembled monolayers of thiolates on metals as a form of nanotechnology. *Chemical Reviews*, *105*(4), 1103–1169. doi:10.1021/cr0300789 PMID:15826011

Lu, L., & Wang, G. (2008). A study on multi-agent supply chain framework based on network economy. *Computers & Industrial Engineering*, *54*(2), 288–300. doi:10.1016/j.cie.2007.07.010

Lynch, C. A. (2003). *Institutional repositones: Essential infrastructure for scholarship in the digital age*. Retrieved from http:// DSpace.uniroma2.it/DSpace/bitstream/2108/261/1/ir.html

Ma, J. K. H., & Hadzija, B. (2012). Basic Physical Pharmacy. Jones & Bartlett Learning.

Machado, M., Ordejón, P., Sánchez-Portal, D., Artacho, E., & Soler, J. M., (submitted for publication). Density functional calculations of planar DNA base-pairs. *Journal of Physical Chemistry B*.

Maggon, K. (2007). Monoclonal antibody gold rush. *Current Medicinal Chemistry*, *14*(18), 1978–1987. doi:10.2174/092986707781368504 PMID:17691940

Maheshwari, S. (October 2011). *Global protein therapeutics market: Beefing up towards futuristic growth*. Pharmaphorum.

Mailänder, V., & Landfester, K. (2009). Interaction of nanoparticles with cells. *Biomacromolecules*, *10*(9), 2379–2400. doi:10.1021/bm900266r PMID:19637907

Makiura, R., Motoyama, S., Umemura, Y., Yamanaka, H., Sakata, O., & Kitigawa, H. (2010). Surface nano-architecture of a metal–organic framework. *Nature Materials*, *9*(7), 565–571. doi:10.1038/nmat2769 PMID:20512155

Manuel, P. (2010). Computational aspects of carbon and boron nanotubes. *Molecules (Basel, Switzerland)*, *15*(12), 8709–8722. doi:10.3390/molecules15128709 PMID:21119566

Marin, M. L., McGilvray, K. L., & Scaiano, J. C. (2008). Photochemical strategies for the synthesis of gold nanoparticles from Au(III) and Au(I) using photoinduced free radical generation. *Journal of the American Chemical Society*, *130*(49), 16572–16584. doi:10.1021/ja803490n PMID:19049456

Markus, M. L., Steinfield, C. W., Wigand, R. T., & Minton, G. (2006). Standards, Collective Action and IS Development-Vertical Information Systems Standards in the Home Mortgage Industry. *MIS Quarterly. Special Issue on Standards and Standardization*, *30*, 439–465.

Martí-Gastaldo, C., Antypov, D., Warren, J. E., Briggs, M. E., Chater, P. A., Wiper, P. V., & Rosseinsky, M. J. et al. (2014). Side-chain control of porosity closure in single- and multiple-peptide-based porous materials by cooperative folding. *Nature Chemistry*, *6*(4), 343–351. doi:10.1038/nchem.1871 PMID:24651203

Martin, C. R., & Kohli, P. (2003). The emerging field of nanotube biotechnology. *National Review*, *2*, 29–37. PMID:12509757

Martínez-Martín, D., Jaafar, M., Pérez, R., Gómez-Herrero, J., & Asenjo, A. (2010). Upper bound for the magnetic force gradient in graphite. *Physical Review Letters*, *105*(25), 257203–1–4. doi:10.1103/PhysRevLett.105.257203 PMID:21231621

Martínez, V., Boldog, I., Gaspar, A. B., Ksenofontov, V., Bhattacharjee, A., Gütlich, P., & Real, J. A. (2010). Spin crossover phenomenon in nanocrystals and nanoparticles of [Fe(3-Fpy)$_2$M(CN)$_4$] (M^{II} = Ni, Pd, Pt) two-dimensional coordination polymers. *Chemistry of Materials*, *22*(14), 4271–4281. doi:10.1021/cm101022u

Martins, S., Sarmento, B., Ferreira, D. C., & Souto, E. B. (2007). Lipid-based colloidal carriers for peptide and protein delivery – liposomes versus lipid nanoparticles. *International Journal of Nanomedicine*, *2*, 595–607. PMID:18203427

Martins, V. (2010). *Validation of an educational game for teaching cardiovascular assessment*. Investigación y Educación en Enfermería Colombia.

Martin, Y., & Wickramasinghe, H. K. (1987). Magnetic imaging by *force microscopy* with 1000Å resolution. *Applied Physics Letters*, *50*(20), 1455–1457. doi:10.1063/1.97800

Mathews, C. K., Baba, M. S., Narasimhan, T. S. L., Balasubramanian, R., Sivaraman, N., Srinivasan, T. G., & Rao, P. R. V. (1992). Vaporization studies on buckminsterfullerene. *Journal of Physical Chemistry*, *96*(9), 3566–3568. doi:10.1021/j100188a002

Mathieu, J. (2007). Blogs, podcasts, and wikis: The new names in information dissemination. *Journal of the American Dietetic Association*, *107*(4), 553–555. doi:10.1016/j.jada.2007.02.027 PMID:17383254

Mattheos, N., Schoonheim-Klein, M., Walmsley, A. D., & Chapple, I. L. C. (2010). Innovative educational methods and technologies applicable to continuing professional development in periodontology. *European Journal of Dental Education*, *14*, 43–52. doi:10.1111/j.1600-0579.2010.00624.x PMID:20415976

McGarr, O. (2009). A Review of Podcasting in Higher Education: Its Influence on the Traditional Lecture. *Australasian Journal of Educational Technology*, *25*(3), 309–321.

McGilvray, D. (2008). *Executing Data Quality Projects: Ten Steps to Quality Data and Trusted Information*. Burlington, MA: Morgan Kaufman.

McGilvray, K. L., Decan, M. R., Wang, D., & Scaiano, J. C. (2006). Facile photochemical synthesis of unprotected aqueous gold nanoparticles. *Journal of the American Chemical Society*, *128*(50), 15980–15981. doi:10.1021/ja066522h PMID:17165719

Meade, O., Bowskill, D., & Lymn, J. S. (2009). Pharmacology as a foreign language: A preliminary evaluation of podcasting as a supplementary learning tool for non-medical prescribing students. *BMC Medical Education*, *9*(1), 74. doi:10.1186/1472-6920-9-74 PMID:20021673

Meade, O., Bowskill, D., & Lymn, J. S. (2011). Pharmacology podcasts: A qualitative study of non-medical prescribing students' use, perceptions and impact on learning. *BMC Medical Education*, *11*(2). Retrieved from http://www.biomedcentral.com/1472-6920/11/2 PMID:21223547

Medical Dictionary. (n.d.). Retrieved august 12, 2014, from http://medicaldictionary.thefreedictionary.com/Antirheumatic+Drugs

Medintz, I. L., Clapp, A. R., Mattoussi, H., Goldman, E. R., Fisher, B., & Mauro, J. M. (2003). Self assembled nanoscale biosensors based on quantum dot FRET donors. *Nature Materials*, *2*(9), 630–638. doi:10.1038/nmat961 PMID:12942071

Meldrum, F. C., Heywood, B. R., & Mann, S. (1992). Magnetoferritin: *In vitro* synthesis of a novel magnetic protein. *Science*, *257*(5069), 522–523. doi:10.1126/science.1636086 PMID:1636086

Melissa Tan, S., & Ladyshewsky, K.,, R., & Gardner, P. (2010). Using blogging to promote clinical reasoning and metacognition in undergraduate physiotherapy fieldwork programs. *Australasian Journal of Educational Technology*, *26*(3), 355–368.

Meng, T., Wang, C. Y., & Wang, S. Y. (2008). First-principles study of a hybrid carbon material: Imperfect fullerenes covalently bonded to defective single-walled carbon nanotubes. *Physical Review B*, *77*, 33415.

Menkiti, M. C., & Onukwuli, O. D. (2010). Coag-flocculation studies of *Moringa oleifera* coagulant (MOC) in brewery effluent: Nephelometric approach. *Journal of American Science*, *6*(12), 788–806.

Mercuri, R., & Meredith, K. (2014). An educational venture into 3D Printing.*Integrated STEM Education Conference (ISEC)*. IEEE.

Meulendijk, M. (2014). *STRIPA*. Unpublished Raw Data.

Meyer, A. D. (2001). Technology transfer into China: Preparing for a new era. *European Management Journal, 19*(2), 140–144. doi:10.1016/S0263-2373(00)00088-8

Meyer, J. C., Geim, A. K., Katsnelson, M. I., Novoselov, K. S., Booth, T. J., & Roth, S. (2007). The structure of suspended graphene sheets. *Nature, 446*(7131), 60–63. doi:10.1038/nature05545 PMID:17330039

Miller, A. D., Bookstaver, P. B., & Norris, L. B. (2009). *Use of Wikis in Advanced Pharmacy Practice Experiences*. Academic Press.

Miller, A. D., Norris, L. B., & Bookstaver, P. B. (2012). Use of wikis in pharmacy hybrid elective courses. *Currents in Pharmacy Teaching and Learning, 4*(4), 256–261. doi:10.1016/j.cptl.2012.05.004

Millett, L. I., & Holden, S. H. (2003, November-December). Authentication and its privacy effects. *IEEE Internet Computing, 7*(6), 54–58. doi:10.1109/MIC.2003.1250584

Minton, A. P. (2000). Effects of excluded surface area and adsorbate clustering on surface adsorption of proteins I. Equilibrium models. *Biophysical Chemistry, 86*(2-3), 239–247. doi:10.1016/S0301-4622(00)00151-4 PMID:11026688

Minton, A. P. (2001). Effects of excluded surface area and adsorbate clustering on surface adsorption of proteins. II. Kinetic models. *Biophysical Journal, 80*(4), 1641–1648. doi:10.1016/S0006-3495(01)76136-X PMID:11259279

Mirkin, C. A., Letsinger, R. L., Mucic, R. C., & Storhoff, J. J. (1996). A DNA-based method for rationally assembling nanoparticles into macroscopic materials. *Nature, 382*(6592), 607–609. doi:10.1038/382607a0 PMID:8757129

Mirk, S. M., Burkiewicz, J. S., & Komperda, K. E. (2010). Student perception of a wiki in a pharmacy elective course. *Currents in Pharmacy Teaching and Learning, 2*(2), 72–78. doi:10.1016/j.cptl.2010.01.002

Mitchell, K. (2011). *A Celebration of Beauty*. Digital Art Guild; Art through Technology. Retrieved March 15, 2014, from http://www.digitalartguild.com/content/view/56/26/

Miyamachi, T., Gruber, M., Davesne, V., Bowen, M., Boukari, S., Joly, L., Scheurer, F., Rogez, G., Yamada, T. K., Ohresser, P., Beaurepaire, E., & Wulfhekel, W. (2012). Robust spin crossover and memristance across a single molecule. *Nature Communications, 3*, 938–1-6.

Mkrtchyan, G. S., & Shmidt, V. V. (1972). Interaction between a cavity and a vortex in a superconductor of the second kind. *Soviet Physics, JETP, 34*, 195–197.

Molina-Sánchez, A., Segura-Ruiz, J., Garro, N., García-Cristóbal, A., Cantarero, A., Iikawa, F., & Rizzi, A. et al. (2012). Inhomogeneous electron distribution in InN nanowires: Influence on the optical properties. *Physica Status Solidi, 9*(c), 1001–1004.

Mondal, A., Li, Y., Herson, P., Seuleiman, M., Boillot, M. L., Rivière, E., & Lescouëzec, R. et al. (2012). Photomagnetic effect in a cyanide-bridged mixed-valence $\{Fe^{II}_2Fe^{III}_2\}$ molecular square. *Chemical Communications, 48*(45), 5653–5655. doi:10.1039/c2cc17835d PMID:22550634

Mondal, A., Li, Y., Seuleiman, M., Julve, M., Toupet, L., Buron-Le Cointe, M., & Lescouëzec, R. (2013). On/off photo-switching in a cyanide-bridged $\{Fe_2Co_2\}$ magnetic molecular square. *Journal of the American Chemical Society, 135*(5), 1653–1656. doi:10.1021/ja3087467 PMID:23321056

Moniruzzaman, M., & Winey, K. I. (2006). Polymer nanocomposites containing carbon nanotubes. *Macromolecules, 39*(16), 5194–5205. doi:10.1021/ma060733p

Monteagudo, P. (2003). *Software educativo para la enseñanza de la semiología clínica del sistema respiratorio*. (Master's Thesis). Ciudad de La Habana: ISCM de La HabanaCentro de Cibernética Aplicada a la Medicina (CECAM).

Moradi, A. (2010). Theory of carbon nanotubes as optical nano waveguides. *Journal of Electromagnetic Analysis and Applications*, *2*(12), 672–676. doi:10.4236/jemaa.2010.212088

Morant-Miñana, M. C. (personal communication).

Muhlen, C., Elverfeldt, D., Bassler, N., Neudorfer, I., Steitz, B., Petri-Fink, A., & Peter, K. et al. (2007). Superparamagnetic iron oxide binding and uptake as imaged by magnetic resonance is mediated by the integrin receptor Mac-1 (CD11b/CD18): Implications on imaging of atherosclerotic plaques. *Atherosclerosis*, *193*(1), 102–111. doi:10.1016/j.atherosclerosis.2006.08.048 PMID:16997307

Mukhina, M. V., Maslov, V. G., Baranov, A. V., Artemyev, M. V., & Fedorov, A. V. (2013). Anisotropic absorption of CdSe/ZnS quantum rods embedded in polymer film. *Advances in Nano Research*, *1*, 153-158.

Mukoyama, T. (2003). Innovation, imitation, and growth with cumulative technology. *Journal of Monetary Economics*, *50*(2), 361–380. doi:10.1016/S0304-3932(03)00005-9

Müller-Meskamp, L., Karthäuser, S., Zandvliet, H. J. W., Homberger, M., Simon, U., & Waser, R. (2010). Field emission resonances at tip/mercaptoalkylferrocene/Au interfaces. *E-Nano Newsletter*, (19), 28-29.

Müller-Meskamp, L., Karthäuser, S., Zandvliet, H. J. W., Homberger, M., Simon, U., & Waser, R. (2009). Field-emission resonances at tip/α,ω-mercaptoalkyl ferrocene/Au interfaces studied by STM. *Small*, *5*(4), 496–502. doi:10.1002/smll.200800802 PMID:19197965

Mundy, A. (2010, July 22). China never investigated tainted Heparin, says probe. *The Wall Street Journal*, p. A7.

Murphy, C. J., & Coffer, J. L. (2002). Quantum dots: A primer. *Applied Spectroscopy*, *56*(1), 16A–27A. doi:10.1366/0003702021954214

Murphy, C. J., Thompson, L. B., Alkilany, A. M., Sisco, P. N., Boulos, S. P., Sivapalan, S. T., & Huang, J. et al. (2010). The many faces of gold nanorods. *The Journal of Physical Chemistry Letters*, *1*(19), 2867–2875. doi:10.1021/jz100992x

Nacci, A., & Cioffi, N. (2011). Special issue: Nano-catalysts and nano-technologies for green organic synthesis. *Molecules (Basel, Switzerland)*, *16*(12), 1452–1453. doi:10.3390/molecules16021452 PMID:21307822

Nagra, K. A. (2012). Building Institutional Repositories in the Academic Libraries. *Community & Junior College Libraries*, *18*(3/4), 137–150. doi:10.1080/02763915.2012.799028

Nair, R. R., Ren, W., Jalil, R., Riaz, I., Kravets, V. G., Britnell, L., & Geim, A. K. et al. (2010). Fluorographene: A two-dimensional counterpart of Teflon. *Small*, *6*(24), 2877–2884. doi:10.1002/smll.201001555 PMID:21053339

Nasibulin, A. G., Anisimov, A. S., Pikhitsa, P. V., Jiang, H., Brown, D. P., Choi, M., & Kauppinen, E. I. (2007a). Investigations of NanoBud formation. *Chemical Physics Letters*, *446*(1-3), 109–114. doi:10.1016/j.cplett.2007.08.050

Nasibulin, A. G., Pikhitsa, P. V., Jiang, H., Brown, D. P., Krasheninnikov, A. V., Anisimov, A. S., & Kauppinen, E. I. et al. (2007b). A novel hybrid carbon material. *Nature Nanotechnology*, *2*(3), 156–161. doi:10.1038/nnano.2007.37 PMID:18654245

Nasibulin, A. G., Shandakov, S. D., Anisimov, A. S., Gonzalez, D., Jiang, H., Pudas, M., & Kauppinen, E. I. et al. (2008). Charging of aerosol products during ferrocene vapor decomposition in N_2 and CO atmospheres. *The Journal of Physical Chemistry C*, *112*(15), 5762–5769. doi:10.1021/jp7118026

Nasto, B. (2011, December). Biotech in China: Special feature on China's emerging biotech industry. *Nature Biotechnology*, 1–16.

Nathan, P., & Chan, A. (2007). Engaging Undergraduates with Podcasting in a Business Subject. In ICT: Providing choices for learners and learning. Singapore.

National Association of Board of Pharmacies (NABP). (2009). *State of the Internet: NABP Position Paper on the Continued Proliferation of Rogue Internet Drug Outlets*. Mount Prospect, IL: NABP.

National Association of Boards of Pharmacy (NABP). (2011). *Internet Drug Outlet Identification Program: Progress Report for State and Federal Regulators*. Mount Prospect, IL: NABP.

National Association of Boards of Pharmacy (NABP). (n.d.). Retrieved from http://www.nabp.net, last accessed on May 28, 2011.

Nazim, M., & Mukherjee, B. (2011). Status of Institutional Repositories in Asian Countries: A Quantitative Study. *Library Philosophy & Practice*, 60-75.

Nederlandse patiënten consumenten federatie. (2014). *Zorgkaart Nederland*. Available from: http://www.zorgkaartnederland.nl/apotheker

Nederlandse Service Apotheek Beheer B.V. (2012). *Analyse Medicatiebeoordeling t.b.v. de KNMP Richtlijn Medicatiebeoordeling*. Available from: www.serviceapotheek.nl

Nel, A., Zhao, Y., & Mädler, L. (2013). Environmental health and safety considerations for nanotechnology. *Accounts of Chemical Research*, *46*(3), 605–606. doi:10.1021/ar400005v PMID:23964654

Neu, J. C., Cañizo, J. A., & Bonilla, L. L. (2002). Three eras of micellization. *Phys. Rev. E*, *66*, 61406.

Neuberger, T., Schöpf, B., Hofmann, H., Hofmann, M., & Rechenberg, B. (2005). Superparammagnetic nanoparticles for biomedical applications. Possibilities and limitations of a new drug delivery system. *Journal of Magnetism and Magnetic Materials*, *293*(1), 483–496. doi:10.1016/j.jmmm.2005.01.064

Neville, A. C. (1977). Symmetry and Asymmetry Problems in Animals. In R. Duncan & M. Weston-Smith (Eds.), *The Encyclopedia of Ignorance: Everything you ever wanted to know about the unknown* (pp. 331–338). Oxford, UK: Paragon Press.

Nia, A., & Zaichkowsky, J. L. (2000). Do counterfeits devalue the ownership of luxury brands? *Journal of Product and Brand Management*, *9*(7), 485–497. doi:10.1108/10610420010351402

Niemeyer, C. M., & Mirkin, C. A. (Eds.). (2004). Nanobiotechnlogy: Concepts, Applications and Perspectives. Weinheim: Wiley-VCH. doi:10.1002/3527602453

Niemeyer, C. M. (2001). Nanoparticles, proteins, and nucleic acids: Biotechnology meets materials science. *Angewandte Chemie International Edition*, *40*(22), 4128–4158. doi:10.1002/1521-3773(20011119)40:22<4128::AID-ANIE4128>3.0.CO;2-S

Nilson Nils, J. (2004). *Inteligencia artificial una nueva síntesis*. Mc GrawHill.

Noachtar, S., & Peters, A. (2009). *Semiology of epileptic seizures: A critical review*. University on Munich Germany.

Noels, K.A. (2001). New orientations in language learning motivation: Towards a model of intrinsic, extrinsic, and integrative orientations and motivation. *Motivation and Second Language Acquisition*, *23*, 43-68.

Noginov, M. A., Zhu, G., Belgrave, A. M., Bakker, R., Shalaev, V. M., Narimanov, E. E., & Wiesner, U. et al. (2009). Demonstration of a spaser-based nanolaser. *Nature*, *460*(7259), 1110–1112. doi:10.1038/nature08318 PMID:19684572

Nolan, P., & Yeung, G. (2001). Big business with Chinese characteristics: Two paths to growth of the firm in China under reform. *Cambridge Journal of Economics*, *25*(4), 443–465. doi:10.1093/cje/25.4.443

Notman, R., Noro, M., O'Malley, B., & Anwar, J. (2006). Molecular basis for dimethylsulfoxide (DMSO) action on lipid membranes. *Journal of the American Chemical Society*, *128*(43), 13982–13983. doi:10.1021/ja063363t PMID:17061853

Novoselov, K. S., Geim, A. K., Morozov, S. V., Jiang, D., Zhang, Y., Dubonos, S. V., & Firsov, A. A. et al. (2004). Electric field effect in atomically thin carbon films. *Science*, *306*(5696), 666–669. doi:10.1126/science.1102896 PMID:15499015

Novoselov, K. S., Jiang, D., Schedin, F., Booth, T. J., Khotkevich, V. V., Morozov, S. V., & Geim, A. K. (2005). Two-dimensional atomic crystals. *Proceedings of the National Academy of Sciences of the United States of America*, *102*(30), 10451–10453. doi:10.1073/pnas.0502848102 PMID:16027370

Nwadiuto Igwe, K. (2014). Open Access Repositories in Academic and Research Institutions for the Realization of Nigeria's Vision 20: 2020. *International Journal Of Information Science & Management*, *12*(1), 33–46.

O'Brien, R. (2001). An Overview of the Methodological Approach of Action Research. In R. Richardson (Ed.), Theory and Practice of Action Research. Brazil: Universidade Federal da Paraíba. Retrieved from http://www.web.ca/~robrien/papers/arfinal.html

O'Sullivan, S., & McGlynn, H. (2010). Use of e-portfolios and web 2.0 tools in the assessment of group work activities in the undergraduate biomedical science classroom. In *EDULEARN10 Proceedings*. Barcelona: IATED. Retrieved from http://library.iated.org/view/OSULLIVAN2010USE

Oberdörster, G., Maynard, A., Donaldson, K., Castranova, V., Fitzpatrick, J., Ausman, K., … Yang, H. (2005). A report from the ILSI Research Foundation/Risk Science Institute Nanomaterial Toxicity Screening Working Group: Particle and fibre toxicology. *Particle and Fibre Toxicology, 2*, 8.

Odell, J., Parunak, V. D., & Bauer, B. (2000). Extending UML for agents. In *Proc. of the Agent-Oriented Information Systems Workshop at the 17th National Conference on Artificial Intelligence*. Academic Press.

OECD. (2009). *The Bioeconomy to 2030: Designing a Policy Agenda*. Organization for Economic Co-Operation and Development.

Okamoto, K., Iyi, N., & Sasaki, T. (2007). Factors affecting the crystal size of the MgAl-LDH (layered double hydroxide) prepared by using ammonia-releasing reagents. *Applied Clay Science*, *37*(1-2), 23–31. doi:10.1016/j.clay.2006.10.008

Okubo, P. G., & Aki, K. (1987). Fractal Geometry in the San Andreas Fault System. *Journal of Geophysical Research*, *92*(B1), 345–355. doi:10.1029/JB092iB01p00345

Oliva, J. M., Allan, N. L., Schleyer, P. R., Viñas, C., & Teixidor, F. (2005). Strikingly long C···C distances in 1,2-di-substituted *ortho*-carboranes and their dianions. *Journal of the American Chemical Society*, *127*(39), 13538–13547. doi:10.1021/ja052091b PMID:16190717

Oliva, J. M., Klein, D. J., Schleyer, P. R., & Serrano-Andrés, L. (2009a). Design of carborane molecular architectures with electronic structure computations: From endohedral and polyradical systems to multidimensional networks. *Pure and Applied Chemistry*, *81*(4), 719–729. doi:10.1351/PAC-CON-08-09-18

Oliva, J. M., Serrano-Andrés, L., Havlas, Z., & Michl, J. (2009b). On the electronic structure of a dianion, a radiacal anion, and a neutral biradical $(HB)_{11}C{-}C{\equiv}C{-}C(BH)_{11}$ carborane dimer. *Journal of Molecular Structure THEOCHEM*, *912*(1-3), 13–20. doi:10.1016/j.theochem.2009.01.033

Olofsson, A. D., Lindberg, J. O., & Hauge, T. E. (2011). Blogs and the Design of Reflective Peer-to-Peer Technology-Enhanced Learning and Formative Assessment. *Campus-Wide Information Systems, 28*(3), 183–194. doi:10.1108/10650741111145715

Olsen, J. E., & Granzin, K. L. (1992). Gaining retailers' assistance in fighting counterfeiting: Conceptualization and empirical test of a helping model. *Journal of Retailing, 68*, 90–109.

Olson, M. (1971). The Logic of Collective Action: Public Goods and the Theory of Groups (rev. ed.). Cambridge, MA: Harvard University Press.

Oomen-Early, J., & Burke, S. (2007). Entering the Blogosphere: Blogs as Teaching and Learning Tools in Health Education. *International Electronic Journal of Health Education, 10*, 186–196.

Open Access. (2006). In *Wikipedia*. Retrieved 26 December 2006 from en.wikipedia.org/wiki/Open_access

Open Doar. (2008). *Introduction to the project*. Retrieved from http://www.opendoar.org/about.html

Oppenländer, T. (2003). Photochemical purification of water and air. Weinheim: Wiley-VCH.

Ordejón, P., Artacho, E., & Soler, J. M. (1996). Mixed approach to incorporate self-consistency into order-N LCAO methods.*Materials Research Society Symposium Proceedings, 408*, 85-90.

Ortega-Villar, N., Thompson, A. L., Muñoz, M. C., Ugalde-Saldívar, V. M., Goeta, A. E., Moreno-Esparza, R., & Real, J. A. (2005). Solid- and solution-state studies of the novel μ-dicyanamide-bridged dinuclear spin-crossover system $\{[(Fe(bztpen)]_2[\mu\text{-}N(CN)_2]\}(PF_6)_3{\cdot}nH_2O$. *Chemistry (Weinheim an der Bergstrasse, Germany), 11*(19), 5721–5734. doi:10.1002/chem.200500171 PMID:16028299

Oxman, A. D. (1994). Checklists for review articles. *BMJ : British Medical Journal, 309*(6955), 648–651. doi:10.1136/bmj.309.6955.648 PMID:8086990

Ozin, G. A., & Arsenault, A. C. (2005). *Nanochemistry*. Cambridge, UK: The Royal Society of Chemistry.

Packard, A. (2001). A 'neural' net that can be seen with the naked eye. In Neuronal coding of perceptual systems (pp. 397-402). World Scientific.

Padgham, L., & Winiko, M. (2002a). Prometheus: A pragmatic methodology for engineering intelligent agents. In *Proceedings of the OOPSLA 2002 Workshop on Agent-Oriented Methodologies* (pp.97-108). Seattle, WA: OOPSLA. doi:10.1145/544741.544749

Padgham, L., & Winiko, M. (2002b). *Prometheus: Engineering intelligent agents*. Tutorial notes. Unpublished.

Pagona, G., Fan, J., Maignè, A., Yudasaka, M., Iijima, S., & Tagmatarchis, N. (2007b). Aqueous carbon nanohorn–pyrene–porphyrin nanoensembles: Controlling charge-transfer interactions. *Diamond and Related Materials, 16*(4-7), 1150–1153. doi:10.1016/j.diamond.2006.11.071

Pagona, G., Mountrichas, G., Rotas, G., Karousis, N., Pispas, S., & Tagmatarchis, N. (2009). Properties, applications and functionalisation of carbon nanohorns. *International Journal of Nanotechnology, 6*(1/2), 176–195. doi:10.1504/IJNT.2009.021715

Pagona, G., Sandanayaka, A. S. D., Araki, Y., Fan, J., Tagmatarchis, N., Charalambidis, G., & Ito, O. et al. (2007a). Covalent functionalization of carbon nanohorns with porphyrins: Nanohybrid formation and photoinduced electron and energy transfer. *Advanced Functional Materials, 17*(10), 1705–1711. doi:10.1002/adfm.200700039

Pagona, G., Sandanayaka, A. S. D., Araki, Y., Fan, J., Tagmatarchis, N., Yudasaka, M., & Ito, O. et al. (2006b). Electronic interplay on illuminated aqueous carbon nanohorn–porphyrin ensembles. *The Journal of Physical Chemistry B, 110*(42), 20729–20732. doi:10.1021/jp064685m PMID:17048875

Pagona, G., Tagmatarchis, N., Fan, J., Yudasaka, M., & Iijima, S. (2006a). Cone-end functionalization of carbon nanohorns. *Chemistry of Materials*, *18*(17), 3918–3920. doi:10.1021/cm0604864

Paolo, B., Anna, P., Paolo, G., Fausto, G., & John, M. (2004). TROPOS: An agent-oriented software development methodology. *Journal of Autonomous Agents and Multi-Agent Systems*, *8*(3), 203–236. doi:10.1023/B:AGNT.0000018806.20944.ef

Parapar. (2007). Manifestaciones oculares de algunas enfermedades reumáticas en el niño. *Hosp. Pedro Borras*. Retrieved April 12, 2014, from http://www.portalesmedicos.com/publicaciones/articles/634/1/Manifestacionesocularesdealgunas-enfermedadesreumaticasenelnino.html

Pareto, V. (1935). *The Mind and Society*. New York: Harcourt, Brace.

Park, C. L., Crocker, C., Nussey, J., Springate, J., & Hutchings, D. (2010). Evaluation of a Teaching Tool--Wiki--in Online Graduate Education. *Journal of Information Systems Education*, *21*(3), 313–321.

Parker, K. R., & Chao, J. T. (2007). Wiki as a teaching tool. *Interdisciplinary Journal of E-Learning and Learning Objects*, *3*, 57–72.

Park, Y. K., Bold, B., Lee, W. K., Jeon, M. H., An, K. H., Jeong, S. Y., & Shim, Y. K. (2011). D-(+)-Galactose-conjugated single-walled carbon nanotubes as new chemical probes for electrochemical biosensors for the cancer marker galectin-3. *International Journal of Molecular Sciences*, *12*(12), 2946–2957. doi:10.3390/ijms12052946 PMID:21686160

Pathak, D., Nordstrom, G., & Kurokawa, S. (2000). Modelling of supply chain: a multi-agent approach.*IEEE International Conference on Systems, Man, and Cybernetics.*

Peelle, B. R., Krauland, E. M., Wittrup, K. D., & Belcher, A. M. (2005). Probing the interface between biomolecules and inorganic materials using yeast surface display and genetic engineering. *Acta Biomaterialia*, *1*(2), 145–154. doi:10.1016/j.actbio.2004.11.004 PMID:16701791

Pendry, J. B., Holden, A. J., Stewart, W. J., & Youngs, I. (1996). Extremely low frequency plasmons in metallic mesostructures. *Physical Review Letters*, *76*(25), 4773–4776. doi:10.1103/PhysRevLett.76.4773 PMID:10061377

Penz, E., & Stottinger, B. (2005). Forget the "real" thing—take the copy! An explanatory model for the volitional purchase of counterfeit products. *Advances in Consumer Research. Association for Consumer Research (U. S.)*, *32*(1), 568–575.

Perdew, J. P., Burke, K., & Ernzerhof, M. (1996). Generalized gradient approximation made simple. *Physical Review Letters*, *77*(18), 3865–3868. doi:10.1103/PhysRevLett.77.3865 PMID:10062328

Perini, A., Bresciani, P., Giunchiglia, F., Giorgini, P., & Mylopoulos, J. (2001). A knowledge level software engineering methodology for agent oriented programming.*Proceedings of the fifth international conference on Autonomous agents* (pp. 648-655).

Perrier, A., Maurel, F., Ciofini, I., & Jacquemin, D. (2011). A theoretical spectroscopy investigation of bifunctional platinum-bridged diarylethenes. *Chemical Physics Letters*, *502*(1-3), 77–81. doi:10.1016/j.cplett.2010.12.044

Persson, P., Laaksolahti, J., & Lönnqvist, P. (2001). Understanding socially intelligent agents—a multilayered phenomenon. *IEEE Transactions on Systems, Man, and Cybernetics. Part A, Systems and Humans*, *31*(5), 349–360. doi:10.1109/3468.952710

Petta, J. R., Slater, S. K., & Ralph, D. C. (2004). Spin-dependent transport in molecular tunnel junctions. *Physical Review Letters*, *93*(13), 136601–1–4. doi:10.1103/PhysRevLett.93.136601 PMID:15524747

Phadtare, A., Bahmani, A., Shah, A., & Pietrobon, R. (2009). Scientific writing: A randomized controlled trial comparing standard and on-line instruction. *BMC Medical Education*, *9*(1), 27. doi:10.1186/1472-6920-9-27 PMID:19473511

Pharmaceutical and Medicine Manufacturing. (n.d.). Bureau of Labor Statistics, U. S. Department of Labor. Retrieved on September 5, 2014, from http://www.bls.gov/oco/cg/cgs009.htm#addinfo

Phillips, W. A. (1972). Tunneling states in amorphous solids. *Journal of Low Temperature Physics*, *7*(3-4), 351–360. doi:10.1007/BF00660072

Pierce, R., & Fox, J. (2012). Vodcasts and Active-Learning Exercises in a "Flipped Classroom" Model of a Renal Pharmacotherapy Module. *American Journal of Pharmaceutical Education*, *76*(10), 196. doi:10.5688/ajpe7610196 PMID:23275661

Pignataro, B. (Ed.). (2010a). Tomorrow's Chemistry Today. Weinheim: Wiley-VCH.

Pignataro, B. (Ed.). (2010b). Ideas in Chemistry and Molecular Sciences: Advances in Nanotechnology, Materials and Devices. Weinheim: Wiley-VCH.

Pilarski, P. P., Alan Johnstone, D., Pettepher, C. C., & Osheroff, N. (2008). From music to macromolecules: Using rich media/podcast lecture recordings to enhance the preclinical educational experience. *Medical Teacher*, *30*(6), 630–632. doi:10.1080/01421590802144302 PMID:18677662

Pino, E., & Encinas, M. V. (2012). Photocatalytic degradation of chlorophenols on TiO_2-325mesh and TiO_2-P25. An extended kinetic study of photodegradation under competitive conditions. *Journal of Photochemistry and Photobiology A Chemistry*, *242*, 20–27. doi:10.1016/j.jphotochem.2012.05.019

Poirier, T., Crouch, M., MacKinnon, G., Mehvar, R., & Monk-Tutor, M. (2009). Updated Guidelines for Manuscripts Describing Instructional Design and Assessment: The IDEAS Format. *American Journal of Pharmaceutical Education*, *73*(3), 55. doi:10.5688/aj730355 PMID:19564998

Pollack, A. (2012, December 30). U.S. clears DNA firm's acquisition by Chinese. *The New York Times*.

Pong, B. K., Trout, B. L., & Lee, J. Y. (2008). Modified ligand-exchange for efficient solubilization of CdSe/ZnS quantum dots in water: A procedure guided by computational studies. *Langmuir*, *24*(10), 5270–5276. doi:10.1021/la703431j PMID:18412382

Ponnurangam, D., & Uma, G. (2005). Fuzzy complexity assessment model for resource negotiation. *Expert Systems with Applications*, *29*(1), 105–119. doi:10.1016/j.eswa.2005.01.008

Poonawalla, T., & Wagner, R. F. (2006). Assessment of a blog as a medium for dermatology education. *Dermatology Online Journal*, *12*(1), 5. PMID:16638373

Prescription Drug Marketing Act (PDMA) of 1987 (P.L. 100-293, 102 Stat. 95).

Pressman, R. S. (2002). *Ingeniería del Software. Un enfoque práctico*. Madrid: McGrawHill.

Prins, F., Monrabal-Capilla, M., Osorio, E. A., Coronado, E., & Zant, H. S. J. (2011). Room-temperature electrical addressing of a bistable spin-crossover molecular system. *Advanced Materials*, *23*(13), 1545–1549. doi:10.1002/adma.201003821 PMID:21449059

Prusinkiewicz, P., Lindenmeyer, A., Hanan, J. S., Fracchia, F. D., Fowler, D. R., de Boer, M. J. M., & Mercer, L. (1991). *The Algorithmic Beauty of Plants (The Virtual Laboratory)* (1st ed.). Springer.

Purchase, H. (2010). Graph Drawing Aesthetics in User-sketched Graph Layouts. In *AUIC'10 Proceedings of the Eleventh Australasian Conference on User Interface*.

Pushkarova, Y., & Kholin, Y. (2014). A procedure for meaningful unsupervised clustering and its application for solvent classification. *Central European Journal of Chemistry*, *12*(5), 594–603. doi:10.2478/s11532-014-0514-6

Puzyn, T., Leszczynska, D., & Leszczynski, J. (2009). Toward the development of *nano-QSARs*: Advances and challenges. *Small*, *5*(22), 2494–2509. doi:10.1002/smll.200900179 PMID:19787675

Puzyn, T., Rasulev, B., Gajewicz, A., Hu, X., Dasari, T. P., Michalkova, A., & Leszczynski, J. et al. (2011). Using nano-QSAR to predict the cytotoxicity of metal oxide nanoparticles. *Nature Nanotechnology*, *6*(3), 175–178. doi:10.1038/nnano.2011.10 PMID:21317892

Quiñones, R., Raman, A., & Gawalt, E. S. (2007). An approach to differentiating between multi- and monolayers using MALDI–TOF MS. *Surface and Interface Analysis*, *39*(7), 593–600. doi:10.1002/sia.2561

Rallo, R., France, B., Liu, R., Nair, S., George, S., Damoiseaux, R., & Cohen, Y. et al. (2011). Self-organizing map analysis of toxicity-related cell signaling pathways for metal and metal oxide nanoparticles. *Environmental Science & Technology*, *45*(4), 1695–1702. doi:10.1021/es103606x PMID:21250674

Rankin, D. W. H., Mitzel, N., & Morrison, C. (2013). *Structural Methods in Molecular Inorganic Chemistry (Inorganic Chemistry: A Textbook Series)* (1st ed.). Wiley.

Ravikumar, R., & Mohsin Khan, I. (2015). Design & Development of a 3D Printer.*Proceedings of 12th IRF International Conference*.

Reines, D., Conaway, R. C., & Conaway, J. W. (1999). Mechanism and regulation of transcriptional elongation by RNA polymerase II. *Current Opinion in Cell Biology*, *11*(3), 342–346. doi:10.1016/S0955-0674(99)80047-7 PMID:10395562

Rejón, C. (2013). *Logic structure of clinical judgment and its relation to medical and psychiatric semiology.* U.S.A. Pub med / U.S. National Library of Medicine.

Resconi, G., & Jain, L. (2004). *Intelligent agents Theory and applications*. Springer Verlag.

Research Unit of the Spanish Society of Rheumatology. (2013). Spanish registry of adverse events of biological therapies in rheumatic diseases. *Biobadaser*. Retrieved August 12, 2014, from http://biobadaser.ser.es/biobadaser/eng/docs/SER_informe_web2012_ENG.pdf

Rich, E., & Knight, K. (1991). *Artificial Intelligence*. Mc GrawHill.

Richter, H., Dereux, A., Grilles, J. M., Guillaume, C., Thiry, P. A., Lucas, A. A., & de Hoffmann, E. et al. (1994). Sublimation of pure C60 fullerene and of C60 adsorbed on MgO or graphite powders. *Berichte der Bunsengesellschaft fur Physikalische Chemie*, *98*(10), 1329–1329. doi:10.1002/bbpc.19940981019

Rieger, R., & Müllen, K. (2010). Forever young: Polycyclic aromatic hydrocarbons as model cases for structural and optical studies. *Journal of Physical Organic Chemistry*, *23*, 315–325.

Rivado-Casas, L., Sampedro, D., Campos, P. J., Fusi, S., Zanirato, V., & Olivucci, M. (2009). Fluorenylidene–pyrroline biomimetic light-driven molecular switches. *The Journal of Organic Chemistry*, *74*(13), 4666–4674. doi:10.1021/jo802792j PMID:19485347

RNCOS. (2013). *In-Vitro Diagnostics Market Analysis to 2017*. RNCOS Industry Research Solutions.

Rodríguez, M. L. G., Carrillo, A. R., & Alvarez, A. M. R. (2010). Aplicación de la filosofía ¿wiki¿ a la actualización del material docente en el área de farmacia y tecnología farmacéutica. *Ars Pharmaceutica*, *51*(4), 187–194.

Rogers, L. (2009). Simulating clinical experience: Exploring Second Life® as a learning tool for nurse education. In *Same places, different spaces. Proceedings ASCILITE Auckland 2009*. Retrieved from http://www.ascilite.org.au/conferences/auckland09/procs/rogers.pdf

Rohrschneider, L. (1966). Eine methode zur chrakterisierung von gaschromatographischen trennflüssigkeiten. *Journal of Chromatography. A*, *22*, 6–22. doi:10.1016/S0021-9673(01)97064-5

Ronald, M., Marie, L., & Antonie, E. (2000). An ABC of drug-Related problems. *Drug Safety*, *22*(6), 415–423. doi:10.2165/00002018-200022060-00001 PMID:10877036

Rosh, A., Jones, K., & Wahl, R. (2009). Receiving: The Use of Web 2.0 to Create a Dynamic Learning Forum to Enrich Resident Education. *Academic Emergency Medicine*, *16*, S274–S275. doi:10.1111/j.1553-2712.2009.00392_2.x

Ros-Rodriguez, J., Encinas, T., Picazo, R.A., Labadia, A., Artalejo, A.R., Gutiérrez-Martin, Y., … Gilabert, J.A. (2011). Use of Wikis as collaborative tools in a B-learning course of Pharmacology. In *INTED 2011 Proceedings*. Valencia: IATED. Retrieved from http://library.iated.org/view/ROSRODRIGUEZ2011USE

Roubeau, O., Alcazar Gomez, J. M., Balskus, E., Kolnaar, J. J. A., Haasnoot, J. G., & Reedijk, J. (2001). Spin-transition behaviour in chains of FeII bridged by 4-substituted 1,2,4-triazoles carrying alkyl tails. *New Journal of Chemistry*, *25*(1), 144–150. doi:10.1039/b007094g

Roy, B., Mukhopadhyay, P., & Biswas, S. (2012). An Analytical Study of Institutional Digital Repositories in India. *Library Philosophy & Practice*, 1-13.

RSNA. (2005). Nanoparticles show promise in cancer detection and treatment. *RSNA News*, (10), 3-5.

Rubesova, E., Berger, F., Wendland, M., Hong, K., Stevens, K., Gooding, C., & Lang, P. (2002). Gd-labeled liposomes for monitoring liposome-encapsulated chemotherapy: Quantification of regional uptake in tumor and effect on drug delivery. *Academic Radiology*, *9*(2), S525–S527. doi:10.1016/S1076-6332(03)80283-0 PMID:12188328

Rugar, D., Mamin, H. J., Guethner, P., Lambert, S. E., Stern, J. E., McFadyen, I., & Yogi, T. (1990). Magnetic force microscopy: General principles and application to longitudinal recording media. *Journal of Applied Physics*, *68*(3), 1169–1183. doi:10.1063/1.346713

Ruiz-Conde, E., & Calderón-Martínez, A. (2014). University institutional repositories: Competitive environment and their role as communication media of scientific knowledge. *Scientometrics*, *98*(2), 1283–1299. doi:10.1007/s11192-013-1159-5

Russell, S., & Norving, P. (1996). *Inteligencia artificial un enfoque moderno*. México: Prentice Hall Hispanoamericana, S.A.

Saleh, S., Taha, M. O., Haddadin, R. N., Marzooqa, D., & Hodali, H. (2011). Preparation of silver- and zinc-doped mullite-based ceramics showing anti-bacterial biofilm properties. *Molecules (Basel, Switzerland)*, *16*(12), 2862–2870. doi:10.3390/molecules16042862

Salicrup, L. A., & Fedorkova, L. (2006). Challenges and opportunities for enhancing biotechnology and technology transfer in developing countries. *Biotechnology Advances*, *24*(1), 69–79. doi:10.1016/j.biotechadv.2005.06.004 PMID:16098701

Salter, B. (n.d.). *Nonarticular Rheumatism in Textbook of Disorders and Injuries of the Musculoskeletal System* (3rd ed.). Lippincott Williams.

Sampedro, D., Migani, A., Pepi, A., Busi, E., Basosi, R., Latterini, L., & Olivucci, M. et al. (2004). Design and photochemical characterization of a biomimetic light-driven Z/E switcher. *Journal of the American Chemical Society*, *126*(30), 9349–9359. doi:10.1021/ja038859e PMID:15281826

Sánchez Costa, J., Lappalainen, K., de Ruiter, G., Quesada, M., Tang, J., Mutikainen, I., & Reedijk, J. et al. (2007). Remarkable steric effects and influence of monodentate axial ligands L on the spin-crossover properties of *trans*-[Fe^{II}(N_4 ligand)L] coplexes. *Inorganic Chemistry*, *46*(10), 4079–4089. doi:10.1021/ic0624017 PMID:17441713

Sánchez-Portal, D., Ordejón, P., Artacho, E., & Soler, J. M. (1997). Density-functional method for very large systems with LCAO basis sets. *International Journal of Quantum Chemistry*, *65*, 453–461. doi:10.1002/(SICI)1097-461X(1997)65:5<453::AID-QUA9>3.0.CO;2-V

Sandars, J. (2006). Twelve tips for using blogs and wikis in medical education. *Medical Teacher*, *28*(8), 680–682. doi:10.1080/01421590601106353 PMID:17594577

Sandars, J., Homer, M., Pell, G., & Croker, T. (2008). Web 2.0 and social software: The medical student way of e-learning. *Medical Teacher*, *30*(3), 308–312. doi:10.1080/01421590701798729 PMID:18608950

Sandler, T. (1992). *Collective Action: Theory and Applications*. Ann Arbor, MI: University of Michigan Press.

Sanvito, S. (2010). Molecular spintronics: The rise of spinterface science. *Nature Physics*, *6*(8), 562–564. doi:10.1038/nphys1714

Saravolats, L. (2007). *The role of Medical Semiology in clinical infectious diseases: Back to the basic*. Lippincott Williams & Wilkins, Inc.

Sato, O., Iyoda, T., Fujishima, A., & Hashimoto, K. (1996). Photoinduced magnetization of a cobalt–iron cyanide. *Science*, *272*(5262), 704–705. doi:10.1126/science.272.5262.704 PMID:8662564

Sato, O., Tao, J., & Zhang, Y. Z. (2007). Control of magnetic properties through external stimuli. *Angewandte Chemie International Edition*, *46*(13), 2152–2187. doi:10.1002/anie.200602205 PMID:17348074

Sawant, S. (2012). Management of Indian institutional repositories. *OCLC Systems & Services*, *28*(3), 130–143. doi:10.1108/10650751211262128

Scaiano, J. C., Stamplecoskie, K. G., & Hallett-Tapley, G. L. (2012). Photochemical Norrish type I reaction as a tool for metal nanoparticle synthesis: Importance of proton coupled electron transfer. *Chemical Communications*, *48*(40), 4798–4808. doi:10.1039/c2cc30615h PMID:22498952

Schellman, J. A. (1975). Macromolecular binding. *Biopolymers*, *14*(5), 999–1018. doi:10.1002/bip.1975.360140509

Schelly, C., Anzalone, G., Wijnen, B., & Pearce, J. M. (2015). Open-Source 3-D Printing Technologies for Education: Bringing Additive Manufacturing to the Classroom. *Journal of Visual Languages and Computing*, *28*, 226–237. doi:10.1016/j.jvlc.2015.01.004

Schilthuizen, M., & Gravendeel, B. (2011). Evolution of Chirality Symposium. *The 13th Congress of the European Society for Evolutionary Biology*. Retrieved August 26, 2011, from http://www.eseb2011.de/

Schnakenberg, L. (1977). *Thermodynamic network analysis of biological systems*. Berlin: Springer. doi:10.1007/978-3-642-96394-0

Scholarly Communication. (2012). In *Wikipedia*. Retrieved March 9, 2012, from http://en.wikipedia.org/w/index.php?title=Scholarly_communication&oldid=486006103

Schöpfel, J., & Prost, H. (2013). Degrees of secrecy in an open environment. The case of electronic theses and dissertations. *ESSACHESS - Journal for Communication Studies, 6*(2).

Schreiber, B. E., Fukuta, J., & Gordon, F. (2010). Live lecture versus video podcast in undergraduate medical education: A randomised controlled trial. *BMC Medical Education*, *10*(1), 68. doi:10.1186/1472-6920-10-68 PMID:20932302

Schuller, J. A., Barnard, E. S., Cai, W., Jun, Y. C., White, J. S., & Brongersma, M. L. (2010). Plamonics for extreme light concentration and manipulation. *Nature Materials*, *9*(3), 193–204. doi:10.1038/nmat2630 PMID:20168343

Schumacher Ralph, H., Klippel John, H., & Robinson Dwight, R. (1988). *Principios de las Enfermedades Reumáticas (12th ed.)*. Arthritis Foundation.

Schwartz, D. K. (2001). Mechanisms and kinetics of self-assembled monolayer formation. *Annual Review of Physical Chemistry*, *52*(1), 107–137. doi:10.1146/annurev.physchem.52.1.107 PMID:11326061

Schwarzenbach, G. (1952). Der chelateffekt. *Helvetica Chimica Acta*, *35*(7), 2344–2359. doi:10.1002/hlca.19520350721

Schwarz, J. A., Contescu, C. I., & Putyera, K. (Eds.). (2004). *Dekker Encyclopedia of Nanoscience and Nanotechnology*. New York, NY: Marcel Dekker.

Science Special Issue. (2002). Supramolecular chemistry and self-assembly. *Science, 295*, 2395-2421.

Seeman, N. C. (2005). Structural DNA nanotechnology: An overview. *Methods in Molecular Biology (Clifton, N.J.)*, *303*, 143–166. PMID:15923682

Segura-Ruiz, J., Garro, N., Cantarero, A., Denker, C., Malindretos, J., & Rizzi, A. (2009). Optical studies of MBE-grown InN nanocolumns: Evidence of surface electron accumulation. *Physical Review B: Condensed Matter and Materials Physics*, *79*(11), 115305–1–9. doi:10.1103/PhysRevB.79.115305

Segura-Ruiz, J., Martinez-Criado, G., Sans, J. A., Tucoulou, R., Cloetens, P., Snigireva, I., & Cantarero, A. et al. (2011). Direct observation of elemental segregation in InGaN nanowires by X-ray nanoprobe.[RRL]. *Physica Status Solidi*, *5*(3), 95–97. doi:10.1002/pssr.201004527

Segura-Ruiz, J., Molina-Sánchez, A., Garro, N., García-Cristóbal, A., Cantarero, A., Iikawa, F., & Rizzi, A. et al. (2010). Inhomogeneous free-electron distribution in InN nanowires: Photoluminescence excitation experiments. *Physical Review B: Condensed Matter and Materials Physics*, *82*(12), 125319–1–9. doi:10.1103/PhysRevB.82.125319

Seker, U. O. S., & Demir, H. V. (2011). Material binding peptides for nanotechnology. *Molecules (Basel, Switzerland)*, *16*(12), 1426–1451. doi:10.3390/molecules16021426 PMID:21307821

Šenolta, L., Vencovskýa, J., Pavelkaa, K., Ospeltb, C., & Gayb, S. (2009). Prospective new biological therapies for rheumatoid arthritis. *Autoimmunity Reviews, 9*(2).

Sepúlveda, B., Angelomé, P. C., Lechuga, L. M., & Liz-Marzán, L. M. (2009). LSPR-based nanobiosensors. *Nano Today*, *4*(3), 244–251. doi:10.1016/j.nantod.2009.04.001

Serpone, N. (2006). Is the band gap of pristine TiO_2 narrowed by anion- and cation-doping of titanium dioxide in second-generation photocatalysts? *The Journal of Physical Chemistry B*, *110*(48), 24287–24293. doi:10.1021/jp065659r PMID:17134177

Serrano-Andrés, L., Klein, D. J., Schleyer, P. R., & Oliva, J. M. (2008). What electronic structures and geometries of carborane mono- and *ortho-*, *meta-*, and *para*-diradicals are preferred? *Journal of Chemical Theory and Computation*, *4*(8), 1338–1347. doi:10.1021/ct800150h PMID:26631709

Serrano-Andrés, L., & Oliva, J. M. (2006). Photochemical window mechanism for controlled atom release in carborane endohedral boxes: Theoretical evidence. *Chemical Physics Letters*, *432*(1-3), 235–239. doi:10.1016/j.cplett.2006.10.077

Seydel, J. K., & Wiese, M. (2002). Drug-Membrane Interactions: Analysis, Drug Distribution, Modeling. Weinheim: Wiley-VCH. doi:10.1002/3527600639

Seyed, M., Rizi, M., Maciej, M., Latek, M., & Armando, G. (2010). *Merging Remote Sensing Data and Population Surveys in Large, Empirical Multi-agent Models: The Case of the Afghan Drug Industry*. Presented during the Third World Social Simulation Congress, Kassel, Germany. Retrieved April, 30, 2011 from: http://www.css.gmu.edu/projects/irregularwarfare/remotesensing.pdf

Shanthi, S., & Priya, K. S. (2012). Photo degradation of dyes from their aqueous solutions of their binary mixture, using TiO_2 as the oxidant with different sources of energy. *Journal of Chemistry and Chemical Engineering*, *6*, 951–955.

Shantikumar, S. (2009). From lecture theatre to portable media: Students' perceptions of an enhanced podcast for revision. *Medical Teacher*, *31*(6), 535–538. doi:10.1080/01421590802365584 PMID:18937140

Sharifi, M., Marschall, R., Wilhelm, M., Wallacher, D., & Wark, M. (2011). Detection of homogeneous distribution of functional groups in mesoporous silica by small angle neutron scattering and in situ adsoprtion of nitrogen or water. *Langmuir*, *27*(9), 5516–5522. doi:10.1021/la2000188 PMID:21480601

Sheetal, O., & Pragati, T. (2010). Kinetics of photocatalytic degradation of methylene blue in a TiO_2 slurry reactor. *Research Journal of Chemistry and Environment*, *14*(4), 9–13.

Shevchenko, V. Y., Madison, A. E., & Shudegov, V. E. (2003). The structural diversity of the nanoworld. *Glass Physics and Chemistry*, *29*(6), 577–582. doi:10.1023/B:GPAC.0000007934.93203.f3

Shilatifard, A., Conaway, J. W., & Conaway, R. C. (1997). Mechanism and regulation of transcriptional elongation and termination by RNA polymerase II. *Current Opinion in Genetics & Development*, *7*(2), 199–204. doi:10.1016/S0959-437X(97)80129-3 PMID:9115429

Shirazi, M., & Soroor, J. (2007). An intelligent agent-based architecture for strategic information system applications. *Knowledge-Based Systems*, *20*(8), 726–735. doi:10.1016/j.knosys.2006.10.004

Shukla, D., Ahearn, W. G., & Farid, S. (2005). Chain amplification in photoreactions of *N*-alkoxypyridinium salts with alcohols: Mechanism and kinetics. *The Journal of Organic Chemistry*, *70*(17), 6809–6819. doi:10.1021/jo050726j PMID:16095300

Shukla, D., Ahearn, W. G., & Farid, S. (2006). Enhancement of chain amplification in photoreactions of *N*-methoxypyridinium salts with alcohols. *Photochemistry and Photobiology*, *82*(1), 146–151. doi:10.1562/2005-06-28-RA-594 PMID:16178662

Singh, P., Gonzalez, M. J., & Manchester, M. (2006). Viruses and their uses in nanotechnology. *Drug Development Research*, *67*(1), 23–41. doi:10.1002/ddr.20064

Singh, R. (2006). Clinical research in China and India: A paradigm shift in drug development. *Drug Discovery Today*, *11*(15/16), 675–676. doi:10.1016/j.drudis.2006.06.009 PMID:16846793

Skiba, D. J. (2008). Nursing education 2.0: Twitter® & tweets. Can you post a nugget of knowledge in 140 characters or less? *Nursing Education Perspectives*, *29*(2), 110–112. PMID:18459627

Slinker, J. D., Muren, N. B., Renfrew, S. E., & Barton, J. K. (2011). DNA charge transport over 34 nm. *Nature Chemistry*, *3*(3), 228–233. doi:10.1038/nchem.982 PMID:21336329

Smirnov, B. M. (1992). Systems of atoms with a short-range interaction. *Uspekhi Fizicheskikh Nauk*, *162*(12), 97–97. doi:10.3367/UFNr.0162.199212b.0097

Smith, B. R., Heverhagen, J., Knopp, M., Schmalbrock, P., Shapiro, J., Shiomi, M., & Lee, S. C. et al. (2007). Localization to atherosclerotic plaque and biodistribution of biochemically derivatized superparamagnetic iron oxide nanoparticles (SPIONs) contrast particles for magnetic resonance imaging (MRI). *Biomedical Microdevices*, *9*(5), 719–727. doi:10.1007/s10544-007-9081-3 PMID:17562181

Snyder, L. R. (1974). Classification of the solvent properties of common liquids. *Journal of Chromatography. A*, *92*(2), 223–230. doi:10.1016/S0021-9673(00)85732-5

Snyder, L. R. (1997). Changing reversed-phase high performance liquid chromatography selectivity: Which variables should be tried first? *Journal of Chromatography. B, Biomedical Sciences and Applications*, *689*(1), 105–115. doi:10.1016/S0378-4347(96)00351-9 PMID:9061486

Som, C., Berges, M., Chaudhry, Q., Dusinska, M., Fernandes, T. F., Olsen, S. I., & Nowack, B. (2010). The importance of life cycle concepts for the development of safe nanoproducts. *Toxicology*, *269*(2-3), 160–169. doi:10.1016/j.tox.2009.12.012 PMID:20025922

Son, S. J., Bai, X., & Lee, S. B. (2007). Inorganic hollow nanoparticles and nanotubes in nanomedicine. Part 1. Drug/gene delivery applications. *Drug Discovery Today*, *12*(15-16), 650–656. doi:10.1016/j.drudis.2007.06.002 PMID:17706547

Sparks, M. A., O'Seaghdha, C., Sethi, S. K., & Jhaveri, K. D. (2011). Embracing the Internet as a Means of Enhancing Medical Education in Nephrology. *American Journal of Kidney Diseases: The Official Journal of the National Kidney Foundation*. doi:10.1053/j.ajkd.2011.06.009

Stamplecoskie, K. G., Fasciani, C., & Scaiano, J. C. (2012). Dual-stage lithography from a light-driven, plasmon-assisted process: A hierarchical approach to subwavelength features. *Langmuir*, *28*(30), 10957–10961. doi:10.1021/la301728r PMID:22803690

Stamplecoskie, K. G., Pacioni, N. L., Larson, D., & Scaiano, J. C. (2011). Plasmon-mediated photopolymerization maps plasmon fields for silver nanoparticles. *Journal of the American Chemical Society*, *133*(24), 9160–9163. doi:10.1021/ja201139z PMID:21615121

Stamplecoskie, K. G., & Scaiano, J. C. (2010). Light emitting diode irradiation can control the morphology and optical properties of silver nanoparticles. *Journal of the American Chemical Society*, *132*(6), 1825–1827. doi:10.1021/ja910010b PMID:20102152

Stanley, W. M. (1937). Crystalline Tobacco-mosaic Virus Protein. *American Journal of Botany*, *24*(2), 59–68. doi:10.2307/2436720

Stewart, D. W., Panus, P. C., & Hagemeier, N. E. (2013). An analysis of student performance with podcasting and active learning in a pharmacotherapy module. *Currents in Pharmacy Teaching and Learning*, *5*(6), 574–579. doi:10.1016/j.cptl.2013.07.004

Stiffler, D., Stoten, S., & Cullen, D. (2011). Podcasting as an Instructional Supplement to Online Learning. *CIN: Computers, Informatics. Nursing*, *29*, TC84–TC88. doi:10.1097/NCN.0b013e3181fc3fdf

Stinaff, E., Scheibner, M., Bracker, A., Ponomarev, I., Korenev, V., Ware, M., & Gammon, D. et al. (2006). Optical signatures of coupled quantum dots. *Science*, *311*(5761), 636–639. doi:10.1126/science.1121189 PMID:16410487

Stoecklein, W., Parkin, S. S. P., & Scott, J. C. (1988). Ferromagnetic resonance studies of exchange-biased Permalloy thin films. *Pysical Review B*, *38*(10), 6847–6854. doi:10.1103/PhysRevB.38.6847 PMID:9945364

Stolberg, H. O., Norman, G., & Trop, I. (2004). Fundamentals of Clinical Research for Radiologists. *AJR*, *183*, 1539–1544. doi:10.2214/ajr.183.6.01831539 PMID:15547188

Stone, J. (2009). *Pearls and Myths in Rheumatology*. London: Springer Dordrecht Heidelberg.

Strichartz, G. R. (Ed.). (1987). Handbook of experimental pharmacology*Local Anesthetics No. 81*. Heidelberg, Germany: Springer. doi:10.1007/978-3-642-71110-7

Suber, P. (2003). *Removing the barriers to research: An introduction to open access for librarians*. Retrieved 26 December 2006 from http://www.earlham.edu/~peters/writing/acrl.htm

Sumer, S. (2008). Web 2.0 and Radiology. *Internet Journal of Radiology, 8*(2). Retrieved from http://www.ispub.com/journal/the_internet_journal_of_radiology/volume_8_number_2_11/article/web_2_0_and_radiology.html

Sun, X., Liu, Z., Welsher, K., Robinson, J. T., Goodwin, A., Zaric, S., & Dai, H. (2008). Nano-graphene oxide for cellular imaging and drug delivery. *Nano Research, 1*(3), 203–212. doi:10.1007/s12274-008-8021-8 PMID:20216934

Sweeney, R. Y., Mao, C., Gao, X., Burt, J. L., Belcher, A. M., Georgiou, G., & Iverson, B. L. (2004). Bacterial biosynthesis of cadmium sulfide nanocrystals. *Chemistry & Biology, 11*(11), 1553–1559. doi:10.1016/j.chembiol.2004.08.022 PMID:15556006

Sycara, K. (1998). Multi-agent Systems. *AI Magazine, 19*(2), 79–92.

Tagmatarchis, N., Maigné, A., Yudasaka, M., & Iijima, S. (2006). Functionalization of carbon nanohorns with azomethine ylides: Towards solubility enhancement and electron-transfer processes. *Small, 2*(4), 490–494. doi:10.1002/smll.200500393 PMID:17193072

Takezawa, N., & Fukushima, K. (1994). Optimal size of a cylindrical insulating inclusion acting as a pinning center for magnetic flux in superconductors. *Physica. C, Superconductivity, 228*(1-2), 149–159. doi:10.1016/0921-4534(94)90186-4

Takezawa, N., & Fukushima, K. (1997). Optimal size of an insulating inclusion acting as a pinning center for magnetic flux in superconductors: Calculation of pinning force. *Physica. C, Superconductivity, 290*(1-2), 31–37. doi:10.1016/S0921-4534(97)01574-8

Tamura, R., & Tsukada, M. (1995). Electronic states of the cap structure in the carbon nanotube. *Physical Review B: Condensed Matter and Materials Physics, 52*(8), 6015–6026. doi:10.1103/PhysRevB.52.6015 PMID:9981793

Tan, M. C., Chow, G. M., & Ren, L. (Ed.) (2010). Nanostructured Materials for Biomedical Applications. Trivandrum, India: Research Signpost.

Tang, J., Sánchez Costa, J., Smulders, S., Molnár, G., Bousseksou, A., Teat, S. J., & Reedijk, J. et al. (2009). Two-step spin-transition iron(III) compound with a wide [high spin-low spin] plateau. *Inorganic Chemistry, 48*(5), 2128–2135. doi:10.1021/ic801973x PMID:19235971

Taniguchi, N. (1974). *Proceedings of the International Conference on Production Engineering, Part II*. Tokyo: Japan-Society of Precision Engineering.

Taranta, A., & Markowitz, M. (1981). *Rheumatic fever: a guide to its recognition, prevention, and cure*. MTP Press.

Tenne, R., & Redlich, M. (2010). Recent progress in the research of inorganic fullerene-like nanoparticles and inorganic nanotubes. *Chemical Society Reviews, 39*(5), 1423–1434. doi:10.1039/B901466G PMID:20419198

Teresawa, M., Takezawa, N., Fukushima, K., Mitamura, T., Fan, X., Tsubakino, H., & Tatara, G. et al. (1998). Flux pinning and flux creep in $La_{2-x}Sr_xCuO_4$ with splayed columnar defects. *Physica. C, Superconductivity, 296*(1-2), 57–64. doi:10.1016/S0921-4534(97)01822-4

The 20th Century Artbook. (1996). San Francisco, CA: Chronicle Books/Phaidon Press.

The American Art Book. (1999). London: Phaidon Press.

The Artbook. (2012). Phaidon Press Inc.

Thorek, D. L. J., Chen, A. K., Czupryna, J., & Tsourkas, A. (2006). Superparamagnetic iron oxide nanoparticle probes for molecular imaging. *Annals of Biomedical Engineering*, *34*(1), 23–28. doi:10.1007/s10439-005-9002-7 PMID:16496086

Tiberj, A., Camara, N., Godignon, P., & Camassel, J. (2011). Micro-Raman and micro-transmission imaging of epitaxial graphene grown on the Si and C faces of 6H-SiC. *Nanoscale Research Letters*, *6*, 478.

Tikhonenko, F. V., Horsell, D. W., Gorbachev, R. V., & Savchenko, A. K. (2008). Weak localization in graphene flakes. *Physical Review Letters*, *100*, 56802.

Timeline of RFID. (2010). Retrieved on September 6, 2014, from http://www.google.com/#q=Timeline+for+RFID&hl=en&sa=X&rlz=1R2SKPB_enUS374&tbs=tl:1,tl_num:100&ei=R9VETKb4CML38AbCyoGTBw&ved=0CIgBEMsBKAQ&fp=1&bav=on.2,or.r_gc.r_pw.&cad=b

Tokmakoff, A., Haynes, D. R., & George, S. M. (1991). Desorption kinetics of C_{60} multilayers from Al_2O_3 (0001). *Chemical Physics Letters*, *186*(4-5), 450–455. doi:10.1016/0009-2614(91)90207-P

Tom, G., Garibaldi, B., Zeng, Y., & Pilcher, J. (1998). Consumer demand for counterfeit goods. *Psychology and Marketing*, *15*(5), 405–421. doi:10.1002/(SICI)1520-6793(199808)15:5<405::AID-MAR1>3.0.CO;2-B

Torrens, F., & Castellano, G. (2013a). *Bundlet* model of single-wall carbon, BC_2N and BN nanotubes, cones and horns in organic solvents. *Journal of Nanomaterials & Molecular Nanotechnology*, *2*, 1000107.

Torrens, F., & Castellano, G. (2013d). Elementary polarizability of Sc/*fullerene*/*graphene* aggregates and di/graphene–cation interactions. *Journal of Nanomaterials & Molecular Nanotechnology*, *S1*, 001.

Torrens, F., & Castellano, G. (2014c). Cluster model expanded to C-nanostructures: Fullerenes, tubes, graphenes and their buds. *Austin Journal of Nanomedicine & Nanotechnology*, *2*(2), 7–1-7.

Torrens, F., & Castellano, G. (in press). Solvents and co-solvents of single-wall carbon nanotubes: New best solvents, superacids, nitric acid and guest–host inclusion complexes. In: E. A. Castro (Ed.), New Developments and Applications of QSAR-QSPR Theory. Trivandrum, India: Research Signpost.

Torrens, F. (2003). Valence topological charge-transfer indices for dipole moments. *Molecules (Basel, Switzerland)*, *8*(1), 169–185. doi:10.3390/80100169 PMID:18007514

Torrens, F. (2004a). Periodic table of carbon nanotubes based on the chiral vector. *Internet Electronic Journal of Molecular Design*, *3*, 514–527.

Torrens, F. (2004b). Valence topological charge-transfer indices for dipole moments: Percutaneous enhancers. *Molecules (Basel, Switzerland)*, *9*(12), 1222–1235. doi:10.3390/91201222 PMID:18007514

Torrens, F. (2004c). Fractal dimension of transdermal-delivery drug models. *Lebanese Science Journal*, *5*(1), 61–70.

Torrens, F. (2005a). Periodic properties of carbon nanotubes based on the chiral vector. *Internet Electronic Journal of Molecular Design*, *4*, 59–81.

Torrens, F. (2005b). Calculations on cyclopyranoses as co-solvents of single-wall carbon nanotubes. *Molecular Simulation*, *31*(2-3), 107–114. doi:10.1080/08927020412331308494

Torrens, F. (2005c). Calculations on solvents and co-solvents of single-wall carbon nanotubes: Cyclopyranoses. *Journal of Molecular Structure THEOCHEM*, *757*(1-3), 183–191. doi:10.1016/j.theochem.2005.03.023

Torrens, F. (2005d). Calculations on solvents and co-solvents of single-wall carbon nanotubes: Cyclopyranose. *Nanotechnology*, *16*(5), S181–S189. doi:10.1088/0957-4484/16/5/009

Torrens, F. (2005e). Some calculations on single-wall carbon nanotubes. *Problems of Nonlinear Analysis in Engineering Systems*, *11*(2), 1–16.

Torrens, F. (2006a). Calculations of organic-solvent dispersions of single-wall carbon nanotubes. *International Journal of Quantum Chemistry*, *106*(3), 712–718. doi:10.1002/qua.20835

Torrens, F. (2006b). Corrigendum: Effect of type, size and deformation on polarizability of carbon nanotubes from atomic increments. *Nanotechnology*, *17*(5), 1541–1541. doi:10.1088/0957-4484/17/5/C01

Torrens, F., & Castellano, G. (2005). Cluster origin of the solubility of single-wall carbon nanotubes. *Computing Letters*, *1*(4), 331–336. doi:10.1163/157404005776611303

Torrens, F., & Castellano, G. (2006). Periodic classification of local anaesthetics (procaine analogues). *International Journal of Molecular Sciences*, *7*(1), 12–34. doi:10.3390/i8010012

Torrens, F., & Castellano, G. (2007a). Cluster nature of the solvation features of single-wall carbon nanotubes. *Current Research in Nanotechnology*, *1*, 1–29.

Torrens, F., & Castellano, G. (2007b). Effect of packing on the cluster nature of C nanotubes: An information entropy analysis. *Microelectronics Journal*, *38*(12), 1109–1122. doi:10.1016/j.mejo.2006.04.004

Torrens, F., & Castellano, G. (2007c). Cluster origin of the transfer phenomena of single-wall carbon nanotubes. *Journal of Computational and Theoretical Nanoscience*, *4*, 588–603.

Torrens, F., & Castellano, G. (2007d). Asymptotic analysis of coagulation-fragmentation equations of carbon nanotube clusters. *Nanoscale Research Letters*, *2*(7), 337–349. doi:10.1007/s11671-007-9070-8

Torrens, F., & Castellano, G. (2008a). Properties of fullerite and other symmetric carbon forms: Similarity laws. *Symmetry*, *19*, 341–370.

Torrens, F., & Castellano, G. (2008b). Fractal dimension of transdermal-delivery drug models: 4-Alkylanilines. *Journal of Liquid Chromatography & Related Technologies*, *31*(15), 2337–2347. doi:10.1080/10826070802281877

Torrens, F., & Castellano, G. (2009a). Asymptotic analysis of coagulation–fragmentation equations. In F. Columbus (Ed.), *Carbon nanotubes: New research* (pp. 7–16). New York, NY: Nova.

Torrens, F., & Castellano, G. (2009b). Topological charge-transfer indices: From small molecules to proteins. *Current Proteomics*, *6*(4), 204–213. doi:10.2174/157016409789973770

Torrens, F., & Castellano, G. (2010a). Fullerite crystal thermodynamic characteristics and the law of corresponding states. *Journal of Nanoscience and Nanotechnology*, *10*(2), 1208–1222. doi:10.1166/jnn.2010.1858 PMID:20352780

Torrens, F., & Castellano, G. (2010b). Cluster nature of the solvent features of single-wall carbon nanohorns. *International Journal of Quantum Chemistry*, *110*(3), 563–570. doi:10.1002/qua.22054

Torrens, F., & Castellano, G. (2011a). (Co-)solvent selection for single-wall carbon nanotubes: *Best* solvents, acids, superacids and guest–host inclusion complexes. *Nanoscale*, *3*(6), 2494–2510. doi:10.1039/c0nr00922a PMID:21331393

Torrens, F., & Castellano, G. (2011b). Using chemical structural indicators for periodic classification of local anaesthetics. *International Journal of Chemoinformatics and Chemical Engineering*, *1*(2), 15–35. doi:10.4018/ijcce.2011070102

Torrens, F., & Castellano, G. (2012). *Bundlet* model for single-wall carbon nanotubes, nanocones and nanohorns. *International Journal of Chemoinformatics and Chemical Engineering*, *2*(1), 48–98. doi:10.4018/IJCCE.2012010105

Torrens, F., & Castellano, G. (2013b). Solvent features of cluster single-wall C, BC_2N and BN nanotubes, cones and horns. *Microelectronic Engineering*, *108*, 127–133. doi:10.1016/j.mee.2013.02.046

Torrens, F., & Castellano, G. (2013c). C-nanostructures cluster models in organic solvents: Fullerenes, tubes, buds and graphenes. *Journal of Chemistry and Chemical Engineering*, *7*, 1026–1035.

Torrens, F., & Castellano, G. (2014a). Cluster *bundlet* model of single-wall C, BC_2N and BN nanotubes, cones and horns. In A. G. Mercader, E. A. Castro, & A. K. Haghi (Eds.), *Nanoscience and computational chemistry: Research progress* (pp. 271–307). Waretown, NJ: Apple Academic.

Torrens, F., & Castellano, G. (2014b). Nanostructures cluster models in solution: Extension to C, BC_2N and BN fullerenes, tubes and cones. In M. Khosrow-Pour (Ed.), *Contemporary advancements in information technology development in dynamic environments* (pp. 221–253). Hershey, PA: IGI Global. doi:10.4018/978-1-4666-6252-0.ch012

Torrens, F., & Castellano, G. (2014d). Cluster solvation models of carbon nanostructures: Extension to fullerenes, tubes and buds. *Journal of Molecular Modeling*, 20, 2263-1-9. PMID:24869779

TPO. (2012). *Nieuwe multidisciplinaire richtlijn polyfarmacie ouderen. Tijdschrift voor praktijkondersteuning*. Available from http://www.tijdschriftpraktijkondersteuning.nl/archief/volledig/id689-nieuwe-multidisciplinaire-richtlijn-polyfarmacie-ouderen.html

Trappey, A., Lu, T., & Fu, D. (2009). Development of an intelligent agent system for collaborative mold production with RFID technology. *Robotics and Computer-integrated Manufacturing*, *25*(1), 42–56. doi:10.1016/j.rcim.2007.06.002

Tufte, E. R. (1983/2001). *The Visual Display of Quantitative Information* (2nd ed.). Cheshire, CT: Graphics Press.

Tufte, E. R. (1990). *Envisioning Information* (2nd ed.). Cheshire, CT: Graphics Press.

Turcu, C., Turcu, Popa, V., & Gaitan, V. (2008). Identification and Monitoring of Patients Using RFID and Agent Technologies: Synergy and Issues. *Electronics and Electrical Engineering*, *6*(86), 17-22.

Türkarslan, Ö., Böyükbayram, A. E., & Toppare, L. (2010). Amperometric alcohol biosensors based on conducting polymers: Polypyrrole, poly(3,4-ethylenedioxythiophene) and poly(3,4-ethylenedioxypyrrole). *Synthetic Metals*, *160*(7-8), 808–813. doi:10.1016/j.synthmet.2010.01.027

U.S. Customs and Border Protection. (2009). *Intellectual property rights—Seizure statistics: Fiscal year 2009*. Retrieved September 12, 2014, from www.cbp.gov/linkhandler/cgov/trade/priority_trade/ipr/pubs/seizure/fy09_stats.ctt/fy09_stats.pdf)

uOttawa. (n.d.). *Study Designs*. Available from http://www.med.uottawa.ca/sim/data/Study_Designs_e.htm

Uptain, S. M., Kane, C. M., & Chamberlin, M. J. (1997). Basic mechanisms of transcript elongation and its regulation. *Annual Review of Biochemistry*, *66*(1), 117–172. doi:10.1146/annurev.biochem.66.1.117 PMID:9242904

Vajpeyi, A. P., Ajagunna, A. O., Tsagaraki, K., Androulidaki, M., & Georgakilas, A. (2009). InGaN nanopillars grown on silicon substrate using plasma assisted molecular beam epitaxy. *Nanotechnology*, *20*(32), 325605–1–5. doi:10.1088/0957-4484/20/32/325605 PMID:19620761

Vallance, P., & Smart, T.G. (2006). The future of pharmacology. *British Journal of Pharmacology*, *147*(S1), S304–7. doi:10.1038/sj.bjp.0706454

Varga-Atkins, T., Dangerfield, P., & Brigden, D. (2010). Developing professionalism through the use of wikis: A study with first-year undergraduate medical students. *Medical Teacher*, *32*(10), 824–829. doi:10.3109/01421591003686245 PMID:20854158

Vert, M., Doi, Y., Hellwich, K., Hess, M., Hodge, P., Kubisa, P., & Schué, F. et al. (2012). Terminology for biorelated polymers and applications (IUPAC Recommendations 2012). *Pure and Applied Chemistry*, *84*(2), 377–410. doi:10.1351/PAC-REC-10-12-04

Viola, I., Birkeland, Å., Solteszova, V., Helljesen, L., Hauser, H., Kotopoulis, S., Nylund, K., Ulvang, D. M., Øye,O. K., Hausken, T., & Gilja, O. H. (2013). *High-Quality 3D Visualization of In-Situ Ultrasonography*. DOI: 10.2312/conf/EG2013/med/001-004

Viola, I., & Gröller, E. (2007). On the Role of Topology in Focus+Context Visualization. In H. Hauser, H. Hagen, & H. Theisel (Eds.), *Topology-based Methods in Visualization: Mathematics and Visualization* (pp. 183–199). Springer. doi:10.1007/978-3-540-70823-0_12

Volatron, F., Catala, L., Rivière, E., Gloter, A., Stéphan, O., & Mallah, T. (2008). Spin-crossover coordination nanoparticles. *Inorganic Chemistry*, *47*(15), 6584–6586. doi:10.1021/ic800803w PMID:18590329

Voliani, V., Ricci, F., Signore, G., Nifosi, R., Luin, S., & Beltram, F. (2011). Multiphoton molecular photorelease in click-chemistry-functionalized gold nanoparticles. *Small*, *7*(23), 3271–3275. doi:10.1002/smll.201101753 PMID:22012898

Vorob'ev, V. S., & Eletskii, A. V. (1995).. . *Теплофизика высоких температур*, *33*, 862–862.

Vorob'ev, V. S., & Eletskii, A. V. (1996). Similarity laws and thermodynamic quantities for fullerite C_{60}. *Chemical Physics Letters*, *254*(3-4), 263–267. doi:10.1016/0009-2614(96)00329-6

Voss-Andreae. (2011). Quantum Sculpture: Art Inspired by the Deeper Nature of Reality. *Leonardo, Journal of the International Society for the Arts, Sciences and Technology*, *44*(1), 14–20.

Wadhavane, P. D., Galian, R. E., Izquierdo, M. A., Aguilera-Sigalat, J., Galindo, F., Schmidt, L., & Luis, S. V. et al. (2012). Photoluminiscence enhancement of CdSe quantum dots: A case of organogel–nanoparticle symbiosis. *Journal of the American Chemical Society*, *134*(50), 20554–20563. doi:10.1021/ja310508r PMID:23214451

Walsh, W., Wellman, A., Walsh, W., & Wellman, M. (2000). Modelling supply chain formation in multi-agent systems. In M. Alexandros, Y. Fredrik, & S. Carles (Eds.), Agent Mediated Electronic Commerce II (LNCS), (vol. 1788, pp. 94–101). Springer-Verlag. doi:10.1007/10720026_5

Wang, X., & Li, J. (2007). Innovation network in harvest. Paper presented at the International Conference on Technology Innovation, Risk Management, and Supply Chain Management.

Wang, F., Wang, J., & Liu, X. (2010). Direct evidence of a surface quenching effect on size-dependent luminescence of upconversion nanoparticles. *Angewandte Chemie International Edition*, *49*(41), 7456–7460. doi:10.1002/anie.201003959 PMID:20809558

Wang, K., Hong, J., Marinova, D., & Zhu, L. (2009). Evolution and governance of the biotechnology and pharmaceutical industry of China. *Mathematics and Computers in Simulation*, *79*(9), 2947–2956. doi:10.1016/j.matcom.2008.09.001

Wang, M., & Li, C. M. (2011). Magnetic properties of all-carbon graphene-fullerene nanobuds. *Physical Chemistry Chemical Physics*, *13*(13), 5945–5951. doi:10.1039/c0cp02433c PMID:21336407

Wang, Q. H., Kalantar-Zadeh, K., Kis, A., Coleman, J. N., & Strano, M. S. (2012). Electronics and optoelectronics of two-dimensional transition metal dichalcogenides. *Nature Nanotechnology*, *7*(11), 699–712. doi:10.1038/nnano.2012.193 PMID:23132225

Wang, R. Y., & Strong, D. M. (1996). Beyond Accuracy: What Data Quality Means to Data Consumers. *Journal of Management Information Systems*, *12*(4), 5–34. doi:10.1080/07421222.1996.11518099

Wang, W., & Richter, C. A. (2006). Spin-polarized inelastic electron tunneling spectroscopy of a molecular magnetic tunnel junction. *Applied Physics Letters*, *89*, 153105–1–3.

Ward, R., Moule, P., & Lockyer, L. (2009). Adoption of Web 2.0 Technologies in Education for Health Professionals in the UK: Where are we and why. *Electronic Journal of E-Learning*, *7*(2), 165–172.

Wasan, S., & Chakravarty, R. (2013). Digital Repositories as Harbingers of Open Access in India: A Study. In T. Ashraf & P. Gulati (Eds.), *Design* (pp. 191–218). Hershey, PA: Development, and Management of Resources for Digital Library Services; doi:10.4018/978-1-4666-2500-6.ch017

Weber, G. (1975). Energetics of ligand binding to proteins. *Advances in Protein Chemistry*, *29*, 1–83. doi:10.1016/S0065-3233(08)60410-6 PMID:1136898

Wee, C. H., Tan, S. J., & Cheok, K. H. (1995). Non-price determinants of intention to purchase counterfeit goods. *International Marketing Review*, *12*(6), 19–46. doi:10.1108/02651339510102949

Wei, L., & Wang, Y. N. (2004). Electromagnetic wave propagation in single-wall carbon nanotubes. *Physics Letters. [Part A]*, *333*(3-4), 303–309. doi:10.1016/j.physleta.2004.10.048

Wei, L., & Yu, J. (2008).Bioinformatics in China: A personal perspective. *PLoS Computational Biology*, *4*, 1–11.

Wen, Y., Xu, J., He, H., Lu, B., Li, Y., & Dong, B. (2009). Electrochemical polymerization of 3,4-ethylenedioxythiphene in aqueous micellar solution containing biocompatible amino acid-based surfactant. *Journal of Electroanalytical Chemistry*, *634*(1), 49–58. doi:10.1016/j.jelechem.2009.07.012

Werner, D., Hashimoto, S., & Uwada, T. (2010). Remarkable photothermal effect of interband excitation on nanosecond laser-induced reshaping and size reduction of pseudospherical gold nanoparticles in aqueous solution. *Langmuir*, *26*(12), 9956–9963. doi:10.1021/la100015t PMID:20210316

Werner, F., Limbach, F., Carsten, N., Denker, C., Malindretos, J., & Rizzi, A. (2009). Electrical conductivity of InN nanowires and the influence of the native indium oxide formed at their surface. *Nano Letters*, *9*(4), 1567–1571. doi:10.1021/nl8036799 PMID:19290610

Wheelan, C. (2011). *Introduction to Public Policy*. New York: W. W. Norton.

White, J. (1997). Mobile agents. In *Software agents*. AAAI Press.

White, R. J., Luque, R., Budarin, V. L., Clark, J. H., & Macquarrie, D. J. (2009). Supported metal nanoparticles on porous materials. Methods and applications. *Chemical Society Reviews*, *38*(2), 481–494. doi:10.1039/B802654H PMID:19169462

White, S. (2000). Competition, capabilities, and the make, buy, or ally decisions of Chinese State-Owned firms. *Academy of Management Journal*, *43*(3), 324–341. doi:10.2307/1556398

Whitesides, G. M. (2005). Nanoscience, nanotechnology and chemistry. *Small*, *1*(2), 172–179. doi:10.1002/smll.200400130 PMID:17193427

Whittemore, R., & Knafl, K. (2005). The integrative review: Updated methodology. *Journal of Advanced Nursing*, *52*(5), 546–553. doi:10.1111/j.1365-2648.2005.03621.x PMID:16268861

Wickline, S., Neubauer, A., Winter, P., Caruthers, S., & Lanza, G. (2006). Applications of nanotechnology to artherosclerosis, thrombosis, and vascular biology. *Arteriosclerosis, Thrombosis, and Vascular Biology*, *26*(3), 435–441. doi:10.1161/01.ATV.0000201069.47550.8b PMID:16373609

WiCON (2012). *Pharma China 2012.* WiCON International Group LLC.

Wiecha, J., Heyden, R., Sternthal, E., & Merialdi, M. (2010). Learning in a Virtual World: Experience With Using Second Life® for Medical Education. *Journal of Medical Internet Research, 12*(1), e1. doi:10.2196/jmir.1337 PMID:20097652

Wigand, R. T., & Benjamin, R. I. (1996) Electronic Commerce: Effects on Electronic Markets. *Journal of Computer-Mediated Communication, 1*(3). Retrieved on September 5, 2014, from http://shum.huji.ac.il/jcmc/jcmc.html

Wigand, R. T., Picot, A., & Reichwald, R. (1997). *Information, Organization and Management: Expanding Markets and Corporate Boundaries*. Chichester, UK: Wiley.

Wigand, R. T., Steinfield, C. W., & Markus, M. L. (2005, Fall). IT Standards Choices and Industry Structure Outcomes: The Case of the United States Home Mortgage Industry. *Journal of Management Information Systems, 22*(2), 165–191.

Willets, K. A., & van Duyne, R. P. (2007). Localized surface plasmon resonance spectroscopy and sensing. *Annual Review of Physical Chemistry, 58*(1), 267–297. doi:10.1146/annurev.physchem.58.032806.104607 PMID:17067281

Williamson, O. (1979). Transaction-Cost Economics: The Governance of Contractual Relations. *The Journal of Law & Economics, 22*(2), 233–261. doi:10.1086/466942

Willner, I., Willner, B., & Tel-Vered, R. (2011). Electroanalytical applications of metallic nanoparticles and supramolecular nanostructures. *Electroanalysis, 23*(1), 13–28. doi:10.1002/elan.201000506

Wooldridge, M. (1997). Agent based software engineering. *IEEE Proceedings of Software Engineering, 144*(1), 26–37.

Wooldridge, M. J., & Jennings, N. (1995). Agent theories, architectures and languages: A survey. *The Knowledge Engineering Review, 10*(2), 115–152. doi:10.1017/S0269888900008122

Wooldridge, M., Jennings, N. R., & Kinny, D. (2000). The gaia methodology for agent-oriented analysis and design. *Journal of Autonomous Agents and Multi-Agent Systems, 3*(3), 285–312. doi:10.1023/A:1010071910869

World Health Organization (WHO). (2010). *Medicines: Counterfeit Medicines*. Fact Sheet No. 175. Retrieved on September 10, 2014, from http://www.who.int/mediacentre/factsheets/fs275/en/index.html

Woulfe, J., Williams, K., & Ryan, G. (2009). Evaluating pharmacy students ' wiki-based collaboration. In R. Atkinson & C. McBeath (Eds.), *Same places, different spaces. Proceedings ascilite Auckland 2009* (pp. 1197–1199). Auckland: The University of Auckland, Auckland University of Technology, and Australasian Society for Computers in Learning in Tertiary Education (ASCILITE). Retrieved from http://www.ascilite.org.au/conferences/auckland09/procs/

Wright, V., & Harvey, A. (1988). *Ilustraciones diagnósticas en reumatología: Pruebas de autoevaluación*. Interamericana McGraw Hill.

Wuister, S. F., de Mello Donegá, C., & Meijerink, A. (2004). Influence of thiol capping on the exciton liminiscence and decay kinetics of CdTe and CdSe quantum dots. *The Journal of Physical Chemistry B, 108*(45), 17393–17397. doi:10.1021/jp047078c

Wu, J., El Hamaoui, B., Li, J., Zhi, L., Kolb, U., & Müllen, K. (2005). Solid-state synthesis of "bamboo-like" and straight carbon nanotubes by thermolysis of hexa-*peri*-hexabenzocoronene–cobalt complexes. *Small, 1*(2), 210–212. doi:10.1002/smll.200400049 PMID:17193432

Wu, X., & Zeng, X. C. (2008). First-principles study of a carbon nanobud. *ACS Nano, 2*(7), 1459–1465. doi:10.1021/nn800256d PMID:19206315

Wu, X., & Zeng, X. C. (2009). Periodic graphene nanobuds. *Nano Letters, 9*(1), 250–256. doi:10.1021/nl802832m PMID:19072128

Xia, Q. V. (2008). Analysis of future directions and growth of biotechnology in China. *Asia Biotech*, *12*, 1–6.

Xing, M., Shen, H., Zhao, W., Liu, Y., Du, Y., Yu, Z., & Chen, X. (2011). dsDNA-coated quantum dots. *BioTechniques*, *50*(4), 259–261. PMID:21548911

Xu, L., Pirollo, K., Tang, W., Rait, A., & Chang, E. (1999). Transferrin-liposome-mediated systemic p53 gene therapy in combination with radiation results in regression of human head and neck cancer xenografts. *Human Gene Therapy*, *10*(18), 2941–2952. doi:10.1089/10430349950016357 PMID:10609655

Yang, J., & Shu, W. (2005). On the choice of technological innovation strategy of Chinese enterprises at present. *China-USA Business Review*, *4*(10), 63–66.

Yanqing, J., Hao, M., Farber, S., John, Y., Peter, D., Richard, M., & Michael, M. (2010). A Distributed, Collaborative Intelligent Agent System Approach for Proactive Post marketing Drug Safety Surveillance. *IEEE Transactions on Information Technology in Biomedicine*, *14*(3), 826–837. doi:10.1109/TITB.2009.2037007 PMID:20007038

Yanqing, J., Hao, Y., John, Y., Shizhuo, Z., Daniel, C., Barth, J., & Michael, M. et al. (2007). A distributed adverse drug reaction detection system using intelligent agents with a fuzzy recognition-primed decision model. *International Journal of Intelligent Systems*, *22*(8), 827–845. doi:10.1002/int.20230

Yanqing, J., Hao, Y., Yen, J., Shizhuo, Z., Massanari, M., & Barth, J. (2005). Team-based multi-agent system for early detection of adverse drug reactions in post marketing surveillance. In *Proc Proceedings of the 24th North American Fuzzy Information Processing Society.*

Younger, P. (2010). Using wikis as an online health information resource. Nursing Standard, 24(36), 49–56.

Yudasaka, M., Iijima, S., & Crespi, V. H. (2008). Single-wall carbon nanohorns and nanocones. *Topics in Applied Physics*, *111*, 605–629. doi:10.1007/978-3-540-72865-8_19

Yu, J., & Wehrly, T. E. (2004). An approach to the residence time distribution for stochastic multi-compartment models. *Mathematical Biosciences*, *191*(2), 185–205. doi:10.1016/j.mbs.2004.06.005 PMID:15363653

Yu, K., & Chen, J. (2014). Graphene-based transparent conductive electrodes. *Material Matters*, *9*(1), 6–13.

Zambonelli, F., Jennings, N. R., & Wooldridge, M. (2001). Organizational rules as an abstraction for the analysis and design of multi agent systems. *International Journal of Software Engineering and Knowledge Engineering*, *11*(3), 303–308. doi:10.1142/S0218194001000505

Zanussi, L., Paget, M., Tworek, J., & McLaughlin, K. (2011). Podcasting in medical education: Can we turn this toy into an effective learning tool? *Advances in Health Sciences Education: Theory and Practice*. doi:10.1007/s10459-011-9300-9 PMID:21544550

Zayat, A. V., & Smolyaninov, I. I. (2003). Near-field photonics: Surface plasmon polaritons and localized surface plasmons. *Journal of Optics A*, *5*(4), 16–50. doi:10.1088/1464-4258/5/4/353

Zeki, S. (1993). Vision of the Brain. Wiley-Blackwell.

Zeki, S. (1999). Art and the Brain. *Journal of Conscious Studies: Controversies in Science and the Humanities*, *6*(6/7), 76–96.

Zhang, B. L., Wang, C. Z., Ho, M., & Chan, C. T. (1993). Melting of carbon cages. *Zeitschrift für Physik D*, *26*(S1), 285–285. doi:10.1007/BF01425692

Zhang, K., Zhang, L. L., Zhao, X. S., & Wu, J. (2010). Graphene/polyaniline nanofiber composites as supercapacitor electrodes. *Chemistry of Materials*, *22*(4), 1392–1401. doi:10.1021/cm902876u

Zhang, X., Coleman, A. C., Katsonis, N., Browne, W. R., van Wees, B. J., & Feringa, B. L. (2010). Dispersion of graphene in ethanol using a simple solvent exchange method. *Chemical Communications*, *46*(40), 7539–7541. doi:10.1039/c0cc02688c PMID:20848021

Zhao, M.-Q., Zhang, Q., Huang, J.-Q., & Wei, F. (2012). Hierarchical nanocomposites derived from nanocarbons and layered double hydroxides – Properties, Synthesis, and applications. *Advanced Functional Materials*, *22*(4), 675–694. doi:10.1002/adfm.201102222

Zhao, X., Xu, S., Wang, L., Duan, X., & Zhang, F. (2010). Exchange-biased $NiFe_2O_4$/NiO nanocomposites derived from NiFe-layered double hydroxides as a single precursor. *Nano Research*, *3*(3), 200–210. doi:10.1007/s12274-010-1023-3

Zhou, W., & Fu, W. (2013). Mesoporous TiO_2: Preparation, doping, and as a composite for photocatalysis. *ChemCatChem*, *5*(4), 885–894. doi:10.1002/cctc.201200519

Zhou, Y. (2007, January). Opportunities in Biopharmaceutical Outsourcing to China. *BioProcess International*, 16–23.

Zhouying, J. (2005). Globalization, technological competitiveness and the 'catch-up' challenge for developing countries: Some lessons of experience. *International Journal of Technology Management and Sustainable Development*, *4*(1), 35–46. doi:10.1386/ijtm.4.1.35/1

Zhu, J., Kase, D., Shiba, K., Kasuya, D., Yudasaka, M., & Iijima, S. (2003). Binary nanomaterials based on nanocarbons: A case for probing carbon nanohorns' biorecognition properties. *Nano Letters*, *3*(8), 1033–1036. doi:10.1021/nl034266q

Zhu, S., & Xu, G. (2010). Single-walled carbon nanohorns and their applications. *Nanoscale*, *2*(12), 2538–2549. doi:10.1039/c0nr00387e PMID:20957266

Zhu, X., & Su, H. (2009). Magnetism in hybrid carbon nanostructures: Nanobuds. *Physical Review B: Condensed Matter and Materials Physics*, *79*(16), 165401–1–5. doi:10.1103/PhysRevB.79.165401

Zhu, Y., Li, W., Li, Q., Li, Y., Li, Y., Zhang, X., & Huang, Q. (2009). Effects of serum proteins on intracellular uptake and cytotoxicity of carbon nanoparticles. *Carbon*, *47*(5), 1351–1358. doi:10.1016/j.carbon.2009.01.026

Zhu, Y., Zhang, X., Zhu, J., Zhao, Q., Li, Y., Li, W., & Huang, Q. et al. (2012). Cytotoxicity of phenol red in toxicity assays for carbon nanoparticles. *International Journal of Molecular Sciences*, *13*(12), 12336–12348. doi:10.3390/ijms131012336 PMID:23202901

Zorginstituut Nederland. (2013). *Fact sheet polyfarmacie*. Available from: https://www.zorginstituutnederland.nl/binaries/content/documents/zinl-www/kwaliteit/meerjarenagenda/agenda-2013/agenda-2013/agenda-2013/zinl%3Aparagraph%5B7%5D/zinl%3Adocuments%5B4%5D/1306-fact-sheet-polyfarmacie/Fact+sheet+polyfarmacie.pdf

Zou, J., Zhong, W., & Ward, R. K. (2006) A novel digital watermarking method for commercial bills based on a class of orthogonal functions. In *IEEE International Symposium on Signal Processing and Information Technology*. Vancouver: IEEE Computer Society. doi:10.1109/ISSPIT.2006.270766

About the Contributors

Floor Aarnoutse is a starting researcher at the department of Information and Computing Sciences at Utrecht University. For the last few years of her academic career she fondly dived in several topics including but not limited to software ecosystems, knowledge management, game production and software product management.

Rupak Chakravarty works as Assistant Professor (Senior Scale) in the Department of Library & Information Science, Panjab University, Chandigarh, India for more than six years. Specialization areas include: Application of ICT in libraries including library automation, Free Open Source Software (FOSS), digital library, digital reference service, library portal development. Other areas of specialization include: Management of Library & Information Centres, Information & Communication, Dewey Decimal Classification, Information Sources, Open Access, Institutional Repositories (IRs), Web 2.0, Library 2.0. Also serving as Mendeley Advisor at Chandigarh region.Worked as Assistant Librarian, Panjab University Library for more than five years as the Reference Librarian and at Periodicals. Publications and Conferences:Over 25 publications in many peer-reviewed journals including International Journals like IFLA journal, conferences both national as-well-as International. Total number of citations is 40 (Google Scholar). Also served/serving as reviewer in many national and international journals. Resources person in many training programmes, refresher courses.LIS Portal: Developed LIS Portal - portal for LIS professionals and students to remain updated and attain new information regarding Library Resources, Open Access Journals, IRs, FOSS and LIS research. This portal is available at http://dlispu.webs.com.

Leong Chan received Ph.D. in technology management from Portland State University, USA. He is an adjunct assistant professor in the department of Engineering and Technology Management, Portland State University. His research interests are in the areas of technology assessment, technology policy, R&D management, and innovation systems.

Liu Dan, Ph.D., associate professor of School of Economics and Management in Beijing University of Posts and Communications.Research areas include Innovation Management, Innovation and Entrepreneurship, Service innovation and management theory and empirical research.

Eduardo César Contreras Delgado is part of systemic studies research department of "Facultad de Sistemas" and General Coordination of Postgraduate Education at "Universidad Autónoma de Coahuila", he works also as research associate of "Instituto Tecnológico de Saltillo". He actually works in the project "Therapy of communication for people with intellectual disabilities", The main purpose of professor

Eduardo is to develop software for people with different disabilities in public special education schools, providing training to therapists and special education teachers on the usage of Information Technologies as a didactic resource and develop software for medical area. His hobby is to play PC games.

Maria João Gomes is an assistant professor at the Department of Curriculum Studies and Educational Technology, Institute of Education, University of Minho - Braga, Portugal. She has a PhD in Education with specialization in Educational Technology and conducts its activities of teaching and research within the area. She taught several undergraduate and postgraduate courses on distance learning and e-learning, educational technology, didactics of informatics, ICT for special needs students and digital games and education. She is the author of over 90 publications and develops her research interests around the themes of the social web, digital portfolios, cyberbullying, internet safety, distance education and e-learning. She is also director of the electronic journal "Educação, Formação & Tecnologias" and coordinator of the Competence Center for Information Technologies and Communication of the University of Minho.

Ângelo Jesus holds a Degree in Pharmacy, and has professional experience in Hospital and Community Pharmacy. He is now a Lecturer at the School of Allied Health Sciences, Polytechnic Institute of Porto, Portugal, where he teaches Pharmacology and Therapeutics. Additionally, he is responsible for International Mobility Programs, and has taught abroad in several locations in Europe. He also has a PhD in Educational Sciences, specialization in Educational Technology from Minho University, and develops his research interests around distance education and e-learning in biomedical settings; Web 2.0 tools in Biomedical Education; Instructional Design, Drug Therapy and Information Counseling. He has recently joined the team of Polytechnic Institute of Porto e-learning Center.

Dinah Mande is an Information Quality (IQ) professional. She has experience in the healthcare and oil industries. Past projects have included data profiling, adapting an IQ methodology for system development from waterfall to agile, information chain mapping and improvement, and coordination of information quality improvement teams. She is an active member of the International Association for Information and Data Quality (IAIDQ) and was the first president of the IAIDQ Little Rock Chapter launched in spring 2010. She has served the IQ discipline in a variety of ways: Program committee member for the ICIQ 2014 and the 2011 and 2010 DGIQ conferences, member of the 2013 IDQ conference planning team and reviewer of a 2010 issue of the International Journal of Information Quality. Dinah earned a Master of Science degree in Information Quality from the University of Arkansas at Little Rock (UALR), Little Rock, Arkansas. She holds a BA in Business Administration from Philander Smith College, Little Rock, Arkansas.

Cassandra Renes is a Researcher at Utrecht University.

Gustavo Julio Puente Salcido was born in Mexico City on march 28th 1954 so he is sixty years old now. He graduated from Medical School, Universidad Nacional Autónoma de México in México City in 1977 and began post grade training in Rheumatology in Centro Médico Nacional IMSS in México City from 1978 to 1982. He have been a board certified Rheumatologist since 1982 and worked from 1982 to 2005 in Rheumatology department of Clínica Hospital 2 IMSS in Saltillo, Coah., Mexico. Now he is retired but still active in solo practice and teaching rheumatology in Medical School in Saltillo and collaborating with experts in informatics to develop a diagnostic tool to identify rheumatic diseases.

Marco Spruit is a researcher, manager and lecturer in the Organization & Information research group at the Information and Computing Sciences department of Utrecht University where he lectures in Business Intelligence and Life Sciences & Health Informatics. He is also the information science curriculum manager. Before, he was a professional software developer for fourteen years in the fields of Information Retrieval & Intelligence, and a product software entrepreneur for SMEs. He is passionate about my Health Intelligence Engineering for Transparancy (HIET) research on data-driven decision-supportive innovations. He applies business intelligence technologies to engineer innovations of organisational processes which support people's performance using an action research approach. It is his strategic goal to contribute to a better world by improving transparency in decision-making processes. He serves on the editorial boards of the international journals on Business Intelligence Research (IJBIR), Decision Analytics, and Computer Information Systems (JCIS).

Francisco Torrens is lecturer in physical chemistry at the Universitat de València. After obtaining a PhD in molecular associations in azines and macrocycles from the Universitat de València, Dr. Torrens undertook postdoctoral research with Professor Rivail at the Université de Nancy I. More recently, Dr. Torrens has collaborated on projects with Professor Tomás-Vert. Major research projects include characterization of the electronic structure of electrically conductive organic materials, theoretical study of new electrically conductive organic materials, protein modelling, electronic correlation, development and applications of high-precision mono and multi referential electronic correlation methods, development and application of high-precision quantum methods, methodological developments and applications of quantum methods, molecular modelling and chemistry-computer applications for the description and prediction of molecular properties, development and integration of a computational system for the discovery of active drugs against AIDS, cancer pandemics and other microbial diseases of high social impact, starting an electronic journal of scientific-technological and gender spreading, computational design, chemical obtaining and biological evaluation of new antiprotozoal and anticancer drugs, prediction of the environmental impact of chemical substances using chemobioinformatics tools, discovery of new inhibitors of tyrosinase: computational design, synthesis, characterization and experimental corroboration, search for new antimicrobial agents that inhibit the synthesis of the bacterial cell wall, minimal active domains of proteins of the Bcl 2 family and analysis of pores of BAX at nanometric scale. His scientific accomplishments include the first implementation in a computer at the Universitat de València of a program for the elucidation of crystallographic structures, and the construction of the first computational-chemistry program adapted to a vector-facility supercomputer in a Spanish university.

Anna Ursyn is a professor and Computer Graphics Area Head at the School of Art and Design, University of Northern Colorado. She combines programming with software and printmaking media, to unify computer generated and painted images, and mixed-media sculptures. Ursyn had over 30 single juried and invitational art shows, participated in over 100 fine art exhibitions, and published articles and artwork in books and journals. Research and pedagogy interests include integrated instruction in art, science, and computer art graphics. Since 1987 she serves as a Liaison, Organizing and Program Committee member of International IEEE Conferences on Information Visualization (iV) London, UK, and Computer Graphics, Imaging and Visualization Conferences (CGIV). She serves as Chair of the Symposium and Digital Art Gallery D-ART iV, 1997-now. Anna has published three books with IGI-Global.

Rolf T. Wigand is Maulden-Entergy Chair and Distinguished Professor Emeritus of Information Science and Business Information Systems, University of Arkansas at Little Rock. He is past director of the Graduate Program in Information Management and founding director, Center for Digital Commerce, both at Syracuse University. Current NSF research grants (2007-2016) continue with standards development research in the mortgage, retail and automotive supply industries; the analysis of virtual organizations, networks in disaster settings and the analysis of online blogs and social media as instruments in collective action. Wigand authored eight books and over 390 articles, chapters, and monographs. His research has appeared in such journals as MIS Quarterly, Journal of MIS, Sloan Management Review, Journal of Information Technology, Journal of Communication, Electronic Markets, International Journal of Information Technology & Decision Making, Journal of Applied Psychology, Knowledge Management: Research & Practice, International Journal of Management, Business & Information Systems Engineering, European Journal of Information Systems.

Index

T

V

W

www.ingramcontent.com/pod-product-compliance
Lightning Source LLC
LaVergne TN
LVHW060407300926
849162LV00006B/1368